The RADON TRANSFORM and LOCAL TOMOGRAPHY

The RADON TRANSFORM and LOCAL TOMOGRAPHY

A.G. Ramm
A.I. Katsevich

CRC Press
Taylor & Francis Group
Boca Raton London New York

CRC Press is an imprint of the
Taylor & Francis Group, an **informa** business

CRC Press
Taylor & Francis Group
6000 Broken Sound Parkway NW, Suite 300
Boca Raton, FL 33487-2742

© 1996 by Taylor & Francis Group, LLC
CRC Press is an imprint of Taylor & Francis Group, an Informa business

First issued in paperback 2019

No claim to original U.S. Government works

ISBN-13: 978-0-367-44867-7 (pbk)
ISBN-13: 978-0-8493-9492-8 (hbk)

Visit the Taylor & Francis Web site at
http://www.taylorandfrancis.com

and the CRC Press Web site at
http://www.crcpress.com

This research was performed under the auspices of the U.S. Department of Energy.

This book was typeset using AMST$_{E}$X

Library of Congress Cataloging-in-Publication Data
Catalog Record available from the Library of Congress.

Dedicated to
Olga and Julia
and
Tanya, Anya, and Gene

PREFACE

The purpose of this book is twofold. On one hand, the book can serve as an introduction to the basic properties of the Radon transform. It can be used as a text: very little background is needed in order to understand this book. Almost all the background facts that are used are explained and partly proved in Chapter 14.

On the other hand, the book presents in a self-contained and systematic way the new theory developed by the authors. This theory deals with the study of the singularities of the Radon transform and its applications to the imaging in tomography. The authors hope that their theory will be widely used by practitioners. Numerical examples show that the algorithms for tomographic imaging proposed in this book are of practical value. They lead to sharp images of the discontinuities of functions. Computationally, these methods are inexpensive and effective. In the presentation of the basic theory of the Radon transform there are many new points, some of which are discussed in the bibliographical notes.

The authors tried to make the book accessible to a broad audience. That is why the necessary background is kept to a minimum and the results used in the book are collected in Chapter 14. A detailed table of contents shows the material presented in the book.

The introductory Chapter 1 explains the contents of the book and its aims and gives several physical motivations to the problems of tomography.

Chapters 2 and 3 contain the basic definitions and facts about the Radon transform and related transforms, for example, X-ray transform, the adjoint Radon transform (backprojection operator), and X_m-transform. The range theorems are discussed in detail and the reader will find many novel points and results in the presentation of the material.

In Chapter 4 the new theory is presented: the singularities of the Radon transform are studied in detail and the asymptotics of the Radon transform near its singular support is found. The asymptotics depends on the geometry of the singular support of the Radon transform. The relation between singularities of f and $\hat{f}$, its Radon transform, is found.

A map which sends the singular support of a piecewise-smooth function $f(x)$ onto the singular support of the Radon transform of f is described. This map turns out to be involutive and is a generalization of the classical Legendre transform known in classical Hamiltonian mechanics, differential geometry and thermodynamics. A relation between the wave front of f and the singular support of $\hat{f}$ is obtained. A duality law, which makes clear a geometrical relation between singsupp f and singsupp $\hat{f}$ for smooth singsupp f is formulated. Asymptotics of the Fourier transform of piecewise-smooth functions is derived in the case of functions of several variables. This result has many applications. It is based on the asymptotics of the Radon transform near its singular support. We study singularities of X-ray transform and give a procedure for recovery of f from three-dimensional X-ray transform data. Formulas describing singularities of R^*f are obtained.

Chapters 5, 6 and 7 present, for the most part, the new theory: methods for finding discontinuities of functions and the sizes of the jumps of these functions across the discontinuities from local tomographic data. Three methods are proposed: generalized local tomography (usual local tomography, as it was known in the literature, did not allow one to find values of jumps), pseudolocal tomography, and geometrical tomography. These methods allow one to obtain sharp images, which carry quantitaive information, using computationally efficient procedures. A class of formulas of local tomography is introduced and a formula optimal with respect to noise stability is found. For more than a decade there was no method for calculating the sizes of the jumps of functions across their discontinuity surfaces given local tomographic data. Several such methods are proposed, described and tested in this monograph. We develop and use systematically the theory of pseudodifferential operators (PDO) with symbols from a special class, introduced by the authors, for a study of tomographic problems of finding discontinuities of f and the sizes of the jumps of f across the discontinuity surfaces. In particular, we obtain the asymptotics of Bf near singular support of f, where B is a PDO with symbol from the above class of symbols, which is of basic interest in tomography.

Numerical examples with simulated data are presented and discussed. Practitioners will get useful new tools for better imaging in tomography.

Chapter 8 deals with incomplete tomographic data. In particular, analytic formulas and methods for inversion of limited-angle data are given.

In Chapter 9 inversion of the cone-beam data is discussed and some new geometries are studied. These geometries arise in γ-tomography. The problems discussed are the practical problems which came from technology and medicine.

In Chapter 10 the Radon transform is defined on some spaces of distributions. Equivalence of several apparently different definitions, used in the literature, is established. A new definition, convenient for practical calculations, is formulated. The other definitions are equivalent to the new one. Examples of applications are considered.

In Chapter 11 the Abel equation is discussed. Most of the material here, except Section 11.4, is standard.

In Chapter 12 a new method for edge, bright spot and straight line detection is developed. This method, proposed by the authors, turned to be very simple to use in applications and very efficient. The method is based on a new idea, which is of statistical nature. The reason for including this method in the book is clear: one needs methods for finding discontinuity curves and other characteristic features in tomographic images.

In Chapter 13 a general test of randomness and a discussion of its possible applications are given. Again it is our goal to give a method for finding discontinuity surfaces of signals from noisy observations of the signal at a discrete set of points. The test of randomness proposed by the authors is very general and examples of its applications to image processing are included in the book. In particular, this test gives an algorithm for finding change surfaces from noisy discrete data in the case of multidimensional signals.

Finally, in Chapter 14 the reader will find the background material, which, as the authors hope, will make this monograph essentially self-contained.

For convenience of the reader the list of notations and index are included, and in the Bibliographical notes one finds comments on the origin of the material presented in this monograph. Basically, the material in Chapters 4–13, with some exceptions, mentioned in the Bibliographical notes, belongs to the authors, and most of the material in Chapters 2 and 3, with some exceptions, is known from the literature.

The authors tried to make the reading of this book as easy as possible and make the book informative for readers with different interests: for researchers in analysis, image processing and mathematics of tomography, for engineers and practitioners who use tomographic image processing, and for students in these fields. The authors hope that all these readers will find that the time spent in reading this book is well spent.

The authors thank Los Alamos National Laboratory and U.S. Department of Energy for support. They are grateful to Dr. V. Faber from LANL for his support.

The new theory, presented in Chapters 4–10, 12–13, and partly in Chapters 2,3, and 14, is based on the series of papers [R], [RK], [RSZ], [RZ], [KR], [K], and [FKR]. The authors thank Academic Press, Pergamon Press, Kluwer Academic Publishers, American Mathematical Society and SIAM in whose journals the above papers have appeared.

Last but not least the authors are grateful to our families for their support and understanding.

AK thanks his wife, Tanya, as well as his parents and brother for their love and support.

AR is especially grateful to L. Ramm, his wife, for her help with TEX files and typing.

He also thanks Professors A. Palanque-Mestre and A. Ruiz, and Drs L. Desbat and A. Zaslavsky for useful comments.

He is grateful to Complutense University, Technion, and the University of Grenoble for hospitality.

Alexander G. Ramm, Manhattan, KS
Alexander I. Katsevich, Los Alamos, NM
September, 1995.

TABLE OF CONTENTS

CHAPTER 1

INTRODUCTION

1.1. Brief description of new
results and the aims of the book

The Radon transform of a function $f(x), x \in \mathbb{R}^n$, is defined to be

$$Rf := \hat{f}(\alpha, p) := \int_{l_{\alpha p}} f(x)ds, \qquad (1.1.1)$$

where $\alpha \in S^{n-1}$, S^{n-1} is the unit sphere in $\mathbb{R}^n$, $p \in \mathbb{R}, l_{\alpha p} := \{x : \alpha \cdot x = p\}$ is a plane, and ds is the Lebesgue measure on this plane. It is assumed that $f(x)$ is integrable over any plane. A natural question to ask is: how can one recover $f(x)$ knowing $\hat{f}(\alpha, p)$? This question, in the case $n = 2$, was posed and solved by Radon in 1917 [Ra]. However, according to [De2, p.3], Uhlenbeck (1925) wrote that Lorentz, the famous Dutch physicist, the author of the theory of electrons and the Nobel prize winner in physics for 1902 (jointly with P. Zeeman), knew the inversion formula for the Radon transform already around the turn of the century. In the mathematical literature the Radon transform was used in the books [J3] and [GS] for construction of fundamental solutions to partial differential equations with constant coefficients. Inversion of the Radon transform was one of the first problems of integral geometry, which is a branch of mathematics dealing with the recovery of functions knowing their integrals over a family of manifolds [LRS].

Radon's paper was not used for a long time. The rapid development of applications of the Radon transform started in the early 1970s. It is not possible to mention all the relevant works in medicine, astronomy, optics, physics, geophysics, and other areas. A vast bibliography is given in [De2], [Nat3]. The most well known applications are in computed tomography (X-ray transmission tomography, emission tomography, and ultrasound tomography). The Nobel prize in physiology and medicine was awarded in 1979 to A. Cormack and G. Hounsfield for their work on applications of tomography to medical diagnostics.

The Radon transform is a particular case of the more general X_m-ray transform, which is defined as follows:

$$X_m f = \int_{\mathbb{R}^n \cap M_m} f\, ds, \qquad (1.1.2)$$

where M_m runs through the set of m-dimensional affine manifolds, $1 \leq m \leq n - 1$, and ds is the m-dimensional Lebesgue measure on $\mathbb{R}^n \cap M_m$. If $m = 1$, we obtain X-ray transform – the integrals of f over straight lines:

$$X f := \int_{-\infty}^{\infty} f(x + \alpha t)dt := g(x, \alpha), \qquad (1.1.3)$$

where $\alpha \in S^{n-1}$ and $x \in \mathbb{R}^n$ runs through a subset of $\mathbb{R}^n$. For example, this subset may be a curve, a surface or other manifold in $\mathbb{R}^n$. If $m = n - 1$, we obtain the Radon transform (1.1.1), that is $X_{n-1} f = R f$.

Let us consider the practically important case $n = 2$. In this case the Radon transform and X-ray transform coincide and, basically, one is given the integrals of a three-dimensional object f along all lines located on a fixed plane through the object. Such data can be collected observing attenuation of X-rays passing through the object (see Section 1.2.1, p. 6). The problem is: given the line integral data, recover f on the plane. This gives a two-dimensional slice of f. Stacking many two-dimensional slices, if necessary, one can recover the three-dimensional object. Unfortunately, the conventional two-dimensional reconstruction is not local: to compute f at a point x one needs to know the integrals of f along all lines on the plane intersecting the support of f, even along the lines far removed from x. Moreover, in practice, it might be impossible to collect the complete data set: for example, if the object is too big. Suppose now that one is interested in the recovery of f not for all $x \in \operatorname{supp} f$, but for x only in some subset $U \subset \operatorname{supp} f$. The subset U will be called the region of interest (ROI). Let us define the local data as the integrals of f only along the lines intersecting the ROI. Clearly, in real medical experiments it is desirable to collect the local data instead of the full data, because this means that X-rays which do not intersect the ROI (e.g., a liver) are shielded, thus reducing the total X-ray dose. However, unless the ROI coincides with the entire cross-section, it is impossible to reconstruct f pointwise inside the ROI. It turns out that some very useful information can, nevertheless, be extracted from the local data. More precisely, one can find locations and values of jumps (sharp variations) of f. As an example, one can think of finding the healthy tissue-tumor interface and of computing the density jump across this interface.

The branch of computed tomography which deals with the analysis of local data is called local tomography. The first algorithm of local tomography was proposed by Russian scientists E. Vainberg, I. Kazak, and V. Kurczaev in 1981. Later on, M. Faingoiz, A. Faridani, F. Keinert, P. Kuchment, E. Ritman, K. Smith, and some others contributed to the development of local tomography.

Very recently, local tomography received another impulse in its development when the authors of this monograph discovered a family of local tomography functions, applied systematically methods of the theory of pseudodifferential operators to tomography, introduced the new notions of pseudolocal and geometrical tomographies, and gave several methods for finding the values of the jumps of f given the local data (previously, it was known how to find only the locations of jumps). The authors consider now the term 'local tomography' as including both the local and pseudolocal tomographies. Currently, local tomography is a rapidly developing field. A considerable part of the monograph is devoted to the theoretical analysis and the description of different algorithms of local tomography.

We decided to include in the monograph also some results on image processing and statistics. Although these results are not directly related to the Radon transform, they are necessary for the analysis of tomographic images. Indeed, inverting the Radon transform data with the help of either regular or local tomography, we obtain a two- or three-dimensional distribution (image) of a certain function f. Quite frequently, this distribution should be analyzed further in order to find some specific features: e.g., locations of sharp variations of f (edges in the image of f), or thin strips of local maxima of f (thin lines), etc. Finding such features is not always an easy task to do, because the initial line integral data may be noisy and, consequently, the resulting images may be corrupted. Therefore, to extract the additional information about f from its image, one frequently has to combine techniques of image processing and statistics. Several such algorithms are described in Chapters 12 and 13.

The range of mathematical and numerical problems related to the Radon transform and computed tomography is extremely wide, and they all cannot be discussed in one monograph. The books by Gelfand, Graev and Vilenkin, by Helgason, and by Natterer, as well as papers by many authors deal with the variety of the mathematical problems arising in the theory of the Radon transform. A much greater number of papers deal with the practical applications of the Radon transform (see [De2] for many references to works in applied areas and [Nat3] for many references related to computational aspects of the inversion of tomographic data).

In the present monograph we decided to emphasize the following problems:

(1) What are the properties of the Radon and X-ray transforms? In particular, from what functional space into what functional space do they act and on what classes of functions are these transforms injective? What are their ranges? How does one obtain the inversion formulas for these transforms? What are the properties of the adjoint operators?

(2) How does one define these transforms on some classes of distributions?

(3) What are the essential properties of $f(x)$ that can be obtained from local tomographic data? What are the algorithms for extracting this information?

(4) What can be found about f from incomplete tomographic data: for example, from limited-angle data? The limited-angle data are the data $\hat{f}(\alpha, p)$ known in a proper cone, that is, not for all angles α.

(5) Given an image of an arbitrary nature, how does one find edges and other features in this image?

Such a selection of problems reflects research interests of the authors: most of the new results presented in this book have been obtained by them (see [R], [RZ], [KR], [RSZ], [FKR], [K]). These results include:

(1) A detailed study of the singularities of Rf, Xf, and R^*f; formulas for the asymptotic behavior of $\hat{f}(\alpha, p)$ near a point $(\alpha, p) \in \hat{S}$, where $\hat{S}$ is the singular support of $\hat{f}$; it is discovered that $S := \operatorname{singsupp} f$ and $\hat{S}$ are in an one-to-one correspondence and there is a simple map which sends $\hat{S}$ onto S: this map is the Legendre transform. The notion of the classical Legendre transform is generalized so that one could define the Legendre transform of a function known on a manifold of codimension greater than one. Geometric properties of $\hat{S}$ are studied and the relation between these properties of S and $\hat{S}$ are discussed. A simple relation between the wave fronts of f and $\hat{f}$ is given. A numerical method for finding S given $\hat{S}$ is described. Asymptotic behavior of $(\mathcal{B}f)(x)$ when x approaches $S := \operatorname{singsupp} f$ is found for a class of pseudodifferential operators $\mathcal{B}$, which is of basic interest in tomography.

(2) New methods for finding discontinuity curves (surfaces) S of f from local tomographic data and the sizes of the jumps of f across S are developed. A new concept, that of pseudolocal tomography, is introduced and its efficiency in finding discontinuities of f from local tomographic data is demonstrated. Estimates of the rate of convergence of a regularized inversion formula for the Radon transform on a class of piecewise-continuous functions are obtained.

(3) New methods for inversion of incomplete tomographic data are described.

(4) New geometries in inversion of cone-beam data used in γ-ray tomography are introduced and studied; methods for exact inversion of cone-beam data in helical tomography are described.

(5) New definition of the Radon transform on various classes of distributions is given; the three earlier definitions (Gelfand-Graev's, Helgason's and Ludwig's), which were considered different in earlier literature (see [GGV], [Hel2], [Lu]), are shown to be equivalent, if properly understood, and equivalent to the new definition; constructive description of the space of test functions is given.

(6) New simple algorithm for edge detection is proposed. This algorithm is justified mathematically and tested on synthetic and real data.

(7) A new fairly general test of randomness is proposed and justified mathematically. Application of the test to edge detection is illustrated by the results of numerical experiments.

It should be noted that many of the above new theoretical results have been implemented numerically. The corresponding algorithms are described and the results of their testing are included in the book.

The traditional material which can be found elsewhere in the literature is often presented in an original way: some derivations are new or simplified, some results are improved. A discussion of the original points in the presentation of the traditional material is given in the Bibliographical notes.

The aims of the book are:

(1) To present the classical properties of the Radon transform in a simple way for a broad audience. Not much background material is required from the reader. Practically all the auxiliary material is presented in Chapter 14. Therefore, the book is essentially self-contained.

(2) To present in a self-contained way the new theories developed by the authors and to demonstrate their practical efficiency.

The book is not an encyclopedia on the Radon transform. The literature cited is not complete (and could not be probably): only the works used in writing this book and a limited number of other works, in which the reader finds further references, are mentioned. Many questions are not discussed at all or discussed insufficiently. For example, the Radon transform on non-Euclidean spaces, on groups, (see [Hel1], [GGV]), inverse problems for transport equation, integral geometry, sampling and numerical reconstruction methods, etc.

The book is written for a broad audience. It can be used as a text for a course on the Radon transform and its applications. It can be used by graduate students and researchers: much new material and some research problems can be found in the book. It can be used by engineers, physicists and radiologists who deal with processing of tomographic data.

For the most part, the references are collected in the Bibliographical notes, where the origin of the results presented in the book are stated.

1.2. Review of some applications of the Radon transform

1.2.1. Applications in medicine and non-destructive evaluation

1.2.1.1. Transport equation and X-ray transmission tomography. We suppose that a beam of particles propagates through a medium. Assume that velocities of particles are the same and equal to v, and we can neglect collisions of particles between themselves. The particles may collide with the fixed atoms in the medium and, in this case, they are absorbed. The scattering is also neglected.

Let $\psi = \psi(x, t)$ be the density of the particles at the spatial location x at the time moment t. Then the transport equation is the differential form of the conservation of particles law:

$$\frac{\partial \psi(x, t)}{\partial t} + v \cdot \nabla \psi = -|v|\sigma(x)\psi + q(x, t), \qquad (1.2.1)$$

where $q(x, t)$ is the source term, and $\sigma(x)$ is the probability of absorption of the particle at the point (x, t). Let $v = |v|\alpha$, where α is a unit vector in the direction of v, and let $x = x_0 + s\alpha$ be the parametric equation of the line L along which the particles propagate. If one considers only the stationary process in which $\frac{\partial}{\partial t} = 0$, then (1.2.1) becomes

$$\frac{\partial \psi(x)}{\partial s} = -\sigma(x)\psi + h(x), \qquad x = x_0 + s\alpha. \qquad (1.2.2)$$

Here $h = q/|v|$; it was assumed that $|v| = \text{const} > 0$ and the source term does not depend on α. In X-ray transmission tomography, $\psi =: I$ is the intensity of an X-ray beam, $h(x) = 0$ (there are no sources of radiation inside the object being interrogated), and $\sigma =: f$ is the X-ray attenuation coefficient. In this case equation (1.2.2) becomes

$$\frac{dI(x)}{ds} = -f(x)I(x), \qquad x = x_0 + s\alpha. \qquad (1.2.3)$$

Integrating (1.2.3) along the line L yields:

$$\frac{I}{I_0} = \exp\left(-\int_L f(x)dx\right), \qquad (1.2.4)$$

where I_0 is the intensity of the beam before it entered the body, and I is the intensity of this beam after it has left the body. Therefore

$$\ln \frac{I_0}{I} = \int_L f(x)dx = g(x_0, \alpha) = \hat{f}(\alpha^\perp, x_0 \cdot \alpha^\perp), \qquad (1.2.5)$$

where $\alpha^\perp$ is a unit vector perpendicular to α. The conclusion is: the observed quantity $\frac{I_0}{I}$ gives the Radon transform of f if $n = 2$ and X-ray transform of f if $n > 2$. Inverting the corresponding transform, one gets the unknown X-ray attenuation coefficient f.

X-ray transmission tomography is popular in medical diagnostics and in many industrial applications, where there is a need for nondestructive evaluation. Transmitting x-rays through an object at many different angles and inverting the resulting line integral data, one obtains the distribution of the X-ray attenuation coefficient f inside the object, from which one can make further conclusions about its inner structure.

1.2.1.2. SPECT. Single photon emission computed tomography (in a short form, SPECT) is based on equation (1.2.2). Now we assume that each volume element of the medium emits the particles so that $h(x)ds$ is the number of particles emitted in a cylinder of length ds and with unit cross section. It is assumed further that the number of the emitted particles does not depend on the direction α. The quantity $\sigma(x)$ has the same meaning as in Section 1.2.1.1. Equation (1.2.2) along a straight line is a linear differential equation. Integration of (1.2.2) over the segment of L between the point where L enters the object and a detector (the latter is supposed to be collimated so as to detect the radiation only along the line L) yields:

$$I(L) = \int_L h(x) \exp\left(-\int_{L(x)} \sigma(y)dy\right) dx. \qquad (1.2.6)$$

Here $L(x)$ is the segment of L between the point x and the detector. Note that $\sigma = 0$ outside the object, that is where $h = 0$. If $n = 2$, the right-hand side of (1.2.6) is called the attenuated Radon transform. The problem is to find h knowing σ and $I(L)$.

SPECT is used in medicine in the case when the patient is given the radiopharmaceutical, and the problem is to find the distribution h of radionuclides inside the body.

REMARK 1.2.1. Suppose that the sources emit particles simultaneously in opposite directions. Suppose also that one counts only such events when two particles reach detectors located on the opposite sides of L simultaneously. Then in formula (1.2.6) the exponential factor is replaced by $\exp\left(-\int_{-\infty}^{\infty} \sigma(y)dy\right)$. If σ is known, equation (1.2.6), under the above assumption, yields the Radon transform of h:

$$\int_L h(x)dx = I(L)\exp\left(\int_L \sigma(y)dy\right).$$

This is the case in positron emission tomography (PET).

REMARK 1.2.2. One cannot determine both functions h and σ from equation (1.2.6) knowing only $I(L)$ for all L. However, under some additional assumptions, one can recover h if σ is known.

An interesting question which is not resolved completely, is the following one:

Problem 1.2.1. Suppose that

$$\int_{\alpha \cdot x=p} f(x)\mu(x,\alpha,p)dx = A(\alpha,p),$$

where $A(\alpha,p)$ is known for all $\alpha \in S^{n-1}, p \in \mathbb{R}^1$, and $\mu > 0$ is a weight function. Under what assumptions on μ is the mapping $T : f \longmapsto A(\alpha,p)$ injective?

It is known that there is $f \in C_0^\infty(\mathbb{R}^2)$, $f \not\equiv 0$, and $0 < \mu(x,\alpha,p) \in C^\infty$ such that $A(\alpha,p) \equiv 0$. On the other hand, some sufficient conditions on μ are known for the map T to be injective.

1.2.2. Applications in geophysics

Consider a monochromatic scalar wavefield

$$[\nabla^2 + k^2 + k^2 v(x)]u = 0, \quad x \in \mathbb{R}^3, \tag{1.2.7}$$

where $v(x)$ is the inhomogeneity in the wave velocity profile which corresponds to an inhomogeneity in the medium, $k > 0$ is the wavenumber. If one looks for u of the form

$$u = A\exp(ik\varphi) = \exp(ik\varphi(x))\sum_{j=0}^{\infty} \frac{A_j}{(ik)^j}, \tag{1.2.8}$$

then, equating in (1.2.7) the coefficients in front of similar terms, one gets

$$(\nabla\varphi)^2 = 1 + v(x), \tag{1.2.9}$$

$$2\nabla\varphi \cdot \nabla A_0 + A_0 \nabla^2\varphi = 0, \tag{1.2.10}$$

$$2 \nabla \varphi \cdot \nabla A_n + A_n \nabla^2 \varphi = -\nabla^2 A_{n-1}, \quad n \geq 1. \tag{1.2.11}$$

Equation (1.2.9) is called the eiconal equation, and (1.2.10) - (1.2.11) are the transport equations.

The eiconal equation (1.2.9) can be solved by the method of characteristics. The characteristics of equation (1.2.9) are the extremals of the functional

$$\int_{\mathcal{L}(s_0,s)} [1 + v(x)]dx = \min_{\mathcal{L}(s_0,s)} := t(s_0,s), \tag{1.2.12}$$

and (1.2.12) is the Fermat principle of minimum of time needed for the signal to reach from one point to another.

Suppose that $v(x)$ is a compactly supported function and $\operatorname{supp} v(x) := D$. The inverse seismic problem, which is of interest in geophysics, consists of finding $v(x)$ given the travel times $t(s_0,s)$, needed for the wave to travel between two points s_0 and s for many pairs $s_0, s \in \Gamma := \partial D$. In other words, given the solutions $t(s_0,s), \forall s, s_0 \in \Gamma$, to the minimization problem (1.2.12), such that $t(s_0,s) = O(|s - s_0|)$ as $s \to s_0$, find $v(x)$.

If $|v(x)| \ll 1$, then one can use the following approximate solution to the above inverse problem. Neglecting $v(x)$, find the rays, that is, the extremals of the functional (1.2.12). If $v = 0$, then the straight lines joining s_0 and s are the extremals of this functional, so that the rays are straight lines. Then find $v(x)$ from the relation

$$\int_{L(s_0,s)} [1 + v(x)]dx = t(s_0,s), \tag{1.2.13}$$

where s_0 and s are arbitrary points on Γ and $L(s_0,s)$ is the straight line joining s_0 and s. But (1.2.13) is just the standard tomography problem: one wants to find $v(x)$ knowing the integrals of $v(x)$ over all straight lines passing through the support of $v(x)$.

If the background is not constant, for example, if equation (1.2.7) is

$$[\nabla^2 + k^2 n_0(x) + k^2 v(x)]u = 0,$$

where $n_0(x) > 0$ is known and $v(x), |v(x)| \ll |n_0(x)|$, is to be found, then in the approximation, similar to the one used above, one takes as rays the extremals of the functional

$$\int_{\mathcal{L}(s_0,s)} n_0(x)dx = \min_{\mathcal{L}(s_0,s)}, \tag{1.2.14}$$

and the problem, analogous to (1.2.13), will be

$$\int_{L(s_0,s)} [n_0(x) + v(x)]dx = t(s_0,s), \quad \forall s_0, s \in \Gamma,$$

where $\mathcal{L}(s_0, s)$ is the extremal of the functional (1.2.14), and $t(s_0, s)$ is the travel time between s_0 and s. This problem can be written as

$$\int_{\mathcal{L}(s_0,s)} v(x)dx = T(s_0, s), \qquad (1.2.15)$$

where $T(s_0, s)$ is known if both $n_0(x)$ and $\mathcal{L}(s_0, s)$ are known:

$$T(s_0, s) := t(s_0, s) - \int_{\mathcal{L}(s_0,s)} n_0(x)dx.$$

Problem (1.2.15) is a typical integral geometry problem, which generalizes the Radon problem: integrals of $v(x)$ are taken not along the straight lines, but along some family of lines, e.g., along the geodesics.

CHAPTER 2

PROPERTIES OF THE RADON
TRANSFORM AND INVERSION FORMULAS

2.1. Definitions and properties of the Radon transform and related transforms

2.1.1. Definition of the Radon transform

Throughout this section (unless specified otherwise) we assume that $f(x)$ is from the Schwartz space S consisting of infinitely differentiable functions which rapidly decay with all their derivatives. More precisely, this space consists of all $f \in C^\infty(\mathbb{R}^n)$ such that

$$\sup_{x \in \mathbb{R}^n} |x|^\beta |\partial_x^\gamma f(x)| < \infty,$$

where $\partial_x = (\frac{\partial}{\partial x_1}, \ldots, \frac{\partial}{\partial x_n})$, β and γ are arbitrary multiindices, i.e., $\beta = (\beta_1, \ldots, \beta_n)$ and β_j are nonnegative integers. The assumptions on f will be considerably relaxed later, in Chapters 3 and 10. There we treat the cases when f is in the Sobolev spaces or in some spaces of distributions.

Define

$$Rf := \hat{f}(\alpha, p) = \int_{l_{\alpha p}} f(x) ds. \qquad (2.1.1)$$

Here $x \in \mathbb{R}^n$,

$$l_{\alpha p} := \{x : \alpha \cdot x = p\}, \quad \alpha \in \mathbb{R}^n, \ p \in \mathbb{R}, \qquad (2.1.2)$$

$\alpha \cdot x$ is the dot product, and ds is the Lebesgue measure on the plane $l_{\alpha p}$. Geometrically, Rf is defined naturally for $\alpha \in S^{n-1}$ and $p \geq 0$, since the plane is uniquely defined by these parameters. Then Rf is extended as an even function $\hat{f}(\alpha, p) = \hat{f}(-\alpha, -p)$ for $S^{n-1} \times \mathbb{R}$. Since the plane $l_{\alpha p}$ remains the same when α and p are multiplied by a nonzero real number, it is natural to consider the point $(\alpha : p)$ as a point in the projective space $\mathbb{RP}_n$. This space consists of all the points $(\alpha : p)$ such that for any $\lambda \in \mathbb{R}, \lambda \neq 0$, the points $(\lambda \alpha : \lambda p)$ are identified with $(\alpha : p)$. We will write (α, p) in place of $(\alpha : p)$.

11

This definition of the Radon transform implies that the Radon transform is extended from $S^{n-1} \times \mathbb{R}$ to $\mathbb{R}^n \times \mathbb{R}$ as a homogeneous function of degree zero:

$$\hat{f}(\alpha, p) = \hat{f}(\lambda\alpha, \lambda p), \quad \lambda \in \mathbb{R}, \ \lambda \neq 0. \tag{2.1.3}$$

and, therefore, can be naturally defined on the projective space $\mathbb{RP}_n$. In particular,

$$\hat{f}(-\alpha, -p) = \hat{f}(\alpha, p). \tag{2.1.3a}$$

This means that $\hat{f}(\alpha, p)$ is a homogeneous function of degree zero. Recall that a homogeneous function of degree m is defined by the identity

$$g(\lambda x) = \lambda^m g(x), \quad \lambda > 0. \tag{2.1.4}$$

Very often the Radon transform is defined differently, namely by the formula

$$R_1 f := \check{f}(\alpha, p) := \int_{\mathbb{R}^n} f(x)\delta(p - \alpha \cdot x)dx, \tag{2.1.5}$$

where δ is the delta-function. Since

$$\delta(\lambda t) = |\lambda|^{-1}\delta(t), \ t \in \mathbb{R}, \ \lambda \neq 0, \tag{2.1.6}$$

definition (2.1.5) implies

$$\check{f}(\lambda\alpha, \lambda p) = |\lambda|^{-1}\check{f}(\alpha, p), \ \lambda \neq 0. \tag{2.1.7}$$

Thus $\check{f}(\alpha, p)$ is a homogeneous function on $\mathbb{R}^n \times \mathbb{R}$ of degree -1. If $\alpha \in S^{n-1}$, where S^{n-1} is a unit sphere in $\mathbb{R}^n$, then $\hat{f} = \check{f}$.

The relation between $\hat{f}$ and $\check{f}$ is given by the formula:

$$\hat{f}(\alpha, p) = |\alpha|\check{f}(\alpha, p), \quad \alpha \in \mathbb{R}^n. \tag{2.1.8}$$

Exercise 2.1.0. Derive (2.1.8).

A natural domain of definition of the Radon transform (2.1.1) is the set of hyperplanes in $\mathbb{R}^n$, i.e., the projective space $\mathbb{RP}_n$ (without the infinite hyperplane H_∞) so that $(\alpha : p)$ must be considered up to a scalar multiple. The space $Z := S^{n-1} \times \mathbb{R}$ is a two-fold covering of $\mathbb{RP}_n \setminus H_\infty$ and we may consider the Radon transform an even function (i.e., $g(-\alpha, -p) = g(\alpha, p)$) defined on Z. When $|\alpha| = 1$, the two definitions coincide. They may be considered as different extensions of the Radon transform from Z to $\mathbb{R}^n \times \mathbb{R}$. We can also extend the Radon transform to this region in some other way. In this book the Radon transform (2.1.1) is used on its natural domain of definition, i.e., on the projective space, or on its two-fold covering Z; it is also used extensively in the sense of the second definition (2.1.5) which is convenient for various analytic derivations.

Since $\hat{f} = \check{f}$ if $|\alpha| = 1$ and since in this book the Radon transform is used mostly on the unit sphere, we decided to use only the notations Rf and $\hat{f}$ when $|\alpha| = 1$, no matter what definition is used at the moment. The notations $R_1 f$ and $\check{f}$ are used only when $|\alpha| \neq 1$.

2.1.2. Some generalizations

Suppose that the integration in (2.1.1) is taken over $m-$ dimensional affine linear space $M_m(x_0) := \{x : x = x_0 + \sum_{j=1}^{m} c_j e_j\}$, where x_0 is a vector in $\mathbb{R}^n$, e_j, $j = 1, \ldots, m$, are m linearly independent vectors, and c_j, $j = 1, \ldots, m$, are arbitrary constants. If $x_0 = 0$, then M_m is a linear m-dimensional subspace of $\mathbb{R}^n$. If $x_0 \neq 0$, then the linear manifold M_m is called an affine space. Define

$$X_m f(x_0, e_1, \ldots, e_m) := \int_{M_m(x_0)} f(x) ds, \qquad (2.1.9)$$

where ds is the m-dimensional Lebesgue measure in M_m. If $m = 1$, formula (2.1.9) can be written as

$$X f(x_0, \alpha) := \int_{-\infty}^{\infty} f(x_0 + \alpha t) dt, \ |\alpha| = 1. \qquad (2.1.10)$$

We will call $X_m f$ the X_m-ray transform of f, and Xf - the X-ray transform of f. As a rule, we write x in place of x_0 in formulas analogous to (2.1.10) in what follows, and we often write $g(x, \alpha)$ for Xf. The physical meaning of X-ray transform is discussed in Chapter 1.

If $n = 2$, then the Radon transform and X-ray transform can be reduced to each other. Namely, if $\alpha = (\alpha_1, \alpha_2)$, $|\alpha| = 1$, then

$$\hat{f}(\alpha, p) = g(\alpha p, \alpha^{\perp}), \qquad (2.1.11)$$

where $\alpha^{\perp} := (-\alpha_2, \alpha_1)$.

If in (2.1.10) we integrate over the ray $[0, \infty)$, we obtain the cone-beam transform:

$$(Df)(x_0, \alpha) := \int_{0}^{\infty} f(x_0 + t\alpha) dt, \ x_0 \in \mathbb{R}^n, \ \alpha \in S^{n-1}.$$

Cone-beam transform is investigated in Chapter 9. In $\mathbb{R}^2$, cone-beam transform is called fan-beam transform.

2.1.3. Simple properties of the Radon transform

Let us give several simple properties of the Radon transform.

(1) R is a linear operator
(2) Suppose $A = (a_{ij})_{i,j=1}^{n}$ is a non-singular matrix of a linear transformation, and $f_A(x) := f(A^{-1}x)$, then

$$\tilde{f}_A(\alpha, p) = |\det A| \check{f}(A'\alpha, p), \qquad (2.1.12)$$

where A' is the transposed matrix, $a'_{ij} = a_{ji}$.

Exercise 2.1.1. Prove (2.1.12).

Hint. Use the substitution $y = A^{-1}x$ and formula (2.1.5).

(3) If $f_\xi(x) := f(x + \xi)$, then

$$\hat{f}_\xi(\alpha, p) = \hat{f}(\alpha, p + \xi \cdot \alpha). \qquad (2.1.13)$$

Exercise 2.1.2. Prove (2.1.13)

Hint. Use the substitution $y = x + \xi$ and formula (2.1.5).

(4) Let $\alpha = (\alpha_1, \ldots, \alpha_n)$, $|\alpha| = 1$. One has

$$R_1 \frac{\partial f}{\partial x_k} = \alpha_k \frac{\partial \check{f}(\alpha, p)}{\partial p}, \quad \frac{\partial(R_1 f)}{\partial \alpha_k} = -\frac{\partial}{\partial p} R_1(x_k f). \qquad (2.1.14)$$

It is implicitly assumed in (2.1.14) that if one differentiates $R_1 f(\alpha, p)$ with respect to α, then $R_1 f(\alpha, p)$ is extended to all $\alpha \in \mathbb{R}^n, |\alpha| \neq 0$, using (2.1.7).

Exercise 2.1.3. Prove (2.1.14)

Hint. Use (2.1.5).

It follows from (2.1.14) that if $P_m(x)$ is a homogeneous polynomial with constant coefficients of degree m and $|\alpha| = 1$, then

$$R_1 (P_m(\partial_x)f) = P_m(\alpha) \frac{\partial^m \check{f}(\alpha, p)}{\partial p^m},$$

$$P_m(\partial_\alpha)(R_1 f) = (-1)^m \frac{\partial^m}{\partial p^m} R_1(P_m(x)f). \qquad (2.1.15)$$

Here $\partial_x = (\frac{\partial}{\partial x_1}, \ldots, \frac{\partial}{\partial x_n})$ and $\partial_\alpha = (\frac{\partial}{\partial \alpha_1}, \ldots, \frac{\partial}{\partial \alpha_n})$. In particular, if Δ is the Laplacian, then

$$R_1(\Delta_x f) = \frac{\partial^2 \check{f}(\alpha, p)}{\partial p^2}, \quad \Delta_\alpha(R_1 f) = \frac{\partial^2}{\partial p^2} R_1(|x|^2 f), \quad |\alpha| = 1. \quad (2.1.16)$$

It is clear that the first equations in (2.1.14)–(2.1.16) hold also in case of the Radon transform defined by (2.1.1).

2.1.4. Radon transform of a convolution

Let us define a convolution $f * g$ of $f(x)$ and $g(x)$:

$$h(x) := (f * g)(x) := \int_{\mathbb{R}^n} f(x - y)g(y)dy. \qquad (2.1.17)$$

Proposition 2.1.1. *One has:*

$$\hat{h}(\alpha,p) = \int_{-\infty}^{\infty} \hat{f}(\alpha,p-t)\hat{g}(\alpha,t)dt := \hat{f} \circledast \hat{g}, \qquad (2.1.18)$$

where $\circledast$ *denotes the one-dimensional convolution with respect to the second argument of the functions* $\hat{f}$ *and* $\hat{g}$.

Proof. One has

$$\hat{h}(\alpha,p) = \int_{\mathbb{R}^n} \delta(p - \alpha \cdot x) \left[\int_{\mathbb{R}^n} f(x-y)g(y)dy \right] dx$$

$$= \int_{\mathbb{R}^n} g(y) \int_{\mathbb{R}^n} \delta[p - \alpha \cdot y - \alpha \cdot (x-y)]f(x-y)dx\, dy$$

$$= \int_{\mathbb{R}^n} g(y)\hat{f}(\alpha,p - \alpha \cdot y)dy := I. \qquad (2.1.19)$$

Let $\alpha \cdot y = t$. Then

$$I = \int_{-\infty}^{\infty} \left(\int_{\alpha \cdot y = t} g(y)dy \right) \hat{f}(\alpha,p-t)dt = \int_{-\infty}^{\infty} \hat{f}(\alpha,p-t)\hat{g}(\alpha,t)dt.$$

The proof is complete. □

2.1.5. The Fourier slice theorem
By this name the following formula is known:

$$F_{p \to t}\hat{f} := \int_{-\infty}^{\infty} \hat{f}(\alpha,p)\exp(ipt)dp = \tilde{f}(t\alpha), \qquad (2.1.20)$$

where

$$\tilde{f}(t\alpha) := (\mathcal{F}f)(t\alpha) := \int_{\mathbb{R}^n} f(x)\exp(it\alpha \cdot x)dx. \qquad (2.1.21)$$

Note that Equation (2.1.20) can be written as:

$$FR = \mathcal{F}, \quad R = F^{-1}\mathcal{F}. \qquad (2.1.20')$$

Proof. Use (2.1.5) to get

$$\int_{-\infty}^{\infty} \exp(ipt) \int_{\mathbb{R}^n} f(x)\delta(p - \alpha \cdot x)dx \, dp$$

$$= \int_{\mathbb{R}^n} f(x) \int_{-\infty}^{\infty} \delta(p - \alpha \cdot x)\exp(ipt)dp \, dx = \int_{\mathbb{R}^n} f(x)\exp(it\alpha \cdot x)dx. \tag{2.1.22}$$

Formula (2.1.20) is proved. □

REMARK 2.1.1. Formula (2.1.20) can be extended from $f \in \mathcal{S}$ to a much larger class of functions and to some classes of distributions. This will be done later, in Chapter 10.

2.1.6. The adjoint operator R^*

Consider the Schwartz space $X := \mathcal{S}(\mathbb{R}^n)$ of functions $f(x)$, and let Y denote the Schwartz space $\mathcal{S}(Z), Z := S^{n-1} \times \mathbb{R}$, of functions $g(\alpha, p)$. The spaces X and Y are equipped with the inner products

$$< f_1, f_2 >: =< f_1, f_2 >_X := \int_{\mathbb{R}^n} f_1(x)\overline{f_2(x)}dx,$$

$$(g_1, g_2) : = (g_1, g_2)_Y := \int_{S^{n-1}} \int_{-\infty}^{\infty} g_1(\alpha, p)\overline{g_2(\alpha, p)}dp \, d\alpha, \tag{2.1.23}$$

where the bar stands for complex conjugation. In this chapter we will consider only real-valued functions f and g, and so complex conjugation will be omitted in what follows. One has, using the definition of the adjoint operator,

$$(Rf, g) =< f, R^*g > . \tag{2.1.24}$$

Thus

$$(Rf, g) = \int_{S^{n-1}} \int_{-\infty}^{\infty} \int_{\mathbb{R}^n} f(x)\delta(p - \alpha \cdot x)g(\alpha, p)dx \, dp \, d\alpha$$

$$= \int_{\mathbb{R}^n} f(x) \int_{S^{n-1}} \int_{-\infty}^{\infty} \delta(p - \alpha \cdot x)g(\alpha, p)dp \, d\alpha \, dx$$

$$= \int_{\mathbb{R}^n} f(x) \int_{S^{n-1}} g(\alpha, \alpha \cdot x)d\alpha \, dx =< f, R^*g > . \tag{2.1.25}$$

Therefore,

$$R^*g := \int_{S^{n-1}} g(\alpha, \alpha \cdot x)d\alpha. \tag{2.1.26}$$

The operator $R : X \to Y$ acts from the space of smooth rapidly decaying functions of x into the space of functions of (α, p). The operator R^* acts from the space of smooth functions of (α, p) into the space of functions of x. The action of these operators on more general spaces and, in particular, the description of the ranges of R and R^*, will be given in Chapters 3 and 10.

2.1.7. Formulas for $R^* R$ and RR^*

Let us denote $|S^{n-1}| = \frac{2\pi^{\frac{n}{2}}}{\Gamma(\frac{n}{2})}$ and $|B_1^n| = \frac{1}{n}|S^{n-1}|$ the area of the unit sphere in $\mathbb{R}^n$ and the volume of the unit ball in $\mathbb{R}^n$, respectively. Here $\Gamma(z)$ is the gamma-function. Recall that $f * g$ denotes convolution in $\mathbb{R}^n$.

Lemma 2.1.1. *One has*

$$R^* R f = |S^{n-2}||x|^{-1} * f, \qquad (2.1.27)$$

or, equivalently,

$$R^* R f = \frac{1}{\gamma} \mathcal{F}^{-1}(|\xi|^{1-n} \tilde{f}(\xi)), \qquad (2.1.28)$$

where $\gamma := 1/[2(2\pi)^{n-1}]$. *If g is even: $g(\alpha, p) = g(-\alpha, -p)$, then*

$$RR^* g = \frac{1}{\gamma} F_{t \to p}^{-1}(|t|^{1-n} F_{p \to t} g). \qquad (2.1.29)$$

Proof of formulas (2.1.27) and (2.1.28). One has, using Equations (2.1.26) and (2.1.5):

$$R^* R f = \int\limits_{S^{n-1}} \int\limits_{\mathbb{R}^n} \delta(\alpha \cdot x - \alpha \cdot y) f(y) dy \, d\alpha = |S^{n-2}| \int\limits_{\mathbb{R}^n} |x - y|^{-1} f(y) dy.$$
$$(2.1.30)$$

Here we have used the identity

$$\int\limits_{S^{n-1}} \delta(\alpha \cdot z) d\alpha = |S^{n-2}||z|^{-1}, \qquad (2.1.31)$$

which can be proved as follows. Denote $\beta = z/|z|$. Then, by (2.1.6),

$$\int\limits_{S^{n-1}} \delta(\alpha \cdot z) d\alpha = |z|^{-1} \int\limits_{S^{n-1}} \delta(\alpha \cdot \beta) d\alpha.$$

Using the spherical coordinate system in $\mathbb{R}^n$ (see Section 14.4.4) with the x_n-axis directed along the vector β, one obtains

$$\int_{S^{n-1}} \delta(\alpha \cdot \beta)d\alpha = \int_0^\pi \int_{S^{n-2}} \delta(\cos\theta_{n-1})(\sin\theta_{n-1})^{n-2}d\gamma \, d\theta_{n-1}$$

$$= |S^{n-2}| \int_{-1}^1 \delta(u)(1-u^2)^{\frac{n-3}{2}} du = |S^{n-2}|. \tag{2.1.32}$$

Here $d\gamma$ is the element of the surface area of S^{n-2}, and the substitution $u = \cos\theta_{n-1}$ was used. Identity (2.1.31) follows from the last two equations. Equation (2.1.28) is proved using the formula for the Fourier transform of a convolution and the identity [GS]

$$\mathcal{F}_{x\to\xi}(|x|^\lambda) = 2^{\lambda+n}\pi^{\frac{n}{2}}\frac{\Gamma\left(\frac{\lambda+n}{2}\right)}{\Gamma\left(\frac{-\lambda}{2}\right)}|\xi|^{-\lambda-n}, \quad \lambda \neq 0, 2, 4, \ldots, \tag{2.1.33}$$

with $\lambda = -1$. Alternatively, Equation (2.1.28) can be proved combining (2.1.20') and the identity $R^* = \gamma^{-1}\mathcal{F}^{-1}|t|^{1-n}F$, which is derived below (see (2.2.18')). Formulas (2.1.27) and (2.1.28) are proved. $\square$

Exercise 2.1.4. Let $h(t)$ be a continuous function defined on the interval $[-1, 1]$, and $\alpha, \beta \in S^{n-1}$. Prove that

$$\int_{S^{n-1}} h(\alpha \cdot \beta)d\beta = |S^{n-2}| \int_{-1}^1 h(t)(1-t^2)^{\frac{n-3}{2}}dt. \tag{2.1.34}$$

Hint. Follow the proof of Equation (2.1.32).

Exercise 2.1.5. Under the assumptions of Exercise 2.1.4, prove that

$$\int_{S^{n-1}} h(\alpha \cdot \beta)Y_l(\beta)d\beta = c(n, l)Y_l(\alpha), \tag{2.1.35}$$

where $Y_l(\beta)$ are the normalized in $L^2(S^{n-1})$ spherical harmonics,

$$c(n, l) = \frac{|S^{n-2}|}{C_l^{\left(\frac{n-2}{2}\right)}(1)} \int_{-1}^1 h(t)C_l^{\left(\frac{n-2}{2}\right)}(t)(1-t^2)^{\frac{n-3}{2}}dt, \tag{2.1.36}$$

and $C_l^{(p)}$ are the Gegenbauer polynomials (see Section 14.4 for the definition and properties of the spherical harmonics and the Gegenbauer polynomials). Formulas (2.1.35) and (2.1.36) are known as the *Funk-Hecke Theorem*.

Hint. One has $h(t) = \sum_{j=0}^{\infty} h_j C_j^{(\frac{n-2}{2})}(t)$, where

$$h_j = \frac{1}{\Lambda_{j,n}} \int_{-1}^{1} h(t) C_j^{(\frac{n-2}{2})}(t)(1-t^2)^{\frac{n-3}{2}} dt,$$

and the constant $\Lambda_{j,n}$ is defined in (14.4.24). Thus, (2.1.35) follows from the formula

$$\int_{S^{n-1}} C_j^{(\frac{n-2}{2})}(\alpha \cdot \beta) Y_l(\beta) d\beta = \frac{|S^{n-2}| \Lambda_{l,n}}{C_l^{(\frac{n-2}{2})}(1)} Y_l(\alpha) \delta_{jl},$$

where $\delta_{jl} = 1$ if $j = l$ and $\delta_{jl} = 0$ if $j \neq l$. This equation is a consequence of (14.4.45) and other properties of the spherical harmonics (see Section 14.4.5).

We now finish the proof of Lemma 2.1.1.

Proof of formula (2.1.29). Define

$$\tilde{g}(\alpha, t) = F_{p \to t} g = \int_{-\infty}^{\infty} g(\alpha, p) \exp(ipt) dp, \qquad (2.1.37)$$

so that

$$g(\alpha, p) = \frac{1}{2\pi} \int_{-\infty}^{\infty} \tilde{g}(\alpha, t) \exp(-ipt) dt. \qquad (2.1.38)$$

First, we prove the identity

$$\mathcal{F}(R^* g) = 2(2\pi)^{n-1} |t|^{1-n} F_{p \to t} g. \qquad (2.1.39)$$

Using (2.1.38) and (2.1.26), one gets for the left-hand side of (2.1.39) the following expression:

$$\int_{\mathbb{R}^n} dx \, \exp(it\alpha \cdot x) \int_{S^{n-1}} d\beta \frac{1}{2\pi} \int_{-\infty}^{\infty} dq \, \exp(-i\beta \cdot xq) \tilde{g}(\beta, q)$$

$$= \frac{1}{2\pi} \int_{S^{n-1}} d\beta \int_{-\infty}^{\infty} dq \, \tilde{g}(\beta, q)(2\pi)^n \delta(t\alpha - \beta q)$$

$$= (2\pi)^{n-1} |t|^{1-n} [\tilde{g}(\alpha, t) + \tilde{g}(-\alpha, -t)] = 2(2\pi)^{n-1} |t|^{1-n} \tilde{g}(\alpha, t).$$
$$(2.1.40)$$

Here we took into account that $\tilde{g}$ is even: $\tilde{g}(\alpha, t) = \tilde{g}(-\alpha, -t)$, if $g(\alpha, p)$ is even, and used the identity

$$\delta(t\alpha - q\beta) = |t|^{1-n}[\delta(t - q)\delta(\alpha - \beta) + \delta(t + q)\delta(\alpha + \beta)].$$

Using the Fourier slice theorem and identity (2.1.39), we finish the proof of (2.1.29):

$$F_{p \to t}(RR^*g) = \mathcal{F}(R^*g) = 2(2\pi)^{n-1}|t|^{1-n}F_{p \to t}g. \tag{2.1.41}$$

☐

2.1.8. Formula for $(R^*g) * f$

One has

$$(R^*g) * f = R^*(g \circledast Rf). \tag{2.1.42}$$

Proof. Using an obvious identity

$$\int_{\mathbb{R}^n} f(y)dy = \int_{-\infty}^{\infty} \int_{\alpha \cdot y = t} f(y)dy\, dt = \int_{-\infty}^{\infty} \hat{f}(\alpha, t)dt$$

and the substitution $t = \alpha \cdot y$, we derive

$$(R^*g) * f = \int_{\mathbb{R}^n} \left[\int_{S^{n-1}} g(\alpha, \alpha \cdot (x - y))d\alpha \right] f(y)dy$$

$$= \int_{S^{n-1}} \int_{\mathbb{R}^n} g(\alpha, \alpha \cdot x - \alpha \cdot y)f(y)dy\, d\alpha$$

$$= \int_{S^{n-1}} \int_{-\infty}^{\infty} \hat{f}(\alpha, t)g(\alpha, \alpha \cdot x - t)dt\, d\alpha = R^*(\hat{f} \circledast g).$$

☐

2.1.9. The Parseval and Plancherel equalities

One has the Plancherel formula:

$$< f, h >:= \int_{\mathbb{R}^n} f(x)h(x)dx$$

$$= \frac{(n-1)!i^n}{(2\pi)^n} \int_{-\infty}^{\infty} \int_{-\infty}^{\infty} \frac{dp_1\, dp_2}{(p_1 - p_2 + i0)^n} \int_{S^{n-1}} d\alpha \hat{f}(\alpha, p_1)\hat{h}(\alpha, p_2). \tag{2.1.43}$$

If $h = f$, then (2.1.43) is called the Parseval formula.

Proof of (2.1.43). Using the Plancherel formula for the Fourier transform in spherical coordinates and then Equation (2.1.20), we get:

$$\int_{\mathbb{R}^n} f(x)h(x)dx = (2\pi)^{-n} \int_{\mathbb{R}^n} \tilde{f}(\xi)\overline{\tilde{h}(\xi)}d\xi$$

$$= (2\pi)^{-n} \int_{S^{n-1}} \int_0^\infty \left[\int_{-\infty}^\infty \hat{f}(\alpha, p_1)e^{ip_1 t}dp_1 \int_{-\infty}^\infty \hat{h}(\alpha, p_2)e^{-ip_2 t}dp_2 \right] t^{n-1} dt\, d\alpha$$

$$= (2\pi)^{-n} \int_{-\infty}^\infty \int_{-\infty}^\infty dp_1 dp_2 \int_{S^{n-1}} d\alpha \int_0^\infty dt\, t^{n-1} e^{it(p_1-p_2)} \hat{f}(\alpha, p_1)\hat{h}(\alpha, p_2)$$

$$= (2\pi)^{-n} \int_{-\infty}^\infty \int_{-\infty}^\infty dp_1 dp_2 \int_{S^{n-1}} d\alpha$$

$$\times \left[\left(\frac{1}{i}\frac{\partial}{\partial p_1} \right)^{n-1} \frac{i}{p_1 - p_2 + i0} \right] \hat{f}(\alpha, p_1)\hat{h}(\alpha, p_2)$$

$$= (2\pi)^{-n} \int_{-\infty}^\infty \int_{-\infty}^\infty dp_1 dp_2 \int_{S^{n-1}} d\alpha \frac{i^n (n-1)!}{(p_1 - p_2 + i0)^n} \hat{f}(\alpha, p_1)\hat{h}(\alpha, p_2).$$
(2.1.44)

Here we have used the formula

$$\int_0^\infty t^{n-1} e^{its} dt = i^n (n-1)! s^{-n} + (-i)^{n-1}\pi\delta^{(n-1)}(s)$$

$$= i^n (n-1)!(s + i0)^{-n},$$
(2.1.45)

which follows immediately by applying the operator $(\frac{1}{i}\frac{d}{ds})^{n-1}$ to the identity

$$\int_0^\infty e^{its} dt = \frac{i}{s + i0}.$$
(2.1.46)

It is known (see Section 14.2) that

$$\frac{1}{s + i0} = \frac{1}{s} - i\pi\delta(s).$$
(2.1.47)

From (2.1.46) and (2.1.47) one gets (2.1.45). □

Recall that

$$\hat{f}(\alpha, p) = \hat{f}(-\alpha, -p), \quad \hat{h}(\alpha, p) = \hat{h}(-\alpha, -p).$$
(2.1.48)

Therefore, if $n = 2m + 1$, one has

$$\int\limits_{-\infty}^{\infty} \int\limits_{-\infty}^{\infty} dp_1 dp_2 \int\limits_{S^{n-1}} d\alpha (p_1 - p_2)^{-2m-1} \hat{f}(\alpha, p_1) \hat{h}(\alpha, p_2) = 0,$$

and formula (2.1.43) takes the form

$$< f, h > = \frac{\pi}{(2\pi)^n} \int\limits_{-\infty}^{\infty} dp \int\limits_{S^{n-1}} d\alpha \hat{f}^{(m)}(\alpha, p) \hat{h}^{(m)}(\alpha, p), \quad n = 2m + 1,$$

$$(2.1.49)$$

where $\hat{f}^{(m)} := \partial^m \hat{f} / \partial p^m$.

If $n = 2m$, then

$$\int\limits_{-\infty}^{\infty} \int\limits_{-\infty}^{\infty} dp_1 dp_2 \int\limits_{S^{n-1}} d\alpha \delta^{(2m-1)}(p_1 - p_2) \hat{f}(\alpha, p_1) \hat{h}(\alpha, p_2) = 0,$$

and formula (2.1.43) takes the form:

$$< f, h > = \frac{i^n (n-1)!}{(2\pi)^n} \int\limits_{-\infty}^{\infty} \int\limits_{-\infty}^{\infty} dp_1 dp_2 \int\limits_{S^{n-1}} d\alpha \frac{\hat{f}(\alpha, p_1) \hat{h}(\alpha, p_2)}{(p_1 - p_2)^n}, \quad n = 2m.$$

$$(2.1.50)$$

Define the operator K_1:

$$K_1 g = \frac{(n-1)!(-i)^n}{(2\pi)^n} (p - i0)^{-n} \circledast g, \quad g \in \mathcal{S}(Z). \qquad (2.1.51)$$

Then, identity (2.1.43) can be written as

$$< f, h > = (Rf, K_1 Rh) = (\hat{f}, K_1 \hat{h}). \qquad (2.1.52)$$

Taking into account that $\hat{f}$ and $\hat{h}$ are even, we get from (2.1.44)

$$< f, h > = \frac{1}{2(2\pi)^n} \int\limits_{-\infty}^{\infty} \int\limits_{-\infty}^{\infty} dp_1 dp_2 \int\limits_{S^{n-1}} d\alpha$$

$$\times \int\limits_{-\infty}^{\infty} dt \, |t|^{n-1} e^{-it(p_1 - p_2)} \hat{f}(\alpha, p_1) \hat{h}(\alpha, p_2).$$

Defining the operator K:

$$K g = \gamma F^{-1}(|t|^{n-1} F g), \quad \gamma := \frac{1}{2(2\pi)^{n-1}}, \quad g \in \mathcal{S}(Z),$$

we also get

$$< f, h > = (Rf, KRh) = (\hat{f}, K\hat{h}). \qquad (2.1.52')$$

For more information on the operator K and its relation with R and R^* see Section 2.2.2 (Equation (2.2.16) and below).

2.1.10. Integrals over a domain
Consider the integral

$$\int_D f(x)dx = \int_{\mathbb{R}^n} \chi_D(x)f(x)dx. \tag{2.1.53}$$

Here D is a bounded domain and

$$\chi_D(x) := \begin{cases} 1, & x \in \bar{D} \\ 0, & x \notin \bar{D} \end{cases} \tag{2.1.54}$$

where $\bar{D}$ is the closure of D.

Applying to (2.1.53) Plancherel formulas (2.1.49) for $n = 2m+1$ and (2.1.50) for $n = 2m$, one obtains

$$\int_D f(x)dx = \frac{\pi}{(2\pi)^n} \int_{-\infty}^{\infty} dp \int_{S^{n-1}} d\alpha \hat{f}^{(m)}(\alpha,p)\hat{\chi}^{(m)}(\alpha,p), \quad n = 2m+1,$$
$$\tag{2.1.55}$$

and

$$\int_D f(x)dx = \frac{i^n(n-1)!}{(2\pi)^n} \int_{-\infty}^{\infty}\int_{-\infty}^{\infty} dp_1\,dp_2 \int_{S^{n-1}} \frac{d\alpha \hat{f}(\alpha,p_1)\hat{\chi}_D(\alpha,p_2)}{(p_1-p_2)^n},$$
$$n = 2m. \tag{2.1.56}$$

Note that

$\hat{\chi}_D(\alpha,p)$ is the area of the section of the domain D

by the plane $\alpha \cdot x = p$.   (2.1.57)

We do not discuss the degree of smoothness of $\hat{\chi}_D(\alpha,p)$ with respect to the p-variable. This is done in Chapter 4. However, if $f(x) \in C_0^\infty(D)$, then $\hat{f}(\alpha,p)$ is smooth and compactly supported. Therefore, one can integrate by parts in (2.1.55) and get

$$\int_D f(x)dx = \frac{\pi(-1)^m}{(2\pi)^n} \int_{-\infty}^{\infty} dp \int_{S^{n-1}} d\alpha \hat{f}^{(n-1)}(\alpha,p)\hat{\chi}_D(\alpha,p), \quad n = 2m+1.$$
$$\tag{2.1.58}$$

2.1.11. Consistency and moment conditions
Suppose the function $g(\alpha,p)$, $\alpha \in S^{n-1}$, $p \in \mathbb{R}$, is given. Under what conditions is this function representable as $g(\alpha,p) = Rf(x)$ for $f(x)$

from a certain class of functions? This is an important question about the range of the Radon transform, which will be discussed in Chapter 3. Here we give some simple necessary conditions for g to be the Radon transform of a function f.

The first necessary condition is given by formula (2.1.3a): the function $g(\alpha, p)$ has to be even

$$g(\alpha, p) = g(-\alpha, -p). \qquad (2.1.59)$$

The second set of necessary conditions, called *the moment conditions*, consists of the following equations:

$$\int_{-\infty}^{\infty} g(\alpha, p) p^m dp = \mathcal{P}_m(\alpha), \quad m = 0, 1, 2, \ldots, \qquad (2.1.60)$$

where $\mathcal{P}_m(\alpha)$ is a restriction to S^{n-1} of a homogeneous polynomial of degree m of $\alpha \in \mathbb{R}^n$. These conditions make sense if

$$\int_{-\infty}^{\infty} |g(\alpha, p)|(1 + |p|)^m dp < \infty, \quad m = 0, 1, 2, \ldots, \qquad (2.1.61)$$

which happens, for example, if $g(\alpha, p)$ is compactly supported or belongs to the Schwartz space with respect to the p-variable.

Let us derive the moment conditions (2.1.60). One has

$$\int_{-\infty}^{\infty} \hat{f}(\alpha, p) p^m dp = \int_{-\infty}^{\infty} \int_{\mathbb{R}^n} \delta(p - \alpha \cdot x) f(x) dx\, p^m dp$$

$$= \int_{\mathbb{R}^n} \int_{-\infty}^{\infty} \delta(p - \alpha \cdot x) p^m dp\, f(x) dx$$

$$= \int_{\mathbb{R}^n} (\alpha \cdot x)^m f(x) dx =: \mathcal{P}_m(\alpha). \qquad (2.1.62)$$

Clearly, $\mathcal{P}_m(\alpha)$ is a homogeneous (of degree m) polynomial of $\alpha \in \mathbb{R}^n$.

Exercise 2.1.6. Can the function $g(\alpha, p) = \exp(-p^2)\cos(2\theta)$ be the Radon transform of a function $f(x)$, $x \in \mathbb{R}^2$, $\alpha = (\cos\theta, \sin\theta)$?

Hint. Check if the moment conditions (2.1.12) are satisfied for all $m = 0, 1, 2, \ldots$.

2.1.12. The Radon transform of spherically symmetric functions

Let $f(x) = f(|x|)$, $x \in \mathbb{R}^n$. Then, using spherical coordinates on the plane $l_{\alpha p}$, one gets:

$$\hat{f}(\alpha, p) = \int\limits_{\alpha \cdot x = p} f(|x|) ds = \int\limits_{S^{n-2}} \int\limits_0^\infty f(\sqrt{p^2 + \rho^2}) \rho^{n-2} d\rho d\alpha$$

$$= |S^{n-2}| \int\limits_0^\infty f(\sqrt{p^2 + \rho^2}) \rho^{n-2} d\rho$$

$$= |S^{n-2}| \int\limits_p^\infty f(s)(s^2 - p^2)^{\frac{n-3}{2}} s\, ds. \tag{2.1.63}$$

In particular, if $n = 2$, then $|S^0| = 2$ and

$$\hat{f}(\alpha, p) = 2 \int\limits_0^\infty f(\sqrt{p^2 + \rho^2}) d\rho = 2 \int\limits_p^\infty f(s) \frac{s\, ds}{\sqrt{s^2 - p^2}}. \tag{2.1.64}$$

If $n = 3$, then $|S^{n-2}| = 2\pi$ and

$$\hat{f}(\alpha, p) = 2\pi \int\limits_0^\infty f(\sqrt{p^2 + \rho^2}) \rho\, d\rho = 2\pi \int\limits_p^\infty f(s) s\, ds. \tag{2.1.65}$$

One can see that if $f(x) = f(|x|)$, then $\hat{f}(\alpha, p) = \hat{f}(p)$, that is $\hat{f}$ does not depend on α. Equations (2.1.63)-(2.1.64) are of the Abel type and can be solved in a closed form (see Sections 11.2, 2.2.5 and 2.2.6).

Exercise 2.1.7. Calculate $R \exp(-|x|^2)$.

Answer.

$$R \exp(-|x|^2) = \pi^{\frac{n-1}{2}} \exp(-p^2), \ n \geq 2.$$

Exercise 2.1.8. Let

$$f(x) = \begin{cases} (1 - |x|^2)^{\lambda-1}, & |x| < 1, \\ 0, & |x| \geq 1, \end{cases} \quad x \in \mathbb{R}^2, \lambda > 1.$$

Prove that

$$\hat{f}(p) = \begin{cases} \frac{\pi^{1/2}\Gamma(\lambda)}{\Gamma(\lambda+\frac{1}{2})}(1 - p^2)^{\lambda-1/2}, & -1 \leq p \leq 1, \\ 0, & |p| \geq 1, \end{cases} \tag{2.1.66}$$

where $\Gamma(\lambda)$ is the gamma-function.

Exercise 2.1.9. Prove the formula:

$$R\{H_m(x_1)H_k(x_2)\exp[-(x_1^2 + x_2^2)]\}$$
$$= \pi^{1/2}(\cos\theta)^m(\sin\theta)^k\exp(-p^2)H_{m+k}(p), \quad (2.1.67)$$

where $\alpha = (\cos\theta, \sin\theta)$ and $H_m(x)$ are the Hermite polynomials (cf. Section 14.4.3).

Hint. Use the Rodrigues formula:

$$\exp(-t^2)H_m(t) = (-1)^m(\partial/\partial t)^m\exp(-t^2)$$

and (2.1.14).

One can use this exercise to calculate the Radon transform of the functions $x_1^{m_1}x_2^{m_2}\exp(-|x|^2)$, $x \in \mathbb{R}^2$. The polynomial $x_1^{m_1}x_2^{m_2}$ can be expressed in terms of linear combinations of the products of Hermite polynomials, and then one can use formula (2.1.67).

2.1.13. Concluding remarks

We finish Section 2.1 by noting that most of the results obtained in this section can be generalized to a class of functions broader than $S(\mathbb{R}^n)$. There are two directions for further study.

(1) To keep $f(x)$ compactly supported or rapidly decaying, but relax the smoothness assumption: consider, for example, $f(x) \in L_0^2(\mathbb{R}^n)$, that is compactly supported L^2 functions.
(2) To study the Radon transform of functions or distributions which can grow at infinity.

The first direction is taken in Chapter 3. The second one is taken in Chapter 10.

2.2. Inversion formulas for R

In this section we assume again that $f \in S(\mathbb{R}^n)$.

2.2.1. The first method

Let us start with an obvious identity which follows from the Fourier slice theorem

$$f = \mathcal{F}^{-1}\mathcal{F}f = \mathcal{F}^{-1}F_{p\to t}\hat{f} = \mathcal{F}^{-1}FRf. \quad (2.2.1)$$

It is clear from the definition of the Radon transform that Rf has compact support if f is compactly supported, and that $\hat{f} \in S(Z)$ if $f \in S(\mathbb{R}^n)$.

Exercise 2.2.1. Prove these claims.

Therefore, FRf is a well defined function of t and α, in fact, of $t\alpha$, which decays rapidly when $|t| \to \infty, t \in \mathbb{R}$. Thus $\mathcal{F}^{-1}FRf$ is well defined in the sense of classical analysis. If one uses spherical coordinates and understands the integrals in the sense of distributions, one gets from (2.2.1) and (2.1.45):

$$f(x) = \frac{1}{(2\pi)^n} \int_{S^{n-1}} d\alpha \int_0^\infty dt\, t^{n-1} \exp(-it\alpha \cdot x) \int_{-\infty}^\infty dp \exp(itp)\hat{f}(\alpha, p)$$

$$= \frac{1}{(2\pi)^n} \int_{S^{n-1}} d\alpha \int_{-\infty}^\infty dp\hat{f}(\alpha, p) \int_0^\infty dt\, t^{n-1} \exp\{it(p - \alpha \cdot x)\}$$

$$= \frac{i^n(n-1)!}{(2\pi)^n} \int_{S^{n-1}} d\alpha \int_{-\infty}^\infty dp\hat{f}(\alpha, p)(p - \alpha \cdot x + i0)^{-n}. \qquad (2.2.2)$$

This is the inversion formula we wanted to prove. Let us formulate the result.

Theorem 2.2.1. *If* $f \in S(\mathbb{R}^n)$, *then*

$$f(x) = \frac{i^n(n-1)!}{(2\pi)^n} \int_{S^{n-1}} \int_{-\infty}^\infty \frac{\hat{f}(\alpha, p)}{(p - \alpha \cdot x + i0)^n} dp\, d\alpha. \qquad (2.2.3)$$

Let us derive some useful corollaries of this theorem. Note that (see Section 14.2.1.2)

$$(p + i0)^{-n} = p^{-n} - \frac{i\pi(-1)^{n-1}}{(n-1)!}\delta^{(n-1)}(p).$$

Since $\hat{f}$ is even, we get

$$\int_{S^{n-1}} \int_{-\infty}^\infty \frac{\hat{f}(\alpha, p)}{(p - \alpha \cdot x)^n} dp\, d\alpha = 0, \quad n = 2m + 1,$$

and

$$\int_{S^{n-1}} \int_{-\infty}^\infty \hat{f}(\alpha, p)\delta^{(n-1)}(p - \alpha \cdot x)dp\, d\alpha = 0, \quad n = 2m.$$

Theorem 2.2.1 and the last two equations prove

Corollary 2.2.1. *If n is odd, then formulas (2.2.3), (2.1.48), and (2.1.45) imply*

$$f(x) = \frac{\pi(-1)^m}{(2\pi)^n} \int_{S^{n-1}} \hat{f}^{(n-1)}(\alpha, \alpha \cdot x) d\alpha, \quad n = 2m + 1. \qquad (2.2.4)$$

If n is even, then

$$f(x) = \frac{(-1)^{\frac{n}{2}}(n-1)!}{(2\pi)^n} \int_{S^{n-1}} \int_{-\infty}^{\infty} \frac{\hat{f}(\alpha, p)}{(p - \alpha \cdot x)^n} dp\, d\alpha, \quad n = 2m, \qquad (2.2.5)$$

which, after integration by parts with respect to p, yields

$$f(x) = \frac{(-1)^{\frac{n}{2}+1}}{(2\pi)^n} \int_{S^{n-1}} \int_{-\infty}^{\infty} \frac{\hat{f}^{(n-1)}(\alpha, p)}{\alpha \cdot x - p} dp\, d\alpha, \quad n = 2m. \qquad (2.2.5')$$

Formulas (2.2.4), (2.2.5), and (2.2.5') are the classical inversion formulas. For smooth $\hat{f}(\alpha, p)$ the singular integral (2.2.5) can be understood as (2.2.5') and the integration by parts is justified.

2.2.2. The second method

Define the operator $\mathcal{I}^a$, called the Riesz potential, by the formula

$$\mathcal{I}^a f := \mathcal{F}^{-1}(|\xi|^{-a} \tilde{f}(\xi)), \quad a < n. \qquad (2.2.6)$$

Similarly, for $g = g(\alpha, p)$, we define

$$\mathcal{I}^a g := F^{-1}(|t|^{-a} F_{p \to t} g). \qquad (2.2.7)$$

Theorem 2.2.2. *If $|a| < n$, then*

$$f = \gamma \mathcal{I}^{-a} R^* \mathcal{I}^{a-n+1} \hat{f}, \quad \gamma := \frac{1}{2(2\pi)^{n-1}}. \qquad (2.2.8)$$

Proof. In spherical coordinates, definition (2.2.6) can be written as

$$\mathcal{I}^a f = \frac{1}{(2\pi)^n} \int_{S^{n-1}} d\alpha \int_0^{\infty} dt\, t^{n-1-a} \exp(-it\alpha \cdot x) \tilde{f}(t\alpha). \qquad (2.2.9)$$

Using the Fourier slice theorem and taking into account that $\hat{f}$ is even, we obtain

$$\mathcal{I}^a f = \frac{1}{2(2\pi)^n} \int_{S^{n-1}} d\alpha \int_{-\infty}^{\infty} dt |t|^{n-1-a} \exp(-it\alpha \cdot x)(F\hat{f})(\alpha, t)$$

$$= \gamma R^* \mathcal{I}^{1+a-n} \hat{f}, \qquad (2.2.10)$$

which is equivalent to (2.2.8). $\quad\square$

Corollary 2.2.2. *Put $a = 0$ in (2.2.8). Then*

$$f = \gamma R^* I^{-n+1} \hat{f}. \qquad (2.2.11)$$

If n is odd, Equation (2.2.11) yields

$$f(x) = (-1)^{\frac{n-1}{2}} \gamma \int_{S^{n-1}} \hat{f}^{(n-1)}(\alpha, \alpha \cdot x) d\alpha, \quad n = 2m + 1. \qquad (2.2.12a)$$

For even n, one gets

$$f = (-1)^{n/2} \gamma R^* \mathcal{H} \hat{f}^{(n-1)}, \quad n = 2m, \qquad (2.2.12b)$$

where $\mathcal{H}$ is the Hilbert transform:

$$\mathcal{H}g := \frac{1}{\pi} \int_{-\infty}^{\infty} \frac{g(q) dq}{q - p} = -\frac{1}{\pi p} \circledast g. \qquad (2.2.13)$$

Comparing (2.2.4) and (2.2.5') with (2.2.12a) and (2.2.12b), respectively, we see that these formulas are identical.

Proof. The passage from (2.2.11) to (2.2.12a) is based on the equation:

$$I^{-n+1} \hat{f} = F^{-1}(|t|^{2m} F_{p \to t} \hat{f}) = (-1)^m \hat{f}^{(2m)}, \; n = 2m + 1.$$

If $n = 2m$, then

$$I^{-n+1} \hat{f} = F^{-1}(|t|^{2m-1} F_{p \to t} \hat{f}) = F^{-1}(\text{sgn} t \, t^{2m-1} F \hat{f})$$

$$= F^{-1}(\text{sgn} t) \circledast F^{-1}(t^{2m-1} F \hat{f}) = \frac{1}{i \pi p} \circledast \left(-\frac{1}{i} \frac{\partial}{\partial p}\right)^{2m-1} \hat{f}$$

$$= (-1)^m \frac{-1}{\pi p} \circledast \hat{f}^{(n-1)} = (-1)^m \mathcal{H} \hat{f}^{(n-1)},$$

where we have used the well-known identities $F^{-1}(Fh \cdot Fg) = h \circledast g$ and

$$F_{p \to t} \frac{1}{p} = i \pi \text{sgn} t, \; F \mathcal{H} g = -i \text{sgn} t \cdot F_{p \to t} g. \qquad (2.2.14)$$

$\square$

Corollary 2.2.3. *One has*

$$f = \gamma(-1)^{\frac{n-1}{2}} \Delta^{\frac{n-1}{2}} R^* \hat{f}, \; n = 2m + 1, \qquad (2.2.15a)$$

and

$$f = \gamma(-1)^{\frac{n}{2}} \Delta^{\frac{n}{2}-1} R^* \mathcal{H} \frac{\partial}{\partial p} \hat{f}, \; n = 2m, \qquad (2.2.15b)$$

where Δ is the Laplacian and $\gamma := 1/[2(2\pi)^{n-1}]$.

Proof. First assume $n = 2m + 1$. Put $a = n - 1$ in (2.2.8) to get $f = \gamma \mathcal{I}^{1-n} R^* \hat{f}$. Since

$$\mathcal{I}^{1-n} h = \mathcal{F}^{-1}(|\xi|^{2m} \tilde{h}(\xi)) = (-\Delta)^m h,$$

this proves (2.2.15a). If $n = 2m$, we put $a = n - 2$ in (2.2.8) to get

$$f = \gamma \mathcal{I}^{2-n} R^* I^{-1} \hat{f} = \gamma(-\Delta)^{m-1} R^* \mathcal{H} \frac{\partial}{\partial p} \hat{f},$$

which proves (2.2.15b). $\square$

Let us introduce the operator K:

$$Kg := \frac{1}{2(2\pi)^{n-1}} I^{-n+1} g = \frac{1}{2(2\pi)^n} \int_{-\infty}^{\infty} dt |t|^{n-1} \exp(-ipt)(Fg)(\alpha, t)$$

$$= \gamma F^{-1}(|t|^{n-1} Fg). \qquad (2.2.16)$$

From (2.2.11) we obtain

$$R^* K R = I, \quad R^{-1} = R^* K, \qquad (2.2.11')$$

where I is the identity operator on $\mathcal{S}(\mathbb{R}^n)$. The operator K can also be represented as

$$K = \begin{cases} \gamma(-1)^{\frac{n-1}{2}} \frac{\partial^{n-1}}{\partial p^{n-1}}, & n \text{ odd}, \\ \gamma(-1)^{\frac{n}{2}} \mathcal{H} \frac{\partial^{n-1}}{\partial p^{n-1}}, & n \text{ even}. \end{cases} \qquad (2.2.16')$$

Exercise 2.2.1. Prove

$$\mathcal{H}^* = -\mathcal{H}, \quad K^* = K, \quad F^* = 2\pi F^{-1}, \quad \mathcal{F}^* = (2\pi)^n \mathcal{F}^{-1}. \qquad (2.2.17)$$

Hint. Use the Fourier transform representation for the first two equations. The last two equations are obvious.

Corollary 2.2.4. *One has*

$$(R^*)^{-1} = KR, \quad KRR^* = I, \tag{2.2.18}$$

where the domain of R^* is assumed to be the space $KRS(\mathbb{R}^n)$.

Proof. Using (2.2.11') and (2.2.17), we get

$$(R^*)^{-1} = (R^{-1})^* = (R^*K)^* = KR.$$

The second equation in (2.2.18) follows easily from the first one. One can also derive this equation from (2.1.29) and (2.2.16). $\square$

Using the last expression in (2.2.16), the first equation in (2.2.18), and the Fourier slice theorem, we also get

$$(R^*)^{-1} = \gamma F^{-1}|t|^{n-1}F, \quad R^* = \gamma^{-1}\mathcal{F}^{-1}|t|^{1-n}F. \tag{2.2.18'}$$

REMARK 2.2.1. Formula (2.1.12) shows that the operator R^{-1} in odd-dimensional spaces is local in the following sense: on the right-hand side of (2.1.12) only the data are used which are integrals of $f(x)$ over the hyperplanes passing through the point x and its arbitrary small neighborhood (in order to compute the derivatives of $\hat{f}$).

Formula (2.2.12b) shows that the operator R^{-1} is non-local in even-dimensional spaces: to calculate $f(x)$, one needs to know integrals of $f(x)$ over *all* hyperplanes, not only the ones which pass through the point x. Practically, this difference is important because local data are often easier to collect and to process. The problem of recovery of some useful information about f from the local tomographic data in even-dimensional spaces is discussed in detail in Chapters 5–7.

2.2.3. Inversion in two-and three-dimensional spaces

Let $n = 2$. Formula (2.2.5') yields:

$$f(x) = \frac{1}{4\pi^2} \int\limits_{S^1} \int\limits_{-\infty}^{\infty} \frac{\hat{f}_p(\alpha, p)}{\alpha \cdot x - p} dp \, d\alpha, \quad x \in \mathbb{R}^2, \tag{2.2.19}$$

where $\hat{f}_p = \partial \hat{f}/\partial p$, and S^1 is the unit sphere in $\mathbb{R}^2$.

Let $n = 3$. Formula (2.2.12a) yields:

$$f(x) = -\frac{1}{8\pi^2} \int\limits_{S^2} \hat{f}_{pp}(\alpha, \alpha \cdot x) d\alpha, \quad x \in \mathbb{R}^3. \tag{2.2.20}$$

Formula (2.2.20) can be written also as follows (cf. (2.2.15a)):

$$f(x) = -\frac{1}{8\pi^2} \Delta_x \int\limits_{S^2} \hat{f}(\alpha, \alpha \cdot x) d\alpha, \quad x \in \mathbb{R}^3, \tag{2.2.21}$$

where Δ_x is the Laplacian acting on the x-variable.

2.2.4. Radon's original inversion formula

Let $n = 2$. Define

$$F(x, q) := \frac{1}{2\pi} \int_{S^1} \hat{f}(\alpha, q + \alpha \cdot x) d\alpha. \tag{2.2.22}$$

Exercise 2.2.2. Prove that

$$F(x, -q) = F(x, q). \tag{2.2.23}$$

Hint. Use the fact that $\hat{f}$ is even.

Let us now derive Radon's inversion formula. Write formula (2.2.19) as:

$$f(x) = \frac{-1}{4\pi^2} \int_{S^1} \int_{-\infty}^{\infty} \frac{\hat{f}_p(\alpha, \alpha \cdot x + q)}{q} dq \, d\alpha = -\frac{1}{2\pi} \int_{-\infty}^{\infty} \frac{F_q(x, q)}{q} dq,$$

$$\tag{2.2.24}$$

where $F_q := \partial F / \partial q$. Using (2.2.23), we get

$$\int_{-\infty}^{0} \frac{F_q(x, q)}{q} dq = \int_{0}^{\infty} \frac{F_q(x, q)}{q} dq. \tag{2.2.25}$$

Combining (2.2.24) and (2.2.25), we prove

Theorem 2.2.3. *Assume $n = 2$. Then the following inversion formula holds:*

$$f(x) = -\frac{1}{\pi} \int_{0}^{\infty} \frac{F_q(x, q)}{q} dq. \tag{2.2.26}$$

Exercise 2.2.3. Prove that

$$f(x) = c(n) \int_{0}^{\infty} \frac{F^{(n-1)}(x, q)}{q} dq, \quad n = 2m, \tag{2.2.27}$$

where $F^{(n-1)}(x, q) = \frac{\partial^{n-1} F(x,q)}{\partial q^{n-1}}$, and

$$c(n) = 2(-1)^{\frac{n}{2}} (2\pi)^{-n} |S^{n-1}|. \tag{2.2.28}$$

Hint. Use (2.2.5') and argue as above.

REMARK 2.2.2. The integrals in (2.2.26) and (2.2.27) are defined as $\lim_{\epsilon \to +0} \int_{\epsilon}^{\infty}$.

2.2.5. Inversion via the spherical harmonics series

Let us denote

$$x^0 := \frac{x}{|x|}, \quad r := |x|, \quad x \in \mathbb{R}^n.$$

Let $Y_l(\alpha)$ be the spherical harmonics orthonormalized in $L^2(S^{n-1})$ (see Section 14.4.5 about the spherical harmonics). One has

$$f(x) = \sum_{l=0}^{\infty} f_l(r)Y_l(x^0), \tag{2.2.29}$$

and, therefore,

$$\hat{f}(\alpha, p) = \sum_{l=0}^{\infty} R\left(f_l(r)Y_l(x^0)\right). \tag{2.2.30}$$

First, assume for simplicity that $n = 2$:

$$R\left(f_l(r)Y_l(x^0)\right)$$

$$= \int_{S^{n-1}} dx^0 \int_0^{\infty} dr\, r^{n-1} f_l(r)Y_l(x^0)\delta(p - r\alpha \cdot x^0)$$

$$= \int_0^{\infty} dr\, r^{n-2} f_l(r) \int_{S^{n-1}} dx^0 Y_l(x^0)\delta\left(\frac{p}{r} - \alpha \cdot x^0\right)\Bigg|_{\substack{n=2 \\ Y_l=\exp(il\vartheta)/\sqrt{2\pi} \\ \alpha=(\cos\theta,\sin\theta)}}$$

$$= \int_p^{\infty} dr\, f_l(r) \int_{-\pi}^{\pi} d\vartheta \frac{\exp(il\vartheta)}{\sqrt{2\pi}} \delta\left(\frac{p}{r} - \cos(\theta - \vartheta)\right)$$

$$= \frac{1}{\sqrt{2\pi}} \int_p^{\infty} dr\, f_l(r) \frac{\left[\exp\left(il \arccos\frac{p}{r}\right) + \exp\left(-il \arccos\frac{p}{r}\right)\right]}{|\sin\left(\arccos\frac{p}{r}\right)|} \exp(il\theta)$$

$$= \sqrt{\frac{2}{\pi}} \int_p^{\infty} dr\, f_l(r) \frac{\cos\left(l \arccos\frac{p}{r}\right)}{\sqrt{1 - \frac{p^2}{r^2}}} \exp(il\theta)$$

$$= \sqrt{\frac{2}{\pi}} \int_p^{\infty} dr\, f_l(r) \frac{T_{|l|}\left(\frac{p}{r}\right)}{\sqrt{1 - \frac{p^2}{r^2}}} \exp(il\theta), \quad p \geq 0, l = 0, \pm1, \pm2, \ldots. \tag{2.2.31}$$

Here

$$T_l(x) = \cos(l \arccos x), \quad -1 \leq x \leq 1, \tag{2.2.32}$$

is the Chebyshev polynomial of the first kind (see Section 14.4.3), and the system

$$\sqrt{\frac{2}{\pi}} T_l(x), \quad l = 0, 1, 2, \ldots, \tag{2.2.33}$$

is the system of orthonormal polynomials in $L^2\left([-1,1];\frac{1}{\sqrt{1-x^2}}\right)$. One gets from (2.2.30) and (2.2.31):

$$\hat{f}(\alpha,p) = \sum_{l=-\infty}^{\infty} \hat{f}_l(p)\frac{\exp(il\theta)}{\sqrt{2\pi}}, \quad \alpha = (\cos\theta, \sin\theta), \qquad (2.2.34)$$

where

$$\hat{f}_l(p) = 2\int_p^{\infty} dr\, f_l(r)\frac{T_{|l|}\left(\frac{p}{r}\right)}{\sqrt{1-\frac{p^2}{r^2}}}. \qquad (2.2.35)$$

If $\hat{f}(\alpha,p)$ is given for all $\alpha \in S^1$ and all $p \in \mathbb{R}$, then one can find $f(x)$ by the formula

$$f(x) = \sum_{l=-\infty}^{\infty} f_l(r)\frac{\exp(il\vartheta)}{\sqrt{2\pi}}, \quad x = r(\cos\vartheta, \sin\vartheta), \qquad (2.2.36)$$

where $f_l(r)$ should be found from the integral Equation (2.2.35). This is an Abel-type integral equation, which can be solved in a closed form (see Section 11.2):

$$f_l(r) = -\frac{1}{\pi}\int_r^{\infty} \frac{T_{|l|}\left(\frac{p}{r}\right)}{\sqrt{p^2-r^2}}\hat{f}_l'(p)\,dp. \qquad (2.2.37)$$

An inversion formula, similar to (2.2.36)-(2.2.37), can be derived for any dimension $n \geq 2$. These formulas will not be used later. We present them and sketch a proof for the completeness of the presentation. One proves that

$$R\big(f_l(r)Y_l(x^0)\big) = \hat{f}_l(p)Y_l(\alpha), \qquad (2.2.38)$$

where the relation between $f_l(r)$ and $\hat{f}_l(p)$ is given by the equation analogous to (2.2.35):

$$\hat{f}_l(p) = \gamma_{nl}\int_p^{\infty} C_l^{\left(\frac{n-2}{2}\right)}\left(\frac{p}{r}\right)\left(1-\frac{p^2}{r^2}\right)^{\frac{n-3}{2}} f_l(r)r^{n-2}\,dr. \qquad (2.2.39)$$

Here $C_l^{\left(\frac{n-2}{2}\right)}(t)$ are the Gegenbauer polynomials, and γ_{nl} are the constants:

$$\gamma_{nl} = \frac{|S^{n-2}|}{C_l^{\left(\frac{n-2}{2}\right)}(1)}. \qquad (2.2.40)$$

Since $C_l^{(\frac{n-2}{2})}(-x) = (-1)^l C_l^{(\frac{n-2}{2})}(x)$, one sees that

$$\hat{f}_l(p) = (-1)^l \hat{f}_l(-p). \tag{2.2.41}$$

Solving Equation (2.2.39) using the method given in Section 11.2 yields

$$f_l(r) = \left(\frac{-1}{2\pi}\right)^{n-1} \frac{\gamma_{nl}}{r} \int\limits_r^\infty \hat{f}_l^{(n-1)}(p) C_l^{(\frac{n-2}{2})}\left(\frac{p}{r}\right)\left(\frac{p^2}{r^2} - 1\right)^{\frac{n-3}{2}} dp. \tag{2.2.42}$$

2.2.6. Inversion in the spherically symmetric case

Comparing Equations (2.1.63)–(2.1.65) with (2.2.35) and (2.2.42), we see that the former are obtained from the latter by setting $l = 0$. Therefore, setting $l = 0$ in (2.2.37), we obtain the solution to (2.1.64):

$$f(r) = -\frac{1}{\pi} \int\limits_r^\infty \frac{\hat{f}'(s)}{\sqrt{s^2 - r^2}} ds. \tag{2.2.43}$$

The solution to (2.1.65) is obvious (it can also be obtained from (2.2.42) putting $l = 0$, $n = 3$, and using that $C_0^\lambda \equiv 1$):

$$f(r) = -\frac{1}{2\pi r}\frac{d}{dp}\hat{f}(p)\Big|_{p=r}. \tag{2.2.44}$$

Since $C_0^{(m)}(t) = 1$, Equation (2.1.39) for $l = 0$ reduces to (2.1.63) and the solution to (2.1.63) is obtained from (2.2.42):

$$f(r) = \left(\frac{-1}{2\pi}\right)^{n-1} \frac{|S^{n-2}|}{r} \int\limits_r^\infty \left(\frac{p^2}{r^2} - 1\right)^{\frac{n-3}{2}} \hat{f}^{(n-1)}(p) dp. \tag{2.2.45}$$

2.3. Singular value decomposition of the Radon transform

Fix any angle $\alpha \in S^{n-1}$ and introduce the operator R_α:

$$(R_\alpha f)(p) = (Rf)(\alpha, p) = \int\limits_{\mathbf{R}^n} f(x)\delta(\alpha \cdot x - p) dx. \tag{2.3.1}$$

The operator R_α^*, which is adjoint to R_α, can be easily found:

$$\int\limits_{\mathbf{R}} (R_\alpha f)(p)g(p) dp = \int\limits_{\mathbf{R}} \int\limits_{\mathbf{R}^n} f(x)\delta(\alpha \cdot x - p)g(p) dx\, dp$$

$$= \int\limits_{\mathbf{R}^n} f(x)g(\alpha \cdot x) dx,$$

therefore, $(R_\alpha^* g)(x) = g(\alpha \cdot x)$. Denote $\omega(p) := (1 - p^2)^{1/2}$.

Lemma 2.3.1. *Let B_1^n be the unit ball in $\mathbb{R}^n$, $n \geq 2$. Then the operator*

$$R_\alpha : \ L^2(B_1^n) \to L^2([-1,1], \omega^{1-n})$$

is continuous.

Proof. Pick any $f \in C_0^\infty(B_1^n)$ and assume f real-valued without loss of generality. Since

$$(R_\alpha f)(p) = \int_{\substack{y \cdot \alpha = 0 \\ |y|^2 + p^2 \leq 1}} f(p\alpha + y) dy,$$

the Cauchy-Schwarz inequality yields

$$|(R_\alpha f)(p)|^2 = \left| \int_{\substack{y \cdot \alpha = 0 \\ |y| \leq \sqrt{1-p^2}}} f(p\alpha + y) dy \right|^2$$

$$\leq (1-p^2)^{\frac{n-1}{2}} |B_1^{n-1}| \int_{y \cdot \alpha = 0} f^2(p\alpha + y) dy,$$

where $|B_1^{n-1}|$ is the volume of the unit ball in $\mathbb{R}^{n-1}$. Dividing by $(1-p^2)^{\frac{n-1}{2}}$ and integrating with respect to p, we complete the proof:

$$\int_{|p| \leq 1} (1-p^2)^{\frac{1-n}{2}} |(R_\alpha f)(p)|^2 dp \leq |B_1^{n-1}| \int_{|p| \leq 1} \int_{y \cdot \alpha = 0} f^2(p\alpha + y) dy \, dp$$

$$= |B_1^{n-1}| \int_{B_1^n} f^2(y) dy.$$

□

Later on we will need the following result.

Lemma 2.3.2. *Fix any $\alpha_1, \alpha_2 \in S^{n-1}$. One has*

$$\left(R_{\alpha_1} R_{\alpha_2}^* g \right)(p) = |B_1^{n-2}| \int_{|t| \leq \sqrt{1-p^2}} g(p \cos \varphi + t \sin \varphi) \omega^{n-2}(\sqrt{p^2 + t^2}) dt,$$

$$(2.3.2)$$

where $\varphi \in [0, \pi]$ is the angle between the vectors α_1 and α_2, that is $\cos \varphi = \alpha_1 \cdot \alpha_2$, $\sin \varphi \geq 0$, and $|B_1^{n-2}|$ is the volume of the unit ball in $\mathbb{R}^{n-2}$.

Proof. Using Equation (2.3.1), the definition of R_α^*, and setting $x = s\alpha_1 + y$, $y \cdot \alpha_1 = 0$, we get

$$
\left(R_{\alpha_1} R_{\alpha_2}^* g\right)(p) = \int_{|x| \le 1} g(\alpha_2 \cdot x)\delta(\alpha_1 \cdot x - p)dx
$$

$$
= \int_{\substack{|y| \le \sqrt{1-p^2} \\ y \cdot \alpha_1 = 0}} g(p \cos \varphi + y \cdot \alpha_2)dy. \tag{2.3.3}
$$

Suppose first that $\alpha_1 = \alpha_2$ or $\alpha_1 = -\alpha_2$. Then $\sin \varphi = 0$, and Equation (2.3.2) holds because

$$
|B_1^{n-2}| \int_{|t| \le \sqrt{1-p^2}} (1 - p^2 - t^2)^{\frac{n-2}{2}} dt = |B_1^{n-1}|(1 - p^2)^{\frac{n-1}{2}},
$$

which is in agreement with Equation (2.3.3). Suppose now that $\alpha_1 \ne \pm\alpha_2$. Representing the vector α_2 and the variable y as

$$
\alpha_2 = \alpha_1 + \alpha', \ \alpha' \cdot \alpha_1 = 0, \ |\alpha'| = \sin \varphi,
$$

$$
y = t\frac{\alpha'}{|\alpha'|} + z, \ z \cdot \alpha_1 = 0, \ z \cdot \alpha' = 0,
$$

we get from (2.3.3):

$$
\left(R_{\alpha_1} R_{\alpha_2}^* g\right)(p) = \int_{|t| \le \sqrt{1-p^2}} g(p \cos \varphi + t \sin \varphi) \int_{\substack{|z| \le \sqrt{1-p^2-t^2} \\ z \cdot \alpha_1 = 0, \ z \cdot \alpha' = 0}} dz \, dt
$$

$$
= |B_1^{n-2}| \int_{|t| \le \sqrt{1-p^2}} g(p \cos \varphi + t \sin \varphi)(1 - p^2 - t^2)^{\frac{n-2}{2}} dt,
$$

which proves Equation (2.3.2) for the case $n > 2$. If $n = 2$, Equation (2.3.2) also holds because $B_1^0 := 1$. □

Let us substitute $C_m^{(\frac{n}{2})}$ for g in (2.3.2). Since $|p| \le 1$, we change variables $t = \sqrt{1 - p^2}u$, denote $p = \cos v$, and use integral (7.316) from [GR] to get

$$
\left(R_{\alpha_1} R_{\alpha_2}^* C_m^{(\frac{n}{2})}\right)(p)
$$

$$
= |B_1^{n-2}|(1 - p^2)^{\frac{n-1}{2}}
$$

$$
\times \int_{-1}^{1} C_m^{(\frac{n}{2})}(\cos v \cos \varphi + u \sin v \sin \varphi)(1 - u^2)^{\frac{n}{2}-1} du \bigg|_{\cos v = p}
$$

$$
= |B_1^{n-2}|\omega^{n-1}(p)\frac{2^{n-1}m!\Gamma^2(\frac{n}{2})}{\Gamma(n+m)} C_m^{(\frac{n}{2})}(p)C_m^{(\frac{n}{2})}(\alpha_1 \cdot \alpha_2),
$$

because $\cos\varphi = \alpha_1 \cdot \alpha_2$. Using Equations (14.4.3), (14.4.23), and formula for the volume of the unit ball in $\mathbb{R}^{n-2}$, we find

$$(R_{\alpha_1} R_{\alpha_2}^* C_m^{(\frac{n}{2})})(p) = \psi_m(\alpha_1 \cdot \alpha_2)\omega^{n-1}(p)C_m^{(\frac{n}{2})}(p), \quad p \in \mathbb{R}, \qquad (2.3.4)$$

where

$$\psi_m(s) = \frac{\pi^{\frac{n-1}{2}}}{\Gamma(\frac{n+1}{2})} \frac{C_m^{(\frac{n}{2})}(s)}{C_m^{(\frac{n}{2})}(1)}, \quad -1 \le s \le 1. \qquad (2.3.5)$$

Consider R_α and R as mappings between the spaces

$$R_\alpha : L^2(B_1^n) \to L^2([-1,1], \omega^{1-n}),$$
$$R : L^2(B_1^n) \to L^2(Z_1, \omega^{1-n}), \qquad (2.3.6)$$

where $Z_1 := S^{n-1} \times [-1,1]$. Since

$$\int_{\mathbb{R}} \omega^{1-n}(p)\left[\int_{\mathbb{R}^n} f(x)\delta(\alpha \cdot x - p)dx\right]g(p)dp = \int_{\mathbb{R}^n} f(x)\omega^{1-n}(\alpha \cdot x)g(\alpha \cdot x)dx,$$

the adjoint operator to R_α, specified by (2.3.6), is

$$(\tilde{R}_\alpha^* g)(x) = \omega^{1-n}(\alpha \cdot x)g(\alpha \cdot x) = [R_\alpha^*(\omega^{1-n}g)](x).$$

Similarly, the adjoint operator to R, specified by (2.3.6), is

$$(\tilde{R}^* g)(x) = \int_{S^{n-1}} \omega^{1-n}(\alpha \cdot x)g(\alpha, \alpha \cdot x)d\alpha = \int_{S^{n-1}} \tilde{R}_\alpha^* g \, d\alpha.$$

Therefore,

$$(R\tilde{R}^* g)(\alpha_1, p) = \int_{S^{n-1}} (R_{\alpha_1} \tilde{R}_{\alpha_2}^* g)(p)d\alpha_2. \qquad (2.3.7)$$

Denoting $h_m = \omega^{n-1}C_m^{(\frac{n}{2})}$, Equation (2.3.4) takes the form

$$R_{\alpha_1} \tilde{R}_{\alpha_2}^* h_m = \psi_m(\alpha_1 \cdot \alpha_2)h_m, \qquad (2.3.8)$$

that is h_m is the eigenfunction of the self-adjoint operator $R_{\alpha_1} \tilde{R}_{\alpha_2}^*$, and $\psi_m(\alpha_1 \cdot \alpha_2)$ is the corresponding eigenvalue. Substituting $g(\alpha, s) = h_m(s)Y_{lk}(\alpha)$ into (2.3.7), where $Y_{lk}, k = 1,\ldots,N(n,l)$, are the spherical harmonics of degree l (see Section 14.4.5) and taking into account

Equation (2.3.8), we find using the Funk-Hecke theorem (2.1.35) (with $Y_l = Y_{lk}$):

$$[R\tilde{R}^*(h_m Y_{lk})](\alpha_1, p) = \int_{S^{n-1}} (R_{\alpha_1} \tilde{R}_{\alpha_2}^* h_m)(p) Y_{lk}(\alpha_2) d\alpha_2$$

$$= h_m(p) \int_{S^{n-1}} \psi_m(\alpha_1 \cdot \alpha_2) Y_{lk}(\alpha_2) d\alpha_2$$

$$= c_n^2(m, l) h_m(p) Y_{lk}(\alpha_1), \qquad (2.3.9a)$$

where

$$c_n^2(m, l) = \frac{|S^{n-2}|}{C_l^{(\frac{n-2}{2})}(1)} \int_{-1}^{1} \psi_m(t) C_l^{(\frac{n-2}{2})}(t) (1 - t^2)^{\frac{n-3}{2}} dt. \qquad (2.3.9b)$$

Equations (2.3.6) and (2.3.9a) show that the functions $h_m(p) Y_{lk}(\alpha)$ are eigenfunctions of the self-adjoint operator

$$R\tilde{R}^* : \quad L^2(Z_1, \omega^{1-n}) \to L^2(Z_1, \omega^{1-n}),$$

and $c_n^2(m, l)$ are the corresponding eigenvalues. Since the functions $h_m Y_{lk}$ form a basis in $L^2(Z_1, \omega^{1-n})$, we conclude that there are no other eigenfunctions. To find $c_n^2(m, l)$, we use (2.3.5) in (2.3.9b) to get

$$c_n^2(m, l) = \frac{\pi^{\frac{n-1}{2}}}{\Gamma(\frac{n+1}{2})} \frac{|S^{n-2}|}{C_m^{(\frac{n}{2})}(1) C_l^{(\frac{n-2}{2})}(1)} \int_{-1}^{1} C_m^{(\frac{n}{2})}(t) C_l^{(\frac{n-2}{2})}(t) (1 - t^2)^{\frac{n-3}{2}} dt.$$
$$(2.3.10)$$

Since $C_m^{(\frac{n}{2})}(t)$ is the polynomial of degree m, the orthogonality property of the Gegenbauer polynomials implies $c_n^2(m, l) = 0$ for $m < l$. Also, $c_n^2(m, l) = 0$ if $m + l$ is odd, because $C_l^{(\cdot)}$ is even (odd) if l is even (odd). Thus the nonzero eigenvalues are

$$c_n^2(m, l) \text{ for } m = 0, 1, 2, \ldots, \ l = m, m - 2, \ldots.$$

Since $\int_{S^{n-1}} Y_{lk}^2(\alpha) d\alpha = 1$, the normalized eigenfunctions of $R\tilde{R}^*$ are

$$g_{mlk}(\alpha, p) = b_m h_m(p) Y_{lk}(\alpha), \quad h_m(p) = (1 - p^2)^{\frac{n-1}{2}} C_m^{(\frac{n}{2})}(p)$$

$$b_m = \left[\int_{-1}^{1} (1 - p^2)^{\frac{1-n}{2}} h_m^2(p) dp \right]^{-1/2} = \left[\frac{\pi 2^{1-n} \Gamma(m + n)}{m! (m + \frac{n}{2}) \Gamma^2(n/2)} \right]^{-1/2}, \qquad (2.3.11)$$

where we have used integral (7.313.2) from [GR]. For m and l such that $c_n^2(m, l) \neq 0$, define

$$f_{mlk}(x) := \frac{1}{c_n(m,l)}(\tilde{R}^* g_{mlk})(x) = \frac{b_m}{c_n(m,l)} \int_{S^{n-1}} C_m^{(\frac{n}{2})}(x \cdot \alpha) Y_{lk}(\alpha) d\alpha.$$

Applying formula (2.1.35) one more time, we derive

$$f_{mlk}(x) = Q_{ml}(|x|) Y_{lk}(x/|x|),$$

$$Q_{ml}(r) = \frac{b_m}{c_n(m,l)} \frac{|S^{n-2}|}{C_l^{(\frac{n-2}{2})}(1)} \int_{-1}^{1} C_m^{(\frac{n}{2})}(rt) C_l^{(\frac{n-2}{2})}(t)(1-t^2)^{\frac{n-3}{2}} dt,$$

$$m = 0, 1, 2, \ldots, \ l = m, m-2, \ldots, \ k = 1, 2, \ldots, N(n,l),$$

where $N(n, l)$ is defined in Section 14.4.5.

Using integral 15 from [PBM, page 563] and Equation (22.5.2) from [AS], we obtain

$$Q_{ml}(r) = \tilde{\tilde{\gamma}}_{ml} r^m P_{\frac{m-l}{2}}^{(-\frac{n}{2}-m,0)}\left(1 - \frac{2}{r^2}\right) = \tilde{\gamma}_{ml} r^l P_{\frac{m-l}{2}}^{(0, l+\frac{n}{2}-1)}(2r^2 - 1)$$

$$= \gamma_{ml} r^l G_{\frac{m-l}{2}}\left(l + \frac{n}{2}, l + \frac{n}{2}, r^2\right),$$

$$m = 0, 1, 2, \ldots, \ l = m, m-2, \ldots, \tag{2.3.12}$$

where $P_j^{(p,q)}(x)$ and $G_j(p, q, x)$ are the Jacobi and shifted Jacobi polynomials, respectively (see [AS]), and $\tilde{\tilde{\gamma}}_{ml}$, $\tilde{\gamma}_{ml}$, γ_{ml} are some constants. In particular, $G_j(p, q, x)$ are orthogonal in $L^2([0,1], (1-x)^{p-q} x^{q-1})$.

The constant γ_{ml} is chosen from the normalization condition:

$$1 = \|f_{mlk}\|_{L^2(B_1^n)}^2 = \int_0^1 r^{n-1} \int_{S^{n-1}} f_{mlk}^2(r\alpha) d\alpha \, dr$$

$$= \gamma_{ml}^2 \int_0^1 r^{n-1+2l} G_{\frac{m-l}{2}}^2\left(l + \frac{n}{2}, l + \frac{n}{2}, r^2\right) dr$$

$$= \frac{\gamma_{ml}^2}{2} \int_0^1 x^{l+\frac{n}{2}-1} G_{\frac{m-l}{2}}^2\left(l + \frac{n}{2}, l + \frac{n}{2}, x\right) dx$$

$$= \gamma_{ml}^2 \frac{\left[\left(\frac{m-l}{2}\right)!\right]^2 \Gamma^2\left(\frac{m+l+n}{2}\right)}{(2m+n)\Gamma^2\left(m + \frac{n}{2}\right)}. \tag{2.3.13}$$

Collecting (2.3.12) and (2.3.13), we find

$$f_{mlk}(x) = \frac{\sqrt{2m+n}\,\Gamma(m+\frac{n}{2})}{(\frac{m-l}{2})!\,\Gamma(\frac{m+l+n}{2})} r^l G_{\frac{m-l}{2}}(l+\frac{n}{2}, l+\frac{n}{2}, |x|^2) Y_{lk}(x/|x|),$$

$$m = 0, 1, 2, \ldots, l = m, m-2, \ldots, k = 1, 2, \ldots, N(n,l),$$
(2.3.14)

Finally, the singular value decomposition of R is given by

$$Rf = \sum_{m=0}^{\infty}\sum_{l=0}^{m}{}' c_n(m,l) \sum_{k=1}^{N(n,l)} <f, f_{mlk}>_{L^2(B_1^n)} g_{mlk},$$
(2.3.15)

where f_{lmk} and g_{lmk} are defined in (2.3.11) and (2.3.14), and the prime near the summation sign denotes that l runs over integers $m, m-2, \ldots$. We see that the multiplicity of each singular value in (2.3.15) is $N(n,l)$, where $N(n,l)$ is the number of linearly independent spherical harmonics of degree l. Decomposition (2.3.15) yields the following inversion formula: if $g = Rf$, then

$$f = \sum_{m=0}^{\infty}\sum_{l=0}^{m}{}' \frac{1}{c_n(m,l)} \sum_{k=1}^{N(n,l)} (g, g_{mlk})_{L^2(Z_1, \omega^{1-n})} f_{mlk}.$$
(2.3.16)

Let us consider the case $n = 2$. From (2.3.10), we obtain using (14.4.33) and integral (3.612.1) from [GR]:

$$c_2^2(m,l) = \frac{\sqrt{\pi}}{\Gamma(3/2)}\frac{2}{m+1}\int_{-1}^{1}\frac{U_m(t)T_l(t)}{\sqrt{1-t^2}}dt$$

$$= \frac{4}{m+1}\int_{0}^{\pi}\frac{\sin((m+1)\theta)}{\sin(\theta)}\cos(l\theta)d\theta = \frac{4\pi}{m+1}$$
(2.3.17)

for $0 \le l \le m$ and $l+m$ even.

We see that $c_2(m,l) \to 0$ as $m \to \infty$, but the rate of decay is relatively slow. Therefore, the inversion of the Radon transform is only weakly ill-posed. Note that the rate of decay of singular values obtained here is in agreement with the asymptotic result obtained in Section 14.7.4 using the results from the theory of pseudodifferential operators (see Equation (14.7.50)).

2.4. Estimates in Sobolev spaces

Let $H_0^s(B_a)$ be the closure of $C_0^\infty(B_a)$ in the norm of $H^s(\mathbb{R}^n)$:

$$\|f\|_s := \left\{\frac{1}{(2\pi)^n}\int_{\mathbb{R}^n}|\tilde{f}(\xi)|^2(1+|\xi|^2)^s d\xi\right\}^{1/2},$$
(2.4.1)

where $H^s(\mathbb{R}^n)$ is the Sobolev space of order s, $-\infty < s < \infty$ (cf. Section 14.1.2). In particular, $\|f\|_0 = \|f\|_{L^2}$. Similarly, $H_0^s(Z_a)$ denotes the closure of $C_0^\infty(Z_a)$ in the norm of $H^s(Z)$:

$$\||g\||_s := \left\{ \frac{1}{2\pi} \int\limits_{S^{n-1}} d\alpha \int\limits_{-\infty}^{\infty} dt(1+t^2)^s |F_{p\to t}g(\alpha,p)|^2 \right\}^{1/2} . \qquad (2.4.2)$$

The main result of this section is

Theorem 2.4.1. *Let* $f \in H_0^s(B_a)$. *Then there exists a constant* $c(s,n) > 0$, *which depends on* s *and* n, *such that the following inequalities hold:*

$$\frac{(2\pi)^{n/2}}{\sqrt{\pi}} \|f\|_s \leq \||\hat{f}\||_{s+\frac{n-1}{2}} \leq c(s,n)\|f\|_s. \qquad (2.4.3)$$

It is clear from these estimates that the Radon transform increases smoothness of f in the scale of Sobolev spaces by $(n-1)/2$ derivatives, where n is the dimension of the space. We also conclude that the Radon transform, which is considered as a mapping $R : H_0^s(B_a) \to H_0^{s+\frac{n-1}{2}}(Z_a)$, has continuous inverse.

If one is interested in the action of the Radon transform in the spaces C^m of continuously differentiable functions, the answer is more complicated: in some directions the Radon transform $\hat{f}(\alpha,p)$ of a piecewise-continuous function may be piecewise-continuous and not smoother. For example, consider the Radon transform $\hat{f}$ of a characteristic function f of a rectangle in the direction α_0 perpendicular to one of its sides. The function $\hat{f}(\alpha_0,p)$ has a jump at some values of p. To be specific, consider the square: $D := \{x : 0 \leq x_1, x_2 \leq 1\}$, $x \in \mathbb{R}^2$, and let $\alpha_0 = (1,0)$. If f is the characteristic function of D, then

$$\hat{f}(\alpha_0,p) = \begin{cases} 0, & p < 0 \text{ or } p > 1, \\ 1, & 0 < p < 1. \end{cases}$$

Therefore, $\hat{f}(\alpha_0,p)$ is not smoother than $f(x)$. In Chapter 4 the singularities of Rf are studied in detail.

Proof of Theorem 2.4.1. Fix any $f \in C_0^\infty(B_a)$. The left inequality in (2.4.3) is proved using (2.1.20):

$$\||\hat{f}\||_{s+\frac{n-1}{2}}^2 = \frac{1}{2\pi} \int\limits_{S^{n-1}} d\alpha \int\limits_{-\infty}^{\infty} dt(1+t^2)^{s+\frac{n-1}{2}} |\tilde{f}(t\alpha)|^2$$

$$\geq \frac{1}{\pi} \int\limits_{S^{n-1}} d\alpha \int\limits_{0}^{\infty} dt\, t^{n-1}(1+t^2)^s |\tilde{f}(t\alpha)|^2 = \frac{(2\pi)^n}{\pi}\|f\|_s^2. \qquad (2.4.4)$$

To prove the right inequality in (2.4.3), we write:

$$\||\hat{f}\||^2_{s+\frac{n-1}{2}} = \frac{1}{\pi} \int\limits_{S^{n-1}} d\alpha \left(\int\limits_0^1 dt(1+t^2)^{s+\frac{n-1}{2}}|F\hat{f}|^2 \right.$$

$$\left. + \int\limits_1^\infty dt(1+t^2)^{s+\frac{n-1}{2}}|F\hat{f}|^2 \right) := \frac{1}{\pi}(I_1 + I_2). \tag{2.4.5}$$

If $t \geq 1$, then $(1+t^2)^{\frac{n-1}{2}} \leq 2^{\frac{n-1}{2}} t^{n-1}$, and, therefore,

$$I_2 \leq 2^{\frac{n-1}{2}} \int\limits_{S^{n-1}} d\alpha \int\limits_1^\infty dt\, t^{n-1}(1+t^2)^s |\tilde{f}(t\alpha)|^2$$

$$\leq 2^{\frac{n-1}{2}} \int\limits_{S^{n-1}} d\alpha \int\limits_0^\infty dt\, t^{n-1}(1+t^2)^s |\tilde{f}(t\alpha)|^2 = 2^{\frac{n-1}{2}} (2\pi)^n \|f\|^2_s. \tag{2.4.6}$$

If $0 \leq t \leq 1$, then

$$I_1 \leq |S^{n-1}| c_1(s,n) \max_{|\xi| \leq 1} |\tilde{f}(\xi)|^2, \quad c_1(s,n) := \int\limits_0^1 (1+t^2)^{s+\frac{n-1}{2}} dt. \tag{2.4.7}$$

Fix any $\chi \in C_0^\infty(\mathbb{R}^n)$ such that $\chi(x) = 1$ if $x \in B_a$. Let us denote $\chi_\xi(x) = \chi(x)\exp(ix \cdot \xi)$, where $\xi \in \mathbb{R}^n$ is a parameter, and let $\tilde{\chi}_\xi$ be the Fourier transform of χ_ξ. Since the support of f is B_a, we get using the Parseval equality for the Fourier transform and the Cauchy-Schwarz inequality:

$$|\tilde{f}(\xi)| = \left| \int\limits_{\mathbb{R}^n} f(x)\chi_\xi(x)dx \right| = \frac{1}{(2\pi)^n} \left| \int\limits_{\mathbb{R}^n} \tilde{f}(\eta)\overline{\tilde{\chi}_\xi(\eta)}d\eta \right|$$

$$= \frac{1}{(2\pi)^n} \left| \int\limits_{\mathbb{R}^n} \tilde{f}(\eta)(1+|\eta|^2)^{s/2}(1+|\eta|^2)^{-s/2}\overline{\tilde{\chi}_\xi(\eta)}d\eta \right|$$

$$\leq \frac{1}{(2\pi)^n} \left(\int\limits_{\mathbb{R}^n} |\tilde{f}(\eta)|^2(1+|\eta|^2)^s d\eta \right)^{\frac{1}{2}} \left(\int\limits_{\mathbb{R}^n} |\tilde{\chi}_\xi(\eta)|^2(1+|\eta|^2)^{-s}d\eta \right)^{\frac{1}{2}}$$

$$= \|f\|_s \|\chi_\xi\|_{-s}. \tag{2.4.8}$$

The norm $\|\chi_\xi\|_{-s}$ is a continuous function of ξ. Therefore, there exists the maximum

$$c_2(s,n) := \max_{|\xi| \leq 1} \|\chi_\xi\|_{-s} < \infty. \tag{2.4.9}$$

Combining (2.4.5)-(2.4.9), we prove the right inequality in (2.4.3), where

$$c^2(s,n) = \frac{2^{\frac{n-1}{2}}(2\pi)^n + |S^{n-1}|c_1(s,n)c_2^2(s,n)}{\pi}. \tag{2.4.10}$$

We have proved estimates (2.4.3) for an arbitrary $f \in C_0^\infty(B_a)$. By considering the closure of $C_0^\infty(B_a)$ in the $H^s(\mathbb{R}^n)$-norm, we see that these estimates are also valid for any $f \in H_0^s(B_a)$. $\square$

REMARK 2.4.1. Let $s > 0$. The norm in the Sobolev space with the negative index can be defined as:

$$\|\psi\|_{H^{-s}(B_a)} := \sup_{\|h\|_{H^s(B_a)}=1} |(\psi,h)_{L^2(B_a)}|.$$

Hence, instead of (2.4.8), one can use a shorter proof:

$$\max_{|t|\leq 1} |\tilde{f}(t\alpha)| \leq \max_{|t|\leq 1} \|\exp(it\alpha \cdot x)\|_{H^{-s}(B_a)} \|f\|_s \leq \tilde{c}\|f\|_s.$$

Exercise 2.4.1. Prove that the norm $\|\chi_\xi\|_{-s}$ is a continuous function of ξ for any fixed $s \in \mathbb{R}$.

Hint. Use (1) the well-known inequality $\left|\|f\| - \|g\|\right| \leq \|f-g\|$, and (2) any order partial derivatives of $\chi_{\xi'}(x)$ with respect to x converge to $\chi_\xi(x)$ in the L^2-norm if $\xi' \to \xi$, because $\chi(x)$ is compactly supported.

REMARK 2.4.2. For nonnegative s, one can easily find an explicit expression for $c(s,n)$. Instead of (2.4.8), we have

$$|\tilde{f}(\xi)| = \left|\int_{\mathbb{R}^n} f(x)e^{ix\cdot\xi}dx\right| \leq |B_a|^{1/2}\left(\int_{\mathbb{R}^n}|f|^2 dx\right)^{1/2}$$

$$= |B_a|^{1/2}\|f\|_0 \leq |B_a|^{1/2}\|f\|_s, \quad s \geq 0. \tag{2.4.11}$$

Taking into account that $|B_a| = a^n|B_1^n| = \frac{a^n}{n}|S^{n-1}|$ and $|S^{n-1}| = 2\pi^{n/2}/\Gamma(n/2)$, and estimating $c_1(s,n)$ (see (2.4.7)), we find

$$c^2(s,n) = \frac{2^{\frac{n-1}{2}}(2\pi)^n + |S^{n-1}|^2 2^{s+\frac{n-1}{2}}\frac{a^n}{n}}{\pi}$$

$$= \frac{2^{\frac{n-1}{2}}}{\pi}\left((2\pi)^n + \frac{4\pi^n}{\Gamma^2(n/2)}2^s\frac{a^n}{n}\right), \quad s \geq 0. \tag{2.4.12}$$

2.5. Inversion formulas for the backprojection operator

2.5.1. Motivation and problem formulation

In applications one may be interested in calculating the integral $I :=$ $\int_D f(x)dx$ in terms of the tomographic data, that is in terms of the Radon transform $Rf := \hat{f}$ of f. For example, in medicine the integral I describes a cumulant property of a tumor (if D is its support), and this property can be used for medical diagnostics. More generally, let us consider the problem of computing a functional $I = <f, h>$, where h is some function (distribution), knowing the Radon transform $\hat{f}$. Let μ be found such that

$$R^*\mu = h, \qquad (2.5.1)$$

where R^* is the operator adjoint to R. Then,

$$I = <f, h> = <f, R^*\mu> = (Rf, \mu) := \int\limits_{S^{n-1}} \int\limits_{-\infty}^{\infty} \hat{f}(\alpha, p)\mu(\alpha, p)dp\, d\alpha.$$

$$(2.5.2)$$

Therefore, if we know the solution to Equation (2.5.1), we can compute I from the tomographic data $\hat{f}$. The purpose of this section is to study Equation (2.5.1): to prove existence of $(R^*)^{-1}$ on a space of smooth and rapidly decaying functions, to give formulas for the inverse operator $(R^*)^{-1}$, and to give conditions for the solution μ to be compactly supported when h is compactly supported. In particular, it is proved that if $x \in \mathbb{R}^n$, n is odd, h is compactly supported: $h(x) = 0$ for $|x| > a$, and an even function μ solves (2.5.1), then $\mu(\alpha, p) = 0$ for $|p| > a$. It is also proved that if n is even, $h \not\equiv 0$ is compactly supported, and μ solves (2.5.1), then μ cannot be compactly supported. These results should be compared with the known results for the Radon transform: for both odd and even n the Radon transform of a sufficiently smooth function f is compactly supported if and only if $\hat{f}$ is compactly supported. The 'if' part is trivial, and the 'only if' part is the known 'hole theorem' (cf. Section 2.7.1 below). Properties of the solution to (2.5.1) with nonsmooth right-hand side are discussed in Chapter 10.

2.5.2. Inversion formulas

Consider the backprojection operator

$$R^*\mu = \int\limits_{S^{n-1}} \mu(\alpha, \alpha \cdot x)d\alpha, \quad R^* : C^\infty(Z) \cap L^1(Z) \to C^\infty(\mathbb{R}^n), \quad (2.5.3)$$

and consider Equation (2.5.1) with $h \in \mathcal{S}(\mathbb{R}^n)$. Suppose the solution μ to (2.5.1) is from the space $C^\infty(Z) \cap L^1(Z)$. Then we can represent

$\mu(\alpha, p)$ as

$$\mu(\alpha, p) = \frac{1}{2\pi} \int_{-\infty}^{\infty} \tilde{\mu}(\alpha, \lambda) \exp(-i\lambda p) d\lambda := F_{\lambda \to p}^{-1} \tilde{\mu}. \qquad (2.5.4)$$

Substituting (2.5.4) into (2.5.1), we find

$$h(x) = \int_{S^{n-1}} \mu(\alpha, \alpha \cdot x) d\alpha = \frac{1}{2\pi} \int_{S^{n-1}} \int_{-\infty}^{\infty} \tilde{\mu}(\alpha, \lambda) \exp(-i\lambda \alpha \cdot x) d\lambda \, d\alpha.$$

$$(2.5.5)$$

Let μ_o and μ_e be the odd and even parts of μ, respectively:

$$\mu_o(\alpha, p) = \frac{\mu(\alpha, p) - \mu(-\alpha, -p)}{2}, \quad \mu_e(\alpha, p) = \frac{\mu(\alpha, p) + \mu(-\alpha, -p)}{2}.$$

Since $\mu_o(\alpha, p) = -\mu_o(-\alpha, -p)$, we obtain $R^* \mu_o = 0$. Thus, by solving Equation (2.5.1) we can recover only the even part of μ. Replacing μ by μ_e in (2.5.5) and using the fact that $\tilde{\mu}_e$ is also even, we get

$$h(x) = \frac{1}{\pi} \int_{S^{n-1}} \int_0^{\infty} \tilde{\mu}_e(\alpha, \lambda) \exp(-i\lambda \alpha \cdot x) d\lambda \, d\alpha$$

$$= \frac{1}{\pi} \int_{S^{n-1}} \int_0^{\infty} (\lambda^{1-n} \tilde{\mu}_e(\alpha, \lambda)) \exp(-i\lambda \alpha \cdot x) \lambda^{n-1} d\lambda \, d\alpha,$$

therefore,

$$\tilde{h}(\lambda \alpha) = \gamma^{-1} \lambda^{1-n} \tilde{\mu}_e(\alpha, \lambda), \; \lambda > 0, \, \alpha \in S^{n-1}, \gamma := \frac{1}{2(2\pi)^{n-1}}, \quad (2.5.6)$$

where $\tilde{h} = \mathcal{F}h$ is the n-dimensional Fourier transform of h. Solving for $\tilde{\mu}_e$ and using the fact that $\tilde{\mu}_e$ is even, we get

$$\tilde{\mu}_e(\alpha, \lambda) = \gamma |\lambda|^{n-1} \tilde{h}(\lambda \alpha), \; \lambda \in \mathbb{R}, \, \alpha \in S^{n-1}. \qquad (2.5.6')$$

The last equation combined with (2.5.4) gives the desired inversion formula. Also, we conclude from (2.5.6') that

(1) given $h \in \mathcal{S}(\mathbb{R}^n)$, μ_e can be uniquely determined; and
(2) $\mu_e \in C^{\infty}(Z) \cap L^1(Z)$, because $\tilde{h}(\lambda \alpha)$ decays faster than any power as $\lambda \to \infty$ and $|\lambda|^{n-1} \tilde{h}(\lambda \alpha)$ is continuous at $\lambda = 0$.

Therefore, we have proved the following theorem.

Theorem 2.5.1. *Consider the equation $R^*\mu = h$, where $h \in S(\mathbb{R}^n)$. There exists a unique even solution $\mu_e \in C^\infty(Z) \cap L^1(Z)$ to this equation, which is given by*

$$\mu_e(\alpha, p) = \frac{\gamma}{2\pi} \int_{-\infty}^{\infty} |\lambda|^{n-1} \tilde{h}(\lambda\alpha) \exp(-i\lambda p) d\lambda. \qquad (2.5.7)$$

The general solution to Equation (2.5.1) is

$$\mu = \mu_e + \mu_o,$$

where μ_o is an arbitrary C^∞ odd function. The null-space of R^* consists of all C^∞ odd functions.

In fact, inversion formula (2.5.7) yields a more precise result than $\mu_e \in C^\infty(Z) \cap L^1(Z)$ if $h \in S(\mathbb{R}^n)$. Indeed, if n is odd, we get $\mu_e \in S(Z)$. If n is even, an integration by parts yields the property $\mu_e(\alpha, p) = O(p^{-n})$ as $|p| \to \infty$.

In Chapter 10 we show that solution formula (2.5.7) is valid for a sufficiently broad class of distributions h, and this formula is a convenient tool for investigating the properties of the solution μ.

An alternative form of the inversion formula can be obtained combining Equation (2.2.16') and Corollary (2.2.4) from Section 2.2:

$$\mu_e = (R^*)^{-1}h = KRh = \begin{cases} \gamma(-1)^{\frac{n-1}{2}} \hat{h}_p^{(n-1)}, & n \text{ odd}, \\ \gamma(-1)^{\frac{n}{2}} \mathcal{H}\hat{h}_p^{(n-1)}, & n \text{ even}, \end{cases} \quad \gamma = \frac{1}{2(2\pi)^{n-1}}, \qquad (2.5.8)$$

where $\hat{h}_p^{(n-1)}$ is the $(n-1)$-st derivative of $\hat{h}(\alpha, p)$ with respect to p.

Exercise 2.5.1. Derive (2.5.8) directly from (2.5.7).

Using the identities

$$R = F^{-1}\mathcal{F}, \quad \lambda^{2m}\tilde{h}(\lambda\alpha) = \int_{\mathbb{R}^n} [(-\Delta)^m h](x) \exp(i\lambda\alpha \cdot x) dx, \qquad (2.5.9)$$

we obtain another pair of inversion formulas

$$\mu_e = (R^*)^{-1}h = \begin{cases} \gamma(-1)^{\frac{n-1}{2}} R(\Delta^{\frac{n-1}{2}}h), & n \text{ odd}, \\ \gamma(-1)^{\frac{n}{2}} \mathcal{H}\frac{\partial}{\partial p} R(\Delta^{\frac{n}{2}-1}h), & n \text{ even}, \end{cases} \quad \gamma = \frac{1}{2(2\pi)^{n-1}}. \qquad (2.5.10)$$

REMARK 2.5.1. One can give an operator-theoretical derivation of formula (2.5.10) for the inverse operator $(R^*)^{-1}$. Namely, assuming that n is odd and representing (2.2.15a) as $\gamma(-1)^{\frac{n-1}{2}}\Delta^{\frac{n-1}{2}}R^*R = I$, where I is the identity operator on $S(\mathbb{R}^n)$, we get

$$(R^*)^{-1} = (R^{-1})^* = \gamma(-1)^{\frac{n-1}{2}}(\Delta^{\frac{n-1}{2}}R^*)^* = \gamma(-1)^{\frac{n-1}{2}}R\Delta^{\frac{n-1}{2}}.$$

Assuming that n is even, we obtain from (2.2.15b) using that $\mathcal{H} = -\mathcal{H}^*$ and $(\partial/\partial p)^* = -\partial/\partial p$:

$$(R^*)^{-1} = (R^{-1})^* = \gamma(-1)^{\frac{n}{2}}\left(\Delta^{\frac{n}{2}-1}R^*\mathcal{H}\frac{\partial}{\partial p}\right)^* = \gamma(-1)^{\frac{n}{2}}\frac{\partial}{\partial p}\mathcal{H}R\Delta^{\frac{n}{2}-1}.$$

Noticing that the operators $\partial/\partial p$ and $\mathcal{H}$ commute, we finish the proof.

The relationship between the supports of h and $\mu = (R^*)^{-1}h$ is described in

Theorem 2.5.2. *If n is even, then Equation (2.5.1) with a compactly supported smooth $h \not\equiv 0$ cannot have a compactly supported solution μ. If n is odd, μ_e is even and solves Equation (2.5.1) with $h(x) = 0$ for $|x| > a$, then $\mu_e(\alpha, p) = 0$ for $|p| > a$.*

Proof. First consider the case of even n. If μ_e and h are compactly supported, then $\tilde{\mu}_e(\alpha, \lambda)$ and $\tilde{h}(\alpha, \lambda)$ are entire functions of λ for all $\alpha \in S^{n-1}$. Then Equation (2.5.6') yields that, for all α, the meromorphic function $\frac{\tilde{\mu}_e(\alpha,\lambda)}{\tilde{h}(\lambda\alpha)}$ of the variable λ is proportional to the function $|\lambda|^{n-1}$, which is not meromorphic if n is even. This contradiction proves that μ, and, therefore, $\mu = \mu_e + \mu_o$, cannot be compactly supported.

If n is odd and $h(x) = 0$ for $|x| > a$, then $\tilde{h}(\alpha, p) = 0$ for $|p| > a$, and Equation (2.5.8) implies that $\mu(\alpha, p) = 0$ for $|p| > a$. $\square$

REMARK 2.5.2. Theorem 2.5.2 and the 'hole theorem' (cf. Section 2.7.1 below) imply that if h is compactly supported and n is even, then Equation (2.5.1) does not have a solution μ_e of the form $\mu_e = Rf$, where f is compactly supported. An alternative proof of this assertion is based on identity (2.1.27). Indeed, substituting $\mu_e = Rf$ into (2.5.1), we get $R^*Rf = h$ and (2.1.28) yields: $2(2\pi)^{n-1}\tilde{f}(\xi)/\tilde{h}(\xi) = |\xi|^{n-1}$. If f and h are compactly supported and n is even, we obtain the contradiction: the meromorphic function $\tilde{f}(\xi)/\tilde{g}(\xi)$ is proportional to the nonmeromorphic function $|\xi|^{n-1}$.

2.6. Inversion formulas for X-ray transform

2.6.1. Definition of X^* and a formula for X^*X

The X-ray transform was defined by formula (2.1.10) for $f \in S(\mathbb{R}^n)$. If

$$Xf = \int_{-\infty}^{\infty} f(x + \alpha t)dt := g(x, \alpha), \quad \alpha \in S^{n-1}, \ x \in \alpha^\perp, \quad (2.6.1)$$

then the function $g(x, \alpha)$ is defined on $T := \alpha^\perp \times S^{n-1}$, where $\alpha^\perp$ is a hyperplane passing through the origin and orthogonal to α. We assume the functions we deal with to be real-valued. Define the scalar product in $L^2(T)$:

$$(g_1, g_2)_{L^2(T)} := \int\limits_{S^{n-1}} \int\limits_{\alpha^\perp} g_1(x, \alpha) g_2(x, \alpha) dx \, d\alpha, \qquad (2.6.2)$$

and the $L^2(T)$-norm is defined as usual: $\|g\|_{L^2(T)}^2 := (g_1, g_2)_{L^2(T)}$.

If $x \in \mathbb{R}^n$, denote $x_\alpha^\perp := x - \alpha(x, \alpha)$, where (x, α) is the inner product in $\mathbb{R}^n$. Clearly, $x_\alpha^\perp$ is the orthogonal projection of x onto the hyperplane $\alpha^\perp$. Then

$$g(x, \alpha) = \int\limits_{-\infty}^{\infty} f(x + \alpha t) dt = \int\limits_{-\infty}^{\infty} f[x - \alpha(\alpha, x) + \alpha(t + (\alpha, x))] dt$$

$$= \int\limits_{-\infty}^{\infty} f(x_\alpha^\perp + \tau \alpha) d\tau. \qquad (2.6.3)$$

Thus, we can consider in the definition (2.6.1) only $x \in \alpha^\perp$.

One has, for $f \in C^\infty(B_a)$,

$$\int\limits_{\alpha^\perp} |g(x, \alpha)|^2 dx = \int\limits_{\alpha^\perp} \left| \int\limits_{-\infty}^{\infty} f(x + \alpha t) dt \right|^2 dx$$

$$\leq \int\limits_{\alpha^\perp} \left[\int\limits_{-\infty}^{\infty} |f(x + \alpha t)|^2 dt \int\limits_{-a}^{a} dt \right] dx = 2a \, \| f \|_{L^2(\mathbb{R}^n)}^2 . \qquad (2.6.4)$$

Integrate (2.6.4) with respect to α and get

$$\| g \|_{L^2(T)} \leq 2a |S^{n-1}| \, \|f\|_{L^2(\mathbb{R}^n)}. \qquad (2.6.5)$$

Therefore, $X : L^2(B_a) \to L^2(T)$ is a bounded operator.

Exercise 2.6.1. Let w be a smooth positive weight function such that

$$c_w := \sup_{\alpha \in S^{n-1}, x \in \alpha^\perp} X(1/w)(x, \alpha) < \infty,$$

where $X(1/w)$ is X-ray transform of $1/w$. Show that if $f \in L^2(\mathbb{R}^n, w)$, then

$$\|g\|_{L^2(\alpha^\perp)}^2 \leq c_w \|f\|_{L^2(\mathbb{R}^n, w)}^2 \qquad (2.6.6)$$

and

$$\|g\|_{L^2(T)}^2 \leq c_w |S^{n-1}| \|f\|_{L^2(\mathbb{R}^n, w)}^2. \tag{2.6.7}$$

Hint. The proof is similar to the proof of formula (2.6.4).

Let us calculate $X^* : L^2(T) \to L^2(B_a)$. One has

$$(Xf, h)_{L^2(T)} = \int_{S^{n-1}} d\alpha \int_{\alpha^\perp} dx \int_{-\infty}^{\infty} dt f(x + \alpha t) h(x, \alpha)$$

$$= \int_{S^{n-1}} d\alpha \int_{\mathbb{R}^n} dy f(y) h(y_\alpha^\perp, \alpha) = \langle f, X^* h \rangle_{L^2(\mathbb{R}^n)}, \tag{2.6.8}$$

where $y_\alpha^\perp := y - \alpha(y, \alpha)$ and

$$(X^* g)(y) := \int_{S^{n-1}} g(y_\alpha^\perp, \alpha) d\alpha. \tag{2.6.9}$$

One has

$$X^* X f = \int_{S^{n-1}} d\alpha \int_{-\infty}^{\infty} d\tau f(x_\alpha^\perp + \alpha \tau) = \int_{S^{n-1}} d\alpha \int_{-\infty}^{\infty} dt f(x + \alpha t)$$

$$= 2 \int_{S^{n-1}} d\alpha \int_0^{\infty} dt \, t^{n-1} (t^{1-n} f(x + \alpha t)) = 2 \int_{\mathbb{R}^n} dy |y|^{1-n} f(x + y), \tag{2.6.10}$$

where we have used the definition $x_\alpha^\perp = x - \alpha(x, \alpha)$ and denoted $t = \tau - (x, \alpha)$.

2.6.2. Inversion formula for X-ray transform

For $\xi \in \alpha^\perp$, we define:

$$\tilde{g}(\xi, \alpha) := \mathcal{F}_{x_\alpha^\perp \to \xi \in \alpha^\perp} g(x_\alpha^\perp, \alpha) := \int_{\alpha^\perp} g(y, \alpha) \exp(i\xi \cdot y) dy.$$

The following theorem can be considered as an inversion formula: it gives the Fourier transform $\tilde{f}(\xi)$ of $f(x)$ in terms of the data $g(x, \alpha)$ for all $\xi \in \alpha^\perp$. Since α is arbitrary, $\tilde{f}(\xi)$ can be calculated for any $\xi \in \mathbb{R}^n$. Inverting $\tilde{f}(\xi)$ by the standard Fourier inversion formula, we obtain $f(x)$.

Theorem 2.6.1. *Let $f \in S(\mathbb{R}^n)$ and $g = Xf$. Then one has*

$$\tilde{f}(\xi) = \tilde{g}(\xi, \alpha), \quad \xi \in \alpha^{\perp}. \tag{2.6.11}$$

Proof. Fix $\alpha \in S^{n-1}$ and $\xi \in \alpha^{\perp}$. We have

$$\int_{\alpha^{\perp}} \exp(i\xi \cdot x_{\alpha}^{\perp}) \int_{-\infty}^{\infty} f(x_{\alpha}^{\perp} + t\alpha) dt dx_{\alpha}^{\perp}$$

$$= \int_{\alpha^{\perp}} \int_{-\infty}^{\infty} \exp(i\xi \cdot (x_{\alpha}^{\perp} + t\alpha)) f(x_{\alpha}^{\perp} + t\alpha) dt dx_{\alpha}^{\perp}$$

$$= \int_{\mathbb{R}^n} \exp(i\xi \cdot y) f(y) dy = \tilde{f}(\xi), \quad \xi \cdot \alpha = 0,$$

which proves (2.6.11). □

Let $\mathcal{I}_{\alpha^{\perp}}^a$ be the $(n-1)$-dimensional Riesz potential (cf. (2.2.6)) acting on the hyperplane $\alpha^{\perp}$. Recall that $\mathcal{I}^a$ denotes the usual Riesz potential acting in $\mathbb{R}^n$.

Theorem 2.6.2. *Let $f \in S(\mathbb{R}^n)$ and $g = Xf$. Then, for any $a, |a| < n$, one has*

$$f = \frac{1}{2\pi |S^{n-2}|} \mathcal{I}^{-a} X^* \mathcal{I}_{\alpha^{\perp}}^{a-1} g. \tag{2.6.12}$$

Proof. First, we prove the identity

$$\int_{S^{n-1}} \int_{\alpha^{\perp}} \varphi(\xi) d\xi \, d\alpha = |S^{n-2}| \int_{\mathbb{R}^n} \frac{\varphi(\xi)}{|\xi|} d\xi. \tag{2.6.13}$$

Transforming the left-hand side of (2.6.13), we derive

$$\int_{S^{n-1}} \int_{\alpha^{\perp}} \varphi(\xi) d\xi \, d\alpha = \int_{S^{n-1}} \int_{\mathbb{R}^n} \varphi(\xi) \delta(\alpha \cdot \xi) d\xi \, d\alpha$$

$$= \int_{\mathbb{R}^n} \varphi(\xi) \int_{S^{n-1}} \frac{1}{2\pi} \int_{-\infty}^{\infty} e^{it\alpha \cdot \xi} dt d\alpha \, d\xi$$

$$= \int_{\mathbb{R}^n} \varphi(\xi) \frac{1}{\pi} \int_{\mathbb{R}^n} |x|^{1-n} e^{ix \cdot \xi} dx \, d\xi.$$

Using identity (2.1.33) with $\lambda = 1 - n$, we obtain (2.6.13). Substituting $\varphi(\xi) = |\xi|^{1-a} \tilde{f}(\xi) e^{-ix \cdot \xi} / (2\pi)^n$ into (2.6.13) and using (2.6.11), we get

$$
\begin{aligned}
(\mathcal{I}^a f)(x) &= \frac{1}{(2\pi)^n |S^{n-2}|} \int\limits_{S^{n-1}} \int\limits_{\alpha^\perp} |\xi|^{1-a} \tilde{f}(\xi) e^{-ix \cdot \xi} d\xi \, d\alpha \\
&= \frac{1}{(2\pi)^n |S^{n-2}|} \int\limits_{S^{n-1}} \int\limits_{\alpha^\perp} |\xi|^{1-a} \tilde{g}(\xi, \alpha) e^{-ix_\alpha^\perp \cdot \xi} d\xi \, d\alpha \\
&= \frac{1}{2\pi |S^{n-2}|} \int\limits_{S^{n-1}} (\mathcal{I}_{\alpha^\perp}^{a-1} g)(x_\alpha^\perp, \alpha) d\alpha = \frac{1}{2\pi |S^{n-2}|} X^* \mathcal{I}_{\alpha^\perp}^{a-1} g,
\end{aligned}
$$

which is equivalent to (2.6.12). $\square$

REMARK 2.6.1. Note that the data $g(x, \alpha)$, $(x, \alpha) \in T$, are overdetermined: one can find f knowing $g(x, \alpha)$, $x \in L$, $\alpha \in S^{n-1}$, where L is a suitable one-dimensional manifold (curve), see Chapter 9.

2.7. Uniqueness theorems for the Radon and X-ray transforms

2.7.1. Uniqueness theorems for the Radon transform

Inversion formulas from Section 2.2 show that if we have a complete data set, then the original function f can be uniquely determined. However, quite frequently in practice, it is either impossible or very difficult to collect the complete data. Hence the important question is: can f be found from some incomplete information? The first result in this direction is the 'hole theorem'

Theorem 2.7.1. *Suppose that $f \in S(\mathbb{R}^n)$ and $\hat{f}(\alpha, p) = 0$ for all (α, p) such that the hyperplanes $l_{\alpha p}$ do not intersect a convex compact set $D \in \mathbb{R}^n$. Then $f(x) = 0$ in $D' := \mathbb{R}^n \setminus D$.*

Proof. First, Equations (2.2.39) and (2.2.42) imply that if $\hat{f}(\alpha, p) = 0$ for $|p| > a$, then $f(x) = 0$ for $|x| > a$. Noticing that for any $x \notin D$, there exists a ball which contains D and does not contain x, we finish the proof. $\square$

The fast decay of f at infinity is necessary, as the following example shows. Let $f = (x_1 + ix_2)^{-\nu}$, where $\nu > 1$. Then $\hat{f}(\alpha, p) = 0$ for all α and all $p \neq 0$, but f does not vanish in the region $|x| > a$ for any a. Indeed, fix any (α, p), $p \neq 0$. On the line $l_{\alpha p} : \alpha \cdot x = p$, the line element ds is proportional to dz, $z = x_1 + ix_2$. So we integrate a continuous function $z^{-\nu}$ from $-\infty$ to $+\infty$, and its antiderivative vanishes at $\pm\infty$. Thus the integral of $z^{-\nu}$ over $l_{\alpha p}$ vanishes and, therefore, $\hat{f}(\alpha, p) = 0$.

Exercise 2.7.1. Assume that $f(x) \in \mathcal{S}(\mathbb{R}^n)$ and $(*) \int_S f(y)ds = 0$ for every sphere S which contains a fixed convex compact set D. Prove that $f(x) = 0$ in $D' = \mathbb{R}^n \setminus D$.

This assertion is similar to the 'hole theorem'. One can assume that $f \in C(\mathbb{R}^n)$ and decays faster than any power of $|x|$. The conclusion remains valid.

Hint. It follows from assumption $(*)$ that

$$\int_{|x-y|\leq R} f(y)dy = \int_{\mathbb{R}^n} f(y)dy = \text{const}$$

for any x and a sufficiently large R, such that the ball $B := \{y : |x-y| \leq R\}$ contains D. Differentiating with respect to x_i and using the Gauss formula, one gets $\int_S y_j f(y)ds = 0$, $1 \leq j \leq n$ where $S = \partial B$. Repeat this argument and get $\int_S P(y)f(y)ds = 0$, where $P(y)$ is an arbitrary polynomial and S is an arbitrary sphere containing D. Since the set of all polynomials is complete in $L^2(S)$, it follows that $f = 0$.

In Theorem 2.7.1 we considered the case when the Radon transform $\hat{f}(\alpha, p)$ was known for all $\alpha \in S^{n-1}$, but not for all $p \in \mathbb{R}$. The case when $\hat{f}(\alpha, p)$ is known for all $p \in \mathbb{R}$, but not for all $\alpha \in S^{n-1}$ is considered in the following

Theorem 2.7.2. *Let $\sigma \subset S^{n-1}$ be a set such that $\mathcal{P}(\xi) \equiv 0$ on σ implies $\mathcal{P}(\xi) \equiv 0$ for any homogeneous polynomial $\mathcal{P}(\xi)$. Assume $f \in C_0^\infty(\mathbb{R}^n)$ and $\hat{f}(\alpha, p) = 0 \, \forall (\alpha, p) \in \sigma \times \mathbb{R}$. Then $f \equiv 0$.*

Proof. Since $\hat{f}(\alpha, p) = 0 \, \forall \alpha \in \sigma$, one has

$$\tilde{f}(t\alpha) = 0 \quad \forall (\alpha, t) \in \sigma \times \mathbb{R}.$$

Since f is compactly supported, $\tilde{f}$ is entire and, therefore, $\tilde{f}$ can be represented as

$$\tilde{f}(\xi) = \sum_{m=0}^{\infty} \mathcal{P}_m(\xi),$$

where $\mathcal{P}_m$ is a homogeneous polynomial of degree $m, m = 0, 1, \ldots$, and the series converges absolutely and uniformly on compact subsets of $\mathbb{C}^n$. Letting $\xi = t\alpha$, we get

$$\sum_{m=0}^{\infty} \mathcal{P}_m(t\alpha) = \sum_{m=0}^{\infty} t^m \mathcal{P}_m(\alpha) = 0 \quad \forall (\alpha, t) \in \sigma \times \mathbb{R},$$

then

$$\mathcal{P}_m(\alpha) = 0 \quad \forall \alpha \in \sigma, \ m = 0, 1, 2, \ldots.$$

By the assumption, this implies $\mathcal{P}_m(\alpha) = 0, \quad m = 0, 1, 2, \ldots$, and Theorem 2.7.2 is proved. $\square$

If $f \in L_0^2(\mathbb{R}^n)$, the conclusion of the above theorem still holds.

2.7.2. Uniqueness theorems for X-ray transform

Since integrals of f over planes can be computed from integrals of f over lines, we immediately obtain the analogue of Theorem 2.7.1.

Theorem 2.7.3. *Let $f \in S(\mathbb{R}^n)$ and integrals of f over all lines not intersecting a convex compact set $D \subset \mathbb{R}^n$ equal zero. Then $f = 0$ outside D.*

The case of a countable number of sources is considered in

Theorem 2.7.4. *Let us assume that $f \in L^2(B_a)$ and $f(x) = 0$ for $|x| > a$, a sequence $x_j \to x$, $|x| > a$, is given, and*

$$g(x_j, \alpha) = 0 \quad \forall \alpha \in S^{n-1}, \, j \geq 1, \qquad (2.7.1)$$

where B_a is a ball with radius a centered at the origin, and g is the X-ray transform of f. Then $f \equiv 0$.

Exercise 2.7.2. Let $f \in C^\infty(B_a)$, $f(x) = 0$ for $|x| > a$. Prove the inequality

$$\int\limits_{S^{n-1}} |g(x, \alpha)|^2 d\alpha \leq \frac{4a}{(|x| - a)^{n-1}} \|f\|_{L^2(B_a)}^2.$$

Conclude that if $f \in L^2(B_a)$ and $f(x) = 0$ for $|x| > a$, then $g(x, \alpha)$ is measurable with respect to α for any fixed $|x| > a$.

Proof of Theorem 2.7.4. Since $x_j \notin B_a$, Exercise 2.7.2 implies that the function $g(x_j, \alpha)$ is well defined for each x_j and almost all α. Moreover, if $x_j \notin B_a$ and $g(x_j, \alpha) = 0$, we conclude that

$$0 = \int\limits_0^\infty f(x_j + \alpha t)dt, \, j \geq 1,$$

for almost all α.

Multiply this equality by $(\alpha \cdot \theta)^{-(n-1)}$, integrate with respect to α over S^{n-1}, and set $y = t\alpha$:

$$0 = \int\limits_{S^{n-1}} \int\limits_0^\infty f(x_j + \alpha t)(\alpha \cdot \theta)^{-(n-1)} dt \, d\alpha = \int\limits_{\mathbb{R}^n} f(x_j + y)(y \cdot \theta)^{-(n-1)} dy$$

$$= \int\limits_{\mathbb{R}^n} f(z)(z \cdot \theta - x_j \cdot \theta)^{-(n-1)} dz = \int\limits_{-\infty}^\infty \hat{f}(\theta, p)(p - t_j)^{-(n-1)} dp,$$

$$t_j := x_j \cdot \theta.$$

Recall that x is a limiting point of the sequence x_j, $x_j \to x$. Choose a sufficiently small neighborhood $\Omega \subset S^{n-1}$ of the point $x/|x|$ such that

$$t_j = x_j \cdot \theta > a, \ j > j_0; \ t = x \cdot \theta > a; \ \forall \theta \in \Omega,$$

where j_0 is a sufficiently large fixed number. Fix any $\theta \in \Omega$. Consider the function h:

$$h(z) := \int_{-\infty}^{\infty} \hat{f}(\theta, p)(p - z)^{-(n-1)} dp, \quad h(t_j) = 0.$$

Since $\hat{f}(\theta, p) = 0$ for $|p| > a$, one has

$$h(z) = \int_{-a}^{a} \frac{\hat{f}(\theta, p)}{(p - z)^{n-1}} dp. \tag{2.7.2}$$

Thus, $h(z)$ is analytic in the domain $z \notin [-a, a]$. Furthermore, $h(z)$ vanishes at the sequence of points $t_j > a, j > j_0$, and this sequence has a limiting point $t_j \to t > a$ in the region where h is analytic. Therefore, $h(z) = 0$ in the region $z \notin [-a, a]$. It follows that the Cauchy integral with the density $\hat{f}(\theta, p)$ vanishes outside the cut $[-a, a]$. By the jump formula, the density must vanish, that is, $\hat{f}(\theta, p) \equiv 0$ for $\theta \in \Omega$. By the Fourier slice theorem, $\tilde{f}(\lambda \theta) = 0$ for $\theta \in \Omega$ and $\lambda \in \mathbb{R}$. Since f is compactly supported, its Fourier transform $\tilde{f}(\xi)$, $\xi \in \mathbb{C}^n$, is an entire function. This function vanishes if $\xi/|\xi| \in \Omega$ and, by the analytic continuation, $\tilde{f}(\xi) = 0$ in $\mathbb{R}^n$. Therefore, $f \equiv 0$ and Theorem 2.7.4 is proved. $\square$

Exercise 2.7.3. Assume $f \in C_0^{\infty}(B_a)$, $g(x, \alpha_j) = 0 \ \forall x \in \mathbb{R}^n$ and for infinitely many α_j. Prove: $f \equiv 0$.

Hint. Use Theorem 2.6.1 to conclude that $\tilde{f}(\xi) \equiv 0$ on the hyperplanes $\xi \cdot \alpha_j = 0$. Therefore, $\tilde{f}$ has zero of infinite order at the origin. Since $\tilde{f}$ is entire, this implies $f \equiv 0$.

Clearly, the uniqueness results presented in this section hold also for the case of fan-beam transform.

2.7.3. Example of the lack of injectivity

An important question is: on what classes of functions $f(x)$ is the operator R injective? This question is studied in Chapters 3, 10, and in Section 2.4. Essentially the situation is this: if $f(x)$ decays sufficiently fast as $|x| \to \infty$, then $Rf = 0$ implies $f = 0$, that is, R in the classical sense (2.1.1) is injective. Our goal in this section is to demonstrate that some restrictions on the rate of growth of $f(x)$ are necessary for R to be injective. In Chapter 10 the injectivity of R will be established on broad classes of distributions. However, we can prove that the definition of R on distributions coincide with the classical one (2.1.1) only on the classes of functions with some decay at infinity, e.g., $\frac{f(x)}{1+|x|} \in L^1(\mathbb{R}^n)$, (see Section 10.6).

In the following lemma the function defined on $\mathbb{C}$ is considered as a function on $\mathbb{R}^2$.

Lemma 2.7.1. *There exists an entire analytic function*

$$f(x_1 + ix_2) = f(z) \not\equiv 0, \quad z = x_1 + ix_2,$$

such that

$$\int_{l_{\alpha p}} |f(z)|ds < \infty \quad \forall \alpha, p, \tag{2.7.3}$$

$$\hat{f}(\alpha, p) = 0 \quad \forall \alpha, p. \tag{2.7.4}$$

Proof. It is sufficient to construct an entire function $g(z) \not\equiv 0$, $z = x_1 + ix_2$, such that

$$\int_{l_{\alpha p}} |g'(z)|dz < \infty \quad \forall \alpha, p \tag{2.7.5}$$

and

$$|g(z)| \to 0 \text{ as } z \to \infty, \quad z \in l_{\alpha p}. \tag{2.7.6}$$

Indeed, if $g(z)$ with these properties is constructed, then $f(x_1, x_2) := g'(z)$, $z = x_1 + ix_2$, is the function in Lemma 2.7.1. Indeed, condition (2.7.3) follows from (2.7.5), and (2.7.4) follows from the Newton-Leibnitz formula and from (2.7.6). Existence of $g(z)$ with the desired properties follows immediately from the Arakelyan theorem [Ar, p.188]. This theorem asserts:

if K is a closed set in the complex plane $\mathbb{C}$, whose boundary is a simple Jordan curve which starts from infinity and ends at infinity, and if $f(z)$ is an arbitrary holomorphic function in the interior of K and continuous in K, then for any $\epsilon > 0$ there exists an entire function $g_\epsilon(z)$ such that

$$|h(z) - g_\epsilon(z)| \le \epsilon(1 + |z|)^{-3}, \quad z \in K. \tag{2.7.7}$$

Let K be the complement to the set

$$\{z : |z| < 2\} \cup \{z : x_1 > 1, \ x_1^{-1} \le x_2 \le 2x_1^{-1}\}, \tag{2.7.8}$$

and let $h(z) = z^{-3}$. Then Arakelyan's theorem implies the existence of an entire function $g_\epsilon(z) := g(z)$ such that

$$|g(z) - z^{-3}| \le \epsilon(1 + |z|)^{-3}, \quad \epsilon > 0, \ z \in K. \qquad (2.7.9)$$

It is clear from (2.7.9) that condition (2.7.6) is satisfied. Condition (2.7.5) is easily verified also: it is obvious that on any straight line, which is not the line $x_2 = 0$, condition (2.7.5) holds. Let us check that it holds also for the line $x_2 = 0$. Let $z \in K \backslash \partial K$ and $d := \frac{1}{2} \operatorname{dist}(z, \partial K) > 0$. Then the Cauchy estimate for the modulus of the derivative of an analytic function yields:

$$|g'(z)| \le \frac{c}{d} \sup |g(\xi)| \le c'|z|^{-2} \text{ as } |z| \to \infty, \ z = x_1. \qquad (2.7.10)$$

Here c and c' are positive constants independent of z, and supremum is taken over the disk $\{\xi : |\xi - z| < d\}$. In this disc estimate (2.7.9) yields:

$$|g(\xi)| \le |\xi|^{-3} + \epsilon|\xi|^{-3} \le c''|z|^{-3}, \quad |\xi| \gg 1, \qquad (2.7.11)$$

while

$$d > \frac{1}{4}|z|^{-1}. \qquad (2.7.12)$$

From (2.7.11) and (2.7.12) one gets (2.7.10). Lemma 2.7.1 is proved. □

An example of the lack of injectivity can be constructed in $\mathbb{R}^n$ for any $n \ge 2$ (see [AG]), and for even n such an example is an immediate consequence of the example constructed in this section.

2.8. Attenuated and exponential Radon transforms

2.8.1. Simplest properties

Let $f(x)$, $x \in \mathbb{R}^2$, be compactly supported. Attenuated Radon transform is defined by the formula

$$(R_\mu f)(\theta, p) = \int_{-\infty}^{\infty} f(p\Theta + t\Theta^\perp) \exp\left(-\int_t^{\infty} \mu(p\Theta + s\Theta^\perp)ds\right)dt,$$

$$= \int_{\mathbb{R}^2} f(x) \exp\left(-\int_0^{\infty} \mu(x + s\Theta^\perp)ds\right)\delta(p - \Theta \cdot x)dx,$$

$$\Theta = (\cos\theta, \sin\theta), \ \Theta^\perp = (-\sin\theta, \cos\theta). \qquad (2.8.1)$$

This transform is used in single photon emission computed tomography (SPECT) (see Section 1.2.1.2). In (2.8.1), $\mu(x)$ is the attenuation coefficient which is assumed to be known, and $f(x)$ is the unknown distribution of radioactive material, which is to be determined knowing $(R_\mu f)(\theta, p)$ for all $\theta \in [0, 2\pi)$ and $p \in \mathbb{R}$.

Suppose now that μ is constant inside the support of f and $\mu = 0$ outside $\operatorname{supp} f$. Let us assume that $\operatorname{supp} f$ is convex. Let $\tau(\theta, p)$ denote a point where the ray $p\Theta + t\Theta^\perp$, $-\infty < t < \infty$, enters $\operatorname{supp} f$. Then

$$\int_t^\infty \mu ds = \int_{\tau(\theta,p)}^\infty \mu ds - \int_{\tau(\theta,p)}^t \mu ds = g_1(\theta, p) + \mu\tau(\theta, p) - \mu t = g_2(\theta, p) - \mu t,$$

where $g_2(\theta, p)$ is a known function. Now Equation (2.8.1) yields

$$(R_\mu f)(\theta, p) = e^{-g_2(\theta, p)}(T_\mu f)(\theta, p), \qquad (2.8.2)$$

where $T_\mu f$, defined by the formula

$$\hat{f}^{(\mu)}(\theta, p) := (T_\mu f)(\theta, p) = \int_{-\infty}^\infty f(p\Theta + t\Theta^\perp)e^{\mu t} dt$$

$$= \int_{\mathbb{R}^2} e^{\mu(x \cdot \Theta^\perp)} f(x)\delta(p - \Theta \cdot x)dx, \qquad (2.8.3)$$

is the exponential Radon transform. Equation (2.8.2) shows that if μ is a known constant, the exponential Radon transform can be computed if the attenuated Radon transform is known.

Let us generalize some of the results obtained in Section 2.1 to the transforms R_μ and T_μ.

The operators adjoint to R_μ and T_μ are given by the formulas:

$$(R_\mu^* g)(x) = \int_{S^1} \exp\left(-\int_0^\infty \mu(x + s\Theta^\perp)ds\right)g(\theta, \Theta \cdot x)d\theta$$

and

$$(T_\mu^* g)(x) = \int_{S^1} e^{\mu(x \cdot \Theta^\perp)} g(\theta, \Theta \cdot x)d\theta. \qquad (2.8.5)$$

Exponential Radon transform of a convolution satisfies the equation:

$$T_\mu(f * g) = (T_\mu f) \circledast (T_\mu g), \qquad (2.8.6a)$$

that is

$$[T_\mu(f * g)](\theta, p) = \int_{-\infty}^\infty \hat{f}^{(\mu)}(\theta, p - s)\hat{g}^{(\mu)}(\theta, s)ds. \qquad (2.8.6b)$$

Formula for $(T^*_{-\mu}g) * f$ takes the form:

$$(T^*_{-\mu}g) * f = T^*_{-\mu}(g \circledast (T_\mu f)), \qquad (2.8.7a)$$

that is

$$[(T^*_{-\mu}g) * f](x) = \int_{S^1} e^{-\mu(x\cdot\Theta^\perp)} \int_{-\infty}^{\infty} g(\theta, \Theta \cdot x - p) \hat{f}^{(\mu)}(\theta, p) dp\, d\theta. \quad (2.8.7b)$$

Formula for $T^*_{-\mu} T_\mu$:

$$(T^*_{-\mu} T_\mu f)(x) = \int_{\mathbb{R}^2} 2\frac{\cosh(\mu|x - y|)}{|x - y|} f(y) dy. \qquad (2.8.8)$$

Exercise 2.8.1. Check formulas (2.8.5)-(2.8.8).

2.8.2. Inversion formulas

Theorem 2.8.1. *Let $f \in C_0^\infty(\mathbb{R}^2)$. Then*

$$f(x) = \frac{1}{4\pi^2} \int_{S^1} e^{-\mu(x\cdot\Theta^\perp)} \int_{-\infty}^{\infty} \frac{\cos(\mu(\Theta \cdot x - p))}{\Theta \cdot x - p} \hat{f}_p^{(\mu)}(\theta, p) dp d\theta, \quad (2.8.9)$$

where $\hat{f}_p^{(\mu)} = \partial \hat{f}^{(\mu)}/\partial p$.

Proof. Equation (2.8.7b) yields an inversion formula if we find g such that $T^*_{-\mu} g = \delta$, where δ is the delta-function. Integrating by parts with respect to p in (2.8.9), we obtain an analogue of (2.8.7b) with

$$g(\theta, p) = \frac{1}{4\pi^2} \frac{\partial}{\partial p}\left(\frac{\cos(\mu p)}{p}\right).$$

Therefore, Equation (2.8.5) implies that we have to prove the identity

$$\frac{1}{4\pi^2} \int_{S^1} e^{-\mu(x\cdot\Theta^\perp)} \frac{\partial}{\partial p}\left(\frac{\cos(\mu p)}{p}\right)\bigg|_{p=\Theta\cdot x} d\theta = \delta(x). \qquad (2.8.10)$$

Let us denote

$$\int_{-\infty}^{\infty} \frac{\cos(\mu p)}{p} e^{ips} dp = \varphi(s). \qquad (2.8.11)$$

Clearly, $\varphi'(s) = \pi i(\delta(s - \mu) + \delta(s + \mu))$. Since $\varphi(s)$ is odd, we get

$$\varphi(s) = \begin{cases} -\pi i, & s < -\mu, \\ 0, & |s| < \mu, \\ \pi i, & s > \mu. \end{cases} \qquad (2.8.12)$$

Using the above equation, we obtain the identity

$$\frac{\partial}{\partial p}\left(\frac{\cos(\mu p)}{p}\right) = \frac{1}{2}\int\limits_{|t|>\mu} |t|e^{itp}dt.$$

Substitution into (2.8.10) yields

$$\frac{1}{8\pi^2}\int\limits_{|t|>\mu}|t|\int\limits_{S^1}e^{-\mu(x\cdot\Theta^\perp)}e^{it\Theta\cdot x}d\theta\,dt$$

$$= \frac{1}{4\pi^2}\int\limits_{\mu}^{\infty}t\int\limits_{S^1}e^{it\Theta\cdot x}\cosh(\mu\Theta^\perp\cdot x)d\theta\,dt$$

$$= \frac{1}{4\pi^2}\int\limits_{0}^{\infty}s\int\limits_{S^1}e^{i\sqrt{s^2+\mu^2}\Theta\cdot x}\cosh(\mu\Theta^\perp\cdot x)d\theta\,ds. \qquad (2.8.13)$$

Let us denote the inner integral in (2.8.13) by $C(\mu)$. Suppose that $C(\mu)$ does not depend on μ. Then, substituting $\mu = 0$ into (2.8.13), we recognize the right-hand side of (2.8.13) to be the delta-function represented as an inverse Fourier transform of 1 in cylindrical coordinate system. It remains to be proved that $C'(\mu) = 0$. Let $|x| = r$. Then

$$C(\mu) = \int\limits_{0}^{2\pi} e^{i\sqrt{s^2+\mu^2}r\cos\theta}\cosh(\mu r\sin\theta)d\theta$$

$$= \int\limits_{0}^{2\pi}\cos\left(\sqrt{s^2+\mu^2}r\cos\theta\right)\cosh(\mu r\sin\theta)d\theta.$$

Therefore,

$$
\begin{aligned}
C'(\mu) = 4r \Bigg[& \int_0^{\pi/2} \sinh(\mu r \sin\theta) \cos\left(\sqrt{s^2 + \mu^2} r \cos\theta\right) \sin\theta d\theta \\
& - \int_0^{\pi/2} \cosh(\mu r \cos\theta) \sin\left(\sqrt{s^2 + \mu^2} r \cos\theta\right) \frac{\mu}{\sqrt{s^2 + \mu^2}} \cos\theta d\theta \Bigg]
\end{aligned}
$$

$$
\begin{aligned}
= 4r \Bigg[& \int_0^1 \sinh(\mu r \sqrt{1-t^2}) \cos\left(\sqrt{s^2 + \mu^2} rt\right) dt \\
& - \int_0^1 \cosh(\mu rt) \sin\left(\sqrt{s^2 + \mu^2} r \sqrt{1-t^2}\right) \frac{\mu}{\sqrt{s^2 + \mu^2}} dt \Bigg].
\end{aligned}
$$

Integrating by parts in one of the integrals on the right-hand side of the last equation, we get $C'(\mu) = 0$. □

REMARK 2.8.1. The integral in (2.8.9) is understood in the Cauchy principal value sense:

$$
f(x) = \lim_{\epsilon \to 0} f_\epsilon(x),
$$

$$
f_\epsilon(x) := \frac{1}{4\pi^2} \int_{S^1} e^{-\mu(x \cdot \Theta^\perp)} \int_{|\Theta \cdot x - p| > \epsilon} \frac{\cos(\mu(\Theta \cdot x - p))}{\Theta \cdot x - p} \hat{f}_p^{(\mu)}(\theta, p) dp d\theta.
$$

Using the above equation, it is possible to extend the applicability of inversion formula (2.8.9) to a more general class of functions – in particular, to piecewise-smooth functions with jump discontinuities. This is done in Section 6.7.

Let us represent $\hat{f}^{(\mu)}$ and f using angular harmonics

$$
\hat{f}^{(\mu)}(\theta, p) = \frac{1}{2\pi} \sum_{l=-\infty}^{\infty} \hat{f}_l^{(\mu)}(p) e^{-il\theta}, \quad f(r\Theta) = \frac{1}{2\pi} \sum_{l=-\infty}^{\infty} f_l(r) e^{-il\theta}.
$$

$$
\tag{2.8.14}
$$

Let us denote also

$$
g_l(\sigma) = \int_{-\infty}^{\infty} \hat{f}_l^{(\mu)}(p) e^{i\sigma p} dp. \tag{2.8.15}
$$

Theorem 2.8.2. *Let $f \in C_0^\infty(\mathbb{R}^2)$. One has*

$$
\hat{f}_l^{(\mu)}(p) = \int_p^\infty 2 \frac{\cosh(\mu\sqrt{r^2 - p^2} - il\arccos(p/r))}{\sqrt{1-(p/r)^2}} f_l(r) dr, \quad p > 0,
$$

$$
\tag{2.8.16}
$$

and

$$f_l(r) = \frac{i^l}{4\pi} \int\limits_{|\sigma|>\mu} |\sigma| \left(\frac{\sigma+\mu}{\sqrt{\sigma^2-\mu^2}}\right)^l J_l(r\sqrt{\sigma^2-\mu^2}) g_l(\sigma) d\sigma, \quad r > 0.$$

$$(2.8.17)$$

Proof. First, let us prove (2.8.16). Substituting (2.8.14) into (2.8.3), we get

$$\frac{1}{2\pi} \sum_{l=-\infty}^{\infty} \hat{f}_l^{(\mu)}(p) e^{-il\theta}$$

$$= \int\limits_0^{2\pi}\int\limits_0^{\infty} e^{\mu r \sin(\alpha-\theta)} \frac{1}{2\pi} \sum_{l=-\infty}^{\infty} f_l(r) e^{-il\alpha} \delta(p - r\cos(\alpha-\theta)) r\, dr\, d\alpha.$$

$$(2.8.18)$$

Simple calculations show that

$$G_l^{(\mu)}(p,r) = \int\limits_0^{2\pi} e^{\mu r \sin\alpha - il\alpha} \delta(p - r\cos\alpha) d\alpha$$

$$= 2\frac{\cosh(\mu\sqrt{r^2-p^2} - il\arccos(p/r))}{\sqrt{r^2-p^2}}, \quad 0 \le p < r,$$

$$G_l^{(\mu)}(p,r) = 0, \quad 0 \le r < p.$$

Substitution of the above equation into (2.8.18) proves (2.8.16). Now, using (2.8.14) in (2.8.9), we get

$$\frac{1}{2\pi} \sum_{l=-\infty}^{\infty} f_l(r) e^{-il\alpha} = \frac{1}{4\pi^2} \int\limits_0^{2\pi} e^{-\mu r \sin(\alpha-\theta)} \int\limits_{-\infty}^{\infty} \frac{\cos\left[\mu(r\cos(\theta)-p)\right]}{r\cos(\theta)-p} \times$$

$$\times \frac{1}{2\pi} \sum_{l=-\infty}^{\infty} (\hat{f}_l^{(\mu)}(p))' e^{-il\theta} dp\, d\theta.$$

Thus, f_l and $\hat{f}_l^{(\mu)}$ are related by the equation

$$f_l(r) = \frac{1}{4\pi^2} \int\limits_0^{2\pi} e^{\mu r \sin\theta - il\theta} \int\limits_{-\infty}^{\infty} \frac{\cos\left[\mu(r\cos\theta-p)\right]}{r\cos\theta-p} (\hat{f}_l^{(\mu)}(p))' dp.$$

Substituting (2.8.15) and using Equations (2.8.11), (2.8.12), we obtain after some transformations

$$f_l(r) = \frac{1}{8\pi^2} \int\limits_{|\sigma|>\mu} |\sigma| g_l(\sigma) \left[\int\limits_0^{2\pi} e^{\mu r \sin\theta - i\sigma r\cos\theta} e^{-il\theta} d\theta\right] d\sigma.$$

Using the integral

$$\int\limits_{0}^{2\pi} e^{\mu r \sin\theta - i\sigma r \cos\theta} e^{-il\theta} d\theta = 2\pi i^l \left(\frac{\sigma + \mu}{\sqrt{\sigma^2 - \mu^2}}\right)^l J_l(r\sqrt{\sigma^2 - \mu^2}),$$

we complete the proof of the theorem. $\square$

2.8.3. Generalized Radon transform

Attenuated Radon transform (2.8.1) is a particular case of the generalized Radon transform

$$\hat{f}^{(\Phi)}(\alpha, p) := (R_\Phi f)(\alpha, p) := \int\limits_{\mathbb{R}^n} f(x)\Phi(x, \alpha, p)\delta(\alpha \cdot x - p)dx,$$

$$\alpha \in S^{n-1}, \ x \in \mathbb{R}^n, \tag{2.8.19}$$

where $\Phi(x, \alpha, p)$ is a smooth, strictly positive measure on the hyperplane $\alpha \cdot x = p$. Only a few properties of R_Φ are known. At present, there is no inversion formula for R_Φ. In general, even if $\Phi(x, \alpha, p) \in C^\infty$, the data $\hat{f}^{(\Phi)}(\alpha, p)$, $(\alpha, p) \in S^{n-1} \times \mathbb{R}^n$, does not determine f uniquely.

Adjoint operator to R_Φ is given by

$$(R_\Phi^* g)(x) = \int\limits_{S^{n-1}} \Phi(x, \alpha, \alpha \cdot x)g(\alpha, \alpha \cdot x)d\alpha. \tag{2.8.20}$$

Now we will show that $R_{1/\Phi}^* R_\Phi$ is an elliptic PDO of order $1 - n$. Substituting (2.8.19) into (2.8.20) and using the well-known identity $(2\pi)^{-1} \int_{-\infty}^{\infty} e^{-ipt} dt = \delta(p)$, we get

$$(R_{1/\Phi}^* R_\Phi f)(x)$$

$$= \int\limits_{S^{n-1}} \frac{1}{\Phi(x, \alpha, \alpha \cdot x)} \int\limits_{\mathbb{R}^n} f(y)\Phi(y, \alpha, \alpha \cdot x)\delta(\alpha \cdot (x - y))dy \, d\alpha$$

$$= \frac{1}{2\pi} \int\limits_{S^{n-1}} \int\limits_{-\infty}^{\infty} \int\limits_{\mathbb{R}^n} \frac{\Phi(y, \alpha, \alpha \cdot x)}{\Phi(x, \alpha, \alpha \cdot x)} f(y)e^{-it\alpha \cdot (x-y)} dy \, dt \, d\alpha$$

$$= \frac{\gamma^{-1}}{(2\pi)^n} \int\limits_{\mathbb{R}^n} \int\limits_{\mathbb{R}^n} |\xi|^{1-n} a(x, y, \xi/|\xi|)f(y)e^{-i\xi \cdot (x-y)} dy \, d\xi, \tag{2.8.21}$$

where $\gamma = 1/[2(2\pi)^{n-1}]$, and

$$a(x, y, \alpha) = \frac{1}{2}\left[\frac{\Phi(y, \alpha, \alpha \cdot x)}{\Phi(x, \alpha, \alpha \cdot x)} + \frac{\Phi(y, -\alpha, -\alpha \cdot x)}{\Phi(x, -\alpha, -\alpha \cdot x)}\right]. \tag{2.8.22}$$

Since $a(x, x, \alpha) = 1$, we see that the principal symbol of the PDO speci-
fied by (2.8.21) and (2.8.22) is $|\xi|^{1-n}/\gamma$, which proves the ellipticity. This
result shows that although f, in general, cannot be recovered uniquely,
one can recover singularities of f. The distinctive feature of computing
$R_{1/\Phi}^{*} R_{\Phi} f = R_{1/\Phi} \hat{f}^{(\Phi)}$ is locality: from (2.8.20) we see that calculation
of $R_{1/\Phi} \hat{f}^{(\Phi)}$ at a point x requires the knowledge of $\hat{f}^{(\Phi)}(\alpha, p)$ only for
α, p satisfying $\alpha \cdot x = p$; that is, we have to know integrals of f over
hyperplanes intersecting x. However, the operator $R_{1/\Phi}^{*} R_{\Phi}$ increases
smoothness, since $|\xi|^{1-n} \to 0$ as $|\xi| \to \infty$ (for $n \geq 2$). Therefore, sin-
gularities of f (e.g., jump discontinuities) will be smoothed to a certain
extent and will not be clearly visible in the image of $R_{1/\Phi}^{*} R_{\Phi} f$. The
method which emphasizes singularities of f and which still uses only the
local data is described in Section 5.7.

2.9. Convergence properties of the inversion formulas on various classes of functions

Let us recall that inversion formula (2.2.19) is understood in the
Cauchy principal value sense:

$$f(x) = \lim_{d \to 0} f_d^c(x),$$

$$f_d^c(x) := \frac{1}{4\pi^2} \int_{S^1} \int_{|\Theta \cdot x - p| > d} \frac{\hat{f}_p(\theta, p)}{\Theta \cdot x - p} dp d\theta. \qquad (2.9.1)$$

As we can see, for any $d > 0$, function f_d^c is the regularized version of the
inversion formula (2.2.19), because the $2d$-interval around the singularity
of the Cauchy kernel is removed. Using definition (2.9.1), we can extend
the range of applicability of inversion formula (2.2.19) to a broader and
practically more important than S or C_0^∞ class of functions: compactly
supported, piecewise-smooth functions. Also, even if f is smooth, it is
interesting to investigate the convergence $f_d^c \to f$ as $d \to 0$.

All these issues are addressed in Chapter 6. However, the main focus
in Chapter 6 is on practical applications. Therefore, we briefly describe
here all the theoretical results from Chapter 6 which are relevant to
the properties of the convergence $f_d^c \to f$. Their proofs are given in
Chapter 6.

First, we assume that the original function f is compactly supported,
piecewise-continuous and bounded. We assume that there exist one-
sided limits of $f(x)$ and of its derivatives (up to a certain order) as
x approaches S (the discontinuity curve of f). This class of functions
includes most, if not all, densities considered in practical tomography.

The convergence $f_d^c \to f$ as $d \to 0$ is investigated in three cases:

(1) On compact sets not intersecting S;
(2) At the points of S; and
(3) In a neighborhood of S.

For the first two cases one has

Theorem 2.9.1. *Suppose $f \in C^2(U)$ for some open set U, $U \subset \mathbb{R}^2$. Then*

$$|f_d^c(x) - f(x)| = O(d) \quad as \ d \to 0, \ x \in U.$$

Moreover, the convergence in the above equation is uniform on all compact subsets of U. If $x_0 \in S$ is fixed and there exists an open neighborhood V of x_0 such that S is smooth in V and f is piecewise C^2 in V, then

$$\left|f_d^c(x_0) - \frac{f_+(x_0) + f_-(x_0)}{2}\right| = O(d|\ln d|), \ d \to 0,$$

where $f_{\pm}(x_0)$ are the limiting values of $f(x)$ as x approaches x_0 from different sides of S along any path nonintersecting S.

In the third case, we establish the existence of a layer of width $O(d)$ around S inside which f_d^c does not converge to f in the *sup*-norm. More precisely, one has

Theorem 2.9.2. *Let $x_0 \in S$, S be smooth in some neighborhood of x_0 and f be piecewise C^2 there. Let n_0 be a unit vectors normal to S at x_0. Then for an arbitrary fixed γ, $\gamma \neq 0$, one has*

$$\lim_{d \to 0}\left[f(x_0 + \gamma dn_0) - f_d^c(x_0 + \gamma dn_0)\right] = D(x_0)\psi(\gamma),$$

where $D(x_0) := \lim_{t \to 0+}\left(f(x_0 + tn_0) - f(x_0 - tn_0)\right)$ and

$$\psi(\gamma) := \frac{2}{\pi^2}\int_0^{\min(1,1/\gamma)}\frac{\arccos(\gamma t)}{(1 - t^2)^{1/2}}dt, \ \gamma > 0; \quad \psi(\gamma) = -\psi(|\gamma|), \ \gamma < 0.$$

The function $\psi(\gamma)$ is strictly positive, monotonically decreasing with $\lim_{\gamma \to 0}\psi(\gamma) = 1/2$ and $\psi(\gamma) = 2/(\pi^2\gamma) + O(\gamma^{-3})$, $\gamma \to \infty$.

Finally, we will present a generalization of Theorem 2.9.1 on convergence $f_d^c \to f$ at the points where f is k times continuously differentiable, $k \geq 0$. Let D^k denote any k-th order derivative, $k \geq 0$. If $k = 0$, we define $D^0 f := f$. Also, C^0 denotes a class of continuous functions.

Theorem 2.9.3. *Suppose there exist $x_0 \in \mathbb{R}^2$ and $R > 0$ such that $f \in C^{k_0}(B(x_0, R))$ for some $k_0 \geq 0$, where $B(x_0, R)$ denotes a ball of radius R centered at x_0. Then*

$$|(\partial^k f_d^c)(x) - (\partial^k f)(x)| = o(1) \quad if \quad k = k_0,$$

$$|(\partial^k f_d^c)(x) - (\partial^k f)(x)| = o(d|\log d|) \quad if \quad k = k_0 - 1,$$

$$|(\partial^k f_d^c)(x) - (\partial^k f)(x)| = O(d) \quad if \quad 0 \leq k \leq k_0 - 2$$

for $x \in B(x_0, R)$. Moreover, the convergence in the above formulas is uniform on any ball $B(x_0, R')$, $0 < R' < R$.

Let us note that the analogues of Theorems 2.9.1 and 2.9.2 can be proved for the exponential Radon transform. This is done in Section 6.7.3.

CHAPTER 3

RANGE THEOREMS AND
RECONSTRUCTION ALGORITHMS

3.1. Range theorems for R on smooth functions

3.1.1. The classical range theorem

Consider the Radon transform as a mapping $R : S(\mathbb{R}^n) \to S(Z)$, where S is the Schwartz space, $Z := S^{n-1} \times \mathbb{R}$, and S^{n-1} is the unit sphere in $\mathbb{R}^n$. It turns out that the range of R can be described constructively. Let $S_e(Z)$ denote the subspace of $S(Z)$ which consists of even functions: $g(\alpha, p) = g(-\alpha, -p)$.

Theorem 3.1.1. *For any g such that:*

$$g(\alpha, p) \in S_e(Z) \tag{3.1.1}$$

and the moment conditions (2.1.60):

$$\int_{-\infty}^{\infty} g(\alpha, p) p^j \, dp = \mathcal{P}_j(\alpha),$$

$\mathcal{P}_j(\alpha)$ *is a restriction to S^{n-1} of a homogeneous polynomial of*

$$\text{degree } j, \ j = 0, 1, 2, \dots, \tag{3.1.2}$$

are satisfied, there exists a unique $f \in S(\mathbb{R}^n)$ such that $\hat{f}(\alpha, p) = g(\alpha, p)$. Moreover, if $g(\alpha, p) = 0$ for $|p| \geq a$, then $f(x) = 0$ for $|x| \geq a$.

Proof. Define

$$f(x) = \mathcal{F}^{-1} F_{p \to \lambda} g(\alpha, p). \tag{3.1.3}$$

Since $g \in S(Z)$, the function $h(\alpha, \lambda) := F_{p \to \lambda} g$ also belongs to $S(Z)$, and h satisfies the estimates:

$$\sup_{|\xi| > \epsilon} (1 + |\xi|)^m |\partial_\xi^j h| \leq c_{jm}(\epsilon), \quad \forall j, m = 0, 1, 2, \dots, \quad \xi := \lambda \alpha, \tag{3.1.4}$$

for any $\epsilon > 0$. Since $\mathcal{F}^{-1}$ is an isomorphism of $S(\mathbb{R}^n)$ onto $S(\mathbb{R}^n)$, the inclusion $f \in S(\mathbb{R}^n)$ will be proved if one proves that $h(\alpha, \lambda) := h(\xi) \in$

$S(\mathbb{R}^n), \xi = \lambda\alpha$. From (3.1.4) it follows that one should prove only that $h(\xi)$ is infinitely differentiable at $\xi = 0$. One has

$$
h(\xi) = \int_{-\infty}^{\infty} g(\alpha, p) \exp(i\lambda p) d\lambda
$$

$$
= \sum_{j=0}^{k} \frac{(i\lambda)^j}{j!} \int_{-\infty}^{\infty} g(\alpha, p) p^j \, dp + \int_{-\infty}^{\infty} \varphi_k(\lambda p)(i\lambda p)^{k+1} g(\alpha, p) dp. \tag{3.1.5}
$$

Here $\lambda = |\xi|$, $\alpha = \xi/\lambda$, and

$$
\varphi_k(\lambda p) := \frac{\exp(i\lambda p) - \sum_{j=0}^{k} \frac{(i\lambda p)^j}{j!}}{(i\lambda p)^{k+1}} = \sum_{j=0}^{\infty} \frac{(i\lambda p)^j}{(j + k + 1)!}.
$$

The last equation shows that $\varphi_k \in C^{\infty}(\mathbb{R})$.

Using (3.1.2), one can write (3.1.5) as

$$
h(\xi) = \sum_{j=0}^{k} P_j(\xi) \frac{i^j}{j!} + \psi_k(\alpha, \lambda), \quad \lambda = |\xi|, \ \alpha = \xi/\lambda, \tag{3.1.6}
$$

where $P_j(\xi)$ are homogeneous polynomials of degree j, and

$$
\psi_k(\alpha, \lambda) := \int_{-\infty}^{\infty} \varphi_k(\lambda p)(i\lambda p)^{k+1} g(\alpha, p) dp = \lambda^{k+1} \eta_k(\alpha, \lambda),
$$

$$
\lambda = |\xi|, \ \alpha = \xi/\lambda, \tag{3.1.7}
$$

$\psi_k(\alpha, \lambda) := \psi_k(\xi) \in C^{\infty}(\mathbb{R}^n \backslash 0)$ and $\eta_k(\alpha, \lambda)$ is an infinitely differentiable function of its arguments. Note that $\alpha_l = \xi_l |\xi|^{-1}, 1 \le l \le n$, and suppose $\alpha_n \ne 0$. Then

$$
\frac{\partial}{\partial \xi_j} = \frac{\partial \lambda}{\partial \xi_j} \frac{\partial}{\partial \lambda} + \sum_{l=1}^{n-1} \frac{\partial \alpha_l}{\partial \xi_j} \frac{\partial}{\partial \alpha_l} = \alpha_j \frac{\partial}{\partial \lambda} + \frac{1}{\lambda} \sum_{l=1}^{n-1} (\delta_{lj} - \alpha_l \alpha_j) \frac{\partial}{\partial \alpha_l}. \tag{3.1.8}
$$

From the above representation of the operator $\partial/\partial \xi_j$ it follows that if one takes any k-th order derivative with respect to ξ on both sides of (3.1.7), one gets

$$
\partial_\xi^k \psi_k(\xi) = \lambda \bar{\eta}(\alpha, \lambda), \quad \lambda = |\xi|, \ \alpha = \xi/\lambda,
$$

for some smooth $\bar{\eta}$. This implies that $\psi_k(\xi)$ is k times continuously differentiable for all ξ and, in particular, for $\xi = 0$. Since k was arbitrary, equations (3.1.6) and (3.1.4) imply $h \in S(\mathbb{R}^n)$. Therefore, $f = \mathcal{F}^{-1}h \in S(\mathbb{R}^n)$.

Let us prove that $\hat{f} = g$. From (3.1.3) and the Fourier slice theorem it follows that $h = F\hat{f}$. Thus $Fg = F\hat{f}$ and, since F is injective, we conclude $g = \hat{f}$. To prove uniqueness of f, suppose there is another function f_1 such that $\hat{f}_1 = g$. Equation (3.1.3) implies $g = F_{\lambda \to p}^{-1}\mathcal{F}f = F_{\lambda \to p}^{-1}\mathcal{F}f_1$. Taking into account that F and $\mathcal{F}$ are injective, we get $f = f_1$. The assertion about the supports of f and g follows from the hole theorem (cf. Section 2.1.13). Theorem 3.1.1 is proved. $\square$

REMARK 3.1.1. Equations (3.1.6) and (3.1.7) imply that

$$\mathcal{P}_j(\xi) = (-i)^j \left(\sum_{m=1}^{n} \xi_m \frac{\partial}{\partial \eta_m} \right)^j h(\eta) \Big|_{\eta=0} .$$

In other words, $\frac{i^j \mathcal{P}_j(\xi)}{j!}$ is the j-th term of the Taylor series of $h(\xi)$ at $\xi = 0$. Since h is the Fourier transform of f, we get also

$$\mathcal{P}_j(\xi) = \int_{\mathbb{R}^n} f(x)(\xi \cdot x)^j dx = i^{-j} \left(\frac{\partial}{\partial t} \right)^j \hat{f}(t\xi) \Big|_{t=0} . \tag{3.1.9}$$

Exercise 3.1.1. Let Π_1^n be the unit cube in $\mathbb{R}^n$. Prove that given any finite number of homogeneous polynomials $\mathcal{P}_j(\xi)$, one can find an $f \in C_0^\infty(\Pi_1^n)$ such that f generates $\mathcal{P}_j(\xi)$ by formula (3.1.2) with $g = \hat{f}$.

Hint. Fix any $w \in C_0^\infty([-1,1])$, $w(t) \geq 0$, $w \not\equiv 0$, and fix a sufficiently large integer M. Let $P_{kM}(t)$ be a sequence of polynomials biorthogonal to the powers t^l, $l = 0, 1, \ldots, M$:

$$\int_{-1}^{1} w(t)P_{kM}(t)t^l dt = \begin{cases} 1, & k = l, \\ 0, & k \neq l, \end{cases} \quad k, l = 0, 1, \ldots, M.$$

The polynomials P_{kM}, $k = 0, 1, \ldots, M$, can be constructed using induction in M. Using (3.1.9), we have to find $f \in \Pi_1^n$ such that

$$\int_{\mathbb{R}^n} f(x)(\xi \cdot x)^j dx = \mathcal{P}_j(\xi)$$

for a finite number of j's. This is equivalent to finding f such that

$$\int_{\mathbb{R}^n} f(x)x_1^{l_1} \cdot \ldots \cdot x_n^{l_n} dx = c_{l_1 \ldots l_n}$$

for some constants $c_{l_1...l_n}$, and for a finitely many multiindices $L = (l_1,...,l_n)$. Check that the function

$$f(x) = w(x_1) \cdot ... \cdot w(x_n) \sum_L c_{l_1...l_n} P_{l_1 M}(x_1) \cdot ... \cdot P_{l_n M}(x_n),$$

where the summation is over the required set of multiindices L, does the job.

REMARK 3.1.2. The important part of the proof of Theorem 3.1.1 is the validity of the inversion formula on $S(\mathbb{R}^n)$. The example, given in Section 2.7.3, shows that the inversion formula $f = \mathcal{F}^{-1}F\hat{f}$ cannot be true without some restrictions on the growth of $f(x)$ at infinity. In particular, the operator R is not injective on $C^\infty(\mathbb{R}^n) \cap D(R)$, where $D(R)$ denotes the set of functions on which the Radon transform (2.1.1) is defined. For instance, $f \in D(R)$ if f is continuous and absolutely integrable over any hyperplane. As the example in Section 2.7.3 shows, for R to be injective on some space X of functions, it is necessary that the functions belonging to X satisfy some restrictions on their growth at infinity. Therefore, the operator $\mathcal{F}^{-1}F$ is not injective on some sets which are larger than the range of R. For example, let $g(\alpha, p)=1$. Then $F1 = 2\pi\delta(\lambda)$ and

$$\mathcal{F}^{-1}F1 = (2\pi)^{-(n-1)} \int_0^\infty d\lambda \lambda^{n-1} \int_{S^{n-1}} d\alpha \delta(\lambda) = 0, \quad n \geq 2.$$

In Chapter 10 the Radon transform is defined on some distribution spaces, and the formula $R^{-1} = \mathcal{F}^{-1}F$ holds on the range of R, which is acting from S'. The above example shows that the range of R on S' cannot contain, for example, the function $g(\alpha, p) = 1$. Members of the range of R must satisfy the moment conditions in some sense, even if R is defined on a distribution space (see, e.g. Section 10.4). The function $g = 1$ does not satisfy these conditions.

3.1.2. What happens if the moment conditions are violated?

First, it is clear from the proof of Theorem 3.1.1 (see Equation (3.1.6)) that if moment conditions (3.1.2) are satisfied only for $0 \leq j \leq m$, then the function $f := \mathcal{F}^{-1}Fg$, $g \in S(Z)$, will not belong to $S(\mathbb{R}^n)$: its Fourier transform $\tilde{f}(\xi)$ will be C^m, rather than C^∞, at the origin.

Suppose now that $g \in S_e(Z)$, but g does not satisfy the moment conditions. Is there a function f such that $\hat{f} = g$? It is proved below that the answer is still 'yes'.

Clearly, $h(\alpha, \lambda) := Fg \in S(Z)$ if $g(\alpha, p) \in S(Z)$. Consider the function

$$f(x) = (\mathcal{F}^{-1}h)(x) = \frac{1}{(2\pi)^n} \int\limits_{S^{n-1}} \int\limits_0^\infty \lambda^{n-1} h(\alpha, \lambda) \exp(-i\lambda\alpha \cdot x) d\lambda d\alpha.$$

(3.1.10)

This function and all its derivatives are well defined and continuous since

$$|\lambda^k h(\alpha, \lambda)| \in L^1(0, \infty), \ k = 0, 1, 2, \ldots.$$

Therefore, $f \in C^\infty(\mathbb{R}^n)$. The following lemma provides an estimate of the behavior of $f(x)$ as $|x| \to \infty$.

Lemma 3.1.1. Let $h \in S(Z)$ and fix $k \geq 0$. Define

$$f(x) = \frac{1}{(2\pi)^n} \int\limits_{S^{n-1}} \int\limits_0^\infty \lambda^k h(\alpha, \lambda) \exp(-i\lambda\alpha \cdot x) d\lambda d\alpha.$$

(3.1.11)

Then $f(x) = O(|x|^{-(k+1)})$ as $|x| \to \infty$.

Proof. Fix $\beta \in S^{n-1}$ and let $x = r\beta$, $r > 0$. For a function φ defined on S^{n-1}, we have

$$\int\limits_{S^{n-1}} \varphi(\alpha) d\alpha = \int\limits_{-1}^1 \int\limits_{\alpha \cdot \beta = t} \varphi(\alpha) d\alpha \frac{dt}{\sqrt{1 - t^2}}$$

$$= \int\limits_{-1}^1 \int\limits_{S_{\beta\perp}^{n-2}} \varphi(t\beta + \sqrt{1 - t^2}\, \omega)(1 - t^2)^{\frac{n-3}{2}} d\omega\, dt,$$

where $S_{\beta\perp}^{n-2}$ is the unit sphere in the hyperplane passing through the origin perpendicular to β. Therefore, Equation (3.1.11) takes the form

$$f(r\beta) = \int\limits_{-1}^1 \int\limits_0^\infty \lambda^k (1 - t^2)^{\frac{n-3}{2}} A_\beta(\lambda, t) e^{-i\lambda tr} d\lambda\, dt,$$

(3.1.12)

where

$$A_\beta(\lambda, t) = \frac{1}{(2\pi)^n} \int\limits_{S_{\beta\perp}^{n-2}} h(t\beta + \sqrt{1 - t^2}\, \omega, \lambda) d\omega.$$

(3.1.13)

Since $h \in \mathcal{S}(Z)$, the function $A_\beta(\lambda, t)$ and all its derivatives with respect to λ decay faster than any power of λ as $\lambda \to \infty$. Also $A_\beta(\lambda, t)$ is C^∞ in t for $|t| \leq \epsilon < 1$. For t in neighborhoods of -1 and 1, an integration by parts gives

$$\int_0^\infty \lambda^k A_\beta(\lambda, t) e^{-i\lambda tr} d\lambda = O(r^{-(k+1)}) \qquad (3.1.14)$$

uniformly in t and β for $0.5 \leq |t| \leq 1$ and $\beta \in S^{n-1}$. Fix any $\psi \in C_0^\infty([-1, 1])$ such that $\psi(t) \equiv 1$ for $|t| \leq 0.5$, and denote $B_\beta(\lambda, t) := (1 - t^2)^{\frac{n-3}{2}} A_\beta(\lambda, t) \psi(t)$. Clearly, $B_\beta(\lambda, t) \in C^\infty([0, \infty) \times [-1, 1])$. From (3.1.12) and (3.1.14), we obtain

$$f(r\beta) = O(r^{-(k+1)}) + \int_0^\infty \int_{-1}^1 \lambda^k B_\beta(\lambda, t) e^{-i\lambda tr} dt \, d\lambda$$

$$= O(r^{-(k+1)}) + \int_0^\infty \int_{-\lambda}^\lambda \lambda^{k-1} B_\beta(\lambda, s/\lambda) e^{-isr} ds \, d\lambda$$

$$= O(r^{-(k+1)}) + \int_{-\infty}^\infty C_\beta(s) e^{-isr} ds, \qquad (3.1.15)$$

where

$$C_\beta(s) := \int_{|s|}^\infty \lambda^{k-1} B_\beta(\lambda, s/\lambda) d\lambda. \qquad (3.1.16)$$

Differentiating $C_\beta(s)$ and using that $B_\beta(\lambda, t) \equiv 0$ in neighborhoods of $t = \pm 1$, we have

$$C_\beta^{(m)}(s) = \int_{|s|}^\infty \lambda^{k-1-m} \frac{\partial^m}{\partial t^m} B_\beta(\lambda, t) \Big|_{t=s/\lambda} d\lambda. \qquad (3.1.17)$$

Clearly, all derivatives $C_\beta^{(m)}(s), m = 0, 1, \ldots, k-1$, exist and are continuous at $s = 0$. Also, for $m = k$, we get

$$C_\beta^{(k)}(s) = \int_{|s|}^\infty \frac{B_\beta^{(k)}(\lambda, s/\lambda)}{\lambda} d\lambda = \int_{|s|}^1 \frac{B_\beta^{(k)}(\lambda, s/\lambda)}{\lambda} d\lambda + (C^\infty\text{-function})$$

$$= \int_{|s|}^1 \frac{B_\beta^{(k)}(0, s/\lambda)}{\lambda} d\lambda + (C\text{-function})$$

$$= c(\beta) \log |s| + (C\text{-function}), \quad s \to 0, \qquad (3.1.18)$$

where $c(\beta)$ is some function which depends smoothly on β, C-function depends continuously on s, and we have denoted $B_\beta^{(k)} = \partial^k B_\beta / \partial t^k$. According to the definitions of B_β and C_β, all the derivatives of C_β exist for $s \neq 0$ and decay faster than any power of s as $|s| \to \infty$. Therefore, integrating by parts k times in (3.1.15) and using Equation (3.1.18), we obtain

$$f(r\beta) = O(r^{-(k+1)}) + \frac{1}{(ir)^k} \int_{-\infty}^{\infty} C_\beta^{(k)}(s) e^{-isr} ds$$

$$= O(r^{-(k+1)}) + \frac{O(1/r)}{(ir)^k} = O(r^{-(k+1)}), \quad r \to \infty,$$

uniformly in $\beta \in S^{n-1}$. $\square$

Exercise 3.1.2. Let $\varphi \in S(\mathbb{R})$. Prove that $\int_{-\infty}^{\infty} \ln |t| \varphi(t) e^{-ist} dt = O(s^{-1}), s \to \infty$.

Hint. Integrate by parts once. The integral which contains $\varphi'(t) \ln |t|$ goes to zero as $s \to \infty$ by the Riemann-Lebesgue lemma. In the second integral, split e^{-ist} into $\sin(st)$ and $\cos(st)$. For one of them,

$$\int_{-\infty}^{\infty} \frac{\cos(st)}{t} \varphi(t) dt = \int_{0}^{\infty} \cos(st) \frac{\varphi(t) - \varphi(-t)}{t} dt \to 0 \quad \text{as } s \to \infty.$$

For the remaining one, use the identity $\int_{-\infty}^{\infty} \frac{\sin t}{t} dt = \pi$.

The result of Exercise 3.1.2 can also be obtained from (14.5.22) - (14.5.25).

Corollary 3.1.1. *Let $g \in S(Z)$. Then $(R^*g)(x) = O(|x|^{-1})$ as $|x| \to \infty$.*

Proof. Since

$$(R^*g)(x) = \int_{S^{n-1}} g(\alpha, \alpha \cdot x) d\alpha = \frac{1}{2\pi} \int_{S^{n-1}} \int_{-\infty}^{\infty} h(\alpha, \lambda) e^{-i\lambda \alpha \cdot x} d\lambda \, d\alpha$$

$$= \frac{1}{2\pi} \int_{S^{n-1}} \int_{0}^{\infty} [h(\alpha, \lambda) + h(-\alpha, -\lambda)] e^{-i\lambda \alpha \cdot x} d\lambda \, d\alpha,$$

we can apply Lemma 3.1.1 with $k = 0$ and finish the proof. $\square$

Now we can prove the main result of this section.

Theorem 3.1.2. *Fix any $g \in S_e(Z)$ and define $f := \mathcal{F}^{-1}Fg$. Then $f \in C^\infty(\mathbb{R}^n)$, $\hat{f} = g$, and f satisfies the properties:*

(1) $f(x) = O(|x|^{-n})$ *as* $|x| \to \infty$;

(2) $f(x) = O(|x|^{-(n+m)})$ *as* $|x| \to \infty$ *for some $m \geq 1$ if and only if*

$$\int_{-\infty}^{\infty} g(\alpha, p)p^j\,dp = \mathcal{P}_j(\alpha), \quad j = 0, 1, \ldots, m-1,$$

where $\mathcal{P}_j$ is a homogeneous polynomial of degree j; and

(3) *if $f(x) = O(|x|^{-(n+m)}), m \geq 0$, as $|x| \to \infty$, then $Q_l(\partial_x)f(x) = O(|x|^{-(n+m+l)}), |x| \to \infty$, for any homogeneous polynomial Q_l of degree l. Here $\partial_x := (\partial_{x_1}, \ldots, \partial_{x_n})$.*

Proof. Property (1) follows immediately from Equation (3.1.10) and Lemma 3.1.1 with $k = n - 1$.

(2) Suppose now that g satisfies the first m moment conditions (3.1.2): for $j = 0, 1, \ldots, m - 1, m \geq 1$. Let us prove that $f(x) = O(|x|^{-(n+m)})$ as $|x| \to \infty$. Exercise 3.1.1 (see also Exercise 10.2.1) implies that there exists $f_m \in C_0^\infty(\mathbb{R}^n)$ such that

$$\int_{-\infty}^{\infty} (g(\alpha, p) - \hat{f}_m(\alpha, p))p^j\,dp \equiv 0, \quad j = 0, 1, \ldots, m-1.$$

Denote $h_m := F(g - \hat{f}_m) = \mathcal{F}(f - f_m)$. Similarly to (3.1.5)-(3.1.7), we get: $h_m(\alpha, \lambda) = \lambda^m \eta_{m-1}(\alpha, \lambda), \eta_{m-1} \in S(Z)$. Therefore, Lemma 3.1.1 with $k = n - 1 + m$ and $h = \eta_{m-1}$ implies: $\mathcal{F}^{-1}h_m = f - f_m = O(|x|^{-(n+m)})$. Thus, $f(x) = O(|x|^{-(n+m)})$ as $|x| \to \infty$.

Conversely, if $f(x) = O(|x|^{-(n+m)}), |x| \to \infty$, for some $m \geq 1$, then Equation (2.1.62) yields

$$\int_{-\infty}^{\infty} g(\alpha, p)p^k\,dp = \int_{\mathbb{R}^n} (\alpha \cdot x)^k f(x)dx, \qquad (3.1.19)$$

provided $\hat{f} = g$ (this assertion is proved below). The integral on the right converges absolutely for $k = 0, 1, \ldots, m - 1$, and it is a homogeneous polynomial of α degree k.

(3) Applying the operator $Q_l(\partial_x)$ on both sides of (3.1.10), with $h = Fg$, and using the homogeneity of degree l of the polynomial Q_l, we get

$$Q_l(\partial_x)f(x) = \frac{1}{(2\pi)^n} \int_{S^{n-1}} \int_0^{\infty} \lambda^{n-1+l} Q_l(-i\alpha)h(\alpha, \lambda)\exp(-i\lambda\alpha \cdot x)d\lambda d\alpha.$$

As in the proof of Assertion (2), we can write $Q_l(-i\alpha)h(\alpha,\lambda) = \lambda^m \eta_{m-1}$. Applying Lemma 3.1.1, we obtain the desired result.

Now let us prove that $\hat{f} = g$. Clearly,

$$\hat{f}(\alpha,p) = \int_0^\infty \int_{S_{\alpha\perp}^{n-2}} f(p\alpha + t\omega)d\omega \, t^{n-2}dt. \qquad (3.1.20)$$

Property (1) of f ensures that the above integral converges absolutely. Since derivatives of f decay even faster than f (according to property (3) with $m = 0$, so that we do not use Assertion 2 in the proof of the relation $\hat{f} = g$), one can differentiate with respect to p under the integral sign in (3.1.20) any number of times. Now let us check that $\hat{f}$ can be differentiated with respect to α. Fix any $\alpha \in S^{n-1}$, and let $\omega \perp \alpha$. It is sufficient to check that the function $\hat{f}(\sqrt{1 - |\omega|^2}\alpha + \omega, p)$ can be differentiated with respect to ω at $|\omega| = 0$. Let us represent $x \in \mathbb{R}^n$ as follows: $x = s\alpha + y$, $y \perp \alpha$. Differentiating the identity

$$\hat{f}(\sqrt{1 - |\omega|^2}\alpha + \omega, p) = \int_{\mathbb{R}^{n-1}} f\left(\frac{p - \omega \cdot y}{\sqrt{1 - |\omega|^2}}\alpha + y\right)dy$$

with respect to $\omega_k, 1 \le k \le n - 1$, and setting $|\omega| = 0$, we get

$$\frac{\partial}{\partial \omega_k}\hat{f}(\sqrt{1 - |\omega|^2}\alpha + \omega, p)\Big|_{|\omega|=0} = -\int_{\mathbb{R}^{n-1}} y_k \frac{\partial}{\partial s}f(s\alpha + y)\Big|_{s=p} dy$$

$$= -R\left[y_k \frac{\partial}{\partial s}f(s\alpha + y)\right](\alpha,p).$$

Since $|f(x)| \le O(|x|^{-n})$, $|\frac{\partial}{\partial s}f(s\alpha + y)| \le O((s^2 + |y|^2)^{-(n+1)/2})$, and, therefore, $|y_k \frac{\partial}{\partial s}f(s\alpha + y)| \le O((s^2 + |y|^2)^{-n/2})$, the integrals on the right-hand sides of the last two equations converge absolutely, and taking the derivative under the integral sign is justified. As a generalization, we get

$$P_m(\partial_\omega)\hat{f}(\sqrt{1 - |\omega|^2}\alpha + \omega, p)\Big|_{|\omega|=0}$$

$$= (-1)^m R\left[P_m(y)\frac{\partial^m}{\partial s^m}f(s\alpha + y)\right](\alpha,p),$$

where $P_m(\omega) = P_m(\omega_1,\ldots,\omega_{n-1})$ is a homogeneous polynomial of degree m. By the assumption, $|\frac{\partial^m}{\partial s^m}f(s\alpha + y)| \le O((s^2 + |y|^2)^{-(n+m)/2})$. Therefore, $R\left(P_m(y)\frac{\partial^m}{\partial s^m}f(s\alpha + y)\right)$ is well defined for any $m \ge 0$. This shows that $\hat{f}(\alpha, p)$ is infinitely differentiable with respect to α. Therefore, $\hat{f} \in C^{\infty}(Z)$.

By the construction, $f = \mathcal{F}^{-1} F g$. Property (1) shows that $f \in L^p(\mathbb{R}^n)$ for any $p > 1$. By the Hausdorff-Young inequality (see Section 14.2.2.1), $\mathcal{F} f(\lambda \alpha) = F g(\alpha, \lambda)$ for almost all $(\alpha, \lambda) \in Z$. Lemma 3.1.2 below implies that $F \hat{f}(\alpha, \lambda) = F g(\alpha, \lambda)$ for almost all $(\alpha, \lambda) \in Z$. Since F is injective and $\hat{f}$ and g are continuous, we conclude that $\hat{f} = g$ for all α and p. The Theorem is proved. $\square$

The above theorem shows that if the first moment condition (3.1.2) ($j = 0$) is satisfied, then $f \in L^1(\mathbb{R}^n)$. Moreover, the more moment conditions are satisfied, the faster f decays at infinity.

Exercise 3.1.3. Find $f \in C^{\infty}(\mathbb{R}^n) \cap L^1(\mathbb{R}^n)$ such that $\hat{f}(\alpha, p)$ does not exist for some (α, p).

Hint. Consider the function in $\mathbb{R}^2$:

$$f(x_1, x_2) = \frac{x_1}{1 + x_1^2} e^{-x_1^2 x_2^2}.$$

Lemma 3.1.2. *Let* $f \in L^s(\mathbb{R}^n), 1 \leq s \leq 2$. *Then for almost all* $\alpha \in S^{n-1}$, *one has*

$$F \hat{f}(\alpha, \lambda) = (\mathcal{F} f)(\lambda \alpha) \quad \text{for almost all } \lambda \in \mathbb{R}. \tag{3.1.21}$$

Proof. If $f \in L^s(\mathbb{R}^n)$, then $f \in \mathcal{S}'(\mathbb{R}^n)$ and the results from Section 10.1 (see Lemma 10.1.2) imply that $F \hat{f} = \mathcal{F} f$ in $\mathcal{S}'(\mathbb{R}^n)$. The Hausdorff-Young inequality implies that $\mathcal{F} f \in L^q$, where $s^{-1} + q^{-1} = 1$. Hence $F \hat{f}(\alpha, \lambda) = \mathcal{F} f(\lambda \alpha)$ for almost all $(\alpha, \lambda) \in Z$. Hence (3.1.21) follows. $\square$

3.2. Range theorem for R on the Sobolev spaces

3.2.1. Introduction

In Section 3.1 it was established that R maps the Schwartz space $\mathcal{S}(\mathbb{R}^n)$ isomorphically onto $\mathcal{S}_e(Z)$, where $\mathcal{S}_e(Z)$ is the space of $C^{\infty}(Z)$ even functions which satisfy moment conditions (3.1.2) and decay sufficiently fast with all their derivatives as $|p| \to \infty$.

Let $B_a \subset \mathbb{R}^n$, $n \geq 2$, be the ball of radius a centered at the origin. In this section we define R on $H_0^s(B_a)$ and find its range. We also prove that R is an isomorphism between $H_0^s(B_a)$ and $H_{0m}^{s+\frac{n-1}{2}}(Z_a)$, where $Z_a := S^{n-1} \times [-a, a]$, and $H_{0m}^{s+\frac{n-1}{2}}(Z_a)$ is a Banach space which is constructed as follows. Let $C_{0m}^{\infty}(Z_a)$ be a subspace of $C_0^{\infty}(Z_a)$ consisting of even functions which satisfy moment conditions (3.1.2). Then $H_{0m}^{s+\frac{n-1}{2}}(Z_a)$ is the closure of $C_{0m}^{\infty}(Z_a)$ in the norm

$$|||g|||_{s+\frac{n-1}{2}} := \left(\int\limits_{S^{n-1}} \int\limits_{-\infty}^{\infty} |Fg|^2 (1+\lambda^2)^{s+\frac{n-1}{2}} d\lambda d\alpha \right)^{1/2}. \qquad (3.2.1)$$

Here, as usual, F denotes the one-dimensional Fourier transform with respect to the p-variable.

For $f \in C_0^{\infty}(B_a)$, define the Radon transform R:

$$\hat{f}(\alpha, p) := Rf := \int\limits_{\mathbb{R}^n} f(x)\delta(p - \alpha \cdot x)dx, \qquad (3.2.2)$$

Unless $s > 0$ is sufficiently large for the imbedding $\imath : H_0^s(B_a) \to C(B_a)$ to be bounded (cf. Section 14.1.2.3), we cannot apply definition (3.2.2) to functions from $H_0^s(B_a)$, because (3.2.2) requires f to be continuous. To define R on $H_0^s(B_a)$ we, first, define R on $C_0^{\infty}(B_a)$ by formula (3.2.2) and, second, prove that so defined linear operator is closable in $H_0^s(B_a)$. Furthermore, we prove that its closure, denoted also by R, is defined on all of $H_0^s(B_a)$ and maps $H_0^s(B_a)$ onto $H_{0m}^{s+\frac{n-1}{2}}(Z_a)$ isomorphically (that is, R is linear, injective, and surjective). In particular, this means that given any $g \in H_{0m}^{s+\frac{n-1}{2}}(Z_a)$, there exists a unique $f \in H_0^s(B_a)$ such that $Rf = g$. By Banach's theorem, R and R^{-1} are both continuous, since they are linear closed operators defined on all of the Banach spaces $H_0^s(B_a)$ and $H_{0m}^{s+\frac{n-1}{2}}(Z_a)$, respectively.

It turns out that the Radon transform Rf of $f \in H_0^s(B_a)$ can be computed explicitly using formula $Rf = F^{-1}\mathcal{F}f$, which was originally derived in Section 2.1.5 for $f \in S(\mathbb{R}^n)$ as a corollary to the Fourier slice theorem (see Remark 3.2.1 and Section 10.1).

REMARK 3.2.1. One has $H_0^s(B_a) \subset S'$. Thus, it is possible to define R on $H_0^s(B_a)$ by defining R on S' via, e.g. the duality approach: $(Rf, h) =< f, R^*h >$, where $f \in S'$, and the test functions h are chosen so that R^*h runs through the whole space S. It is interesting to note that the definition based on the closure process coincides with the definition based on duality (see Exercise 10.1.2).

Let us formulate the basic result.

Theorem 3.2.1. *The operator $R : H_0^s(B_a) \to H_{0m}^{s+\frac{n-1}{2}}(Z_a)$ is an isomorphism.*

3.2.2. Proof of Theorem 3.2.1
The proof of Theorem 3.2.1 is based on two lemmas.

Lemma 3.2.1. *One has:*

(1) *The operator R, defined on $C_0^\infty(B_a)$, is closable in $H_0^s(B_a)$;*

(2) *The closure of R is defined on the whole space $H_0^s(B_a)$ and maps $H_0^s(B_a)$ into $H_{0m}^{s+\frac{n-1}{2}}(Z_a)$; and*

(3) *R is correctly solvable. This means that*

$$\|f\|_s \le c \||Rf\||_{s+\frac{n-1}{2}} \quad \forall f \in H_0^s(B_a),$$

where $c > 0$ is a constant independent of f.

Corollary 3.2.1. *One has*

(1) *The range of R is closed in $H_{0m}^{s+\frac{n-1}{2}}(Z_a)$.*

(2) *The null-space of R is trivial: $N(R) = \{0\}$.*

Lemma 3.2.2. *The operator $R : H_0^s(B_a) \to H_{0m}^{s+\frac{n-1}{2}}(Z_a)$ is surjective.*

From Lemma 3.2.1, Corollary 3.2.1 and Lemma 3.2.2 it follows that $R : H_0^s(B_a) \to H_{0m}^{s+\frac{n-1}{2}}(Z_a)$ is a linear continuous bijection of $H_0^s(B_a)$ onto $H_{0m}^{s+\frac{n-1}{2}}(Z_a)$, i.e., an isomorphism of $H_0^s(B_a)$ onto $H_{0m}^{s+\frac{n-1}{2}}(Z_a)$. This is the conclusion of Theorem 3.2.1.

Proof of Lemma 3.2.1. Define R on $C_0^\infty(B_a)$ by formula (3.2.2). Let us prove that R is closable in $H_0^s(B_a)$. This means that

$$\{f_n \in C_0^\infty, \ \|f_n\|_s \to 0, \ \||Rf_n - g\||_{s+\frac{n-1}{2}} \to 0\} \text{ implies } g = 0.$$

The above claim follows from estimate (2.4.1):

$$c_1\|f\|_s \le \||Rf\||_{s+\frac{n-1}{2}} \le c_2\|f\|_s, \quad \forall f \in C_0^\infty(B_a), \qquad (3.2.3)$$

where c_1, c_2 are positive constants independent of f. Suppose that $f_n \in C_0^\infty(B_a)$, $\|f_n\|_s \to 0$, and $\||Rf_n - g\||_{s+\frac{n-1}{2}} \to 0$. Then the right inequality in (3.2.3) implies that $\||Rf_n\||_{s+\frac{n-1}{2}} \to 0$, so $g = 0$ and R is closable. Since $C_0^\infty(B_a)$ is dense in $H_0^s(B_a)$, the closure of R, which we denote also by R, is defined on all of $H_0^s(B_a)$ and maps $H_0^s(B_a)$ into $H_{0m}^{s+\frac{n-1}{2}}(Z_a)$. Correct solvability of R follows from the left inequality in (3.2.3). $\square$

Proof of Lemma 3.2.2. Fix any $g \in H_{0m}^{s+\frac{n-1}{2}}(Z_a)$. There exists a sequence $g_n \in C_{0m}^\infty(Z_a)$ such that $\||g - g_n\||_{s+\frac{n-1}{2}} \to 0$ as $n \to \infty$. Denoting $f_n := R^{-1}g_n$ and using estimate (3.2.3), we get:

$$c_1\|f_m - f_n\|_s^2 \le \||g_m - g_n\||_{s+\frac{n-1}{2}}^2, \quad c_1 > 0.$$

Therefore, $f_n, n = 1, 2, \ldots,$ is a Cauchy sequence in $H^s(\mathbb{R}^n)$. Since $H^s(\mathbb{R}^n)$ is a Banach space, there exists $f \in H^s(\mathbb{R}^n)$ such that $\|f - f_n\|_s \to 0$ as $n \to \infty$. According to Theorem 3.1.1, $f_n \in C_0^\infty(B_a)$. Therefore, $f \in H_0^s(B_a)$. According to the definition of R on $H_0^s(B_a)$, we find:

$$Rf = \lim_{n \to \infty} Rf_n = \lim_{n \to \infty} g_n = g.$$

Hence $\hat{f} = g$, and Theorem 3.2.2 is proved. $\quad\square$

3.2.3. The range theorem in terms of the Fourier coefficients

Let us derive another range theorem. It will be formulated in terms of the Fourier coefficients of the function $g = \hat{f}$. Assume for simplicity that $n = 2$. Without essential difficulties the argument extends to the case $n > 2$. Consider R as a mapping

$$R: \ H_0^s(B_1) \to H_0^{s+\frac{1}{2}}(Z_1),$$

where B_1 is the unit ball in $\mathbb{R}^2$, and $Z_1 := S^1 \times [-1, 1]$. Let $\Theta = (\cos\theta, \sin\theta), -\pi < \theta < \pi$. For $g \in C_0^\infty(Z_1)$ we can write

$$g(\Theta, p) = \sum_{l=-\infty}^{\infty} g_l(p) \exp(il\theta). \tag{3.2.4}$$

Assume that (3.1.1) holds. Then one can easily prove the relation

$$g_l(p) = (-1)^l g_l(-p), \tag{3.2.5}$$

if one takes into account that the map $\Theta \to -\Theta$ is equivalent to the map $\theta \to \theta + \pi$, and

$$\exp[il(\theta + \pi)] = (-1)^l \exp(il\theta).$$

Assume that the moment conditions (3.1.2) hold. Then

$$\int_{-1}^{1} p^j g(\Theta, p)\, dp = \sum_{l=-\infty}^{\infty} \exp(il\theta) \int_{-1}^{1} g_l(p) p^j\, dp = \mathcal{P}_j(\cos\theta, \sin\theta), \quad j \geq 0,$$

where $\mathcal{P}_j(x_1, x_2)$ is a homogeneous polynomial of degree j. Therefore,

$$c_{lj} := \int_{-1}^{1} g_l(p) p^j\, dp = 0 \text{ if } |l| > j \text{ or } |l| + j \text{ is odd.} \tag{3.2.6}$$

Equation (3.2.6) implies that $g_l(p)$ is orthogonal in $L^2[-1,1]$ to all polynomials of degree $j < |l|$. Therefore, we get from (3.2.4) that

$$g(\Theta, p) = \sum_{l=-\infty}^{\infty} \sideset{}{'}\sum_{m \geq |l|} g_{lm} P_m(p) \exp(il\theta), \qquad (3.2.7)$$

where the primed sum means summation over m which are of the same parity as $|l|$, so that $m = |l|, |l| + 2, ..., P_m(p)$ are the Legendre polynomials, and

$$g_{lm} = \frac{2m + 1}{2} \int_{-1}^{1} g_l(p) P_m(p) dp \ \text{if } m \geq |l| \text{ and } |l| + m \text{ is even.} \quad (3.2.8)$$

Note that the sum in (3.2.7) is an even function since

$$P_m(-p) = (-1)^m P_m(p) = (-1)^{|l|} P_m(p), \quad m \equiv |l| \pmod 2. \quad (3.2.9)$$

We have proved the following lemma.

Lemma 3.2.3. *Any even function $g \in C_0^\infty(Z_1)$, which satisfies moment conditions (3.1.2), is of the form (3.2.7).*

We want to prove that any function $g \in H_0^{s+\frac{1}{2}}(Z_1)$, which is of the form (3.2.7), is the Radon transform of a function $f \in H_0^s(B_1)$.

To prove this assertion, one checks that $g \in C_0^\infty(Z_1)$, defined by (3.2.7), is even and satisfies conditions (3.1.2). Considering a sequence $C_0^\infty(Z_1) \ni g_n \to g \in H_0^{s+\frac{1}{2}}(Z_1)$ and applying Theorem 3.2.1, we prove the desired claim. Evenness of g follows directly from (3.2.7). Let us check that the function g, defined in (3.2.7), satisfies conditions (3.1.2). One has

$$\int_{-1}^{1} p^j g(\Theta, p) dp = \sum_{l=-\infty}^{\infty} \sideset{}{'}\sum_{m \geq |l|} g_{lm} \exp(il\theta) \int_{-1}^{1} P_m(p) p^j dp.$$

Since

$$b_{mj} := \int_{-1}^{1} P_m(p) p^j dp = 0 \quad \text{if } j < m \text{ or } j + m \text{ is odd}, \qquad (3.2.10)$$

we get

$$\int_{-1}^{1} p^j g(\Theta, p) dp = \sum_{l=-\infty}^{\infty} \sideset{}{'}\sum_{m=|l|}^{j} g_{lm} b_{mj} \exp(il\theta)$$

$$= \sideset{}{'}\sum_{|l| \leq j} \sum_{m=|l|}^{j} g_{lm} b_{mj} \exp(il\theta). \qquad (3.2.11)$$

Consider l such that $|l| + j$ is odd. Since indices m in the summation on the right-hand side of (3.2.11) are chosen so that $|l| + m$ is even, we get that $m + j$ is also odd and, in virtue of (3.2.10), $b_{mj} = 0$. Therefore, we conclude that the right-hand side of (3.2.11) is a linear combination of exponentials $\exp(il\theta)$ with l such that $|l| \leq j$ and $|l| + j$ is even. Clearly, the function $\exp(il\theta)$, where $|l| \leq j$ and $|l| + j$ is even, can be considered as a homogeneous polynomial of degree j, because in this case we have

$$\exp(il\theta) = \exp(il\theta)(\cos^2\theta + \sin^2\theta)^{\frac{j-|l|}{2}}, \qquad (3.2.12)$$

and the right-hand side in (3.2.12) is obviously a homogeneous polynomial of degree j of the variables $\cos\theta$ and $\sin\theta$. Equations (3.2.11) and (3.2.12) imply the desired result:

$$\int_{-1}^{1} p^j g(\Theta, p)\, dp = P_j(\Theta),$$

where $P_j(\Theta)$ is a homogeneous polynomial of degree j. We have proved

Theorem 3.2.2. *Assume that $g \in H_0^{s+\frac{1}{2}}(Z_1)$ is of the form:*

$$g(\Theta, p) = \sum_{l=-\infty}^{\infty} {\sum_{m \geq |l|}}' g_{lm} P_m(p) \exp(il\theta).$$

Then there exists a unique $f \in H_0^s(B_1)$, such that $\hat{f} = g$.

3.3. Range theorems for R^*

Let X be a complete linear metrizable topological vector space of functions on which KR (see (2.2.11') and (2.5.8)) is a continuous linear map from X onto $Y := KRX$. Assume that the identity $R^*KR = I$ holds on X, and Y is complete. Then the map KR is injective and the map $(KR)^{-1} : Y \to X$ is linear and defined on a complete space Y. By Banach's theorem, the map $(KR)^{-1}$ is continuous. Therefore, the operator $R^* = (KR)^{-1}$ and $R^* : Y \to X$ is a continuous isomorhism of Y onto X. We have proved

Lemma 3.3.1. *The operator $R^* : Y \to X$ is a continuous isomorphism, provided that X and Y are complete, KR is continuous, and $R^*KR = I$ on X.*

In particular, let $X = \mathcal{S}(\mathbb{R}^n)$. Then, as it was established in Section 3.1.1, the space $RX := \mathcal{S}_m(Z)$ consists of $\mathcal{S}(Z)$ even functions which satisfy moment conditions (3.1.2). If n is odd and $X = \mathcal{S}$, then the topologies on $Y := KRS$ and X are the usual topologies of the Schwartz space, and Y is complete. If n is even, then the topology of $Y := \mathcal{H}\partial^{n-1}\mathcal{S}_m$ is defined so that the convergence $g_n \to g$ in Y is equivalent to the convergence of $\mathcal{H}g_n \to \mathcal{H}g$ in $\mathcal{S}(Z)$. Then Y is complete and Lemma 3.3.1 is applicable.

Let us study the action of the operator ∂^{n-1} on $\mathcal{S}_m(Z)$. Fix any $h \in \mathcal{S}_m(Z)$. Put $g = h$ in (3.1.2) and integrate by parts $n-1$ times:

$$\int_{-\infty}^{\infty} h_p^{(n-1)}(\alpha, p) p^{k+n-1} dp = \mathcal{P}_k(\alpha), \qquad (3.3.1)$$

where $h_p^{(n-1)} := \partial_p^{n-1}h$, $\partial_p^{n-1} := \partial^{n-1}/\partial p^{n-1}$, and $\mathcal{P}_k$ in (3.3.1) differs by a constant factor from $\mathcal{P}_k$ in (3.1.2). Denoting $l = k + n - 1$, we obtain that functions from $\partial_p^{n-1}\mathcal{S}_m(Z)$ satisfy the conditions

$$\int_{-\infty}^{\infty} g(\alpha, p) p^l dp = \mathcal{P}_{l-n+1}(\alpha), \quad l = n - 1, n, n + 1, \ldots, \qquad (3.3.2)$$

where $\mathcal{P}_k$ is a homogeneous polynomial of degree k. Now fix any g such that $g = h_p^{(n-1)}$, $h \in \mathcal{S}_m(Z)$. Since $g \in \mathcal{S}(Z)$, we get

$$\int_{-\infty}^{\infty} g(\alpha, p) p^l dp = \text{const} \int_{-\infty}^{\infty} h_p^{(n-1-l)}(\alpha, p) dp = 0, \ 0 \le l \le n - 2. \ (3.3.3)$$

Fix now any $g \in \mathcal{S}(Z)$ such that conditions (3.3.2) and (3.3.3) are satisfied. Define

$$h(\alpha, p) = \int_{-\infty}^{p} g(\alpha, t) \frac{(p - t)^{n-2}}{(n - 2)!} dt. \qquad (3.3.4)$$

Clearly, $h_p^{(n-1)} = g$. It is also clear that $h(\alpha, p)$ and all its derivatives decay faster than any negative power of p as $p \to -\infty$. Using (3.3.3), we find

$$\int_{-\infty}^{p} g(\alpha, t) t^l dt = - \int_{p}^{\infty} g(\alpha, t) t^l dt, \ 0 \le l \le n - 2.$$

Equation (3.3.4) now implies that $h(\alpha, p)$ and all its derivatives decay faster that any power of p as $p \to +\infty$. Hence $h \in \mathcal{S}(Z)$. Conditions (3.3.2) ensure that h satisfies moment conditions (3.1.2). Also, it is clear that $g = \partial_p^{n-1}h$ is even (odd) if and only if n is odd (even). The set of all h such that $h_p^{(n-1)} = g \in \mathcal{S}(Z)$ can be represented as $h = h_0 + a(\alpha)P(p)$, where $h_0 \in \mathcal{S}(Z)$ and $P(p)$ is any polynomial of degree $\le n-2$. Therefore, there exists a unique $h \in \mathcal{S}(Z)$ such that $h_p^{(n-1)} = g$. We have proved

Lemma 3.3.2. *The operator ∂_p^{n-1} maps $S_m(Z)$ isomorphically into the space of functions $g \in S(Z)$, which satisfy conditions (3.3.2) and (3.3.3), and which are even (odd) if n is odd (even).*

Combining Lemmas 3.3.1, 3.3.2, definition (2.2.16') of K, and Theorem 3.1.1, proves

Theorem 3.3.1. *Let $S_{sm}(Z)$ be the space of $S(Z)$-functions which are even (odd) if n is odd (even), and which satisfy the shifted moment conditions*

$$\int_{-\infty}^{\infty} g(\alpha, p) p^j \, dp = \begin{cases} 0, & 0 \le j \le n-2, \\ \mathcal{P}_{j-n+1}(\alpha), & j \ge n-1. \end{cases} \qquad (3.3.5)$$

Let n be odd. Then the mapping $R^* : S_{sm}(Z) \to S(\mathbb{R}^n)$ is an isomorphism. For even n, the mapping $R^* : \mathcal{H}S_{sm}(Z) \to S(\mathbb{R}^n)$ is an isomorphism. Here $\mathcal{H}S_{sm}(Z)$ is the image of $S_{sm}(Z)$ under the action of the Hilbert transform acting on the p-variable.

Let $\mathcal{D}_{sm}(Z)$ be the space of $C_0^\infty(Z)$-functions which are even (odd) if n is odd (even), and which satisfy (3.3.5). Let n be odd. Then the mapping $R^* : \mathcal{D}_{sm}(Z) \to \mathcal{D}$ is an isomorphism. For even n, the mapping $R^* : \mathcal{H}\mathcal{D}_{sm}(Z) \to \mathcal{D}$ is an isomorphism. Here $\mathcal{D} := C_0^\infty(\mathbb{R}^n)$.

Suppose now that one wants to study the range of R^* on spaces different from $Y = KRS(\mathbb{R}^n)$ and $Y = KRC_0^\infty(\mathbb{R}^n)$. In this case, unfortunately, there are no constructive results similar to those from Theorem 3.3.1. The only available result is based on Theorem 2.5.1: namely, let X and Y be the spaces of functions on Z and $\mathbb{R}^n$, respectively, and one wants to check whether $h \in Y$ is in the range of $R^* : X \to Y$. Theorem 2.5.1 implies that, first, one has to compute the function μ:

$$\mu(\alpha, p) := \frac{1}{2(2\pi)^n} \int_{-\infty}^{\infty} |\lambda|^{n-1} \tilde{h}(\lambda \alpha) \exp(-i\lambda p) d\lambda, \qquad (3.3.6)$$

where $\tilde{h} = \mathcal{F}h$. If the function μ, defined by (3.3.6), belongs to X, then the function h is in the range of $R^* : X \to Y$ and $(R^*)^{-1}h = \mu$. This result has a drawback: the condition for h to belong to the range of R^* is stated in terms of the integral operator (3.3.6) acting on h, rather than in terms of generic properties of h, e.g. smoothness and decay rate of h. The decay rate of $h = R^*\mu$ is not possible to describe only in terms of smoothness and decay rate of μ (see Corollary 3.1.1 and Exercise 3.3.1).

Exercise 3.3.1. Let $\mu \in C_0^\infty(Z_1)$, $\mu(\alpha, p) \ge 0$ and $\mu(\alpha, p) > c > 0$ if $|p| < \frac{1}{2}$. Prove that

$$c_1 |x|^{-1} \le R^*\mu \le c_2 |x|^{-1} \text{ as } |x| \to \infty, \ c_1, c_2 > 0.$$

In particular, $R^*\mu$ is not compactly supported.

Hint. Let $|x| = r$ and let the x_n axis be directed along x. Let K_r be the double cone $r|\cos\theta_{n-1}| \leq 1$, where θ_{n-1} is the angle between x and α. Then

$$c_1 r^{-1} \leq \int_{S^{n-1} \cap K_{2r}} \mu(\alpha, r\cos\theta_{n-1}) d\alpha$$

$$\leq \int_{S^{n-1} \cap K_r} \mu(\alpha, r\cos\theta_{n-1}) d\alpha \leq c \int_{S^{n-1} \cap K_r} d\alpha \leq c_2 r^{-1}.$$

The result of Exercise 3.3.1 is in agreement with Corollary 3.1.1, which asserts that, in general, $(R^*g)(x) = O(|x|^{-1})$ as $|x| \to \infty$.

3.4. Range theorem for X-ray transform

Let Xf be the X-ray transform of $f \in S(\mathbb{R}^n)$:

$$Xf = \int_{-\infty}^{\infty} f(x + \alpha t) dt := g(x, \alpha). \tag{3.4.1}$$

Let us recall (cf. Section 2.6.1) that the natural domain of definition of the function Xf is $T := \alpha^\perp \times S^{n-1}$, where $\alpha^\perp$ is the plane orthogonal to α and passing through the origin. Indeed, denoting $x^\perp := x - \alpha(x, \alpha)$ (thus $x^\perp \in \alpha^\perp$), we obtain from (3.4.1) by a change of variables:

$$g(x, \alpha) = \int_{-\infty}^{\infty} f(x^\perp + \alpha t) dt. \tag{3.4.2}$$

This implies that $g(x, \alpha) = g(x^\perp, \alpha)$, and so one can consider $g(x, \alpha)$ as a function defined on T.

Fix any $\alpha \in S^{n-1}$ and consider for $y \perp \alpha$:

$$P_m(y) := \int_{\alpha^\perp} (x \cdot y)^m g(x, \alpha) dx, \quad m = 0, 1, \ldots. \tag{3.4.3}$$

We have

$$P_m(y) = \int_{-\infty}^{\infty} \int_{\alpha^\perp} f(x + \alpha t)(x \cdot y)^m dx\, dt = \int_{\mathbb{R}^n} f(z)(x \cdot y)^m dz, \tag{3.4.4}$$

so that $\mathcal{P}_m(y)$ is a homogeneous polynomial of degree m and this polynomial does not depend on α.

If $f = 0$ for $|x| > a$, then $g(x, \alpha) = 0$ for $|x| > a$ and $x \perp \alpha$. Therefore, for a function $g(x, \alpha)$ to be the X-ray transform of f, $f = 0$ for $|x| > a$, it is necessary that the following two conditions hold:

$$g(x, \alpha) = 0 \quad \text{for} \quad |x| > a, x \perp \alpha, \tag{3.4.5}$$

and

moment conditions (3.4.3) hold, where $\mathcal{P}_m(y)$ is a homogeneous polynomial of degree m independent of α. $\tag{3.4.6}$

Assume now that

$$g \in \mathcal{S}(T). \tag{3.4.7}$$

The basic result of this section is

Theorem 3.4.1. *If conditions (3.4.5)-(3.4.7) hold, then there is an $f \in \mathcal{S}(\mathbb{R}^n)$, $f(x) = 0$ for $|x| > a$, such that $Xf = g$.*

Proof. Note that if g is of the form (3.4.1) (or (3.4.2)), then

$$\int_{x \cdot \beta = p, x \in \alpha^{\perp}} g(x, \alpha) dx = \int_{x^{\perp} \cdot \beta = p} \int_{-\infty}^{\infty} f(x^{\perp} + \alpha t) dt \, dx^{\perp}$$

$$= \int_{x \cdot \beta = p} f(x) dx = \hat{f}(\beta, p) \text{ for } \beta \perp \alpha. \tag{3.4.8}$$

Therefore, one can use the following approach:

(1) Given $g(x, \alpha)$, which satisfies conditions (3.4.5)-(3.4.7), define $\psi(\beta, p)$ by the formula

$$\psi(\beta, p) = \int_{x \cdot \beta = p, x \in \alpha^{\perp}} g(x, \alpha) dx \text{ for } \beta \perp \alpha, \beta \in S^{n-1}; \tag{3.4.9}$$

(2) Check that so defined function $\psi(\beta, p)$ does not depend on the choice of α, and that $\psi(\beta, p)$ satisfies all the conditions of Theorem 3.1.1;

(3) Use Theorem 3.1.1 to infer that $Xf = g$, where f is such that $\psi = Rf$, $f \in \mathcal{S}(\mathbb{R}^n)$, and $f = 0$ for $|x| > a$.

Let us use the approach described above. First, we prove that the function $\psi(\beta, p)$, defined by (3.4.9), does not depend on α. Indeed, we get using (3.4.4)

$$
\int_{-\infty}^{\infty} p^m \psi(\beta, p) dp = \int_{-\infty}^{\infty} p^m \int_{x \cdot \beta = p, x \in \alpha^\perp} g(x, \alpha) dx dp
$$

$$
= \int_{x \in \alpha^\perp} (x \cdot \beta)^m g(x, \alpha) dx = \mathcal{P}_m(\beta).
$$

$$(3.4.10)$$

Since, by the assumption, $\mathcal{P}_m(\beta)$ is independent of α, the integral on the left-hand side of (3.4.10) does not depend on α. Since m was arbitrary, we conclude that $\psi(\beta, p)$ does not depend on α. By the assumption, $\mathcal{P}_m(\beta)$ is a homogeneous polynomial of degree m. Therefore, Equation (3.4.10) implies also that $\psi(\beta, p)$ satisfies moment conditions (3.1.2). It follows from Equation (3.4.9) that

$$
\psi(-\beta, -p) = \psi(\beta, p)
$$

and, since (3.4.7) holds,

$$
\psi(\beta, p) \in \mathcal{S}(Z).
$$

If (3.4.5) holds, then Equation (3.4.9) implies

$$
\psi(\beta, p) = 0 \text{ for } |p| \geq a.
$$

Thus we have checked that $\psi(\beta, p)$ satisfies all the assumptions of Theorem 3.1.1. Therefore, there exists $f \in \mathcal{S}(\mathbb{R}^n)$, $f = 0$ for $|x| \geq 0$, such that $\psi = Rf$. Equation (3.4.9) shows that $\psi(\beta, p)$, $\beta \perp \alpha$, is the $(n-1)$-dimensional Radon transform of $g(x, \alpha)$ with respect to x in the plane $\alpha^\perp$ for a fixed α. Similarly, Equation (3.4.8) implies that $\psi(\beta, p)$, $\beta \perp \alpha$, is the $(n-1)$-dimensional Radon transform of $Xf(x, \alpha)$ with respect to x in the plane $\alpha^\perp$ for a fixed α. Since $\alpha \in S^{n-1}$ is arbitrary and the $(n-1)$-dimensional Radon transform is injective, we conclude that $g = Xf$. $\square$

3.5. Numerical solution of the equation $Rf = g$ with noisy data

3.5.1. Introduction

Consider the equation

$$
Rf = g. \tag{3.5.1}
$$

Assume that $f \in H_0^s(B_1)$, $B_1 \subset \mathbb{R}^2$, and

$$\sup_{\Theta, p} |g_\delta(\Theta, p) - g(\Theta, p)| < \delta, \tag{3.5.2}$$

where the supremum is taken over $\Theta \in S^1$ and $p \in [-1, 1]$, and $\delta > 0$ is a small known number.

Recall that $R : H_0^s(B_1) \to H_0^{s+\frac{1}{2}}(Z_1)$ is an isomorphism, as it follows from Theorem 3.2.1. The problem of solving Equation (3.5.1) in $H_0^s(B_1)$ numerically is ill posed, since the operator R increases smoothness. Equation (3.5.1), with g replaced by g_δ, may not be solvable: g_δ may not belong to the range of R. Even if it is solvable, the function $R^{-1}g_\delta$ may differ much from $R^{-1}g$.

In this section we want to describe several possible methods for solving Equation (3.5.1) with noisy data stably. This means that we want to describe methods for computing $f_\delta(x)$ such that

$$\|f_\delta - f\| \le \varphi(\delta) \to 0 \text{ as } \delta \to 0. \tag{3.5.3}$$

Here and everywhere below in this section $f = R^{-1}g$ and $\| \cdot \| := \| \cdot \|_{L^2(B_1)}$.

The filtered backprojection algorithm, which can also be considered as a technique for regularizing the Radon transform inversion formula, is described in Section 3.6.

3.5.2. Regularization 1

The basic idea of this section is to use the standard passage to the normal equation $R^*Rf = R^*g$. Consider the equation

$$(\nu I + R^*R)f_{\nu,\delta} = R^*g_\delta. \tag{3.5.4}$$

The operator R^*R is a positive elliptic pseudo-differential operator with the kernel $2|x - y|^{-1}$, and $\nu > 0$ is a parameter. Therefore, Equation (3.5.4) is uniquely solvable in $L^2(B_1)$, since $R^*g_\delta \in L^2(B_1)$ for any $g_\delta \in L^\infty(Z_1)$. We want to find $\nu = \nu(\delta)$ such that the solution $f_{\nu(\delta),\delta}(x) := f_\delta(x)$ to (3.5.4) will satisfy the estimate

$$\|f_\delta - f\| \le \varphi(\delta) \to 0 \quad \text{as} \quad \delta \to 0. \tag{3.5.5}$$

Note that

$$f_{\nu,\delta} = (\nu I + R^*R)^{-1}R^*g_\delta := QR^*g_\delta. \tag{3.5.6}$$

Thus, with $T := R^*R$, $T > 0$ on $L^2(B_1)$, one has

$$\|f_{\nu\delta} - f\| \le \|QR^*g_\delta - R^{-1}g\| \le \|QR^*(g_\delta - g)\| + \|(QR^* - R^{-1})g\|$$

$$\le \frac{2\pi^{3/2}\delta}{\nu} + \|[(\nu I + T)^{-1}T - I]f\| \tag{3.5.7}$$

where we have used the estimate $||R^*(g_\delta - g)||_{L^2(B_1)} \leq 2\pi^{3/2}\delta$.

Since T is a positive selfadjoint operator on $L^2(B_1)$, we can write

$$(\nu I + T)^{-1}Tf - f = \int_0^{||T||} [(\nu + t)^{-1}t - 1]dE_t f = -\nu \int_0^{||T||} (\nu + t)^{-1}dE_t f,$$

$$(3.5.8)$$

where E_t is the resolution of the identity corresponding to T, and $||T||$ is the operator norm of T. Let us assume that

$$\int_0^{||T||} t^{-2}d(E_t f, f) \leq c < \infty, \qquad (3.5.9)$$

where $c > 0$ is a constant. This assumption means that $f \in D(T^{-1})$, that is, f is in the range of T. The operator T is elliptic, of order -1, with the symbol $\text{const}|\xi|^{-1}$. The right-hand side of (3.5.7) is majorized by the function $\text{const}(\frac{\delta}{\nu} + \nu)$, which attains its minimum as a function of ν at $\nu(\delta) = \delta^{\frac{1}{2}}$ and this minimum is $\text{const}\sqrt{\delta}$. Therefore, if $f_\delta := f_{\nu(\delta),\delta}$ where $\nu(\delta) = \delta^{\frac{1}{2}}$, then

$$||f_\delta - f|| \leq c_1\sqrt{\delta}. \qquad (3.5.10)$$

Exercise 3.5.1. Prove that if a linear equation $Rf = g$ is solvable, then it is equivalent to the equation $R^*Rf = R^*g$.

Hint. Let $g = Rf_1$. Then $R^*R(f - f_1) = 0$. Thus

$$(R^*R(f - f_1), f - f_1) = ||R(f - f_1)||^2 = 0,$$

so

$$Rf = Rf_1 = g.$$

Since the operator $T = R^*R$ in $L^2(B_1)$ with the kernel $2|x - y|^{-1}$ is compact and selfadjoint in $L^2(B_1)$, its resolution of the identity is an integral operator whose kernel can be written explicitly in terms of eigenfunctions and eigenvalues of T and one can calculate f_δ in terms of these eigenfunctions and eigenvalues if these are known analytically, for example. On the other hand, one can calculate f_δ by solving numerically the variational problem for which (3.5.4) is the Euler's equation (a necessary condition for a minimum):

$$||Rf - g_\delta||^2_{L^2(Z_1)} + \delta^{1/2}||f||^2_{L^2(B_1)} = \min.$$

This is the standard regularization approach.

3.5.3. Regularization 2

The basic idea of this section is to use the standard inversion formula, but to calculate stably the derivative of $g(\Theta, p)$ using the method of Section 14.8.1. Consider the inversion formula

$$f(x) = \frac{1}{4\pi^2} \int_{S^1} \int_{-1}^{1} \frac{g_p(\Theta, p)}{\Theta \cdot x - p} dp\, d\theta, \qquad (3.5.11)$$

where $g = Rf$, $g_p := \frac{\partial g}{\partial p}$,

$$g(\Theta, p) = 0 \quad \text{for} \quad |p| > 1. \qquad (3.5.12)$$

Assume that $f \in C^2(B_1)$. Then

$$g \in C^2(Z_1), \quad g_p \in C^1(Z_1). \qquad (3.5.13)$$

If $g_\delta(\Theta, p) \in L^\infty(Z_1)$ is given such that

$$\sup_{\Theta, p} |g_\delta(\Theta, p) - g(\Theta, p)| < \delta, \qquad (3.5.14)$$

and a constant $M > 0$ is known such that

$$\sup_{\Theta, p} |\partial g / \partial p| \le M, \qquad (3.5.15)$$

then one can use the results developed in Section 14.8.1 and calculate stably an estimate G_δ of g_p by the formula

$$G_\delta(\Theta, p) = \frac{g_\delta(\Theta, p + h(\delta)) - g_\delta(\Theta, p - h(\delta))}{2h(\delta)}, \qquad (3.5.16)$$

where

$$h(\delta) = \sqrt{\frac{2\delta}{M}}. \qquad (3.5.17)$$

The error of this estimate is (see Section 14.8.1):

$$\sup_{\Theta, p} |G_\delta - g_p| \le \sqrt{2M\delta}. \qquad (3.5.18)$$

Thus, given g_δ, one calculates G_δ by formula (3.5.16) and

$$f_\delta(x) := \frac{1}{4\pi^2} \int_{S^1} \int_{-1}^{1} \frac{G_\delta(\Theta, p)}{\Theta \cdot x - p} dp\, d\theta \qquad (3.5.19)$$

The function f_δ satifies the estimate

$$\|f_\delta - f\|_{L^2(B_1)} \le c\delta^{\frac{1}{2}}, \qquad (3.5.20)$$

as follows from (3.5.18) and the boundedness of the singular integral operator with the Cauchy kernel in L^2.

3.5.4. Regularization 3

The basic idea of this section is to use the range theorem 3.2.2.

Given $g_\delta(\Theta, p)$ satisfying (3.5.12) and (3.5.14), calculate the even part of g_δ

$$g_{e\delta}(\Theta, p) := \frac{g_\delta(\Theta, p) + g_\delta(-\Theta, -p)}{2}. \qquad (3.5.21)$$

Note that

$$|g_{e\delta} - g| < \frac{|g_\delta(\Theta, p) - g(\Theta, p)|}{2} + \frac{|g_\delta(-\Theta, -p) - g(-\Theta, -p)|}{2} \le \delta. \qquad (3.5.22)$$

Next, calculate

$$g_{e\delta l}(p) = \frac{1}{2\pi} \int_{-\pi}^{\pi} g_{e\delta}(\Theta, p) \exp(-il\theta) d\theta \qquad (3.5.23)$$

and

$$c_{\delta l m} = \frac{2m + 1}{2} \int_{-1}^{1} g_{e\delta l}(p) P_m(p) dp, \quad 0 \le m < |l|. \qquad (3.5.24)$$

Finally, define

$$\gamma_{\delta l}(p) := g_{e\delta l}(p) - \sum_{m=0}^{|l|-1} c_{\delta l m} P_m(p) \qquad (3.5.25)$$

and

$$G_\delta(\Theta, p) := \sum_{l=-\infty}^{\infty} \gamma_{\delta l}(p) \exp(il\theta). \qquad (3.5.26)$$

By Theorem 3.2.2, the function $G_\delta(\Theta, p)$, defined by (3.5.26), is in the range of R, and by (3.5.22):

$$|\gamma_{\delta l}(p) - g_l(p)| < \delta, \qquad (3.5.27)$$

where

$$g(\Theta, p) = \sum_{l=-\infty}^{\infty} g_l(p) \exp(il\theta). \qquad (3.5.28)$$

Thus $G_\delta(\Theta, p)$ is a Fourier series for $g(\Theta, p)$ with perturbed coefficients. One can use the result of Section 14.8.2 and approximate stably g given G_δ:

$$\|g - G_{1\delta}\| < \varphi(\delta) \qquad (3.5.29)$$

where
$$\varphi(\delta) \to 0 \quad \text{as} \quad \delta \to 0, \tag{3.5.30}$$

and
$$G_{1\delta}(\Theta, p) := \sum_{l=-N(\delta)}^{N(\delta)} \gamma_{\delta l}(p) \exp(il\theta). \tag{3.5.31}$$

The number $N(\delta)$ is chosen by the algorithm described in Section 14.8.2. Similarly, one can calculate $G_{2\delta}$, which gives a stable approximation of g_p.

One can now apply the inversion formula (3.5.11) with $G_{2\delta}$ in place of g_p and obtain a stable approximation to f, for instance, in the $L^2(B_1)$-norm.

3.6. Filtered backprojection algorithm

3.6.1. Derivation of the algorithm

In Section 3.6 we will describe one of the most popular reconstruction algorithms: filtered backprojection (FBP) algorithm. The idea of the algorithm is based on Equation (2.1.42):

$$W_\epsilon * f = R^*(w_\epsilon \circledast \hat{f}), \quad \hat{f} = Rf, \quad W_\epsilon = R^* w_\epsilon, \tag{3.6.1}$$

where $w_\epsilon, \epsilon \to 0$, is a sequence of sufficiently smooth functions. Let us find w_ϵ so that $W_\epsilon \to \delta$ as $\epsilon \to 0$. Here δ is the delta-function and the convergence $W_\epsilon \to \delta, \epsilon \to 0$, is understood in the distributional sense. If W_ϵ is sufficiently close to the delta-function, then $W_\epsilon * f$ is a sufficiently good approximation to f (at least at the points where f is smooth). Frequently, W_ϵ is referred to as a point-spread function.

Let W_ϵ be radial. A typical choices of W_1 is:

$$W_1 = \mathcal{F}^{-1}\tilde{W}_1, \quad \tilde{W}_1(\xi) = \begin{cases} 1 - q|\xi|, & |\xi| < 1, \\ 0, & |\xi| > 1, \end{cases} \tag{3.6.2}$$

where $\mathcal{F}^{-1}$ denotes two-dimensional inverse Fourier transform, and q, $0 \le q \le 1$, is fixed. Define $W_\epsilon(x) := \epsilon^{-2}W_1(x/\epsilon)$. Using Equations (2.1.39) with $n = 2$ and (3.6.1), we find

$$\tilde{w}_1(\sigma) = (Fw_1)(\sigma) = \begin{cases} \frac{|\sigma|}{4\pi}(1 - q|\sigma|), & |\sigma| < 1, \\ 0, & |\sigma| > 1, \end{cases}$$

and

$$w_1(p) = (F^{-1}\tilde{w}_1)(p) = \frac{1}{2\pi} \int_{-1}^{1} \frac{|\sigma|}{4\pi}(1 - q|\sigma|)e^{-i\sigma p} d\sigma$$

$$= \frac{1}{(2\pi)^2} \int_{0}^{1} \sigma(1 - q\sigma)\cos(\sigma p) d\sigma.$$

Finally, we obtain

$$w_\epsilon(p) = \frac{1}{\epsilon^2} w_1(p/\epsilon), \ w_1(p) = \frac{1}{(2\pi)^2}(u(p) - qv(p)),$$

$$u(p) = \begin{cases} \frac{\cos p - 1}{p^2} + \frac{\sin p}{p}, & p \neq 0, \\ \frac{1}{2}, & p = 0, \end{cases}$$

$$v(p) = \begin{cases} \frac{2\cos p}{p^2} + \left(1 - \frac{2}{p^2}\right)\frac{\sin p}{p}, & p \neq 0, \\ \frac{1}{3}, & p = 0. \end{cases} \quad (3.6.3)$$

3.6.2. The parallel beam protocol

In the case of the parallel beam protocol, we are given the data

$$\hat{f}(\theta_j, p_k), \ \theta_j = 2\pi\frac{j-1}{N_\theta}, \ p_k = k\Delta p,$$

$$j = 1, \ldots, N_\theta, \ k = -P, -P+1, \ldots, P.$$

Calculation of $W_\epsilon * f$ according to (3.6.1) can be made in two steps.

Step 1. For each $j, j = 1, \ldots, N_\theta$, compute discrete convolutions

$$h_{j,l} = \Delta p \sum_{k=-P}^{P} w_\epsilon(p_l - p_k)\hat{f}(\theta_j, p_k), \ l = -P, -P+1, \ldots, P. \quad (3.6.4)$$

Step 2. For each reconstruction point x, compute discrete backprojection

$$(W_\epsilon * f)(x) \approx \frac{2\pi}{N_\theta} \sum_{j=1}^{N_\theta} \frac{h_{j,k}(p_{k+1} - \Theta_j \cdot x) + h_{j,k+1}(\Theta_j \cdot x - p_k)}{\Delta p},$$

where $\Theta_j = (\cos\theta_j, \sin\theta_j)$, and index k, which depends on j and x, is chosen so that $p_k \leq \Theta_j \cdot x < p_{k+1}$.

The filter w_ϵ in (3.6.4) depends on the choice of a point-spread function. In particular, one can take the filter as in (3.6.3). Let us estimate the operation count of the algorithm. Convolutions (3.6.4) require $O(N_\theta P \log P)$ operations if one uses Fast Fourier Transform (FFT). The backprojection step requires $O(N_x N_\theta)$ operations, where N_x is the number of reconstruction points. If $N_x = O(P^2)$, the total operation count is $O(N_\theta P^2)$.

3.6.3. The fan beam protocol

In this case, an x-ray source moves over a circle of radius A, and an object is contained inside the circle of radius R, $R < A$. Most frequently, x-ray detectors lie on an arc centered at the x-ray source. Let $a =$

$A(\cos\alpha,\sin\alpha)$ be a position of the source, and let φ be the angle between vector $-a$ and an x-ray $L(\alpha,\varphi)$ (see Figure 3.6.1). Using parallel beam coordinates (θ,p), Equation (3.6.1) takes the form:

$$(W_\epsilon * f)(x) = \int_0^{2\pi} \int_{-R}^{R} w_\epsilon(\Theta \cdot x - p)\hat{f}(\theta,p)dp\,d\theta. \tag{3.6.5}$$

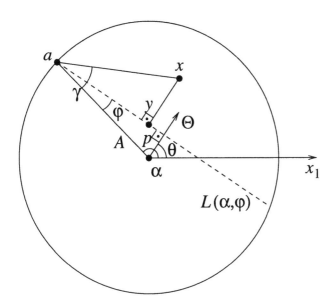

FIGURE 3.6.1. The geometry of the fan beam protocol.

Clearly,

$$p = A\sin\varphi, \quad \theta = \alpha + \varphi - \frac{\pi}{2}. \tag{3.6.6}$$

Fix any reconstruction point x and let γ be the angle between vectors $x - a$ and $-a$:

$$\gamma = \arcsin\left(\frac{(x_1 - A\cos\alpha)A\sin\alpha - (x_2 - A\sin\alpha)A\cos\alpha}{A|x - a|}\right)$$

$$= \arcsin\left(\frac{x_1\sin\alpha - x_2\cos\alpha}{\left[(x_1 - A\cos\alpha)^2 + (x_2 - A\sin\alpha)^2\right]^{0.5}}\right), \quad x = (x_1, x_2),$$

and let y be the orthogonal projection of x onto the ray $L(\alpha,\varphi)$. Note that γ does not depend on φ. Using the triangle axy, we conclude

$$\Theta \cdot x - p = (x - y) \cdot \Theta = |x - a|\sin(\gamma - \varphi). \tag{3.6.7}$$

Substituting (3.6.6) and (3.6.7) into (3.6.5), we obtain

$$(W_\epsilon * f)(x) = A \int_0^{2\pi} \int_{-\varphi_{max}}^{\varphi_{max}} w_\epsilon(|x - a| \sin(\gamma - \varphi)) \hat{f}(\alpha, \varphi) \cos \varphi d\varphi \, d\alpha,$$

$$a = A(\cos \alpha, \sin \alpha), \quad \varphi_{max} = \arcsin(R/A). \tag{3.6.8}$$

Note that the integral with respect to φ is of the convolution type. However, the convolution kernel $w_\epsilon(|x - y| \sin(\gamma - \varphi))$ depends also on x and α. Therefore, we have to compute the convolution for all possible pairs x, α, which would make the algorithm based on Equation (3.6.8) very inefficient. To overcome this difficulty, we use the scaling property of the mollifier w_ϵ: $w_\epsilon(p) = \epsilon^{-2} w_1(p/\epsilon)$ (see (3.6.3)). Equation (3.6.8) now becomes

$$(W_\epsilon * f)(x) = A \int_0^{2\pi} |x - a|^{-2} \int_{-\varphi_{max}}^{\varphi_{max}} w_{\epsilon/|x-a|}(\sin(\gamma - \varphi)) \hat{f}(\alpha, \varphi) \cos \varphi d\varphi \, d\alpha.$$

$$\tag{3.6.9}$$

If the diameter R of the circle containing the object is much less than A, $R \ll A$, we have $\epsilon/|x - a| \approx \epsilon/A$, and the approximate analogue of Equation (3.6.9) is

$$(W_\epsilon * f)(x) \approx A \int_0^{2\pi} |x - a|^{-2} \int_{-\varphi_{max}}^{\varphi_{max}} w_{\epsilon/A}(\sin(\gamma - \varphi)) \hat{f}(\alpha, \varphi) \cos \varphi d\varphi \, d\alpha.$$

$$\tag{3.6.10}$$

Assume now that we are given the fan-beam data

$$\hat{f}(\alpha_j, \varphi_k), \quad \alpha_j = 2\pi \frac{j - 1}{N_\alpha}, \quad j = 1, \ldots, N_\alpha,$$

$$\varphi_k = k \Delta \varphi, \quad k = -K, -K + 1, \ldots, K, \quad \Delta \varphi = \frac{\varphi_{max}}{K}. \tag{3.6.11}$$

Similarly to the parallel beam case, calculation of $W_\epsilon * f$ according to (3.6.10) can be made in two steps.

Step 1. For each $j, j = 1, \ldots, N_\alpha$, compute discrete convolutions

$$h_{j,l} = \Delta \varphi \sum_{k=-K}^{K} w_{\epsilon/A}(\sin(\varphi_l - \varphi_k)) \hat{f}(\alpha_j, \varphi_k), \quad l = -K, -K + 1, \ldots, K.$$

$$\tag{3.6.12}$$

Step 2. For each reconstruction point x, compute discrete backprojection

$$(W_\epsilon * f)(x) \approx A \frac{2\pi}{N_\alpha} \sum_{j=1}^{N_\alpha} \frac{h_{j,k}(\varphi_{k+1} - \gamma(j, x)) + h_{j,k+1}(\gamma(j, x) - \varphi_k)}{|x - A(\cos \alpha_j, \sin \alpha_j)|^2 \Delta \varphi},$$

$$\tag{3.6.13}$$

where

$$\gamma(j,x) = \arcsin\left(\frac{x_1 \sin \alpha_j - x_2 \cos \alpha_j}{\left[(x_1 - A\cos\alpha_j)^2 + (x_2 - A\sin\alpha_j)^2\right]^{0.5}}\right),$$

$$x = (x_1, x_2),$$

and the index k, which depends on j and x, is chosen so that $\varphi_k \leq \gamma(j,x) < \varphi_{k+1}$.

3.7. Other reconstruction algorithms

3.7.1. Fourier algorithm

The Fourier algorithm is based on the Fourier slice theorem (2.1.20) and the Fourier transform inversion formula:

$$\tilde{f}(t\Theta) = \int_{-\infty}^{\infty} \hat{f}(\theta, p)e^{ipt}\,dp, \quad f(x) = \frac{1}{(2\pi)^2}\int_{\mathbb{R}^2} \tilde{f}(\xi)e^{-ix\cdot\xi}\,d\xi, \quad \xi, x \in \mathbb{R}^2.$$

Suppose we are given the parallel beam data

$$\hat{f}(\theta_j, p_k), \; \theta_j = 2\pi\frac{j-1}{N_\theta}, \; p_k = k\Delta p,$$

$$j = 1, \ldots, N_\theta, \; k = -P, -P+1, \ldots, P-1.$$

Let G_{PN_Θ} denote the grid in polar coordinates:

$$G_{PN_\Theta} := \{\xi : \; \xi = t_l\Theta_j, \, j = 1, \ldots, N_\theta,$$
$$t_l = \Delta t l, \, l = -P, -P+1, \ldots, P-1\},$$

where $\Delta t := \pi/(\Delta p P)$. The standard Fourier algorithm operates as follows

Step 1. For $j = 1, \ldots, N_\theta$, one computes

$$\tilde{f}(t_l\Theta_j) \approx \Delta p \sum_{k=-P}^{P-1} \hat{f}(\theta_j, p_k)e^{ip_k t_l},$$

$$t_l = \Delta t l, \, l = -P, -P+1, \ldots, P-1.$$

Step 2. For each grid point $m = (m_1, m_2)$, $|m_1| + |m_2| < P$, set

$$\tilde{f}(\Delta t m) \approx \tilde{f}_m := \tilde{f}(t_l\Theta_j),$$

where $t_l\Theta_j$ is the closest to $\Delta t m$ point on the grid G_{PN_Θ}.

Step 3. Compute inverse Fourier transform:

$$f(x) \approx \frac{(\Delta t)^2}{(2\pi)^2} \sum_{|m|<P} \tilde{f}_m e^{-i\Delta t m\cdot x}.$$

It is essential to use fast Fourier transforms (FFT) on Steps 1 and 3, because otherwise the algorithm will not be efficient. If one uses FFT, the first step requires $O(N_\theta P \log P)$ operations, the third step requires $O(P^2 \log P)$ operations. Since the second step can be neglected, the total operation count is $O(P(N_\theta + P) \log P)$ or $O(P^2 \log P)$ if $P \sim N_\theta$. Therefore, the standard Fourier algorithm is more effective than the FBP algorithm, because the latter requires $O(P^3)$ operations (if $P \sim N_\theta$). Unfortunately, the accuracy of the standard Fourier algorithm is very poor. This drawback is partly removed by improved Fourier algorithms (see e.g. [Nat3]) which use more sophisticated techniques for interpolation from the grid in polar coordinates to the rectangular grid. Note that the total operation count of the improved algorithms is still $O(P^2 \log P)$.

3.7.2. Algebraic reconstruction algorithms

There exist many different modifications of algebraic reconstruction algorithms. We will describe only one of them. Let $f(x), x \in \mathbb{R}^2$, be a compactly supported function. The problem is to recover f given its Radon transform data g_j:

$$\int_{L_j} f(x)dx = g_j, \ j = 1, \dots, N, \qquad (3.7.1)$$

where L_j are lines intersecting supp f. Let $\cup_{m=1}^M S_m$ be a pairwise disjoint partition of supp f by sufficiently small rectangles S_m. Assume that f is constant inside each S_m. Define

$$a_{jm} := \text{length}(L_j \cap S_m).$$

Note that since any fixed line L_j intersects very few rectangles S_m (about $M^{1/2}$), most of a_{jm} are zeros. Equations (3.7.1) can now be written as

$$(a_j, \varphi) = g_j, \ j = 1, \dots, N,$$
$$\varphi := (f_1, \dots, f_M), \ a_j := (a_{j1}, \dots, a_{jM}). \qquad (3.7.2)$$

Here f_m is the value of f inside S_m, and $(\cdot, \cdot)$ denotes the scalar product in $\mathbb{R}^M$. In the matrix form, equations (3.7.2) can be written as

$$A\varphi = G, \ A := (a_{jm})_{\substack{j=1,\dots,N \\ m=1,\dots,M}}, \ G := (g_1, \dots, g_N)^T,$$

where 'T' denotes transposition. Since the number of equations N and the number of unknowns M are usually very large, and the matrix A of the system is sparse, it is convenient to solve this system using iterative methods.

Let us describe the Kaczmarz method. Fix ω, $0 < \omega < 2$. The iteration step consists of computing

$$\varphi_j = \varphi_{j-1} + \frac{\omega}{||a_j||^2}(g_j - (a_j, \varphi_{j-1}))a_j, \; j = 1, \ldots, N,$$

$$\varphi_0 = \varphi^{(k)}, \; ||a_j||^2 = \sum_{m=1}^{M} a_{jm}^2. \qquad (3.7.3)$$

The $(k+1)$-st approximation is then $\varphi^{(k+1)} := \varphi_N$. Recall that there are only $O(M^{1/2})$ nonzero elements in each vector a_j. Therefore, for each j, one updates only $O(M^{1/2})$ elements of φ_{j-1} and this requires $O(M^{1/2})$ operations. Thus, each iteration step (3.7.3) requires $O(NM^{1/2})$ operations. If $M = N_\theta P$, where $N_\theta = O(P)$, and $N = O(P^2)$, the total number of operations per each iteration becomes $O(P^3)$, which is the same as for the FBP algorithm. Thus the algebraic algorithm is relatively slow. The advantage of the algorithm is in its flexibility.

If the initial guess $\varphi^{(0)}$ belongs to the linear span of $a_j, j = 1, \ldots, N$ (e.g., if $\varphi^{(0)} = 0$), then algoritm (3.7.3) converges to a generalized solution to (3.7.2).

CHAPTER 4

SINGULARITIES OF THE RADON TRANSFORM

4.1. Introduction

In this chapter we study the singularities of the Radon transform of piecewise-smooth compactly supported functions. We describe a recipe for finding $S := \text{singsupp } f$, given $\text{singsupp } \hat{f}$, and the asymptotics of $\hat{f}$ in a neighborhood of $\hat{S}$, where $\hat{S}$ is the dual variety with respect to S. We define below the notion of dual variety. These results are of independent interest, but also are of use in several chapters below. They are used in this chapter for finding asymptotics of the Fourier transform for large values of the argument in multidimensional cases.

Let us formulate the basic assumptions. Let $D \subset \mathbb{R}^n$ be a compact domain, $\partial D := S = \cup_{j \in \mathcal{J}} S_j$ be its piecewise-smooth boundary, $\mathcal{J}$ is a set of indices, $S_{\{1,2,\ldots,m\}} := S_1 \cap S_2 \cap \cdots \cap S_m$, $S_{\mathcal{J}'} := \cap_{j \in \mathcal{J}'} S_j$, $S_{\{j\}} := S_j$. We assume throughout that the S_j are in general position. This means that if a point $x \in S_{\{1,2,\ldots,m\}}$, $m \leq n$, then the set of normals to the surfaces S_j at the point x is a linearly independent set of vectors. Assume that $f(x) = \chi_D(x)\phi(x)$, where $\chi_D(x)$ is the characteristic function of D, and $\phi(x)$ is a smooth function not vanishing on S. If $x \in S$ and S is a smooth hypersurface in a neighborhood of the point x, then the tangent hyperplane $l_{\alpha p}$ to S at the point x is defined, and we have $(\alpha : p) = (G_{x_1}, \ldots, G_{x_n} : (\sum_{i=1}^n x_i G_{x_i}))$. If $x \in S_{\{1,\ldots,m\}}, m > 1$, then we say that $l_{\alpha p}$ is tangent (nontransversal) to S at the point x if and only if $l_{\alpha p}$ contains the $(n-m)$-dimensional tangent space to $S_{\{1,\ldots,m\}}$ at the point x. This is equivalent to saying that α is a linear combination of the normals to $S_j, 1 \leq j \leq m$, at the point x. We write the indices $1, \ldots, m$ for convenience only: they might be replaced by $j_1, \ldots, j_m$. Denote by $\hat{S}_{\mathcal{J}'}$, $\mathcal{J}' \subset \mathcal{J}$, and call it the dual variety to $S_{\mathcal{J}'}$, the set of all $(\alpha : p) \in \mathbb{RP}_n$ such that $l_{\alpha p}$ is tangent to S at a point $x \in S_{\mathcal{J}'}$, and by $\hat{S}$ the dual variety to S. If S_j can be locally described by an equation $x_n = g(x'), x' = (x_1, \ldots, x_{n-1})$, with $g \in C^k$, then we say that S_j is C^k smooth. Similarly we define C^∞ and real analytic surfaces. The domain D is called C^k if all the hypersurfaces S_j are C^k smooth. We assume that $k \geq 2$. For simplicity we will formulate the results assuming that S_j are C^∞, but the results and proofs remain valid without essential changes in the cases of $C^k, k \geq 2$ and real analyticity. We often write the equation of the hyperplane as $x_n = \beta \cdot x' - q$, where the inhomogeneous coordinates β and q are defined by the formulas $\beta := -\frac{\alpha'}{\alpha_n}, q := -\frac{p}{\alpha_n}$, and we assume that $\alpha_n \neq 0$.

The basic results we prove in this chapter are:

1) singsupp $\hat{f} = \hat{S}$;

2) S and $\hat{S}$ are in an one-to-one correspondence, and there is an explicitly described involutive map, the Legendre map, which sends $\hat{S}$ onto S. By the one-to-one correspondence we mean that S defines $\hat{S}$ uniquely and vice versa. This does not imply that each point of S is mapped onto a unique point of $\hat{S}$;

3) There is a duality law: S and $\hat{S}$ are envelopes of a family of the planes tangent to them, this family depends on two parameters, x_n and q, and if we choose $x_n = g(x')$ as the parametrization of the family and find its envelope, we get the equation of $\hat{S}$, while if we choose $q = h(\beta)$ as the parametrization of the family and find its envelope, we get the equation of S; here $x_n = g(x')$ is the local equation of S, and $q = h(\beta)$ is the local equation of $\hat{S}$;

4) Asymptotics of $\hat{f}$ in a neighborhood of a point on $\hat{S}$ is given;

5) Asymptotics of the Fourier transform of piecewise-smooth compactly supported functions is obtained in multidimensional case;

6) A relation between the wavefront sets of f and $\hat{f}$ is described;

7) A stable numerical procedure to calculate the Legendre transform of $\hat{f}$, given noisy values of $\hat{f}$, is obtained;

8) Singularities of Xf, the X-ray transform of f, are described and a procedure for finding S from some X-ray transform data is given.

4.2. Singular support of the Radon transform

Theorem 4.2.1. *Let us assume that $\phi \in C^\infty(\mathbb{R}^n)$, $f = \chi_D(x)\phi(x)$ and $\phi \neq 0$ on S. Then singsupp $\hat{f} = \hat{S}$.*

Proof. We have to prove that:

(1) In a neighborhood of a point $(\alpha : p) \notin \hat{S}$ the function

$$\hat{f}(\alpha, p) := \int_{l_{\alpha p}} f(x) ds \qquad (4.2.1)$$

is C^∞ smooth; and

(2) In a neighborhood of a point $(\alpha : p) \in \hat{S}$ the function $\hat{f}(\alpha, p)$ is not C^∞ smooth.

The second statement follows from formula (4.5.7), which is proved in Section 4.5. Here we prove the first statement.

Without loss of generality, assume that $\alpha_n \neq 0$ and write $x_n = \beta \cdot x' - q$ the equation of the hyperplane $l_{\alpha p}$. Let $\tilde{D}$ denote the integration region $D \cap l_{\alpha p}$ in (4.2.1) and D' the orthogonal projection of D onto $\mathbb{R}^{n-1}, x' = (x_1, \ldots, x_{n-1})$. One can write $\tilde{D} = \cup D_l$, where $D_s \cap D_l = \varnothing$ if $s \neq l$. If D_l does not have common points with the boundary S, then clearly the integrals over D_l are as smooth with respect to β and q as the data, that is, as the function ϕ.

If D_l has common points with S, then we want to show that the smoothness of the integral $I_l := \int_{D_l} \phi(x) ds$ is $\min(k_\phi, k_D)$, where k_ϕ and k_D are the integers which measure the smoothness of the data ϕ and D respectively. For example, if $\phi \in C^{k_\phi}$ then we say that k_ϕ is the smoothness of ϕ, and k_D is defined similarly.

The integral I_l is taken over a small neighborhood of a point at the boundary S. Suppose that this point belongs to $S_{\{1,\ldots,m\}}$. Since the surfaces S_i are assumed to be in general position, there exists a coordinate transformation $x_i \to u_i$, which is k_D smooth, such that the equations of the boundary of the projection of the integration domain onto $\mathbb{R}^n$ are of the form $u_i = 0, 1 \leq i \leq m$, and the dependence on β and q is in the integrand. The integrand depends on β and q through ϕ, which is k_ϕ smooth, and through the Jacobian of the coordinate transformation $x_i \to u_i$, which is k_D smooth. Therefore, the integral I_l is $\min(k_\phi, k_D)$ smooth. Statement (1) is proved. Theorem 4.2.1 is proved (up to the proof of Statement (2), which is given in Section 4.5 as we explained above). $\square$

4.3. The relation between S and $\hat{S}$

In this section we introduce the map which sends S onto $\hat{S}$. This map, the Legendre map L, is involutive, that is $L^2 = L$, so L maps $\hat{S}$ onto S. In this sense S and $\hat{S}$ are dual varieties.

We start with the definition of the Legendre map. Let $g(x') \in C^2$ and $g_{,ij} := \frac{\partial^2 g}{\partial x_i \partial x_j}$. Fix a point y' and assume that $\det g_{,ij}(y') \neq 0$. Let U be a neighborhood of y' in $\mathbb{R}^{n-1}$. Consider the equation

$$\nabla g = \beta, \qquad (*)$$

which is a system of $n-1$ equations with $n-1$ unknowns. Denote $\bar{\beta} := g(y')$. Since $\det g_{,ij}(y') \neq 0$, one can uniquely solve $(*)$ in a neighborhood V of $\bar{\beta}$ and the solution $x'(\beta) \in C^{k-1}(V)$ if $g \in C^k(U)$.

Definition 4.3.1. The Legendre map $Lg := h(\beta)$ is defined in V by the formula $h := \beta \cdot x'(\beta) - g(x'(\beta))$.

Lemma 4.3.1. *If* $g \in C^k(U)$ *and* $\det g_{,ij} \neq 0$, $k \geq 2$, *then* $h \in C^k(V)$.

Proof. Denote $g_{,i} := \frac{\partial g}{\partial x_i}$, $h_{,i} := \frac{\partial h}{\partial \beta_i}$ and assume summation over the repeated indices. Then $h_{,i} = x_i(\beta) + \beta_j x_{j,i} - g_{,j} x_{j,i} = x_i(\beta) \in C^{k-1}(V)$, where we have used equation $(*)$. The conclusion of Lemma 4.3.1 follows. $\square$

Exercise 4.3.1. Assume that $\det g_{,ij} \neq 0$. Prove

$$g_{,ij} h_{,jk} = \delta_{ik}. \tag{4.3.1}$$

Hint. Differentiate the equation $h_{,j}(\beta) = x_j(\beta)$.

Exercise 4.3.2. Calculate Lg for $g = a \cdot x' - q$, where $a \in \mathbb{R}^{n-1}$ and $q \in \mathbb{R}$.

Hint. Lg is defined only at the point $\beta = a$ and $Lg = q$ at $\beta = a$.

Exercise 4.3.3. Let $g = a_{ij} x_i x_j + b_i x_i + c := (Ax, x) + (b, x) + c$, where the matrix A is symmetric nonsingular constant matrix, b_i and c are constants. Calculate Lg.

Hint. $2Ax + b = \beta$, $x(\beta) = \frac{1}{2}A^{-1}(b - \beta)$, $h(\beta) = \frac{1}{4}(A^{-1}(\beta - b), \beta - b) - c$.

Exercise 4.3.4. Let $g(x) = x^3/3$, $x \geq 0$. Calculate $h = Lg$.

Hint. $g' = x^2 = \beta$, $x = \beta^{1/2}$, $h(\beta) = \frac{2}{3}\beta^{3/2}$. We see that h is less smooth than g in this example. The reason is the degeneracy of $g''(0)$.

Exercise 4.3.5. Let $n = 2$. Calculate Lg, where $g = |x_1|$.

Hint. We have $g = -x_1$ for $x_1 < 0$, $L(-x_1)$ is defined at $\beta = -1$ only, so $q = 0$ (see Exercise 4.3.2). Similarly, $L(x_1)$ is defined at $\beta = 1$ only, and $L(x_1) = 0$ at $\beta = 1$, so $q = 0$. In the space $\mathbb{RP}_2$ the images of the points $\beta = -1$, $q = 0$ and $\beta = 1$, $q = 0$ are the straight lines $((\alpha_2, \alpha_2) : 0)$ and $((\alpha_2, -\alpha_2) : 0)$.

Exercise 4.3.6. Suppose D is a polygon in $\mathbb{R}^2$ or a polyhedron in $\mathbb{R}^n$. Find the image of $\hat{S}$ of $S := \partial D$ under the Legendre map.

Hint. The faces of S are mapped into the vertices of $\hat{S}$ in $\mathbb{R}^n$. The vertices of S are mapped into the faces of $\hat{S}$ in $\mathbb{R}^n$.

Recall that an operator L is called involutive if $L^2 = I$, where I is the identity operator.

Lemma 4.3.2. *The Legendre map is involutive.*

Proof. We have $Lh = \beta(x') \cdot x' - h(\beta(x'))$, where $\nabla_\beta h = x'$. Moreover, $\nabla_\beta h = x' + \beta_j x_{j,i} - g_{,j} x_{j,i} = x'$. Therefore, $Lh = \beta(x') \cdot x' - h = \beta(x') \cdot x' - x'\beta(x') \cdot x' + g(x') = g(x')$. $\square$

Theorem 4.3.1. *Assume that* $\det g_{,ij}(y') \neq 0$ *and let* $x_n = g(x')$ *be the local equation of* S *in a neighborhood* U *of the point* $y' \in S$. *Then the local equation of* $\hat{S}$ *in a neighborhood* V *of the point* $(\bar{\beta}, \bar{q})$ *is*

$$q = h(\beta), \quad \text{where} \quad h(\beta) = Lg. \tag{4.3.2}$$

The point $(\bar{\beta}, \bar{q})$ *is the point such that the plane* $l_{\bar{\beta}\bar{q}}$ *with the equation* $x_n = \bar{\beta} \cdot x' - \bar{q}$ *is tangent to* S *at the point* $(y', g(y')) \in S$. *Conversely, if* $q = h(\beta)$ *is the local equation of* $\hat{S}$ *in a neighborhood* V *of the point* $(\bar{\beta}, \bar{q})$, *and* $\det g_{,ij}(y') \neq 0$, *then the local equation of* S *in a neighborhood* U *of the point* $(y', g(y'))$ *is* $x_n = g(x')$, *where* $g = Lh$ *and* $y' = \nabla_\beta h(\bar{\beta})$.

Proof. If the plane $l_{\bar{\beta}\bar{q}}$ is tangent to S at the point $(y', g(y'))$, then

$$\nabla_{y'} g(y') = \bar{\beta}, \quad g(y') = \bar{\beta} \cdot y' - \bar{q}. \tag{4.3.3}$$

By the assumption, the first equation defines uniquely $y'(\beta)$ in a neighborhood of $\bar{\beta}$, and $q = \beta \cdot y'(\beta) - g(y'(\beta)) := h(\beta) := Lg$ is the local equation of $\hat{S}$ in a neighborhood V of the point $(\bar{\beta}, \bar{q})$, that is, Equation (4.3.2). The converse statement follows from the involutivity of the Legendre map, see Lemma 4.3.2. □

Theorem 4.3.1 gives a description of the one-to-one map, the Legendre map, which sends S onto $\hat{S}$ and $\hat{S}$ onto S.

A numerical implementation of a method for finding the discontinuity surface S of $f(x)$, given $\hat{f}(\alpha, p)$, is given in Chapter 7. The idea of the method is based on Theorem 4.3.1.

Let us discuss now the situation when the non-degeneracy condition $\det g_{,ij}(y') \neq 0$ does not hold. Assume that S is locally, in an open set $U \subset \mathbb{R}^{n-1}$, a hypersurface with the equation $x_n = g(x')$, $g \in C^2(U)$.

Lemma 4.3.3. *If* m *principal curvatures of* S *vanish in* U, *and* $n - 1 - m$ *principal curvatures do not vanish in* U, *then the Legendre map* $h = Lg$ *is defined on a* C^1 *submanifold in* $\mathbb{R}^{n-1}$ *of codimension* m.

Proof. In Exercise 4.3.2 we had an example when the Legendre map $h(\beta)$ was defined on a submanifold of dimension $< n - 1$. By the assumption, $\text{rank} \, g_{,ij} = n - 1 - m$, and we may assume that the principal nonzero minor of the matrix $g_{,ij}$ corresponds to $1 \leq i, j \leq n - 1 - m$. Therefore, the dimension of the range of the map $x' \to \nabla g(x')$ is $n - 1 - m$. Since the Legendre map is defined on this range, it is defined on a C^1 submanifold in $\mathbb{R}^{n-1}$ of codimension m. □

Since the classical Legendre map is defined on an open set in, say, $\mathbb{R}^{n-1}$, we introduce the generalized Legendre map, which is defined on a C^2 submanifold $M \subset \mathbb{R}^{n-1}$ of codimension $m - 1, 1 \leq m < n$. Let

$\beta \in \mathbb{R}^{n-1}$ and $y = (y_1, \ldots, y_{n-m})$ be local coordinates on M. Consider a C^2 function $g(x')$, $x' \in \mathbb{R}^{n-1}$ as a function on M, so that $x' = x'(y)$. For a given β, find $y(\beta)$ from the equation

$$\nabla_y[\beta \cdot x'(y) - g(x'(y))] = 0. \tag{4.3.4}$$

Let us denote the generalized Legendre map (GLM), or the generalized Legendre transform (GLT), of $g(x'(y))$ by Lg.

Definition 4.3.2. $Lg := h(\beta)$ is the function $h(\beta) := \beta \cdot x'(y(\beta)) - g(x'(y(\beta)))$.

If $m = n$ then M is a union of a set of isolated points $\bar{x}$, and we assume that this set is finite. To each such a point $\bar{x}$, there corresponds $Lg := \beta \cdot \bar{x}' - g(\bar{x}')$ and the equation $q = \beta \cdot \bar{x}' - g(\bar{x}')$ defines a hyperplane in $\mathbb{RP}_n$. Thus, each point $\bar{x}$ is mapped by the Legendre map onto a hyperplane in $\mathbb{RP}_n$.

The generalized Legendre map has the following properties:

1) If M is an open set in $\mathbb{R}^{n-1}$, then GLM coincides with the Legendre map from Definition 4.3.1;

2) If $\bar{\beta} \in \mathbb{R}^{n-1}$ and $y(\bar{\beta})$ is a nondegenerate critical point of the function $\beta \cdot x'(y) - g(\bar{x}'(y))$ on M, then $h(\beta) = Lg$ is defined on an open neighborhood of $\bar{\beta}$ in $\mathbb{R}^{n-1}$, and $Lh = g(x')$ is defined on M in a neighborhood of the point $x'(y(\bar{\beta}))$. The GLM is involutive.

4.4. The envelopes and the duality law

In this section we derive the relation between S and $\hat{S}$ by a geometric argument which makes the duality relation between S and $\hat{S}$ quite transparent.

The starting point is the observation that S is the envelope of the family of the tangent to S planes. The equation of this family can be written as

$$\beta \cdot x' - x_n - q = 0, \tag{4.4.1}$$

where $q = h(\beta)$ is the equation which describes the family of planes tangent to S. To find the envelope of this family, one has to eliminate parameter $\beta \in \mathbb{R}^{n-1}$ from the equations:

$$\beta \cdot x' - x_n - h(\beta) = 0, \quad \nabla h = x'. \tag{4.4.2}$$

If this is done, then the equation of S is $x_n = g(x')$, where $g(x') := Lh$, and Lh is the Legendre transform of h.

Similarly, $\hat{S}$ is the envelope of the family of the tangent to $\hat{S}$ planes. The equation of this family is again (4.4.1), but the parametrization is given by the equation $x_n = g(x')$, where the last equation is a local equation of S. Eliminating parameter x' from the equations

$$\beta \cdot x' - g(x') - q = 0, \quad \nabla g = \beta, \tag{4.4.3}$$

we obtain the equation of $\hat{S}$ of the form $q = h(\beta)$, where $h(\beta) := Lg$, and Lg is the Legendre transform of g.

Thus, we have proved the following duality law:

Theorem 4.4.1. *The sets S and $\hat{S}$ are dual sets: each is the envelope of the family of tangent planes (4.4.1), and this family is defined by the parametrization $q = h(\beta)$, for the family tangent to S, and $x_n = g(x')$ for the family tangent to $\hat{S}$. Here the equation $q = h(\beta)$ is the local equation of $\hat{S}$, and $x_n = g(x')$ is the local equation of S. The functions h and g are the Legendre transforms of each other.*

Our argument shows that the Legendre map is involutive: $L^2 = I$, where I is the identity map.

4.5. Asymptotics of Rf near S

In this section we want to find the asymptotic behavior of $\hat{f}(\alpha, p)$ when the point $(\alpha : p)$ approaches a point $(\bar{\alpha} : \bar{p}) \in \hat{S}$. We assume that the data are piecewise-smooth, that is, $\phi \in C^{\infty}$ and $S_j \in C^{\infty}$. The arguments and the results remain valid without essential changes for piecewise real-analytic data and for piecewise C^k, $k > \max(2, \frac{n+m-2}{2})$ data, where n is the dimension of the space and m is the codimension of the component of S at which the point $\bar{x}$ is situated. Here $\bar{x}$ is the point at which the plane $l_{\bar{\alpha}\bar{p}}$ is tangent to S. We denote $y_+ := \max(y, 0)$, $y_- := \max(-y, 0)$, so $y = y_+ - y_-$. By I we denote the number of negative eigenvalues of a symmetric real-valued matrix. We assume that S is connected. If S is a union of disjoint varieties, then each of its connected components can be treated separately and independently of the others. For example, if $f(x)$ has several disjoint discontinuity surfaces, then each of these enters independently to the set singsupp f. We also assume that the point $(\bar{\alpha} : \bar{p})$ is generic. This means that the function $z := \bar{\alpha} \cdot x - \bar{p}$ has only nondegenerate (Morse-type) critical points on S. These critical points are the points at which the plane $l_{\bar{\alpha}\bar{p}}$ is tangent to S. Almost all points $(\bar{\alpha} : \bar{p})$ are generic as it follows from the Morse theory.

Let $\bar{x} \in S_{\{1,\ldots,m\}}$, $1 \le m < n$. Assume that the plane $l_{\bar{\alpha}\bar{p}}$ is tangent to S at the point $\bar{x}$. Let $x_i = g_i(x_{m+1}, \ldots, x_n)$, $1 \le i \le m$, be the equations of S_i, and

$$z = \sum_{i=1}^{m} \bar{\alpha}_i g_i(x_{m+1}, \ldots, x_n) + \sum_{j=m+1}^{n} \bar{\alpha}_j x_j - \bar{p}. \tag{4.5.1}$$

As usual, we denote $g_{,i} := \frac{\partial g}{\partial x_i}$, etc. Let

$$H := \det z_{,jk} = \det \left(\sum_{i=1}^{m} \bar{\alpha}_i g_{i,jk}(\bar{x}_{m+1}, \dots, \bar{x}_n) \right), \quad m+1 \le j, k \le n, \tag{4.5.2}$$

and

$$J := \det \left(\sum_{i=1}^{m} \bar{\alpha}_i g_{i,jk}(x_{m+1}, \dots, x_n) \right), \quad m+1 \le j, k \le n. \tag{4.5.3}$$

If $(\bar{\alpha} : \bar{p})$ is generic, then $J = H \ne 0$ at $\alpha = \bar{\alpha}$ and $x = \bar{x}$. We assume that $\bar{\alpha}_i \ne 0$, $1 \le i \le m$, and define

$$\Xi := |H|^{-1/2} \frac{(2\pi)^{(n-m)/2}(1 + |\tilde{\beta}|^2)^{1/2}}{|\bar{\alpha}_2 \dots \bar{\alpha}_m| \Gamma((n+m)/2)}, \tag{4.5.4}$$

where $\tilde{\beta} := -\frac{\tilde{\alpha}}{\bar{\alpha}_1}, \tilde{\alpha} := (\bar{\alpha}_2, \dots, \bar{\alpha}_n)$. If $m = 1$, then we define $|\bar{\alpha}_2 \dots \bar{\alpha}_m|$ $:= 1$. Note that Ξ depends on $\bar{x}$ (through H) and on $(\bar{\alpha} : \bar{p})$.

Lemma 4.5.1. *If $(\bar{\alpha} : \bar{p})$ is generic and $l_{\bar{\alpha}\bar{p}}$ is tangent to S at the point $\bar{x} \in S_{\{1,\dots,m\}}$, then, in a neighborhood of $(\bar{\alpha} : \bar{p})$, the variety $\hat{S}$ is a smooth hypersurface.*

Proof. If $l_{\alpha p}$ is tangent to S at the point $x \in S_{\{1,\dots,m\}}$ and N_i is the normal to S_i at x, then

$$\alpha = \sum_{i=1}^{m} c_i N_i, \quad p = \sum_{i=1}^{m} c_i x_i - \sum_{i=1}^{m} c_i \sum_{j=m+1}^{n} x_j g_{i,j}(x_{m+1}, \dots, x_n).$$

On the other hand, $p = \sum_{i=1}^{n} \alpha_i x_i$. Differentiating with respect to x_i the above formulas for p, we get $\alpha_i = c_i$, $1 \le i \le m$, and

$$\alpha_j = -\sum_{i=1}^{m} c_i g_{i,j}(x) = -\sum_{i=1}^{m} \alpha_i g_{i,j}(x), \quad j > m, \tag{4.5.5}$$

$$p = \sum_{i=1}^{m} \alpha_i x_i - \sum_{i=1}^{m} \alpha_i \sum_{j=m+1}^{n} x_j g_{i,j}(x). \tag{4.5.6}$$

Since $H \ne 0$, Equation (4.5.5) allows us to find $x_j = x_j(\alpha_{m+1}, \dots, \alpha_n)$ in a neighborhood of $\bar{\alpha}$ and (4.5.6) yields $p = p(\alpha_1, \dots, \alpha_n)$, where $p(\alpha_1, \dots, \alpha_n)$ is a smooth function in a neighborhood of $\bar{\alpha}$. Lemma 4.5.1 is proved. $\square$

We shall say that a function $r(\alpha, p)$ is as smooth as the data allow, if it is in C^k, $k := \min(k_\phi, k_D)$, where k_ϕ is defined by the condition $\phi \in C^{k_\phi}(\mathbb{R}^n)$ and k_D is defined by the condition $D \in C^{k_D}$ (that is, all the components S_j of S are C^{k_D} smooth) and we assume that $k_D > \max(2, \frac{n+m-2}{2})$, and that $1 \le m < n$ in Theorem 4.5.1.

Theorem 4.5.1. *If $(\bar{\alpha} : \bar{p})$ is generic, $\alpha_1 > 0$, and I is the number of negative eigenvalues of matrix (4.5.2), then*

$$\hat{f}(\alpha, p) = \begin{cases} y_+^{\frac{n+m-2}{2}} r_1 + r_2, & \text{if } I(n+m-1) \text{ is even} \\ y^{\frac{n+m-2}{2}} (\log |y|) r_1 + r_2, & \text{if } I(n+m-1) \text{ is odd.} \end{cases} \quad (4.5.7)$$

Here $y = \pm(p - \bar{p})$, where the minus sign corresponds to the case when I and $n + m$ are of the same parity, and the plus sign corresponds to the case when they are of different parity or $I = 0$. The functions $r_j, j = 1, 2$, are as smooth as the data allow,

$$r_1(\bar{\alpha} : \bar{p}) = \begin{cases} \phi(\bar{x})(-1)^\mu \Xi, & \text{if } I(n+m-1) \text{ is even,} \\ \phi(\bar{x})(-1)^\mu \Xi \pi^{-1}, & \text{if } I(n+m-1) \text{ is odd,} \end{cases} \quad (4.5.8)$$

and the number μ is determined as follows:

$$\mu = \begin{cases} \dfrac{n+m-I}{2}, & I \text{ and } n+m \text{ are of the same parity and } I > 0, \\[2mm] \dfrac{I}{2}, & I > 0 \text{ is even and } n+m \text{ is odd,} \\[2mm] \dfrac{I+1}{2}, & I \text{ is odd and } n+m \text{ is even,} \\[2mm] 0, & I = 0. \end{cases}$$

REMARK 4.5.1. *If $m = n$, then the variety $S_{\{1,...,n\}}$ reduces to a discrete set of points, generally, and the quantity (4.5.4) is not well defined. In this case the behavior of $\hat{f}$ near $\hat{S}$ depends on the geometry of S in a neighborhood of $\bar{x}$. Since we assume that S_j are in general position, the cases of cusps are excluded, and the formula*

$$\hat{f}(\alpha, p) = y_+^{n-1} r_1 + r_2, \quad y := \bar{p} - p,$$

holds for almost all α. The limiting value of r_1 depends on the chosen $(\bar{\alpha}, \bar{p})$. Take, for instance, $n = m = 2$. According to our assumption, the boundary S does not have a cusp at $\bar{x}$, that is, the limiting tangent straight lines to S from the left and right of $\bar{x}$ form a nonzero angle. Then, for every fixed α, except for the values of α corresponding to normals to these two limiting tangent lines, one has:

$$\hat{f}(\alpha, p) = (\bar{p} - p)_+ r_1(\alpha, p) + r_2(\alpha, p),$$

where the behavior of $r_1(\alpha, p)$, as α approaches each of its exceptional values, depends on the geometry of S near $\bar{x}$. For example, if the boundary S coincides with the tangent lines in a neighborhood of $\bar{x}$, then r_1

goes to infinity as α approaches each of the exceptional values and p approaches $\bar{p}$. In the last equation, we have chosen $\bar{p} - p$ rather than $p - \bar{p}$ because we assumed that for $p > \bar{p}$ the line $l_{\bar{\alpha}p}$ does not intersect D.

Proof. The proof is relatively long. In the case $n = 2$ a simple proof and a somewhat different approach are presented in Section 4.6. Let us consider a curve $\alpha = \bar{\alpha}, p = p$, so that p is a parameter which fixes the position of the point on the curve. We want to show that along this curve formula (4.5.7) holds. The above curve is transversal to $\hat{S}$ at the point $(\bar{\alpha} : \bar{p})$. Let U be a neighborhood of $\bar{x}$ and u_i, $1 \le i \le n$, be the local coordinates in U such that $u_i = 0$ is the equation of S_i, $1 \le i \le m$ and $u_i > 0$ in D. Consider the function (4.5.1) on $S_{\{1,...,m\}}$. We assume that this function has a nondegenerate critical point at $\bar{x}$. By Morse's lemma, the coordinates $u_{m+1}, \ldots, u_n$ can be chosen so that

$$z = \sum_{j=m+1}^{n-I} u_j^2 - \sum_{j=n-I+1}^{n} u_j^2, \quad u_1 = \cdots = u_m = 0, \qquad (4.5.9)$$

where I is the number of the negative eigenvalues of H and $u_i > 0$, $1 \le i \le m$ in D. Note that on $S_{\{1,...,m\}}$ one has:

$$| \det z_{,jk}(\bar{x})| = | \det(z_{,u_s u_t} u_{s,j} u_{t,k} + z_{,u_s} u_{s,jk})| = | \det z_{,u_s u_t}|| \det u_{s,j}|^2$$
$$= 2^{n-m} | \det u_{s,j}|^2, \quad m+1 \le j,k,s,t \le n,$$

where we have used the equation $z_{,u_s}(\bar{x}) = 0$, which holds since $\bar{x}$ is a critical point of z and summation over the repeated indices is understood. Thus

$$| \det u_{s,j}(\bar{x})| = 2^{\frac{m-n}{2}} |H|^{\frac{1}{2}} > 0, \quad m+1 \le j,s \le n. \qquad (4.5.10)$$

Since $\det u_{s,j}(x)$ is a continuous function, it follows from (4.5.10) that $\det u_{s,j}(x)$, $m+1 \le j,s \le n$, does not vanish in a small neighborhood U of $\bar{x}$. Therefore, u_j, $m+1 \le j \le n$, are functions of x_j, $m+1 \le j \le n$, only, and x_j, $m+1 \le j \le n$, are functions of u_j, $m+1 \le j \le n$, only. In particular, $x_{j,u_i} = 0$ if $j > m$ and $i \le m$. At a point $x \notin S$, one has $z_{,u_i} = z_{,x_k} x_{k,u_i} = \bar{\alpha}_i$, $1 \le i,k \le m$. Indeed, $u_k = x_k - g_k(x_{m+1}, \ldots, x_n)$, $1 \le k \le m$, so $x_k = u_k + g_k(x_{m+1}, \ldots, x_n)$, and $x_{k,u_i} = \delta_{ik} + g_{k,x_j} x_{j,u_i} = \delta_{ik}$, $1 \le i,k \le m$, $j > m$. Thus, $z_{,u_i} = \bar{\alpha}_i$.

Let us assume without loss of generality that $\bar{\alpha}_j > 0$, $1 \le j \le m$: one can always choose a coordinate system in which these inequalities hold. Let $v_i := \bar{\alpha}_i u_i$, $2 \le i \le m$. Then:

$$z = \bar{\alpha}_1 u_1 + \sum_{i=2}^{m} v_i + \sum_{j=m+1}^{n-I} u_j^2 - \sum_{j=n-I+1}^{n} u_j^2 \ge 0. \qquad (4.5.11)$$

If $m = 1$, then the first sum in (4.5.11) does not appear. On the plane $l_{\bar{\alpha}p}$ one has $z = p - \bar{p}$. Consider the integral

$$\hat{f}(\bar{\alpha}, p) = \int_{l_{\bar{\alpha}p}} f(x)ds. \qquad (4.5.12)$$

The integration is actually taken over the domain $D \cap l_{\bar{\alpha}p}$, and this region is locally (in U) described by the inequalities $u_i > 0$, $1 \le i \le m$, or $v_i > 0$, $2 \le i \le m$, and $u_1 \ge 0$, that is:

$$z - \sum_{i=2}^{m} v_i - \sum_{j=m+1}^{n-I} u_j^2 + \sum_{j=n-I+1}^{n} u_j^2 \ge 0. \qquad (4.5.13)$$

Only the part of the integral (4.5.12) taken over a small neighborhood U of $S \cap (D \cap l_{\bar{\alpha}p})$ should be studied, since the rest of the integral is a smooth function of $(\alpha : p)$. Write the equation of $l_{\bar{\alpha}p}$ as $x_1 = \tilde{\beta} \cdot \tilde{x} - \tilde{q}$, where $\tilde{x} := (x_2, \dots, x_n)$, $\tilde{q} := -\frac{p}{\bar{\alpha}_1}$, $\tilde{\beta} := (-\frac{\bar{\alpha}_2}{\bar{\alpha}_1}, \dots, -\frac{\bar{\alpha}_n}{\bar{\alpha}_1})$. Then the integral

$$I_0 := \int_U f(x)ds = (1 + |\tilde{\beta}|^2)^{1/2} \int_U f(u, z)|\det x_{i,u_k}| du_2 \dots du_n$$

$$= (1 + |\tilde{\beta}|^2)^{1/2} \omega_m \int_U f(v, u, w, z)|\det x_{j,u_l}| dv du dw, \qquad (4.5.14)$$

where $x_{j,u_l}, m + 1 \le j, l \le n$, is the Jacobian and, by (4.5.10),

$$|\det x_{i,u_k}| = 2^{\frac{n-m}{2}}|H|^{-\frac{1}{2}}, \quad \omega_m := \frac{1}{\bar{\alpha}_2 \dots \bar{\alpha}_m},$$

$$v := (v_2, \dots, v_m), \quad u := (u_{m+1}, \dots, u_{n-I}), \quad w := (u_{n-I+1}, \dots, u_n). \qquad (4.5.15)$$

Note that starting from the formula for I_0 and everywhere below u is defined by formula (4.5.15). Taking U small and using smoothness of ϕ, where $f := \phi \chi_D(x)$, we can write the main singular term of I_0 as

$$I_1 := 2^{\frac{n-m}{2}}|H|^{-\frac{1}{2}}\omega_m(1 + |\tilde{\beta}|^2)^{1/2}\phi(\tilde{x})I_2 = \phi(\tilde{x})\Xi\frac{\Gamma(\frac{n+m}{2})}{\pi^{\frac{n-m}{2}}}I_2, \qquad (4.5.16)$$

where $I_2 := \int_U dv du dw$.

For simplicity we calculate only the leading singular term of $\hat{f}$. If we would not take the limiting value of ϕ out of the integral (4.5.14), we would get an integral whose integrand is as smooth as the data allow, and this integral produces the functions r_j in formula (4.5.7), which are also as smooth as the data allow. Here we use the following claim:

Let $f(x, z, y) \in C^k$ with respect to all its arguments. Then, for $a > -1$, we have

$$\int\limits_0^z x^a f(x, z, y) dx = \frac{z^{a+1}}{a+1} F(z, y), \quad F(z, y) \in C^k, F(0, y) = f(0, 0, y).$$

Using this claim, one can integrate in the case when in place of dv_2 one has ϕdv_2, where $\phi \in C^k$.

If we are interested in the leading singular term of $\hat{f}$, it is sufficient to calculate the limiting value of r_1, which is done below.

We assume that U is small: $|w| < \epsilon$, $|u| < 2\epsilon$, $|z| << \epsilon$, $v_i > 0$ and (4.5.13) holds in U. Denote $\xi_j := \sum_{i=j}^m v_i$ and consider I_2 for $I > 0$:

$$I_2 = \int\limits_{|w|<\epsilon} dw \int\limits_{|u|^2<z+|w|^2} du$$

$$\times \int\limits_0^{z-|u|^2+|w|^2} dv_m \int\limits_0^{z-|u|^2+|w|^2-\xi_m} dv_{m-1} \cdots \int\limits_0^{z-|u|^2+|w|^2-\xi_3} dv_2$$

$$= \frac{1}{\Gamma(m)} \int\limits_{|w|<\epsilon} dw \int\limits_{|u|^2<z+|w|^2} du(z - |u|^2 + |w|^2)^{m-1} := \frac{1}{\Gamma(m)} I_3.$$
$$\hspace{11cm}(4.5.17)$$

If $I = 0$ ($I = n - m$), then the integration with respect to w (u) is absent. Using the spherical coordinates with $|w| = r, |u| = \rho$, we obtain from (4.5.17):

$$I_3 = \int\limits_0^\epsilon dr\, r^{I-1} \int\limits_0^{(z+r^2)_+^{1/2}} d\rho\, \rho^{n-m-I-1} (z + r^2 - \rho^2)^{m-1} |S^{I-1}||S^{n-I-m-1}|$$

$$:= |S^{I-1}||S^{n-I-m-1}| I_4. \hspace{5cm}(4.5.18)$$

Recall that

$$|S^{n-1}| = \frac{2\pi^{n/2}}{\Gamma(n/2)}. \hspace{5cm}(4.5.19)$$

Consider I_4. Let $r^2 := t, \rho^2 = s$. Then

$$I_4 = \frac{1}{4} \int\limits_0^{\epsilon^2} dt\, t^{\frac{I-2}{2}} \int\limits_0^{(z+t)_+} ds\, s^{\frac{n-I-m-2}{2}} (z + t - s)_+^{m-1} := \frac{1}{4} I_5. \quad (4.5.20)$$

Let $s := \nu(z+t)_+$. Then

$$I_5 = \int_0^{\epsilon^2} dt\, t^{\frac{I-2}{2}} \int_0^1 d\nu\, \nu^{\frac{n-I-m-2}{2}} (1-\nu)^{m-1}(z+t)_+^{\frac{n+m-I-2}{2}}$$

$$:= B(\frac{n-m-I}{2}, m)I_6, \tag{4.5.21}$$

where formula (14.4.6) was used,

$$I_6 := \int_0^{\epsilon^2} dt\, t^{\frac{I-2}{2}}(z+t)_+^{\frac{n+m-I-2}{2}},$$

and $B(x,y)$ is the beta-function. Formulas (4.5.16)-(4.5.21) imply:

$$I_1 = \phi(\bar{x})2^{\frac{n-m}{2}}|H|^{-\frac{1}{2}}\omega_m(1+|\tilde{\beta}|^2)^{1/2}\frac{1}{\Gamma(m)}$$
$$\times \frac{1}{4}|S^{I-1}||S^{n-I-m-1}|B(\frac{n-m-I}{2}, m)I_6. \tag{4.5.22}$$

Use (4.5.19) and formula (14.4.7): $B(x,y) = \frac{\Gamma(x)\Gamma(y)}{\Gamma(x+y)}$, assume $I \neq 0$, and get:

$$I_1 = \phi(\bar{x})\Xi\frac{1}{B(\frac{I}{2}, \frac{n+m-I}{2})}I_6. \tag{4.5.23}$$

If $I = 0$, then

$$I_1 = \phi(\bar{x})\Xi\frac{1}{B(m, \frac{n-m}{2})}\int_0^{z_+} ds\, s^{\frac{n-m-2}{2}}(z-s)^{m-1}. \tag{4.5.24}$$

Let us study integrals (4.5.23) and (4.5.24) and consider the following cases:

(1) $I = 0$,
(2) $I > 0, I$ even, $n+m$ even,
(3) $I > 0, I$ even, $n+m$ odd,
(4) $I > 0, I$ odd, $n+m$ odd,
(5) $I > 0, I$ odd, $n+m$ even.

Case 1: If $I = 0$, then let $s = zt, z > 0$, and get:

$$\int_0^{z_+} ds\, s^{\frac{n-m-2}{2}}(z-s)^{m-1} = z_+^{\frac{n+m-2}{2}}B(m, \frac{n-m}{2}). \tag{4.5.25}$$

Thus, if $I = 0$, then (4.5.23) and (4.5.25) imply that the leading singularity of I_1 is:

$$I_1 \sim z_+^{\frac{n+m-2}{2}} r_1, \quad r_1(\bar{\alpha}, \bar{p}) = \phi(\bar{x})\Xi, \qquad (4.5.26)$$

where Ξ is defined in (4.5.4). This corresponds to the first line in (4.5.8) with $\mu = 0$ and $y_+ = (p - \bar{p})_+$.

Cases 2 and 4: In these cases $\frac{n+m-I-2}{2}$ is an integer. Using the formula $u_+ = u + u_-$, we get:

$$I_6 = \int_0^{\epsilon^2} dt\, t^{\frac{I-2}{2}} (z + t)_+^{\frac{n+m-I-2}{2}} = \nu(z) + \eta(z), \qquad (4.5.27)$$

where $\eta(z)$ denotes, here and below, a smooth function of z, and

$$\nu(z) := \int_0^{\epsilon^2} dt\, t^{\frac{I-2}{2}} (z + t)_-^{\frac{n+m-I-2}{2}}, \quad \nu(z) = 0, \ z \geq 0. \qquad (4.5.28)$$

If $z = -|z| < 0$, we get:

$$\nu(z) = \int_0^{|z|} dt\, t^{\frac{I-2}{2}} (|z| - t)^{\frac{n+m-I-2}{2}} = |z|^{\frac{n+m-2}{2}} B(\frac{I}{2}, \frac{n+m-I}{2})$$

$$= (-1)^{\frac{n+m-2}{2}} z_-^{\frac{n+m-2}{2}} B(\frac{I}{2}, \frac{n+m-I}{2}). \qquad (4.5.29)$$

From (4.5.29) and (4.5.23) we get:

$$I_1 = \phi(\bar{x})\Xi(-1)^{\frac{n+m-2}{2}} z_-^{\frac{n+m-2}{2}} + \eta(z). \qquad (4.5.30)$$

This corresponds to the first line in (4.5.8), $y_+ = (\bar{p} - p)_+$ and $z_- = y_+$.

Case 3: Integrate by parts $I/2$ times in I_6 and get:

$$I_6 = \int_{-0}^{\epsilon^2} dt\,\delta(t)(z + t)_+^{\frac{n+m-2}{2}} (-1)^{I/2} B(\frac{I}{2}, \frac{n+m-I}{2})$$

$$= z_+^{\frac{n+m-2}{2}} (-1)^{I/2} B(\frac{I}{2}, \frac{n+m-I}{2}), \qquad (4.5.31)$$

where $\delta(t)$ is the delta-function. From (4.5.31) and (4.5.23) we get:

$$I_1 = z_+^{\frac{n+m-2}{2}} (-1)^{I/2} \Xi + \eta(z). \qquad (4.5.32)$$

This corresponds to the first line in (4.5.8) and $y_+ = (p - \bar{p})_+$.

Case 5: Integrate by parts $\frac{n+m-l-1}{2}$ times in I_6 and get:

$$I_6 = (-1)^{\frac{n+m-l-1}{2}} \frac{\Gamma(\frac{l}{2})\Gamma(\frac{n+m-l}{2})}{\Gamma(\frac{1}{2})\Gamma(\frac{n+m-1}{2})} I_7 + \eta(z), \qquad (4.5.33)$$

where

$$I_7 := \int_{-0}^{\epsilon^2} dt\, t^{\frac{n+m-2}{2}} (tz + t^2)_+^{-\frac{1}{2}}. \qquad (4.5.34)$$

Use in (4.5.34) the known formulas:

$$\int x^k (xz + x^2)^{-\frac{1}{2}} dx = \frac{x^{k-1}(xz + x^2)^{\frac{1}{2}}}{k}$$
$$- \frac{(2k-1)z}{2k} \int x^{k-1}(xz + x^2)^{-\frac{1}{2}} dx,$$

$$\int (xz + x^2)^{-\frac{1}{2}} dx = \log\left(2(xz + x^2)^{\frac{1}{2}} + 2x + z\right) + c.$$

Then (4.5.33), (4.5.34), and (4.5.23) yield:

$$I_1 = z^{\frac{n+m-2}{2}} (\log|z|) r_1 + \eta(z), \qquad (4.5.35)$$

where

$$r_1(\bar{\alpha}, \bar{p}) = \phi(\bar{x})\Xi(-1)^{\frac{l+1}{2}} \pi^{-1}. \qquad (4.5.36)$$

This corresponds to the second line in (4.5.8) and $y = p - \bar{p}$.

Theorem 4.5.1 is proved. $\square$

4.6. Singularities of the Radon transform: an alternative approach

In this section we develop an alternative approach to the study of singularities of the Radon transform. This will be done in two-dimensional case. This approach in principle allows us to consider degenerate critical points of the function z on the variety M. Let $x_2 = \beta \cdot x_1 - q$. In this section we write x for x_1 and y for x_2. Assume that $y = g(x)$ and that the line $l_{\bar{\beta}\bar{q}}$ is tangent to S at the point $(\bar{x}, \bar{y})$. We want to study the behavior of $\hat{f}$ in a neigborhood of the point $(\bar{\beta}, \bar{q})$. As before, β and q are the inhomogeneous coordinates defining the line $l_{\beta q}$. We consider three cases.

The first (generic) case: $g(x)$ is smooth in a neighborhood of $\bar{x}$ and $g^{(2)}(\bar{x}) \neq 0, g'(\bar{x}) = \bar{\beta}$.

The second case: $g'(\bar{x}) = \bar{\beta}$, $g^{(j)}(\bar{x}) = 0, 2 \leq j \leq k - 1$, $g^{(k)}(\bar{x}) \neq 0$. We consider two subcases: 2a) k is even, and 2b) k is odd.

The second case was not considered in Section 4.5, where the critical points were assumed nondegenerate (Morse-type), as in the first case.

The third case: $g(x)$ is not differentiable at $\bar{x}$. This case corresponds to the corner or spike point at $(\bar{x}, g(\bar{x}))$.

We start with the first and second cases and use the formula:

$$\hat{f}(\beta, q) = \int_{\phi(\beta,q)}^{\psi(\beta,q)} f(x, \beta x - q)dx, \qquad (4.6.1)$$

where $\phi(\beta, q)$ and $\psi(\beta, q)$ are the x-coordinates of the points of intersection of the line $l_{\beta q}$ with S. These coordinates can be found solving the equation $g(x) = \beta x - q$. We understand formula (4.6.1) in the following sense. Generally, the line $l_{\beta q}$ may have finitely many points of intersection with S. We consider only those which lie in a small neighborhood of the critical point. Depending on the situation (details are discussed below) there are one or two points of intersection in a small neighborhood of the critical point. If there is one point of intersection, then one of the limits of integration in (4.6.1) does not contribute to the singularity associated with the investigated critical point directly, but possibly contributes through the contribution by another critical point which lies on the same straight line $l_{\beta q}$. In the latter case, this limit of integration can be obtained as a point of intersection near another critical point, or an inflection point of S which lies on $l_{\beta q}$. In the case 2a) the above line intersects S in just two points in a neighborhood of the point $(\bar{x}, g(\bar{x}))$.

Use the coordinate system in which the origin is at the point $(\bar{x}, g(\bar{x}))$, the y-axis is the normal to S at the origin, and the straight line $l_{\bar{\beta}\bar{q}}$ is tangent to S at the point $(\bar{x}, g(\bar{x}))$. Hence $l_{\bar{\beta}\bar{q}}$ is the line with $\bar{\beta} = \bar{q} = 0$ and $l_{\bar{\beta}\bar{q}}$ coincides with the x-axis. Then, in a small neighborhood of the origin, the local equation of S is

$$y = \frac{g^{(k)}(0)}{k!}x^k + O(|x|^{k+1}). \qquad (*)$$

If k is even, a straight line $y = \beta x - q$, for sufficiently small β, intersects S in just two points in a small neighborhood of $\bar{x}$.

If k is odd, then the local equation of S has the same form $(*)$. The function y has an inflection point at $(\bar{x}, g(\bar{x}))$. The analysis in both cases 2a) and 2b) is similar to a considerable extent.

Also, if k is odd, there is only one point of intersection of the straight line $l_{\bar{\beta}q}$ with S in a small neighborhood of the point $(\bar{x}, g(\bar{x}))$. The x

coordinate of this point is the unique small solution to Equation (4.6.2) (see below). By small solution we mean the solution in a neighborhood of zero.

We fix $\beta = \bar{\beta}$ and denote $Q := q - \bar{q}, \xi := x - \bar{x}$. The equation $g(x) = \bar{\beta} x - q$ can be written as

$$Q = \xi^k(-\epsilon_k)W(\xi), \quad W(0) = \frac{|g^{(k)}(\bar{x})|}{k!} \neq 0, \qquad (4.6.2)$$

where $\epsilon_k := \text{sgn} g^{(k)}(\bar{x})$, we have used the equation $\bar{\beta} = 0$, and $W(\xi)$ is a smooth function in a neighborhood of zero. In the case 2b) there is only one point of intersection of the straight line $l_{\bar{\beta} q}$ with S in a small neighborhood of the point $(\bar{x}, g(\bar{x}))$.

Consider the case 2a), when there are two small solutions to (4.6.2). Let us assume $k = 2$. The general case is treated quite similarly. Write (4.6.2) as

$$\xi = \pm \nu v(\xi), \quad v(0) = [|g^{(2)}(\bar{x})|/2]^{-1/2} \neq 0, \qquad (4.6.3)$$

where $v(\xi) := W^{-1/2}(\xi)$, $\nu := (\pm Q)^{1/2}$, and the sign is chosen so that $(-\epsilon)(\pm Q) > 0$ whith $\epsilon := \epsilon_2$. By the implicit function theorem, Equation (4.6.3) defines smooth solutions $\xi_\pm = \pm \xi(\nu)$, and

$$\xi_\pm = \pm[|g^{(2)}(\bar{x})|/2]^{-1/2}\nu + O(\nu^2). \qquad (4.6.4)$$

The leading term of the singularity of $\hat{f}$, according to formula (4.6.1), is

$$f(\bar{x}, g(\bar{x}))[\psi(\bar{\beta}, q) - \phi(\bar{\beta}, q)], \qquad (4.6.5)$$

and, by (4.6.4),

$$\psi(\bar{\beta}, q) - \phi(\bar{\beta}, q) = 2^{3/2}|g^{(2)}(\bar{x})|^{-1/2}[\epsilon(q - \bar{q})]_+^{1/2}.$$

Thus, the leading singularity of $\hat{f}$ is given by the term:

$$\hat{f} \sim f(\bar{x}, g(\bar{x}))2^{3/2}|g^{(2)}(\bar{x})|^{-1/2}[\epsilon(q - \bar{q})]_+^{1/2}. \qquad (4.6.6)$$

For $\hat{f}$ one can get the representation:

$$\hat{f} = [\epsilon(q - \bar{q})]_+^{1/2}r_1 + r_2, \qquad (4.6.7)$$

where r_1, r_2 are smooth functions of (β, q) in a neighborhood of the point $(\bar{\beta}, \bar{q})$, and

$$r_1(\bar{\beta}, \bar{q}) = f(\bar{x}, g(\bar{x}))2^{3/2}|g^{(2)}(\bar{x})|^{-1/2}. \qquad (4.6.8)$$

Indeed, in (4.6.5) the functions ψ and ϕ are as smooth as the data allow.

We could keep the β variable in a neighborhood of $\bar{\beta}$ and get r_1 and r_2 as smooth functions of both variables β and q.

If $k > 2$ is even, then there are two real-valued solutions, one positive and one negative, similar to (4.6.4), of the equation similar to (4.6.3). The leading term of the singularity of $\hat{f}$ generically is of the order $[\pm(q-\bar{q})]_+^{1/k}$, and the analysis is similar to the case $k = 2$.

Consider the case 2b) and assume $k = 3$. The general case is treated quite similarly.

If $k = 3$, then (4.6.2) can be written as

$$\xi = \nu v(\xi), \quad v(0) = [|g^{(2)}(\bar{x})|/2]^{-1/2} \neq 0, \tag{4.6.9}$$

where $v(\xi) := -\epsilon_3 W^{-1/3}(\xi)$, $W(0) \neq 0$, $\nu := Q^{1/3}$.

By the implicit function theorem, Equation (4.6.9) defines the unique small smooth solution $\xi = \xi(\nu)$, and

$$\xi = -\epsilon_3 W^{-1/3}(0)\nu + O(\nu^2). \tag{4.6.10}$$

The difference with the case 2a) is that only one of the limits in the integral (4.6.1) corresponds now to the small solution to (4.6.2). Suppose that this is the lower limit $\phi(\beta, q)$. Then the input of this limit to the leading singularity of $\hat{f}$ is $-f(\bar{x}, g(\bar{x}))\phi(\bar{\beta}, q) = (p - \bar{p})^{\frac{1}{3}}r_1 + r_2$, where $r_1(\bar{\alpha}, \bar{p}) = -(\alpha_2)^{-\frac{1}{3}}\epsilon_3 W^{-1/3}(0)f(\bar{x}, g(\bar{x}))$, and we assume $\alpha_2 \neq 0$.

In principle, this approach can be used for the multidimensional problems also.

Consider the third case. In this case $\bar{x}$ solves the equation $g_1(x) = g_2(x)$, where $y = g_j(x), j = 1, 2$, are the equations of the two curves S_j, which are the parts of S, and which intersect at the point with x-coordinate $\bar{x}$. In this case we have

$$q = \beta\bar{x} - g_1(\bar{x}), \quad g_1(\bar{x}) = g_2(\bar{x}). \tag{4.6.11}$$

We assume that $\bar{x}$ is an isolated point of the intersection of S_1 and S_2. Equation (4.6.11) is an equation of a straight line in the (β, q) space. Therefore, the image of the point $(\bar{x}, g_1(\bar{x})) \in S$, which is a point of nondifferentiability of S, is the straight line in the (β, q) space. In other words, according to the remark below Definition 4.3.2, if we consider in this case a function $g(x)$ on the variety M, which consists of one point $(\bar{x}, g_1(\bar{x})) \in S$, then the Legendre map of the function $g(x)$ on this variety is the straight line in the (β, q) space.

The equation of this straight line is (4.6.11). If $(\bar{\alpha} : \bar{p})$ is any finite point on this straight line, $\bar{\alpha} := (\bar{\alpha}_1, \bar{\alpha}_2), \bar{\beta} := -\frac{\bar{\alpha}_1}{\bar{\alpha}_2}, \bar{p} = -\bar{\alpha}_2\bar{q}, \bar{\alpha}_2 \neq 0$, then $\hat{f} = y_+r_1 + r_2$, $y_+ := (\bar{p} - p)_+$, provided that the limiting tangent lines to S at the point $(\bar{x}, g(\bar{x}))$ form a nonzero angle, that is, S does not have a cusp at this point. The functions r_j are as smooth as the data allow, but only outside a small neighborhoods of the exceptional values of α described in Remark 4.5.1

4.7. Asymptotics of the Fourier transform

4.7.1. Introduction

In this section we study the asymptotics of the Fourier transform of a function of the type $f(x) = \phi(x)\chi_D(x)$, where ϕ is smooth, $\phi \neq 0$ on ∂D, D is a bounded domain in $\mathbb{R}^n$, ∂D is a union of hypersurfaces S_j, $j \in \mathcal{J}$, where $\mathcal{J}$ is a finite set of indices. The assumptions in this section are the same as in Section 4.5. The hypersurfaces S_j are assumed to be smooth and in general position. The word 'smooth' means 'belonging to C^n'.

Consider $\tilde{f} := \mathcal{F}f(t\alpha), \alpha \in S^{n-1}, t > 0$. We want to derive an asymptotics of $\tilde{f}$ as $t \to \infty$ valid for almost all α. Examples (one of which is given later in this section), show that the asymptotics, in general, does not hold for all α.

Fix $\alpha \in S^{n-1}$ and consider the function $z := \alpha \cdot x$ as a function on S. Let $\bar{x}$ be a stationary point of this function, so that the hyperplane $l_{\alpha\bar{p}}, \bar{p} := \alpha \cdot \bar{x}$, is not transversal to S at the point $\bar{x}$. Let us assume that $\bar{x}$ belongs to a component of S, which is a smooth variety of codimension $m, m = m(\alpha, \bar{x})$. Each stationary point $\bar{x}$ contributes to the asymptotics of $\tilde{f}$, and the sum of these contributions over all the stationary points corresponding to the fixed α, gives the desired asymptotics. We want to find the first term of this asymptotics.

By formula (2.1.20), we can calculate the asymptotics of $\tilde{f}$ as the asymptotics of one-dimensional Fourier integrals of the Rf. Such an asymptotics is known, and the formulas we use are collected in Section 14.5. The basic result on the singularities of the Radon transform, that we use in this calculation, is Theorem 4.5.1.

So the strategy of our derivation is quite simple. In the next subsection the result is formulated and proved.

4.7.2. Statement and proof of the result

The quantities Ξ and I, used in the following theorem, are defined in Section 4.4, formula (4.5.4) and Theorem 4.5.1 respectively, while $m = m(\alpha, \bar{x})$ is defined in Section 4.7.1.

Theorem 4.7.1. Let $\phi(x) \in C^k(\mathbb{R}^n), k \geq n, \phi \neq 0$ on ∂D. Then, for almost all $\alpha \in S^{n-1}$, one has

$$\widetilde{\phi\chi_D}(t\alpha) = t^{-\frac{n+m}{2}} \sum \exp(it\alpha \cdot x)\phi(x)\Xi c + o\left(t^{-\frac{n+m}{2}}\right) \qquad (4.7.1)$$

as $t \to +\infty$, where $c := \exp\left(i\pi\frac{n+m-2I}{4}\right)$, and the summation is taken over all the critical points of the function $\alpha \cdot x$ on the hypersurface $S := \partial D$.

REMARK 4.7.1. If $m = 1$ and $f(x)$ is replaced by $b(x)f(x)$, where $b(x) \in C^k(D), k \geq \max(2, 1 + n + \nu)$ and in a neighborhood of ∂D the function $b(x)$ equals to $[\rho(x, \partial D)]^\nu$, where $\rho(x, \partial D)$ is the distance from x to ∂D, then one has for almost all $\alpha \in S^{n-1}$ and $\nu > -1$:

$$\widetilde{b\phi\chi_D}(t\alpha) = t^{-\frac{n+1}{2}-\nu}(2\pi)^{\frac{n-1}{2}}\Gamma(\nu+1)\sum \exp(it\alpha \cdot x)\phi(x)\Xi c_\nu$$

$$+ o\left(t^{-\frac{n+1}{2}-\nu}\right) \qquad (4.7.1')$$

as $t \to +\infty$, where $c_\nu := \exp\left[\dfrac{i\pi}{2}\left(\nu + \dfrac{n+1-2I}{2}\right)\right].$

REMARK 4.7.2. It is easy to give an example of a domain $D \subset \mathbb{R}^n$ and direction $\alpha \in S^{n-1}$ such that $\tilde\chi_D(t\alpha)$ has the asymptotic behavior different from (4.7.1). Take $n = 2$, $D = \{(x_1, x_2) \in \mathbb{R}^2 : 1 \geq x_2 \geq x_1^2\}$, $\alpha = (0, 1)$. Then

$$R_\alpha(p) = \begin{cases} 0, & p \notin [0, 1], \\ 2p^{\frac{1}{2}}, & p \in [0, 1] \end{cases}$$

and, by Lemma 14.5.5,

$$\tilde\chi_D(t\alpha) = t^{-1}\left[e^{it}(-i) + o(1)\right].$$

Here the decay rate as $t \to \infty$ is different from the one given by formula (4.7.1) in the case $n = 2$, $m = 1$, that is $O(t^{-\frac{3}{2}})$. For any direction α', $\alpha' \neq \alpha$, in a small neighborhood of α, one has decay $O(t^{-\frac{3}{2}})$ as follows from formula (4.7.1). In fact, any domain whose boundary contains a plane face, yields a similar example.

Proof of Theorem 4.7.1. By formula (2.1.20),

$$\tilde f(t\alpha) = \int_{-\infty}^{\infty} \hat f(\alpha, p)\exp(ipt)dp. \qquad (4.7.2)$$

Substitute (4.5.7) into (4.7.2). As $t \to \infty$, the main term of the asymptotics of $\tilde f(t\alpha)$ is the same as that of the the integral

$$I_1 := r_1(\alpha, \bar p)\int_{-\infty}^{\infty} y_+^{\frac{n+m-2}{2}}\exp(ipt)dp, \qquad (4.7.3)$$

if $I(n + m - 1)$ is even, or that of the integral

$$I_2 := r_1(\alpha, \bar p)\int_{-\infty}^{\infty} y^{\frac{n+m-2}{2}}\log(|y|)\exp(ipt)dp, \qquad (4.7.4)$$

if $I(n + m - 1)$ is odd, where α is the same as $\bar{\alpha}$ in formula (4.5.7), $y :=$ $\pm(p - \bar{p})$, the minus (plus) sign is taken when $I > 0$ and $n + m$ are of the same (different) parity, and plus sign is taken when $I = 0$. Asymptotics of I_1 and I_2 are calculated by formulas (14.5.28) and (14.5.32). Applying these formulas to various integrals below we always assume (as is the case under our assumptions), that the integrand vanishes near infinity. In all cases the point which contributes to the main term of the asymptotics is the zero point, which can always be made the lower limit of integration (after a change of variables in some cases).

There are several cases to consider.

Case 1. If $I > 0$ and $m + n$ are of the same parity, then $I(n + m - 1)$ is even, $y = p - \bar{p}, \mu = \frac{n+m-I}{2}$, and we have:

$$\tilde{f}(t\alpha) \sim \phi(\bar{x})\Xi(-1)^{\frac{n+m-I}{2}} \int_{-\infty}^{\infty} (p - \bar{p})_+^{\frac{n+m-2}{2}} \exp(ipt)dp$$

$$\sim \phi(\bar{x})\Xi\Gamma(\frac{n+m}{2})t^{-\frac{n+m}{2}}e^{it\alpha \cdot x}e^{\frac{i\pi(n+m-2I)}{4}},$$

which is formula (4.7.1).

Case 2. If $I = 0$, then $y = \bar{p} - p, \mu = 0$, and

$$\tilde{f}(t\alpha) \sim \phi(\bar{x})\Xi\Gamma(\frac{n+m}{2})t^{-\frac{n+m}{2}}e^{it\alpha \cdot x}e^{\frac{i\pi(n+m)}{4}},$$

which is again formula (4.7.1).

Case 3. If $I > 0$ and $n + m$ are of different parity and I is even, then $I(n + m - 1)$ is even, $\mu = I/2, y = p - \bar{p}$ and

$$\tilde{f}(t\alpha) \sim \phi(\bar{x})\Xi(-1)^{\frac{I}{2}} \int_{-\infty}^{\infty} (p - \bar{p})_+^{\frac{n+m-2}{2}} \exp(ipt)dp$$

$$\sim \phi(\bar{x})\Xi\Gamma(\frac{n+m}{2})t^{-\frac{n+m}{2}}e^{it\alpha \cdot x}e^{\frac{i\pi(n+m+2I)}{4}},$$

which is formula (4.7.1), if we take into account that, for I even, $e^{i\pi 2I/4}$ $= e^{-i\pi 2I/4}$.

Case 4. If $I > 0$ and $n + m$ are of different parity and I is odd, then $I(n + m - 1)$ is odd, $\mu = \frac{I+1}{2}, y = p - \bar{p}$ and

$$\tilde{f}(t\alpha) \sim \phi(\bar{x})\Xi\frac{1}{\pi}(-1)^{\frac{I+1}{2}} \int_{-\infty}^{\infty} (p - \bar{p})_+^{\frac{n+m-2}{2}} \log(|p - \bar{p}|) \exp(ipt)dp$$

$$\sim \phi(\bar{x})\Xi\Gamma(\frac{n+m}{2})t^{-\frac{n+m}{2}}e^{it\alpha \cdot x}e^{\frac{i\pi(n+m-2I)}{4}},$$

which is formula (4.7.1). Theorem 4.7.1 is proved. $\square$

Exercise 4.7.1. Let S be a smooth strictly convex hypersurface, that is, its Gaussian curvature is strictly positive. Then $m = 1, I = 0$, there exist exactly two stationary points $\bar{x}_\pm(\alpha)$ for any $\alpha \in S^{n-1}$. Prove that in this case formula (4.7.1') can be written as:

$$\widetilde{\phi\chi_D}(t\alpha) = t^{-\frac{n+1+2\nu}{2}}[\exp(it\alpha \cdot \bar{x}_+)a_+ + \exp(it\alpha \cdot \bar{x}_-)a_- + o\,(1)] \quad (4.7.1'')$$

as $t \to +\infty$, where

$$a_\pm := (2\pi)^{\frac{n-1}{2}}\Gamma(\nu+1)\exp\left(\pm i\pi\frac{n+1+2\nu}{4}\right)\phi(\bar{x}_\pm)\mathcal{K}_\pm^{-\frac{1}{2}},$$

$\mathcal{K}_\pm$ are the Gaussian curvatures of S at the points $\bar{x}_\pm$, inner normal to S at the point $\bar{x}_+$ is directed along α, and ν is the same as in Remark 4.7.1.

Hint. Assume first that $\nu = 0$, take one of the stationary points, say $\bar{x}_+ \in S$, put the origin at $\bar{x}_+$, and let x_n be oriented along the inner normal to S at the point $\bar{x}_+$, so that the equation of S in the local coordinates is $x_n = g(x')$, where $x' = (x_1, \ldots, x_{n-1})$, $g(0) = 0$, $\nabla g(0) = 0$ and $g(x') = \sum_{j=1}^{n-1}\frac{x_j^2}{2R_j} + o(|x'|^2)$ as $x' \to 0$. Here R_j are the radii of curvature of S at the point $\bar{x}_+$. The plane $\bar{\alpha} \cdot x = p$ intersects S over a variety whose equation is $g(x') = p$, and $\bar{p} = 0$. If $\alpha = e_n$, where e_n is the unit vector along x_n-axis, then $\hat{f}(e_n, p) = \int_{g(x')<p} f(x')dx'$. As $p \to 0$, one gets the main term of the asymptotics of the Radon transform:

$$\hat{f} \sim f(0)\int\limits_{\sum_{j=1}^{n-1}\frac{x_j^2}{2R_j}<p} dx' = f(0)2^{\frac{n-1}{2}}p^{\frac{n-1}{2}}\mathcal{K}^{-\frac{1}{2}}\int\limits_{\sum_{j=1}^{n-1}y_j^2<1} dy.$$

Here $\mathcal{K} := (R_1 \ldots R_{n-1})^{-1}$ is the Gaussian curvature of S at the point $\bar{x}_+$. Thus, using the value $\frac{2\pi^{\frac{n-1}{2}}}{(n-1)\Gamma(\frac{n-1}{2})}$ for the volume of the unit ball in $\mathbb{R}^{n-1}$, we get

$$\hat{f} \sim f(0)2^{\frac{n-1}{2}}\mathcal{K}^{-\frac{1}{2}}\frac{2\pi^{\frac{n-1}{2}}}{(n-1)\Gamma(\frac{n-1}{2})}p^{\frac{n-1}{2}} + o(p^{\frac{n-1}{2}}) \text{ as } p \to 0.$$

In an arbitrary coordinate system whose x_n axis is directed as the inner normal to S, this main term is

$$f(\bar{x}_+)2^{\frac{n-1}{2}}\mathcal{K}^{-\frac{1}{2}}\frac{2\pi^{\frac{n-1}{2}}}{(n-1)\Gamma(\frac{n-1}{2})}(p-\bar{p})_+^{\frac{n-1}{2}},$$

where $\bar{p} = \bar{\alpha} \cdot \tilde{x}_+$. One has:

$$\int_{-\infty}^{\infty} (p - \bar{p})_+^{\frac{n-1}{2}} \exp(ipt) dp \sim \exp(i\bar{p}t) \exp(i\pi \frac{n+1}{4}) \Gamma(\frac{n+1}{2}) t^{-\frac{n+1}{2}}$$

$$+ o(t^{-\frac{n+1}{2}}) \text{ as } t \to \infty.$$

Since $\frac{2\Gamma(\frac{n+1}{2})}{(n-1)\Gamma(\frac{n-1}{2})} = 1$, we get the formula for a_+. Formula for a_- is obtained similarly, but, in the coordinate system we use, the asymptotics of $\hat{f}$ in a neighborhood of the point $\bar{p} = \bar{\alpha} \cdot \tilde{x}_-$ contains a factor $(\bar{p}-p)_+^{\frac{n-1}{2}}$ rather than $(p-\bar{p})_+^{\frac{n-1}{2}}$, that corresponded to $\tilde{x}_+$. This leads to the phase factor $\exp(-i\pi \frac{n+1}{4})$ rather than $\exp(i\pi \frac{n+1}{4})$. If $\nu > -1$, then the factor $(p - \bar{p})_+^{\nu + \frac{n-1}{2}}$ appears in place of $(p - \bar{p})_+^{\frac{n-1}{2}}$, and $\int_{|y|<1}(1 - |y|^2)^{\nu/2} dy$ - in place of $\int_{|y|<1} dy$. This leads to formula (4.7.1″).

See also [LRZ], where another proof is given.

4.8. Wave front sets

In this section we study the relation between wave front sets of f and $\hat{f}$. The definition of the wave front set of f, $WF(f)$, is given in Chapter 14.

We assume that f is piecewise-smooth, $f = \chi_D(x)\phi(x)$, where ϕ does not vanish on S and is C^∞-smooth. Then singsupp $f = S$.

Theorem 4.8.1. *If $(x, \xi) \in WF(f)$, then $x \in S$ and $(\xi : \xi \cdot x) \in \hat{S}$. Conversely, if $(\bar{\alpha} : \bar{p}) \in \hat{S}$ is a point of $\hat{S}$, at which the plane $l_{\alpha p}$ in $\mathbb{RP}_n$ with the equation $\alpha \cdot x = p$ is tangent to $\hat{S}$, then $x \in S$ and $(x, \xi) \in WF(f)$, where $\xi = \bar{\alpha}$.*

Proof. If $(x, \xi) \in WF(f)$, then $x \in S$ and ξ is directed along the normal to S if S is a hypersurface in a neighborhood of x. Therefore, in this case $\xi = \alpha$, $p = \alpha \cdot x$, and the plane $l_{\alpha p}$ is tangent to S at the point x. Thus, $(\xi : \xi \cdot x) \in \hat{S}$, as claimed. If $x \in S_{\{1,...,m\}}$, then ξ is a linear combination of the m normals to S_j, $1 \le j \le m$, $\alpha = \xi$, and the rest of the argument is as above.

Conversely, if $(\bar{\alpha} : \bar{p}) \in \hat{S}$, and the plane $l_{\alpha p}$ is tangent to $\hat{S}$ at the point $(\bar{\alpha} : \bar{p})$, then $(x', -1)$ is directed along the normal to $\hat{S}$ at the point $(\bar{\alpha} : \bar{p})$, or, in nonhomogeneous coordinates, at the point $(\bar{\beta}, \bar{q})$, $\bar{\beta} := -\frac{\bar{\alpha}'}{\bar{\alpha}_n}$, $\bar{q} := -\frac{\bar{p}}{\bar{\alpha}_n}$, $\bar{\alpha}_n \ne 0$. Thus $\nabla h(\bar{\beta}) = x'$, $x_n = \bar{\beta} \cdot x' - \bar{q}$, $\bar{q} = h(\bar{\beta})$. Therefore, $(x', x_n) = x \in S$, and $\xi = \bar{\alpha}$ is directed along the normal to S at the point x. This means that $(x, \xi) \in WF(f)$. $\square$

4.9. Singularities of X-ray transform

4.9.1. Introduction

In this section we study the singularities of X-ray transform and describe a constructive procedure for finding S, the discontinuity surface of f, from some X-ray transform data, which are specified below.

First, let us note that the results of Sections 4.2 – 4.6 can be used for a study of singularities of Xf. Indeed, let l_m be an m-dimensional affine space in $\mathbb{R}^n$, $m < n$, and let $l_m \subset l_{m+1}$. The restriction of $X_m f$ to l_m is the Radon transform of the restriction of f to l_{m+1}. Therefore, the results concerning the singularities of the Radon transform are applicable. Let $m = 1$, that is, consider X-ray transform. Fix an l_2, a two-dimensional affine space, denote $u_j, j = 1, 2$, the coordinates on l_2, and let $w(u_1, u_2)$ be the restriction of f onto l_2. The singularities of w can be found from the line integrals of w along the lines belonging to l_2. Since we can take as l_2 an arbitrary plane, we can find singularities of f on an arbitrary plane in $\mathbb{R}^n$.

However, the knowledge of the full X-ray transform data is not necessary for the unique reconstruction of f. Much less data will suffice (see Chapter 9).

We consider in this chapter the practically important case of the data corresponding to a family of straight lines which pass through a smooth curve Λ in $\mathbb{R}^n$. The parametric equation of Λ is $\mathbf{x} = \mathbf{x}(t) = (\mathbf{x}_i(t))$, $i = 1, \ldots, n$, $0 \le t \le 1$.

Denote by S_Λ^* the set of straight lines in $\mathbb{R}^n$ having nonvoid intersection with Λ and not transversal to S. We assume that S_Λ^* is given by the equation $t = H(b)$, where $b = (b_1 : \cdots : b_n)$ are the homogeneous coordinates in $\mathbb{RP}_{n-1}$, and $H(b)$ is a smooth function, homogeneous of degree zero, so that $H(b)$ does not depend on the choice of homogeneous coordinates b. Parameter t fixes a point $\mathbf{x}(t) \in \Lambda$, and b fixes the tangent to S straight line through $\mathbf{x}(t)$ in the direction b. The hypersurface S is defined by the equation $\mathfrak{g}(x) = 0$ with $\operatorname{grad} \mathfrak{g}(x) \ne 0$ on S. The problem is : *given the functions $H(b)$ and $\mathbf{x}_i(t)$, find the hypersurface S.* Note that S is the *envelope* of S_Λ^*. We give a *constructive procedure for finding the envelopes of some families of straight lines in $\mathbb{R}^n$.*

Fix $x \in S_\Lambda^*$, and let $g(x, \alpha) := g(\alpha)$ be defined by (1.1.2). Let $\hat{S}_x \subset S^{n-1}$ be a subset of those $\alpha \in S^{n-1}$ for which the straight line $\{x + \alpha t\}_{t \in \mathbb{R}}$ is tangent to S. If $f = \phi(x)\chi_D(x), \phi \in C^\infty$, then $\operatorname{singsupp} g(\alpha) = \hat{S}_x$. The proof of this claim is similar to the proof of Theorem 4.2.1.

4.9.2. Description of the procedure

The following lemma will be useful.

Lemma 4.9.1. *Suppose that in a domain $\Delta \subset \mathbb{R}^\nu$ every solution of* *the system*

$$\phi_k(z) = 0, \quad k = 1, \ldots, K, \quad z = (z_1, \ldots, z_\nu) \in \Delta, \qquad (4.9.1)$$

satisfies the equation

$$\psi(z) = 0. \qquad (4.9.2)$$

Suppose that $\phi_k(z), \psi(z) \in C^2(\Delta)$ and that the matrix

$$\frac{\partial \phi(z)}{\partial z} = \left(\frac{\partial \phi_k(z)}{\partial z_j} \right)_{\substack{k=1,\ldots,K \\ j=1,\ldots,\nu}}$$

has the maximal possible rank $\min(K, \nu)$ for all $z \in \Delta$. Then, for every z satisfying (4.9.1), there exists $\lambda(z) = (\lambda_k(z))_{k=1,\ldots,K}$ such that

$$\frac{\partial \psi(z)}{\partial z} = \lambda(z) \frac{\partial \phi(z)}{\partial z}, \quad \frac{\partial \psi(z)}{\partial z} = \left(\frac{\partial \psi(z)}{\partial z_j} \right)_{j=1,\ldots,\nu}. \qquad (4.9.3)$$

Proof. If $K \geq \nu$, the claim is evident. Let $K < \nu$, and z_0 satisfy (4.9.1). Then, by the implicit function theorem, there exists a neighborhood U of z_0 and a system of local coordinates $y = (y_1, \ldots, y_\nu)$ in U such that $y_k = \phi_k$, $k = 1, \ldots, K$, and, by the assumption, there exist C^1-functions $\zeta_k(y)$, $k = 1, \ldots, K$, such that

$$\psi(z) = \sum_{k=1}^{K} y_k \zeta_k(y) \qquad (4.9.4)$$

From (4.9.4) one gets (4.9.3) with $\zeta_k(y(z)) := \lambda_k(z)$. $\square$

We now describe an algebraic procedure for calculating $x \in S$ given the data

$$\{b_i, \ t = H(b), \ \frac{\partial H(b)}{\partial b_i}, \ \mathbf{x}(t) = (x_i(t)), \ \dot{\mathbf{x}}(t) = (\dot{x}_i(t)), i = 1, \ldots, n\}.$$

Here $\dot{x}_i(t) = \frac{dx_i(t)}{dt}$. Below the summation over the repeated indices is assumed, and the symbol $G_{,j}$ means the derivative with respect to the j-th variable. Note that if $(t, b) \in S_\Lambda^*$, $t = H(b), b \in \mathbb{R}^n$, then there exists $\tau \in \mathbb{R}$ such that

$$\phi_i := x_i(t) + \tau b_i - x_i = 0, \quad 1 \leq i \leq n, \qquad (4.9.5)$$

$$\phi_{n+1} := g_{,i}(x) b_i = 0, \qquad (4.9.6)$$

$$\phi_{n+2} := g(x) = 0, \qquad (4.9.7)$$

$$\psi := t - H(b) = 0. \qquad (4.9.8)$$

Thus, if we apply Lemma 4.9.1 to equations (4.9.5) – (4.9.8), then $K = n + 2$, $z := (x_1, x_2, \ldots, x_n, \tau, t, b_1, \ldots, b_n)$, $\nu = 2n + 2$. By Lemma 4.9.1 there exist $\lambda = (\lambda_1, \ldots, \lambda_n)$ and $\mu, \nu \in \mathbb{R}$ such that

$$-\lambda_i + \mu b_j g(x)_{,ij} + \nu g(x)_{,i} = 0; \tag{4.9.9}$$

$$\lambda_i b_i = 0; \tag{4.9.10}$$

$$\lambda_i \dot{x}_i(t) = 1; \tag{4.9.11}$$

$$\tau \lambda_i + \mu g(x)_{,i} = -H_{,i}. \tag{4.9.12}$$

After multiplying (4.9.9) by b and using (4.9.6) and (4.9.10), one gets

$$-\lambda \cdot b + \mu b_j g(x)_{,ij} b_i + \nu g(x)_{,i} b_i = \mu b_j g(x)_{,ij} b_i = 0. \tag{4.9.13}$$

Let us assume that the matrix $g(x)_{,ij}$ is either positive or negative definite.

Then, since $b \neq 0$, Equation (4.9.13) implies $\mu = 0$. From (4.9.9) – (4.9.12) we get

$$-\tau \nu g(x)_{,i} = H(b)_{,i}, \tag{4.9.14}$$

$$\tau \lambda_i = -H(b)_{,i},$$

and

$$\tau = -H(b)_{,i} \dot{x}_i(t). \tag{4.9.15}$$

Consider the rank of the matrix

$$(\phi_{i,j}) := \Im = \begin{pmatrix} \tau E & -E & \dot{x}^t & b^t \\ g(x)_{,i} & b_j g(x)_{,ij} & 0 & 0 \\ 0 & g(x)_{,i} & 0 & 0 \end{pmatrix}.$$

Here the columns came from the differentiation with respect to b_i, x_i, t and τ respectively, b^t stands for the transposed vector, and E is the unit $n \times n$ matrix. Since $\min(K, \nu) = n + 2$, we need to prove that $r(\Im_1) = n + 2$, where $r(\Im_1)$ is the rank of the matrix

$$\Im_1 = \begin{pmatrix} \tau E & -E \\ g(x)_{,i} & b_j g(x)_{,ij} \\ 0 & g(x)_{,i} \end{pmatrix}.$$

We have $r(\Im_1) = r(\Im_2)$, where

$$\Im_2 = \begin{pmatrix} 0 & -E \\ g(x)_{,i} + \tau b_j g(x)_{,ij} & b_j g(x)_{,ij} \\ \tau g(x)_{,i} & g(x)_{,i} \end{pmatrix}.$$

Consider two possibilities:

(i) τ, defined by (4.9.15), vanishes identically; and

(ii) τ does not vanish.

Our considerations are of local nature. Therefore, the case when $\tau = 0$ at some $t = t_0$ but does not vanish for other t in a neighborhood of t_0, may be omitted from consideration because the set of these τ has measure zero.

(i) If $\tau \equiv 0$, then it follows from (4.9.14) that $H(b) = \text{const}$; thus by (4.9.8), all the straight lines of S_Λ^* cross the line Λ at a common point. By (4.9.5) the hypersurface S degenerates to this point.

(ii) If $\tau \neq 0$, then $r(\mathfrak{J}_2) = n + 2$ if the two vectors $\mathfrak{g}(x)_{,i}$ and $b_j \mathfrak{g}(x)_{,ij}$ are not parallel. We have $b_j \mathfrak{g}(x)_{,ij} b_i \neq 0$ if $b \neq 0$, $\mathfrak{g}(x)_{,i} b_i = 0$ by (4.9.6), and $\mathfrak{g}(x)_{,i}$ is a nonzero vector, as it follows from (4.9.14), (4.9.15), and the assumption $\tau \neq 0$. So, the two vectors $\mathfrak{g}(x)_{,i}$ and $b_j \mathfrak{g}(x)_{,ij}$ are not parallel, and we conclude that $r(\mathfrak{J}) = n + 2$. We have proved:

Theorem 4.9.1. *Consider the n-dimensional family S_Λ^* of straight lines given by (4.9.8). Calculate τ by formula (4.9.15). If $\tau \equiv 0$, then all the straight lines of S_Λ^* cross the line Λ in a common point, and S coincides with this point. Otherwise (4.9.5) is a parametric equation of S.*

Exercise 4.9.1. Show that for $n = 2$ the procedure described in Theorem 4.9.1 reduces to the calculation of the Legendre transform described in Theorem 4.3.1.

Hint. See [RZ5]

4.10. Stable calculation of the Legendre transform

4.10.1. Introduction

A method for finding S given $\hat{S}$ can be based on the calculation of the Legendre transform of the function $h(\beta)$, which is the local equation $q = h(\beta)$ of $\hat{S}$. In practice $\hat{f}$ is measured with errors, so $h(\beta)$ is known with errors. The problem is: how does one calculate stably the Legendre transform of a function given the noisy values of this function? In the remaining sections we answer this question.

4.10.2. The Legendre transform

4.10.2.1. Let $f(x)$ be a vector-function with values in a Euclidean space, x being an element of $\mathbb{R}^{n-1}$. Denote the second and third derivatives of this function by $f''(x)$ and $f'''(x)$; these are bilinear and three-linear forms on the tangent bundle of $\mathbb{R}^{n-1}$, respectively. Define the norm of an m-linear form $H(\zeta_1, \ldots, \zeta_m)$, $\zeta_1, \ldots, \zeta_m \in \mathbb{R}^{n-1}$, by the formula

$$\|H(\zeta_1, \ldots, \zeta_m)\| = \sup_{\zeta_i \in \mathbb{R}^{n-1}, \|\zeta_i\| \leq 1} |H(\zeta_1, \ldots, \zeta_m)|.$$

Denote by $B[x; r]$ the ball of radius r with center x in $\mathbb{R}^{n-1}$. For any set $U \subset \mathbb{R}^{n-1}$ we define

$$U_\epsilon = \{x \in U : B[x, \epsilon] \subset U\} \quad \text{for} \quad \epsilon > 0. \tag{4.10.1}$$

Let $U \subset \mathbb{R}^{n-1}$ be an open set, $g(x) \in C^2(U)$, $h(x) = \operatorname{grad} g(x)$. Assume that there exist constants m, M, $0 < m \leq M < \infty$, such that

$$m\|x - y\| \leq \|h(x) - h(y)\| \leq M\|x - y\| \quad \forall x, y \in U, \tag{4.10.2}$$

and the norm is the Euclidean norm in $\mathbb{R}^{n-1}$ throughout the section. In this case $h(x)$ is a diffeomorphism of U onto an open set $V = h(U) \subset \mathbb{R}^{n-1}$, and the Legendre transform $Lg(\beta)$ of $g(x)$ is defined for $\beta \in V$ (see Section 4.6).

Consider the problem of calculating the Legendre transform $Lg(\beta)$ assuming that the function $g(x)$ is given with some error. Assume that the function

$$g_\delta(x) = g(x) + \psi(x), \quad \sup_{x \in U} |\psi(x)| \leq \delta \tag{4.10.3}$$

is known, where $\psi(x) \in C(U)$. The question is: *how does one calculate stably $Lg(\beta)$ given $g_\delta(x)$?* The function $g_\delta(x)$ is not assumed differentiable, so $\operatorname{grad} g_\delta(x)$ is not defined, in general. Even if it is defined, it may differ from $\operatorname{grad} g(x)$ very much.

The following example shows that a small perturbation of $g(x)$ may change V, the domain of definition of the Legendre transform: for instance, an open subset W of V may disappear.

Example 4.10.1. Let $U = \{x = (x_1, x_2) \in \mathbb{R}^2 : \|x\| < 2\}$, $g(x) :$ $U \to \mathbb{R}$ is given by $g(x) = (x_1 - 1)^2 - x_2^2$. Let T denote the subset of U given by $\{x_1 = x_2, x_1 \geq 1\}$. Define a neighborhood T_ϵ of T in the following way. Let $\rho_\epsilon(\phi) \geq 0$ be a smooth function defined on $[0, 2\pi]$ and vanishing outside $\{\phi : |\phi - \pi/4| < \epsilon\}$, such that $\rho_\epsilon(\pi/4) = 1 + \epsilon$, and $\pi/4$ is the point of maximum of the function $\rho_\epsilon(\phi)$. Now let T_ϵ be given in polar coordinates (r, ϕ) by the inequality $R > 2 - \rho_\epsilon(\phi)$.

Note first that U and $U \setminus \bar{T}_\epsilon$ are homeomorphic. Moreover, there exists a diffeomorphism $\chi_\epsilon : U \setminus \bar{T}_\epsilon \to U$ for which the points $x \in U \setminus \bar{T}_{2\epsilon}$ are fixed points. One can define χ_ϵ, for example, by the formula $\chi_\epsilon(r, \phi) = (r + \rho_\epsilon(\phi)\sigma_\epsilon(r - 2 + \rho_\epsilon(\phi)), \phi)$, where $\sigma_\epsilon(r)$ is a smooth monotonous function such that $\sigma_\epsilon(r) = 0$ for $r < -\epsilon$ and $\sigma_\epsilon(r) = 1$ for $r > 0$.

Secondly, note that the function $g_\epsilon(x) := g(\chi_\epsilon^{-1}(x))$ satisfies the inequality $|g_\epsilon(x) - g(x)| \leq c_1 \epsilon$. Indeed, one can check that $\max_{x \in T_{2\epsilon}} |g(x)| \leq c\epsilon$, and the claim follows from the definition of χ_ϵ and $f_\epsilon(x)$ with $c_1 = 2c$.

Finally, we claim that $g_\epsilon(x)$ has no critical points in U. Indeed, suppose that x_0 is a critical point of $g_\epsilon(x)$. Then, since χ_ϵ is a diffeomorphism, one concludes that $\chi_\epsilon^{-1}(x_0)$ is a critical point of $g(x)$, i.e., $\chi_\epsilon^{-1}(x_0) = (1, 1)$. But by the definition of $\chi_\epsilon(x)$, this equation has no solutions.

Therefore, the domain of definition of the Legendre transform of $g_\epsilon(x)$ contains neither the point $\beta = 0$, nor a neighborhood W of this point.

4.10.2.2. Some lemmas. We need several lemmas in which the notation (4.10.1) is used.

Lemma 4.10.1. *Let $e_1, \ldots, e_{n-1}$ be the canonical orthogonal basis in $\mathbb{R}^{n-1}$, let $t(\delta) = \sqrt{2\delta/M}$, and*

$$h_\delta(x) = \sum_{i=1}^{n-1} \frac{g_\delta(x + t(\delta)e_i) - g_\delta(x - t(\delta)e_i)}{2t(\delta)} e_i.$$

Then for every $x \in U_{t(\delta)}$ the following estimate holds:

$$\|h_\delta(x) - h(x)\| \le \sqrt{2M(n-1)\delta}.$$

Proof. The conclusion of this lemma follows from the results in Section 14.8.1.

Lemma 4.10.2. *For every positive t such that U_t is not empty, one has $h(U_t) \supset V_{Mt}$, where $V = h(U)$.*

Proof. Choose an arbitrary $\beta_0 \in V_{Mt}$, and let $x_0 = h^{-1}(\beta_0)$. Obviously

$$h(U \cap B[x_0; t]) \subset B[\beta_0; Mt] \subset V.$$

Let us prove that $B[x_0; t] \subset U$. Suppose the contrary. Then there exist a sequence of $x_j \in U \cap B[x_0; t]$, $j = 1, 2, \ldots$, and $x \in \mathbb{R}^{n-1} \setminus U$ such that $\lim_{j\to\infty} x_j = x$. The sequence $h(x_j)$ is contained in the ball $B[\beta_0; Mt]$ which is a compact subset of $\mathbb{R}^{n-1}$. Without loss of generality we can assume that the sequence $h(x_j)$ converges to some $\beta \in B[\beta_0; Mt]$. Since $\beta \in V$, one can define $y = h^{-1}(\beta) \in U$, and get

$$\|y - x_j\| \le \frac{1}{m}\|h(y) - h(x_j)\| \to 0 \quad \text{as } j \to \infty.$$

Thus, $\lim_{j\to\infty} x_j = y \in U$. Since $\lim_{j\to\infty} x_j = x \notin U$, one gets a contradiction. The lemma is proved. $\square$

Lemma 4.10.3. *Under the assumptions (4.10.2) and (4.10.3) one has*

$$h_\delta(U_{t(\delta)}) \supset V_{(1+\sqrt{n-1})\sqrt{2M\delta}}.$$

Proof. Choose an arbitrary $\beta \in V_{(1+\sqrt{n-1})\sqrt{2M\delta}}$, and consider the mapping $\sigma : x \to h^{-1}(\beta - h_\delta(x) + h(x))$. Since $B[\beta; \sqrt{2(n-1)M\delta}] \subset V_{\sqrt{2M\delta}} = V_{Mt(\delta)}$, by Lemma 4.10.2 one has $h^{-1}\left(B[\beta; \sqrt{2(n-1)M\delta}]\right) \subset U_{t(\delta)}$. On the other hand, by Lemma 4.10.1, we have

$$\sup_{x \in U_{t(\delta)}} \|h_\delta(x) - h(x)\| \le \sqrt{2M(n-1)\delta},$$

and hence $\sigma(U_{t(\delta)}) \subset h^{-1}\left(B[\beta; \sqrt{2(n-1)M\delta}]\right)$.

So, σ maps continuously a homeomorphic image of the ball

$$h^{-1}\left(B[\beta; \sqrt{2(n-1)M\delta}]\right)$$

onto itself. By the Brouwer theorem, σ has a fixed point $x_\delta \in U_{t(\delta)}$. Clearly $h_\delta(x_\delta) = \beta$. This proves Lemma 4.10.3. $\square$

Lemma 4.10.4. *For every $\beta \in V_{(1+\sqrt{n-1})\sqrt{2M\delta}}$ the set $h_\delta^{-1}(\beta)$ is nonempty, and one has*

$$h_\delta^{-1}(\beta) \subset B\left[h^{-1}(\beta); \frac{\sqrt{2M(n-1)\delta}}{m}\right].$$

Proof. The set $h_\delta(\beta)$ is nonempty by Lemma 4.10.4. If $x \in h_\delta^{-1}(\beta)$, then $x = h^{-1}(\beta - h_\delta(x) + h(x))$, and we have

$$\|x - h^{-1}(\beta)\| = \|h^{-1}(\beta - h_\delta(x) + h(x)) - h^{-1}(\beta)\|$$
$$\le \frac{1}{m}\|\beta - h_\delta(x) + h(x) - \beta\|$$
$$= \frac{1}{m}\|h_\delta(x) - h(x)\| \le \frac{\sqrt{2M(n-1)\delta}}{m}.$$

Lemma 4.10.4 is proved. $\square$

4.10.2.3. The main result. Let us define the function $L_\delta g_\delta(\beta)$ as follows: for every $\beta \in V_{(1+\sqrt{n-1})\sqrt{2M\delta}}$, choose any $x \in U_{t(\delta)}$ such that $h_\delta(x) = \beta$, and set

$$L_\delta g_\delta(\beta) = <\beta, x> -g_\delta(x).$$

Although $L_\delta g_\delta(\beta)$, so defined, may be multivalued, Theorem 4.10.1 below shows that all its values lie in the ball centered at $Lg(\beta)$ with radius $\delta(1 + M^2 m^{-2}(n-1))$.

Theorem 4.10.1. *For every* $\beta \in V_{(1+\sqrt{n-1})\sqrt{2M\delta}}$ *one has*

$$|L_\delta g_\delta(\beta) - Lg(\beta)| < \delta(1 + M^2 m^{-2}(n-1)).$$

Proof. Obviously

$$\left| < \beta, x > -g_\delta(x) - \left(< \beta, h^{-1}(\beta) > -g(h^{-1}(\beta)) \right) \right|$$
$$\leq \left| < \beta, x > -g - \left(< \beta, h^{-1}(\beta) > -g(h^{-1}(\beta)) \right) \right| + |g_\delta(x) - g(x)|.$$

The second term on the right hand side is majorized by δ. To estimate the first term, use the inequality

$$|g(x) - g(y) - < h(y), y - x > | \leq \frac{M}{2}\|y - x\|^2$$

with $y = h^{-1}(\beta)$. One has

$$\left| < \beta, x > -g - \left(< \beta, h^{-1}(\beta) > -g(h^{-1}(\beta)) \right) \right| \leq \frac{M}{2}\|h^{-1}(\beta) - x\|^2.$$

By Lemma 4.10.4,

$$\|x - h^{-1}(\beta)\| \leq \frac{\sqrt{2M(n-1)\delta}}{m}.$$

Therefore,

$$|L_\delta g_\delta(\beta) - Lg(\beta)| < \delta(1 + M^2 m^{-2}(n-1)).$$

Theorem 4.10.1 is proved. □

4.10.2.4. An iterative process. We conclude this section with a discussion of an *iterative process for an approximate calculation* of the solution to the equation $h_\delta(x) = \beta$. This iterative process is of the form

$$x^{j+1} = x^j - A_\delta \left[h_\delta(x^j) - \beta \right], \quad j = 0, 1, \ldots, \qquad (4.10.4)$$

where

$$\|A_\delta - A\| \leq \eta(\delta) \to 0 \quad \text{as } \delta \to 0,$$
$$A = H(x_0), \quad H(x) = [g_{ik}(x)]^{-1},$$
$$g_{,ik} = \frac{\partial^2 g}{\partial x_i \partial x_k}, \quad \|h_\delta(x) - h(x)\| \leq c\delta^{1/2}.$$

Assume that

$$x = h^{-1}(\beta), \quad x^0 \in B[x; r_0] \subset U_\delta,$$
$$\|y - x - A[h(y) - h(x)]\| \leq q\|y - x\|, \quad y \in B[x; r_0], \quad 0 < q < 1.$$

One has

$$
\begin{aligned}
\|x^{j+1} - x^j\| &= \|x^j - x - A_\delta[h_\delta(x^j) - g(x)]\| \\
&\leq \|x^j - x - A[g(x^j) - g(x)]\| \\
&\quad + \|(A_\delta - A)[g(x^j) - g(x)]\| + \|A_\delta[g_\delta(x^j) - g(x^j)]\| \\
&\leq \|x^j - x\| + M\eta(\delta)\|x^j - x\| + c[\|A\| + \eta(\delta)]\delta^{1/2}.
\end{aligned}
$$

Denote $\rho_j = \|x^j - x\|$, $q(\delta) = q + M\eta(\delta)$, $\gamma(\delta) = c[\|A\| + \eta(\delta)]\delta^{1/2}$, then $\rho_{j+1} \leq q(\delta)\rho_j + \gamma(\delta)$. Since $q(\delta) < 1$ for sufficiently small $\delta > 0$, all the sequence $\{x^j\}$ lies in a ball $B[x; r_0]$, provided $r_0 \geq \gamma(\delta)/(1 - q(\delta))$. One has

$$\rho_j \leq \frac{\gamma(\delta)}{1 - q(\delta)} + \rho_0 q^j(\delta).$$

Note that $q_0 \leq r_0$. Choose minimal j such that

$$r_0 q^j(\delta) \leq \frac{\gamma(\delta)}{1 - q(\delta)}. \qquad (4.10.5)$$

Let J denote this minimal j. Then $\rho_J \leq 2\gamma(\delta)/(1 - q(\delta))$. This is the stopping rule for the iterative process (4.10.4) with the error estimate

$$\|x - x^J\| \leq \frac{2\gamma(\delta)}{1 - q(\delta)}.$$

We summarize:

if the initial approximation x_0 is sufficiently close to the solution of the equation $h(x) = \beta$, namely $\|x_0 - x\| \leq r_0$, and $B[x, r_0] \subset U_\delta$, then the iterative process (4.10.4) with the stopping rule (4.10.5) allows one to calculate this solution with the accuracy of order $\sqrt{\delta}$.

4.10.3. Calculation of the generalized Legendre transform

We want to show that the generalized Legendre transform can be calculated stably given the noisy data. In this section we assume that the generalized Legendre transform is defined on an open domain.

4.10.3.1. Definition. Let $\mathcal{M} \subset \mathbb{R}^{n-1}$ be a variety given by the equations

$$x_i = f_i(x_1, \ldots, x_{n-k}), \quad i = n - k + 1, \ldots, n - 1, \qquad (4.10.6)$$

where $1 \leq k \leq n-1$, $f_i(x_1, \ldots, x_{n-k}) \in C^2(U)$, $U \subset \mathbb{R}^{n-k}$ is a domain. Set $\tilde{x}_1 = (x_1, \ldots, x_{n-k})$, $\tilde{x}_2 = (x_{n-k+1}, \ldots, x_{n-1})$, then one can write (4.10.6) as $\tilde{x}_2 = f(\tilde{x}_1)$, $f = (f_{n-k+1}, \ldots, f_{n-1})$. Now, let $g : \mathcal{M} \to \mathbb{R}$ be a C^2-smooth function. The variables $\tilde{x}_1$ are the coordinates in $\mathcal{M}$, so we may assume that g is a function of $\tilde{x}_1$. Let $V_1 \subset \mathbb{R}^{n-k}$, $V_2 \subset \mathbb{R}^{k-1}$ be open sets such that for every $(\tilde{\beta}_1, \tilde{\beta}_2) \in V_1 \times V_2$ the system

$$\text{grad}_{\tilde{x}_1}[g(\tilde{x}_1) - <\tilde{\beta}_2, f(\tilde{x}_1)>] = \tilde{\beta}_1 \qquad (4.10.7)$$

is uniquely solvable in U. Here $< \cdot, \cdot >$ stands for inner product in $\mathbb{R}^{k-1}$. We assume that $\sup_{\tilde{\beta}_2 \in V_2} \|\tilde{\beta}_2\| < \infty$, i.e., V_2 is bounded, and use the notation $N = \sup_{\tilde{\beta}_2 \in V_2} \|\tilde{\beta}_2\|$. Let, for $\tilde{\beta}_2 \in V_2$,

$$h(\tilde{x}_1, \tilde{\beta}_2) = \text{grad}_{\tilde{x}_1}[g(\tilde{x}_1) - <\tilde{\beta}_2, f(\tilde{x}_1)>].$$

Assume that there exist constants m and M, $0 < m \leq M < \infty$, such that for every $\tilde{\beta}_2, \tilde{\gamma}_2 \in V_2$ and $\tilde{x}_1, \tilde{y}_1 \in U$ one has:

$$m\|\tilde{x}_1 - \tilde{y}_1\| \leq \|h(\tilde{x}_1, \tilde{\beta}_2) - h(\tilde{y}_1, \tilde{\beta}_2)\|,$$
$$|h(\tilde{x}_1, \tilde{\beta}_2) - h(\tilde{y}_1, \tilde{\gamma}_2)\| \leq M\|(\tilde{x}_1, \tilde{\beta}_2) - (\tilde{y}_1, \tilde{\gamma}_2)\|. \qquad (4.10.8)$$

In this case, for every $\tilde{\beta}_2 \in V_2$, the map $\tilde{h}(\cdot, \tilde{\beta}_2) : \tilde{x}_1 \to h(\tilde{x}_1, \tilde{\beta}_2)$, with $\tilde{\beta}_2$ fixed, is a diffeomorphism of U on a domain containing V_1, and the generalized Legendre transform $\tilde{L}g(\beta)$ is given by the formula

$$\tilde{L}(\tilde{\beta}_1, \tilde{\beta}_2) = <\tilde{\beta}_1, h(\cdot, \tilde{\beta}_2)^{-1}(\tilde{\beta}_1)> + <\tilde{\beta}_2, f(\tilde{h}(\cdot, \tilde{\beta}_2)^{-1}(\tilde{\beta}_1))>$$
$$- g(h(\cdot, \tilde{\beta}_2)^{-1}(\tilde{\beta}_1)). \qquad (4.10.9)$$

4.10.3.2. Results. The argument below is similar to the one in Section 4.10.2. Assume that the functions $g(\tilde{x}_1)$ and $f_i(\tilde{x}_1)$ are known with some error, i.e., given are the functions

$$g_\delta(\tilde{x}_1), \ f_\delta(\tilde{x}_1) \in C(U), \ f_\delta(\tilde{x}_1) = (f_{i,\delta}(\tilde{x}_1))_{i=n-k+1,\ldots,n-1},$$

such that $\sup_{\tilde{x}_1 \in U} |g_\delta(\tilde{x}_1) - g(\tilde{x}_1)| < \delta$ and $\sup_{\tilde{x}_1 \in U} \|f_\delta(\tilde{x}_1) - f(\tilde{x}_1)\| < \delta$.

The cases $k < n-1$ and $k = n-1$ must be considered separately. First we deal with the case $k < n-1$, $k \geq 1$. As in Lemma 4.10.2, consider the approximate gradient

$$h_\delta(\tilde{x}_1, \tilde{\beta}_2) = \sum_{i=1}^{n-k} \frac{J_{\tilde{\beta}_2,\delta}(\tilde{x}_1 + t(\delta)e_i) - J_{\tilde{\beta}_2,\delta}(\tilde{x}_1 - t(\delta)e_i)}{2t(\delta)} e_i,$$

where $J_{\tilde\beta_2,\delta}(\tilde x_1) = g_\delta(\tilde x_1) - <\tilde\beta_2, f_\delta(\tilde x_1)>$, and $t(\delta) = \sqrt{2\delta/M}$. It follows from Lemma 4.10.1 that

$$\left\| h_\delta(\tilde x_1,\tilde\beta_2) - \tilde h(\tilde x_1,\tilde x_1\beta_2)\right\| \leq \sqrt{2M(n-k-1)\delta}. \qquad (4.10.10)$$

Choose an arbitrary $\tilde\beta_2 \in V_2$ and $t > 0$. Applying Lemma 4.10.2, one proves the inclusion

$$h(\cdot,\tilde\beta_2)(U_t) \supset V_{2,Mt}. \qquad (4.10.11)$$

It follows from Lemma 4.10.3 that for every $\tilde\beta_2 \in V_2$ one has

$$h(\cdot,\tilde\beta_2)(U_{t(\delta)}) \supset V_{2,1+\sqrt{n-k-1})\sqrt{2M\delta}}. \qquad (4.10.12)$$

And, finally, it follows from Lemma 4.10.4 that for every $\beta = (\tilde\beta_1,\tilde\beta_2) \in V_1 \times V_2$, one has

$$\varnothing \neq h_\delta(\cdot,\tilde\beta_2)^{-1}(\tilde\beta_1) \subset B\left[\tilde g(\cdot,\tilde\beta_2)(\tilde\beta_1); \frac{\sqrt{2M(n-k-1)\delta}}{m}\right]. \qquad (4.10.13)$$

Define the approximation of the generalized Legendre transform:

$$(L_\delta g_\delta)(\tilde\beta_1,\tilde\beta_2) = <\tilde\beta_1, h_\delta(\cdot,\tilde\beta_2)^{-1}(\tilde\beta_1)> + <\tilde\beta_2, f_\delta(h_\delta(\cdot,\tilde\beta_2)^{-1}(\tilde\beta_1))>$$
$$- g_\delta(h_\delta(\cdot,\tilde\beta_2)^{-1}(\tilde\beta_1)). \qquad (4.10.14)$$

We may formulate now the generalization of Theorem 4.10.1. It shows that formula (4.10.14) allows one to calculate the generalized Legendre transform with an error of the same order δ as the noise level. Thus, this yields a stable method for calculating the generalized Legendre transform.

Theorem 4.10.2. *Let $1 \leq k < n-1$. Then for every*

$$\beta \in V_{(1+\sqrt{n-k-1})\sqrt{2M\delta}}$$

one has:

$$|L_\delta g_\delta(\beta) - Lg(\beta)| \leq \delta\left(1 + N + \frac{\nu(n-k-1)M}{m^2}\right), \qquad (4.10.15)$$

where $\nu = M_2(g) + NM_2(f)$, $M_2(g) = \sup_{x\in U}\|g''(x)\|$, and $M_2(f) = \sup_{x\in U}\|f''\|$.

Proof. Choose $\beta \in V_{(1+\sqrt{n-k-1})\sqrt{2M\delta}}$, and let $\tilde x_1, \tilde y_1$ be such that $\tilde\beta_1 = h_\delta(\tilde x_1,\tilde\beta_2)$, $\tilde\beta_1 = h(\tilde y_1,\tilde\beta_2)$. Using formulae (4.10.7), (4.10.9), (4.10.13),

(4.10.14), one gets

$$|L_\delta g_\delta(\beta) - Lg(\beta)|$$
$$= |<\tilde{\beta}_1, \tilde{x}_1 - \tilde{y}_1> + <\tilde{\beta}_2, f_\delta(\tilde{x}_1) - f(\tilde{y}_1)>$$
$$\quad - g_\delta(\tilde{x}_1) + g(\tilde{y}_1)|$$
$$\leq |<h(\tilde{y}_1, \tilde{\beta}_2), \tilde{x}_1 - \tilde{y}_1> + <\tilde{\beta}_2, f(\tilde{x}_1) - f(\tilde{y}_1)>$$
$$\quad - g(\tilde{x}_1) + g(\tilde{y}_1)|$$
$$\quad + |<\tilde{\beta}_2, f_\delta(\tilde{x}_1) - f(\tilde{x}_1)>| + |g_\delta(\tilde{x}_1) - g(\tilde{x}_1)|$$
$$\leq |g(\tilde{x}_1) - f(\tilde{y}_1) - <h(\tilde{y}_1), \tilde{x}_1 - \tilde{y}_1>|$$
$$\quad + \left|<\tilde{\beta}_2, f(\tilde{x}_1) - f(\tilde{y}_1) - (\tilde{x}_1 - \tilde{y}_1)\frac{\partial f(\tilde{y}_1)}{\partial \tilde{y}_1}>\right| + N\delta + \delta$$
$$\leq \delta(1 + N) + \frac{M_2(g)}{2}\|\tilde{x}_1 - \tilde{y}_1\|^2 + NM_2(f)\|\tilde{x}_1 - \tilde{y}_1\|^2$$
$$\leq \delta(1 + N + \frac{(M_2(g) + NM_2(f))M(n - k - 1)}{m^2}),$$

and this proves (4.10.15). $\square$

We now proceed to the case $k = n - 1$, which needs special treatment. In this situation we have $\dim \mathcal{M} = 0$, so $\mathcal{M}$ consists of isolated points. Without loss of generality one can assume that $\mathcal{M}$ consists of a single point $\mathcal{M} = \{x_0\}$. Let $g(x_0) = a$, where, as above, $g : \mathcal{M} \to \mathbb{R}$. According to Definition 4.3.2, the generalized Legendre transform in this case is the function $L(\beta) = \beta \cdot x_0 - a$. In the case $k = n - 1$ the problem of calculation of the generalized Legendre transform from noisy data is posed as follows. Suppose that instead of x_0 and a (i.e., instead of the function $g(x)$) given are a vector $x_{0,\delta} \in \mathbb{R}^{n-1}$ and a number $a_\delta \in \mathbb{R}$ (i.e., a function $g_\delta(x)$) such that

$$\|x_0 - x_{0,\delta}\| < \delta, \quad |a_\delta - a| < \delta.$$

Let $V \subset \mathbb{R}^{n-1}$ be a bounded domain, $N = \max_{\beta \in V} \|\beta\|$. Then one can define $L_\delta g_\delta(\beta)$ by the formula $L_\delta g_\delta(\beta) = \beta \cdot x_{0,\delta} - a_\delta$. One easily proves the following analogue of Theorem 4.10.2.

Theorem 4.10.2'. *Suppose $k = n - 1$. Then in the above notations one has*

$$|L_\delta g_\delta(\beta) - Lg(\beta)| < \delta(N + 1).$$

In other words, in this degenerate case also, the generalized Legendre transform may be calculated with error of the same order as the accuracy of the data.

4.10.4. A sufficient condition for (4.10.2)

Let $g(x) \in C^4(\bar{U})$. Assume that $M_m = \sup_{x \in U} \|g^{(m)}(x)\| < \infty$, $m = 2, 3$.

Suppose also that $g''(x)$, considered as an operator acting in $\mathbb{R}^{n-1}$, is invertible for all $x \in U$ and that

$$m_2 = \left(\sup_{x \in U} \|[g''(x)]^{-1}\| \right)^{-1} < \infty.$$

Proposition 4.10.1. *Let* $U \subset \mathbb{R}^{n-1}$ *be a convex bounded domain such that* $\operatorname{diam} U = \sup_{x,y \in U} \|x - y\| < \infty$. *If* $m_2 - (M_3/2) \operatorname{diam} U > 0$, *then* $g(x)$ *is a diffeomorphism of* U *onto* V *and condition (4.10.2) holds with* $m = m_2 - (M_3/2) \operatorname{diam} U$, $M = M_2$.

Proof. Let $h(x) = \operatorname{grad} g(x)$. Then

$$\|h(x) - h(y) - (x - y)g''(y)\| \leq \frac{M_3}{2} \|x - y\|^2.$$

Therefore,

$$\|h(x) - h(y)\| \geq \|(x - y)g''(y)\| - \frac{M_3}{2}\|x - y\|^2$$

$$\geq \left(m_2 - \frac{M_3}{2}\|x - y\|^2 \right) \|x - y\| \geq m\|x - y\|,$$

with $m = m_2 - (M_3/2) \operatorname{diam} U$. The inequality

$$\|h(x) - h(y)\| \leq M_2 \|x - y\|$$

is obvious. This completes the proof. $\square$

CHAPTER 5

LOCAL TOMOGRAPHY

5.1. Introduction

The major disadvantage of the inversion formula (2.2.19) is its *nonlocality*: for the reconstruction of the original function f at a point x one needs to know its Radon transform $\hat{f}(\theta, p)$ for all (θ, p). In other words, one needs to know the integrals of f along all lines intersecting the support of f, even for the lines which are far from the point x. Moreover, in practice, it might be impossible to collect the complete data set – for example, if the object is too big. Suppose now that one is interested in the recovery of f not for all $x \in \text{supp} f$, but for x only in some subset $U \subset \text{supp} f$. Suppose also that one wants not to recover $f(x)$, $x \in U$, pointwise, but to find only locations and values of sharp variations of f (discontinuities or singular support of f). In this case it is possible to propose a technique which recovers this information from only the local data: reconstruction at a given point requires the knowledge of integrals of f along lines intersecting an arbitrary small neighborhood of x. Moreover, these techniques even allow reconstruction of discontinuities of f with better resolution than regular inversion formulas. These techniques are called 'local tomography'.

The basic idea of local tomography is to compute not the original function f, but the result of action of an elliptic pseudodifferential operator (PDO) $\mathcal{B}$ on f, $\mathcal{B}f$. If the operator $\mathcal{B}$ is appropriately chosen, then calculation of $\mathcal{B}f$ is local. Also, if the order of $\mathcal{B}$ is positive (in practice, the order of $\mathcal{B}$ equals 1), then $\mathcal{B}$ is essentially equivalent to taking derivatives and, therefore, discontinuities of f are emphasized in $\mathcal{B}f$. In Section 5.2 we introduce a family of local tomography functions $\psi = \mathcal{B}f$ and show how to construct PDO $\mathcal{B}$, so that calculation of $\mathcal{B}f$ is local. Since, by construction, $\mathcal{B}$ is elliptic, wavefronts of f and ψ are the same, $WF(f) = WF(\psi)$. In Section 5.3 we indicate how to choose the representative of this family which is most noise stable. In Section 5.9 we derive the asymptotics of $\mathcal{B}f$ in a neighborhood of discontinuities of f under the assumption that the amplitude of $\mathcal{B}$ satisfies certain condition at infinity. Using these results, we develop two closely related algorithms for finding values of jumps of f based on local to-mography. One of the algorithms works with already computed values of local tomography function. The second algorithm works in (θ, p) domain and finds values of jumps, basically in one step. These algorithms and the results of their numerical testing are reported in Sections 5.4 and 5.5. Finally, generalizations of local tomography to the exponential and generalized Radon transforms are described in Sections 5.6 and 5.7, respectively. In Section 5.8, local tomography for the limited-angle problem is developed.

5.2. A family of local tomography functions

5.2.1. Definition of a family. Basic property

Multiply Equation (2.1.20) by $(2\pi)^{-n}t^{n-1+m}b(\alpha)\exp(-it\alpha\cdot x)$, where $m > 0$ is an integer,

$$b \in C^\infty(S^{n-1}), \qquad \min_{\alpha \in S^{n-1}} b(\alpha) > 0, \quad b(\alpha) = b(-\alpha), \qquad (5.2.1)$$

and integrate over $(0, \infty)$ with respect to t and over S^{n-1} with respect to α to get:

$$\psi(x) := \mathcal{B}f = \mathcal{F}^{-1}\left(t^m b(\alpha)\tilde{f}(t\alpha)\right)$$

$$= \frac{1}{(2\pi)^n} \int_{S^{n-1}} \int_0^\infty t^{n-1+m} \exp(-it\alpha \cdot x)\tilde{f}(t\alpha)b(\alpha)dt d\alpha$$

$$= \frac{1}{(2\pi)^n} \int_{S^{n-1}} b(\alpha) \int_{-\infty}^\infty \hat{f}(\alpha, p) \int_0^\infty t^{n-1+m} \exp\{it(p - \alpha \cdot x)\}dt dp d\alpha$$

$$= \frac{1}{(2\pi)^n} \int_{S^{n-1}} b(\alpha) \int_{-\infty}^\infty \hat{f}(\alpha, p)\left[i^{n+m}(m+n)!(p - \alpha \cdot x)^{-n-m}\right.$$

$$\left. + (-i)^{n-1+m}\pi\delta^{(n-1+m)}(p - \alpha \cdot x)\right] dp d\alpha,$$

$$(5.2.2)$$

where we have used Equation (2.1.45). *The basic idea of our derivation is very simple: if $n + m$ is an odd integer, then the distribution $(p - \alpha \cdot x)^{-n-m}$ annihilates $\hat{f}(\alpha, p)$ and formula (5.2.2) yields a 'local tomography' formula:*

$$\psi(x) = \frac{\pi i^{n-1+m}}{(2\pi)^n} \int_{S^{n-1}} b(\alpha)\frac{\partial^{n-1+m}\hat{f}(\alpha, p)}{\partial p^{n-1+m}}\bigg|_{p=\alpha\cdot x} d\alpha. \qquad (5.2.3)$$

This formula is the result, which gives the basis for local tomography. Take, for example, $n = 2$ and $m = 1$. Then (5.2.3) yields:

$$\psi(x) = \frac{-1}{4\pi} \int_{S^1} b(\alpha)\hat{f}_{pp}(\alpha, \alpha \cdot x)d\alpha, \quad x \in \mathbb{R}^2. \qquad (5.2.4)$$

By definition (5.2.2), $\psi(x) = \mathcal{F}^{-1}\{t^m b(\alpha)\tilde{f}(t\alpha)\}$. We could take any odd integer m if $n = 2$. We have chosen $m = 1$, the smallest odd positive integer, because this choice minimizes the order of the derivative in formula (5.2.3). Increasing m enhances the sharpness of the discontinuity surfaces and, at the same time, increases the degree of ill-posedness of the problem, since the larger m is, the higher order derivative in formula (5.2.3) will appear. So, there is a trade-off between enhancing the sharpness of the image of the discontinuity and increasing the ill-posedness of the imaging problem, its sensitivity to noise in the data.

We will prove that $WF(f) = WF(\psi)$. This implies that singsupp ψ = singsupp f.

Lemma 5.2.1. *If assumptions (5.2.1) hold then $WF(f) = WF(\psi)$. In particular, singsupp f = singsupp ψ.*

Proof. Take any function $\varphi \in C^\infty(\mathbb{R})$, which is strictly positive outside the origin, vanishes with all derivatives at the origin, and such that $\varphi(t) = 1$, $|t| > c$, for some $c > 0$. Then

$$\psi = \mathcal{F}^{-1}\{\varphi(t)t^m b(\alpha)\tilde{f}(t\alpha)\} + \mathcal{F}^{-1}\{(1-\varphi(t))t^m b(\alpha)\tilde{f}(t\alpha)\} = B_1 f + B_2 f.$$

The operator B_1 has symbol $\varphi(t)t^m b(\alpha)$, $t = |\xi|$, $\alpha = \xi/t$. Assumptions (5.2.1) imply that B_1 is an elliptic PDO of order m. Therefore, B_1 preserves wavefronts. The operator B_2 is smoothing, i.e. $B_2 f \in C^\infty(\mathbb{R}^n)$, which proves the first assertion of the Lemma. Since singsupp f is the natural projection of $WF(f)$ onto its first component $\mathbb{R}^n$, the last conclusion of the Lemma follows. $\square$

REMARK 5.2.1. The operator $f \to \mathcal{F}^{-1}(t^m \mathcal{F} f) := Af$ can be written as a composition of two operators each of which preserves $WF(f)$. This argument is valid for any positive integer n and any integer m. However, the operator A will be local only if $n + m$ is odd. For example, if $n = 2$ and $m = 1$, then $Af = (-\Delta)Tf$, $Tf := \text{const} \int_{\mathbb{R}^2} \frac{f(y)dy}{|x-y|}$, where Δ is the Laplacian. In general, if m is odd, then $Af = (-\Delta)^{m_1}Tf$, where $m = 2m_1 + 1$.

In Section 5.1.4 an elementary proof of the relation $WF(f) = WF(\psi)$ is given. The elementary proof does not use any results from the theory of elliptic PDO.

Formula (5.2.3) yields a function whose singular support is the same as that of $f(x)$ (because $WF(\psi) = WF(f)$). Function (5.2.3) can be calculated from the data $\hat{f}(\alpha, p)$ by means of local (with respect to p) operations: differentiation with respect to p and weighted averaging with respect to α. In this sense one calls (5.2.3) a 'local tomography' formula. In order to find the singular support of $f(x)$, given $\hat{f}(\alpha, p)$, one calculates $\psi(x)$ by formula (5.2.3), and then singsupp ψ = singsupp f.

One can use the choice of $b(\alpha)$ to optimize the formula in some sense (for example, to improve the resolution ability of the method for finding singularities of $f(x)$, based on formula (5.2.3)).

REMARK 5.2.2. Let

$$Bf := \mathcal{F}^{-1}[b(x,t,\alpha)\tilde{f}]$$

be a pseudodifferential operator with a smooth symbol $b(x,t,\alpha), t := |\xi|$, $\alpha := \xi/|\xi|, \xi \in \mathbb{R}^n$. We define

$$A\hat{f} := Bf,$$

and investigate the following problem:

When is A a local operator on $\hat{f}$ in the sense of local tomography?

In other words,

When is A an integral operator with a kernel supported on local to-mographic data, that is in the region:

$$|\alpha \cdot x - p| < d,$$

where d is a given small positive number, and x is a point in the space?

One has

$$Bf = \int_{S^{n-1}} d\alpha \int_{-\infty}^{\infty} ds \hat{f}(\alpha, s) a(x, \alpha, \alpha \cdot x - s) =: A\hat{f}. \qquad (5.2.5)$$

Here

$$a(x, \alpha, p) := \frac{1}{(2\pi)^n} \int_{0}^{\infty} dt t^{n-1} b(x, t, \alpha) \exp(-itp), \qquad (5.2.6)$$

and

$$A\hat{f} = R^*(\hat{f} \circledast a).$$

Our basic result is the following

Theorem 5.2.1. *The operator A is local if and only if the function*

$$Q(x, t, \alpha) := b(x, t, \alpha)t_+^{n-1} + b(x, -t, -\alpha)t_-^{n-1}$$

is an entire function of t of exponential type $\leq d$, where d does not depend on $x \in \mathbb{R}^n$ and $\alpha \in S^{n-1}$. If, in addition, $b(x, t, \alpha)$ is a hypoelliptic symbol, then $WF(A\hat{f}) = WF(f)$.

Proof of this result is given in [R 22]. In this remark we gave a formula for writing a general PDO as an operator on tomographic data. If the symbol $b(x; t, \alpha) = 1$, then $Bf = f$, and formulas (5.2.5) and (5.2.6) give a simple derivation of the inversion formula for the Radon transform by the method of Section 2.2.1.

5.2.2. An elementary proof of the relation
$WF(f) = WF(\psi)$

The purpose of this section is to derive the conclusion of Lemma 5.2.1 without using the results from the theory of PDO. The definition of the wave front of f, $WF(f)$ can be found in Section 14.2.3. For convenience of the reader, we recall this definition. Let D be a domain in $\mathbb{R}^n$, x and ξ are vectors in $\mathbb{R}^n$, $|\xi| > 0$. The wave front $WF(f)$ is a subset of $D \times (\mathbb{R}^n \setminus 0)$, which is defined via its complementary set.

We say that the point (x_0, ξ_0) does not belong to $WF(f)$ if there exist $\epsilon > 0$, a function ϕ with properties $\phi \in C_0^\infty(\mathbb{R}^2)$, $\phi(x_0) \neq 0$, and constants c_N such that the inequalities $|(\mathcal{F}(\phi f))(\xi)| < c_N(1 + |\xi|^2)^{-N}$ hold for all $N > 0$ and all $\xi \neq 0$, $\left|\frac{\xi}{|\xi|} - \frac{\xi_0}{|\xi_0|}\right| < \epsilon$.

Let $f \in L^2(B_a)$, $(x_0, \xi_0) \notin WF(f)$, and $m > -n$. Let us prove that $(x_0, \xi_0) \notin WF(\psi)$. The converse is proved similarly. Let K be a cone with vertex at the origin and ϕ be a function such that $\phi \in C_0^\infty(U)$, $\phi(x) = 1$ for all $x \in U_1$, $x_0 \in U_1 \subset U$, and we have

$$|\mathcal{F}(\phi f)| < c_N(1 + |\xi|^N)^{-1} \qquad (5.2.7)$$

for all $N > 0$ and $\xi \in K$. By c_N and by c we denote below various constants. Let $K_1 \subset K$ be a cone with vertex at the origin, the same axis of symmetry as that of K and half of the opening of K, $\eta \in C_0^\infty(U)$, $\eta = 1$ in U_1, $x_0 \in U_1 \subset U$. We want to prove that

$$|\mathcal{F}(\eta \mathcal{B} f)| \leq c_N(1 + |\xi|^N)^{-1} \quad \text{for all } N > 0, \xi \in K_1. \qquad (5.2.8)$$

Since $\mathcal{B}f = \mathcal{B}[\phi f] + \mathcal{B}[(1-\phi)f]$ and $\mathcal{B}[(1-\phi)f] \in C^\infty(U_1)$, it is sufficient to prove (5.2.8) with ϕf in place of f. One has

$$q(\xi) := \mathcal{F}[\eta \mathcal{B}(\phi f)] = \int \zeta(\xi - \xi')b(\xi'/|\xi'|)|\xi'|^m \mathcal{F}(\phi f)(\xi')d\xi', \quad \int := \int_{\mathbb{R}^n},$$

where $\mathcal{F}(\phi f)$ satisfies (5.2.7) and $\zeta := \mathcal{F}\eta$. Let $\xi \in K_1$, $\xi' \in \mathbb{R}^n \setminus K$. Then $c|\xi| < |\xi - \xi'|$, $c|\xi'| < |\xi - \xi'|$, $c > 0$. Note that $|\mathcal{F}(\phi f)(\xi)| < c \; \forall \xi \in \mathbb{R}^n$. Using Equation (5.2.1) and the fact that $|\zeta(\xi)|$ decays faster than any power of $|\xi|$, one gets:

$$|q(\xi)| \leq c_N \int_K |\xi'|^m (1 + |\xi - \xi'|^N)^{-1}(1 + |\xi'|^{N+n+1+m})^{-1}d\xi'$$

$$+ c_N \int_{\mathbb{R}^n \setminus K} |\xi'|^m |\psi(\xi')|(1 + |\xi - \xi'|^{N+n+1+m})^{-1}d\xi'$$

$$\leq c_N(1 + |\xi|^N)^{-1} + c_N \int_{\mathbb{R}^n \setminus K} (1 + |\xi'|^{n+1})^{-1}(1 + |\xi - \xi'|^N)^{-1}d\xi'$$

$$\leq c_N(1 + |\xi|^N)^{-1}.$$

The proof is complete. It is well known that the relation $WF(f) = WF(\mathcal{B}f)$ implies the relation $\text{singsupp} f = \text{singsupp} \mathcal{B}f$, which is of interest for us.

5.3. Optimization of noise stability

In Section 5.2 the family of local tomography functions (5.2.3) is proposed. Taking $n = 2$ and $m = 1$ in (5.2.3), we get Equation (5.2.4).

Our purpose in this section is to choose among all formulas (5.2.4), in other words, among all b satisfying conditions (5.2.1), the one which is optimal in the sense of robustness towards noise in the data. More precisely, assume that the data $\hat{f}(\theta, p)$ are collected with some noise $\nu(\Theta, p)$. Given the noisy data $\hat{f} + \nu$, one calculates

$$\omega(x) = -\frac{1}{4\pi} \int\limits_{S^1} b(\Theta)[\hat{f}_{pp}(\theta, \Theta \cdot x) + \nu_{pp}(\Theta, \Theta \cdot x)]d\theta. \qquad (5.3.1)$$

We want to choose $b(\Theta)$ so that

$$\overline{|\omega(x) - \psi(x)|^2} = \min, \qquad (5.3.2)$$

where ψ is defined in formula (5.2.4). Here the bar stands for statistical average, and we assume that

$$\overline{\nu(\Theta, p)} = 0, \quad \overline{\nu(\Theta, p)\nu(\Theta', p')} = R(\Theta, \Theta', p, p'), \qquad (5.3.3)$$

where ν is real-valued and $R(\Theta, \Theta', p, p')$ is the covariance function of the noise. Let us assume that the noise is homogeneous, and

$$R(\Theta, \Theta', p, p') = R_1(\Theta - \Theta', p - p'). \qquad (5.3.4)$$

Note that

$$\overline{\nu_{pp}(\Theta, p)\nu_{p'p'}(\Theta', p')} = \frac{\partial^4 R}{\partial p^2 \partial p'^2} = R_1^{(4)}(\Theta - \Theta', p - p'), \qquad (5.3.5)$$

where $R_1^{(4)}$ denotes the fourth order derivative of R_1 with respect to the second argument.

If f is piecewise-smooth, then the function ψ is not in $L_{loc}^1(\mathbb{R}^2)$ (see formula (5.4.4) from Section 5.4). In this case, the functions ψ in (5.2.3) and (5.2.4) should be understood as distributions. For example, $\psi = \mathcal{F}^{-1}(|\xi|\mathcal{F}f)$ is a well defined distribution for such an f, and it coincides with function (5.2.4) for smooth f. However, in practice, one computes not ψ, but $\psi * W$, where $W(x)$ is a mollifier with small support. In this case, Equations (2.1.42) and (5.2.4) yield

$$(\psi * W)(x) = -\frac{1}{4\pi} \int_{S^1} b(\Theta)(\hat{f} \circledast \hat{W})(\theta, \Theta \cdot x) d\theta, \quad \hat{W} = RW,$$

where '$\circledast$' denotes convolution with respect to p only. Thus, the formula for computing $\psi * W$ is of exactly the same form as Equation (5.2.4): the only difference is that $\hat{f}$ is replaced by $\hat{f} \circledast \hat{W}$. Similarly, if we compute not $\omega(x)$ by (5.3.1), but $(\omega * W)(x)$, we should replace both $\hat{f}$ and ν by $\hat{f} \circledast \hat{W}$ and $\nu \circledast \hat{W}$. One can easily check that if the correlation function of ν satisfies (5.3.4), then correlation function of $\nu \circledast \hat{W}$ also depends only on the differences $\Theta - \Theta'$ and $p - p'$.

It is clear from (5.3.1) and (5.3.2) that the optimal $b(\Theta)$ solves the following optimization problem:

$$\int_{S^1} \int_{S^1} \overline{\nu_{pp}(\Theta, \Theta \cdot x)\nu_{p'p'}(\Theta', \Theta' \cdot x)} b(\Theta)b(\Theta') d\theta\, d\theta' = \min, \qquad (5.3.6)$$

where b satisfies constraints (5.2.1) and normalization condition

$$\frac{1}{2\pi} \int_0^{2\pi} b(\Theta) d\theta = 1. \qquad (5.3.7)$$

Equation (5.3.6) can be rewritten as

$$\int_{S^1} \int_{S^1} R_1^{(4)}(\Theta - \Theta', (\Theta - \Theta') \cdot x) b(\Theta)b(\Theta') d\theta\, d\theta' = \min. \qquad (5.3.8)$$

Let us introduce a weight function

$$w(x) > 0, \quad \int_{R^2} w\, dx = 1. \qquad (5.3.9)$$

If we are interested in the optimal estimate not at a particular point x, but globally, with the weight $w(x)$, then (5.3.8) is replaced by

$$\int_{S^1} \int_{S^1} b(\Theta)b(\Theta') \left(\int_{R^2} w(x) R_1^{(4)}(\Theta - \Theta', (\Theta - \Theta') \cdot x) dx \right) d\theta\, d\theta' = \min. \qquad (5.3.10)$$

Thus (5.3.10) should be solved under constraints (5.2.1) and (5.3.7). Let us denote

$$A(\Theta - \Theta') := \int_{R^2} w(x) R_1^{(4)}(\Theta - \Theta', (\Theta - \Theta') \cdot x) dx \qquad (5.3.11)$$

The function b, minimizing (5.3.11), solves equation

$$Ab := \int_{S^1} A(\Theta - \Theta') b(\Theta') d\theta' = \lambda b(\Theta), \qquad (5.3.12)$$

where λ is a constant. According to (5.2.1), we are interested only in positive solutions to (5.3.12). Note that there are two cases when it is a priori known that Equation (5.3.12) is solvable and the solution is positive:

(1) $A(\Theta - \Theta') > 0 \ \forall \Theta, \Theta' \in S^1$, or
(2) $A(\Theta - \Theta') = A(|\Theta - \Theta'|)$.

In the first case, the Krein-Rutman theorem implies that Equation (5.3.12) has a unique (up to a scalar factor) positive eigenfunction such that the corresponding eigenvalue λ is positive. This eigenvalue is the largest eigenvalue of the linear compact operator A considered in the space of continuous on S^1 functions. Condition (5.3.7) plays the role of a normalization condition.

In the second case, we have $A(|\Theta - \Theta'|) = A(2 - 2\cos(\theta - \theta'))$, where $\Theta = (\cos\theta, \sin\theta)$. Therefore, $A(|\Theta - \Theta'|) =: Q(\theta - \theta')$, where $Q(\theta) = Q(-\theta)$ is a 2π-periodic function of θ. The integral equation

$$Qb := \int_0^{2\pi} Q(\theta - \theta') b(\theta') d\theta' = \lambda b(\theta) \qquad (5.3.13)$$

has a complete in $L^2([0, 2\pi])$ system of mutually orthogonal eigenfunctions $\exp(-im\theta)$, $m = 0, \pm 1 \pm 2, \ldots$, and the corresponding eigenvalues are Q_m, the Fourier coefficients of Q,

$$Q_m := \int_0^{2\pi} Q(\theta) \exp(im\theta) d\theta. \qquad (5.3.14)$$

In particular, the only eigenfunction of the operator Q which does not change sign on the interval $[0, 2\pi]$, is the eigenfunction $b(\theta) = \text{const} \neq 0$. One can always assume that this constant is positive, since eigenfunctions are defined up to a constant factor. The normalization condition (5.3.7) requires that this constant be 1.

Let us summarize the results.

Theorem 5.3.1. *If $A(\Theta - \Theta') = A(|\Theta - \Theta'|)$ then the optimal local tomography formula is formula (5.2.4) with $b(\Theta) \equiv 1$, and this formula is unique under the assumption that noise has covariance function satisfying Equation (5.3.4).*

REMARK 5.3.1. Theorem 5.3.1 gives a proof of an optimality of the local tomography formula with $b(\Theta) \equiv 1$ for a wide class of noises. However, it is clear that if $A(\Theta - \Theta') \neq A(|\Theta - \Theta'|)$, then the eigenfunction of the equation, analogous to (5.3.12), is not necessarily $b(\Theta) = \text{const}$.

In general, one has to solve the equation

$$Ab := \int_{S^1} A(\Theta, \Theta')b(\Theta')d\theta' = \lambda b(\Theta), \qquad (5.3.15)$$

where

$$A(\Theta, \Theta') := \int_{R^2} \frac{\partial^4 R(\Theta, \Theta', p, p')}{\partial p^2 \partial p'^2}\Big|_{\substack{p=\Theta \cdot x \\ p'=\Theta' \cdot x}} w(x)dx. \qquad (5.3.16)$$

If (5.3.16) has a unique positive eigenfunction b, then formula (5.2.4) with this $b(\Theta)$ yields an optimal local tomography formula. A sufficient condition for A to have the unique positive eigenfunction corresponding to a positive eigenvalue λ, is positivity of $A(\Theta, \Theta')$, that is $A(\Theta, \Theta') > 0 \; \forall \Theta, \Theta' \in S^1$.

5.4. Algorithm for finding values of jumps of a function using local tomography

5.4.1. Derivation of the algorithm. Basic result

Let us set $b(\alpha) \equiv 1$ in (5.2.4). Then we get the local tomography function f_Λ:

$$f_\Lambda(x) := -\frac{1}{4\pi} \int_{S^1} \hat{f}_{pp}(\theta, \Theta \cdot x)d\theta. \qquad (5.4.1)$$

Using (5.2.2), we see that f_Λ satisfies

$$f_\Lambda(x) = \mathcal{F}^{-1}(|\xi|\tilde{f}(\xi)) = (-\Delta)^{1/2}f, \qquad (5.4.2)$$

where $\Lambda := (-\Delta)^{1/2}$ is the square root of the negative Laplacian, which explains the notation f_Λ. Let $W_\epsilon(x)$ be a sequence of sufficiently smooth mollifiers with the properties:

(a) $W_\epsilon(x)$ is a radial function, $W_\epsilon(x) := W_\epsilon(|x|)$;
(b) $W_\epsilon(x) = 0$, $|x| \geq \epsilon$, $W_\epsilon(x) > 0$, $|x| < \epsilon$; and
(c) $W_\epsilon(x) = \epsilon^{-2}W_1(x/\epsilon)$, $\int_{|x| \leq 1} W_1(x)dx = 1$.

In practice one computes not f_Λ, but its mollification $W_\epsilon * f_\Lambda$. Using (5.4.1), properties (a) – (c), and Equation (2.1.42), we get

$$f_{\Lambda\epsilon}(x) := (W_\epsilon * f_\Lambda)(x) = -\frac{1}{4\pi} \int_{S^1} \int_{\mathbb{R}} w_\epsilon(\Theta \cdot x - p) \hat{f}_{pp}(\theta, p) dp d\theta,$$

where $w_\epsilon = RW_\epsilon$, and an integration by parts yields

$$f_{\Lambda\epsilon}(x) = -\frac{1}{4\pi} \int_{S^1} \int_{\mathbb{R}} w_\epsilon''(\Theta \cdot x - p) \hat{f}(\theta, p) dp d\theta. \qquad (5.4.3)$$

From Equation (5.4.2) we see that f_Λ is the result of the action of a PDO with the symbol $|\xi|$ on f. Let S be a discontinuity curve of f. Pick an arbitrary $x_0 \in S$, so that S is smooth in a neighborhood of x_0. Let $n(x_s)$ be a unit vector perpendicular to S at $x_s \in S$ and $D(x_s)$ be the jump of f at x_s: $D(x_s) = \lim_{t\to 0+}[f(x_s + tn(x_s)) - f(x_s - tn(x_s))]$. Using the results obtained in Section 5.9.3 (see Theorem 5.9.3), the asymptotic formula for f_Λ in a neighborhood of x_0 becomes

$$f_\Lambda(x_s + hn(x_s)) = \frac{D(x_s)}{\pi}\left[\frac{1}{h} + c_1(x_s)\ln|h| + c_2(x_s)\mathrm{sgn}(h) + c_3(x_s, h)\right].$$
$$(5.4.4)$$

Substituting (5.4.4) into the definition of $f_{\Lambda\epsilon}$, using properties (a) – (c) of W_ϵ, and considering $\epsilon \to 0$, we prove

Theorem 5.4.1. *Let W_ϵ be a sequence of mollifiers with properties (a) – (c) (see below Equation (5.4.2)). Pick any $x_0 \in S$ such that S is smooth in a neighborhood of x_0 and $D(x_0) \neq 0$. Denote $f_{\Lambda\epsilon} := W_\epsilon * f_\Lambda$ and $n_0 := n(x_0)$. Then one has*

$$f_{\Lambda\epsilon}(x_0 + qen_0) = \frac{D(x_0)}{\pi}\frac{\psi(q)}{\epsilon}(1 + O(\epsilon)) + \psi_\epsilon(x_0, q) + O(\epsilon \ln \epsilon), \quad \epsilon \to 0,$$
$$(5.4.5)$$

where $\psi_\epsilon(x_0, q) = O(\ln \epsilon)$ is an even function of q, and

$$\frac{\partial}{\partial h}f_{\Lambda\epsilon}(x_0 + hn_0)\bigg|_{h=0} = \frac{D(x_0)}{\pi}\frac{\psi'(0)}{\epsilon^2}(1 + O(\epsilon)), \quad \epsilon \to 0. \qquad (5.4.6)$$

The function $\psi(q)$ used in (5.4.5) and (5.4.6) equals:

$$\psi(q) = \int_{q-1}^{q+1} \frac{w_1(q-t)}{t} dt. \qquad (5.4.7)$$

Proof of Theorem 5.4.1 is given in Sections 5.4.2 and 5.4.3. The reader who is not interested in technical details may skip these sections.

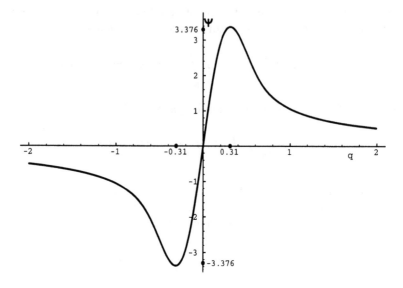

FIGURE 5.4.1.   The graph of $\psi(q)$

In the case of the mollifier

$$W_1(x) = \frac{m+1}{\pi}(1-x^2)^m, \quad m = 8, \tag{5.4.8}$$

the graphs of functions ψ and ψ' are presented in Figures 5.4.1 and 5.4.2, respectively.

5.4.2. Proof of Theorem 5.4.1 in the case of the locally flat S

Fix $x_0 \in S$ and a sufficiently small $\epsilon > 0$ such that S is flat inside ϵ - neighborhood of x_0: $n(x_s) = n(x_0) =: n_0$, $|x_s - x_0| \leq \epsilon$. Using Equation (5.4.4), we obtain

$$f_{\Lambda\epsilon}(x_0 + hn_0) = \int_{\mathbb{R}^2} W_\epsilon((h-t)n_0 + (x_0 - x_s))$$

$$\times \frac{D(x_s)}{\pi}\left[\frac{1}{t} + c_1(x_s)\ln|t| + c_2(x_s)\mathrm{sgn}(t) + c_3(x_s,t)\right]dx_s\,dt. \tag{5.4.9}$$

Let $A(x_s)$ be a smooth function. Then:

$$\int_{\mathbb{R}} W_\epsilon((h-t)n_0 + (x_0 - x_s))A(x_s)dx_s$$

$$= \int_{\mathbb{R}} W_\epsilon((h-t)n_0 + (x_0 - x_s))\big(A(x_0) + A'(x_0)(x_s - x_0) + O(\epsilon^2)\big)dx_s$$

$$= A(x_0)w_\epsilon(h-t) + \chi_\epsilon(h-t), \tag{5.4.10}$$

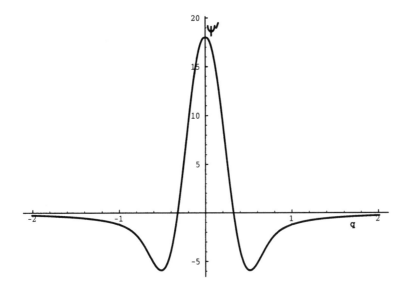

FIGURE 5.4.2. The graph of $\psi'(q)$

where we have used that x_0 is fixed, W_ϵ is radial, and denoted

$$\chi_\epsilon(h - t) := \int_{\mathbb{R}} W_\epsilon((h - t)n_0 + (x_0 - x_s))O(\epsilon^2)dx_s. \qquad (5.4.11)$$

A similar result holds for $A(x_s, t)$, where A is a smooth function of x_s and a bounded function of t. Using Equation (5.4.10) in (5.4.9) yields

$$f_{\Lambda\epsilon}(x_0 + hn_0)$$

$$= \frac{D(x_0)}{\pi} \int_{h-\epsilon}^{h+\epsilon} w_\epsilon(h - t)\left[\frac{1}{t} + c_1(x_0) \ln|t| + c_2(x_0)\mathrm{sgn}(t) + c_3(x_0, t)\right]dt$$

$$+ \int_{h-\epsilon}^{h+\epsilon} \chi_\epsilon(h - t)\frac{1 + O(t \ln|t|)}{t}dt. \qquad (5.4.12)$$

Since $w_\epsilon = RW_\epsilon$, we get

$$w_\epsilon(p) = \int_{\mathbb{R}} W_\epsilon(\sqrt{p^2 + t^2})dt = \frac{1}{\epsilon^2} \int_{\mathbb{R}} W_1\left(\sqrt{\frac{p^2 + t^2}{\epsilon^2}}\right)dt$$

$$= \frac{1}{\epsilon} \int_{\mathbb{R}} W_1\left(\sqrt{\frac{p^2}{\epsilon^2} + t^2}\right)dt = \frac{1}{\epsilon}w_1\left(\frac{p}{\epsilon}\right), \qquad (5.4.13)$$

and substitution into (5.4.12) gives

$$f_{\Lambda \epsilon}(x_0 + q\epsilon n_0) = \frac{D(x_0)}{\pi} \int_{q-1}^{q+1} w_1(q - p)$$

$$\times \left[\frac{1}{\epsilon p} + c_1(x_0) \ln |\epsilon p| + c_2(x_0)\mathrm{sgn}(p) + c_3(x_0, \epsilon p) \right] dp + O(\epsilon \ln \epsilon),$$

$$(5.4.14)$$

where we put $h = q\epsilon$, $q = $ const, $\epsilon \to 0$. In Equation (5.4.14) we have used the relation

$$\int_{h-\epsilon}^{h+\epsilon} \chi_\epsilon(h - t) \frac{1 + O(t \ln |t|)}{t} dt = O(\epsilon \ln \epsilon). \qquad (5.4.15)$$

Clearly, it is sufficient to prove that $\int_{h-\epsilon}^{h+\epsilon} \chi_\epsilon(h - t) \frac{dt}{t} = O(\epsilon \ln \epsilon)$. Differentiating on both sides of (5.4.11) and using the scaling property of the mollifier, we compute

$$\chi_\epsilon'(p) := \int_{\mathbb{R}} \frac{\partial}{\partial p} W_\epsilon(pn_0 + (x_0 - x_s))O(\epsilon^2)dx_s$$

$$= \int_{\mathbb{R}} \frac{\partial}{\partial q} W_1(qn_0 + (\frac{x_0}{\epsilon} - u))O(1)du \bigg|_{q=\frac{p}{\epsilon}} = O(1), \quad \epsilon \to 0. \qquad (5.4.16)$$

Also, $\chi_\epsilon(-\epsilon) = \chi_\epsilon(\epsilon) = 0$. Therefore,

$$\int_{h-\epsilon}^{h+\epsilon} \chi_\epsilon(h - t)\frac{dt}{t} = \int_{h-\epsilon}^{h+\epsilon} \chi_\epsilon'(h - t) \ln |t| dt$$

$$= \int_{h-\epsilon}^{h+\epsilon} O(1) \ln |t| dt = O(\epsilon \ln \epsilon), \quad \epsilon \to 0.$$

Let us assume that the following condition is satisfied

$$\int_{q-1}^{q+1} w_1(q - t)t^{-1}dt \neq 0, \quad 0 < |q| \leq 1. \qquad (5.4.17)$$

In this case, we readily establish for any Q, $0 < Q < \infty$, the following estimate:

$$\max_{|q| \leq Q < \infty} \frac{\int_{q-1}^{q+1} w_1(q - t)\mathrm{sgn}(t)dt}{\int_{q-1}^{q+1} w_1(q - t)t^{-1}dt} < \infty.$$

Exercise 5.4.1. Prove the above relation using (5.4.17).

The following exercise shows that assumption (5.4.17) is not very restrictive.

Exercise 5.4.2. Prove that (5.4.17) holds if the mollifier $w_1(q)$ is even, nonnegative, and nonincreasing on $[0, 1]$.

Note that evenness and nonnegativity of $w_1(q)$ follow from properties (a) – (c) (see below (5.4.2)).

After simple transformations, Equation (5.4.14) yields

$$
f_{\Lambda\epsilon}(x_0 + q\epsilon n_0)
$$
$$
= \frac{D(x_0)}{\pi} \left(\frac{\psi(q)}{\epsilon}(1 + O(\epsilon)) + c_1(x_0)\psi_{1\epsilon}(q) + \psi_{3\epsilon}(x_0, q) \right) + O(\epsilon \ln \epsilon),
$$
$$
\tag{5.4.18}
$$

where

$$
\psi(q) := \int\limits_{q-1}^{q+1} \frac{w_1(q-p)}{p}\, dp, \quad \psi_{1\epsilon}(q) = \int\limits_{q-1}^{q+1} w_1(q-p)\ln|\epsilon p|\, dp,
$$

$$
\psi_{3\epsilon}(x_0, q) = \int\limits_{q-1}^{q+1} w_1(q-p)c_3(x_0, \epsilon p)\, dp. \tag{5.4.19}
$$

Equations (5.9.29) imply

$$
\psi_{3\epsilon}(x_0, q) - \psi_{3\epsilon}(x_0, 0) = O(\epsilon \ln \epsilon), \quad |q| < \infty, \quad \epsilon \to 0. \tag{5.4.20}
$$

Since $w_1(q)$ is even, $\psi_{1\epsilon}(q)$ is also even in q. Denoting

$$
\psi_\epsilon(x_0, q) = \frac{D(x_0)}{\pi}(c_1(x_0)\psi_{1\epsilon}(q) + \psi_{3\epsilon}(x_0, 0)),
$$

and using Equations (5.4.18), (5.4.20), we obtain

$$
f_{\Lambda\epsilon}(x_0 + q\epsilon n_0) = \frac{D(x_0)}{\pi}\frac{\psi(q)}{\epsilon}(1 + O(\epsilon)) + \psi_\epsilon(x_0, q) + O(\epsilon \ln \epsilon), \quad \epsilon \to 0,
$$

which proves Equation (5.4.5).

Recall that in our derivation we assume $h = q\epsilon, q = \text{const}, \epsilon \to 0$. Now, using (5.9.28) and (5.4.19), we find

$$\frac{\partial}{\partial h} f_{\Lambda\epsilon}(x_0 + hn_0)$$

$$= \int_{\mathbb{R}^2} W_\epsilon((h-t)n_0 + (x_0 - x_s))$$

$$\times \frac{D(x_s)}{\pi}\left[-\frac{1}{t^2} + \frac{c_1(x_s)}{t} + 2c_2(x_s)\delta(t) + \frac{\partial}{\partial t}c_3(x_s,t)\right]dx_s\,dt$$

$$= \frac{D(x_0)}{\pi}\int_{h-\epsilon}^{h+\epsilon} w_\epsilon(h-t)\left[-\frac{1}{t^2} + \frac{c_1(x_0)}{t} + 2c_2(x_0)\delta(t) + \frac{\partial}{\partial t}c_3(x_0,t)\right]dt$$

$$+ \int_{h-\epsilon}^{h+\epsilon} \chi_\epsilon(h-t)\frac{1+O(t)}{t^2}dt$$

$$= \frac{D(x_0)}{\pi}\left(\frac{\psi'(q)}{\epsilon^2}(1+O(\epsilon)) + c_1(x_0)\psi'_{1\epsilon}(q) + O(\ln\epsilon)\right) + O(\ln\epsilon),$$

$$h = q\epsilon,\ \epsilon \to 0, \tag{5.4.21}$$

because, similarly to (5.4.16), we get

$$\chi''_\epsilon(p) = \int_{\mathbb{R}} \frac{\partial^2}{\partial p^2}W_\epsilon(pn_0 + (x_0 - x_s))O(\epsilon^2)dx_s = O(\epsilon^{-1}),\ \epsilon \to 0,$$

and

$$\int_{h-\epsilon}^{h+\epsilon} \chi_\epsilon(h-t)\frac{dt}{t^2} = -\int_{h-\epsilon}^{h+\epsilon} \chi''_\epsilon(h-t)\ln|t|dt$$

$$= \int_{h-\epsilon}^{h+\epsilon} O(\epsilon^{-1})\ln|t|dt = O(\ln\epsilon),\ \epsilon \to 0. \tag{5.4.22}$$

Since $w_1(q)$ is even, we get from (5.4.19) $\psi'(0) \neq 0$ and $\psi'_{1\epsilon}(0) = 0$, and Equation (5.4.21) yields

$$\frac{\partial}{\partial h} f_{\Lambda\epsilon}(x_0 + hn_0)\bigg|_{h=0} = \frac{D(x_0)}{\pi}\frac{\psi'(0)}{\epsilon^2}(1 + O(\epsilon)),\quad \epsilon \to 0,$$

and Equation (5.4.6) is also proved. □

5.4.3. Proof of Theorem 5.4.1 in case of the convex S

Introduce the cartesian coordinate system, the origin of which coincides with x_0, and x_1-axis points in the same direction as n_0. Let $x_1 = \varphi(x_2)$ be the local equation of S in a neighborhood of x_0. Suppose, for example, that n_0 points from the center of curvature of S towards x_0. The case when n_0 points in the opposite direction can be considered analogously. Clearly, $\varphi(0) = \varphi'(0) = 0$ and $\varphi''(0) = -R_0^{-1} \neq 0$, where R_0 is the radius of curvature of S at x_0. If $x_s = (\varphi(x_2), x_2)$ is a point on S, then

$$n(x_s) = \frac{1}{\sqrt{1 + (\varphi'(x_2))^2}}(1, -\varphi'(x_2)). \qquad (5.4.23)$$

For each point $x \in \mathbb{R}^2$, which is sufficiently close to x_0, we have a unique representation

$$x = x_s + hn(x_s), \quad x_s \in S. \qquad (5.4.24)$$

Using (5.4.24), we introduce the curvilinear cordinate system, the origin of which coincides with x_0, the first coordinate is h, and the second coordinate is the directed distance $u(x_s)$ from x_0 to x_s along S:

$$u(x_s) = \int_0^{x_2} \sqrt{1 + (\varphi'(x_2))^2}\,dx_2, \quad x_s = (\varphi(x_2), x_2). \qquad (5.4.25)$$

First, we find the Jacobian J of the transformation $(x_1, x_2) \to (h, u)$. In the coordinate form, Equation (5.4.24) is equivalent to

$$x_1 = \varphi(y) + h\frac{1}{\sqrt{1 + (\varphi'(y))^2}},$$
$$x_2 = y + h\frac{-\varphi'(y)}{\sqrt{1 + (\varphi'(y))^2}}, \qquad (5.4.26)$$

where $y = y(u)$ is chosen so that

$$u = \int_0^y \sqrt{1 + (\varphi'(t))^2}\,dt. \qquad (5.4.27)$$

Everywhere below in this section, we drop the arguments of φ and y, assuming that they are y and u, respectively. Derivatives of φ and y are taken with respect to y and u, respectively. Differentiation of (5.4.26) with respect to h and u yields

$$\frac{\partial x_1}{\partial h} = \frac{1}{\sqrt{1 + (\varphi')^2}}, \quad \frac{\partial x_1}{\partial u} = \left(1 - h\frac{\varphi''}{(1 + (\varphi')^2)^{3/2}}\right)\varphi'y';$$
$$\frac{\partial x_2}{\partial h} = -\frac{\varphi'}{\sqrt{1 + (\varphi')^2}}, \quad \frac{\partial x_2}{\partial u} = \left(1 - h\frac{\varphi''}{(1 + (\varphi')^2)^{3/2}}\right)y'.$$

Hence,

$$\det(J) = \frac{y'}{\sqrt{1+(\varphi')^2}}\left(1 - h\frac{\varphi''}{(1+(\varphi')^2)^{3/2}}\right)(1+(\varphi')^2)$$

$$= y'\sqrt{1+(\varphi')^2}\left(1 - h\frac{\varphi''}{(1+(\varphi')^2)^{3/2}}\right).$$

Differentiating (5.4.27) with respect to y, we get $\frac{\partial u}{\partial y} = \sqrt{1+(\varphi')^2}$. Therefore, $\frac{\partial y}{\partial u} = (1+(\varphi')^2)^{-1/2}$, and

$$\det(J) = 1 - h\frac{\varphi''}{(1+(\varphi')^2)^{3/2}}. \tag{5.4.28}$$

Using (5.4.26), let us find the distance d between two points in the coordinates h and u. If a point is on the x_1-axis, then

$$(\tilde{h}, 0)_{\text{in } (x_1,x_2)\text{-coordinates}} \longleftrightarrow (\tilde{h}, 0)_{\text{in } (h,u)\text{-coordinates}}.$$

Hence

$$d^2 = (\tilde{h} - x_1)^2 + x_2^2$$

$$= \left(\tilde{h} - (\varphi + h\frac{1}{\sqrt{1+(\varphi')^2}})\right)^2 + \left(y - h\frac{\varphi'}{\sqrt{1+(\varphi')^2}}\right)^2$$

$$= \tilde{h}^2 - 2\tilde{h}(\varphi + h\frac{1}{\sqrt{1+(\varphi')^2}}) + \varphi^2 + \frac{2h\varphi}{\sqrt{1+(\varphi')^2}} + \frac{h^2}{1+(\varphi')^2}$$

$$+ y^2 - \frac{2yh\varphi'}{\sqrt{1+(\varphi')^2}} + \frac{h^2(\varphi')^2}{1+(\varphi')^2}$$

$$= (\tilde{h} - h)^2 + y^2 + 2\left[\tilde{h}h - \tilde{h}\varphi + \frac{-\tilde{h}h + \varphi h - yh\varphi'}{\sqrt{1+(\varphi')^2}}\right] + \varphi^2. \tag{5.4.29}$$

Suppose the points $(\tilde{h}, 0)$ and (x_1, x_2) are within the distance $O(\epsilon)$ from the origin. Then (5.4.27) yields $y = u + O(u^3)$. Therefore, $\varphi = O(\epsilon^2)$, $(1+(\varphi')^2)^{-1/2} = 1 + O(\epsilon^2)$, and substituting into (5.4.29), we obtain

$$d^2 = (\tilde{h} - h)^2 + u^2(1 + O(\epsilon)^2)$$

$$+ 2[\tilde{h}h - \tilde{h}\varphi + (-\tilde{h}h + \varphi h - yh\varphi')(1 + O(\epsilon^2))] + O(\epsilon^4)$$

$$= (\tilde{h} - h)^2 + u^2 + 2[(h - \tilde{h})\varphi - uh\varphi'] + O(\epsilon^4).$$

Note that the term in brackets on the right-hand side of the last equation is of order $O(\epsilon^3)$. Using that $\varphi(y) = -\frac{1}{2R_0}y^2 + O(y^3)$ and $y = u + O(u^3)$, we finally get

$$d^2 = (\tilde{h} - h)^2 + u^2\left(1 + \frac{\tilde{h} + h}{R_0}\right) + O(\epsilon^4). \tag{5.4.30}$$

Let us consider the leading term of the convolution $W_\epsilon * f_\Lambda \sim W_\epsilon * \frac{D(x_s)}{\pi h}$. Other less singular terms of f_Λ can be considered analogously. Since W_ϵ is radial: $W_\epsilon(x) := W_\epsilon(|x|)$, we have in the curvilinear coordinate system

$$f_{\Lambda\epsilon}(x_0 + \tilde{h}n_0) \sim \int_{\mathbb{R}^2} W_\epsilon\left(\sqrt{(\tilde{h} - h)^2 + u^2\left(1 + \frac{\tilde{h} + h}{R_0}\right) + O(\epsilon^4)}\right)$$

$$\times \frac{D(u)}{\pi h}\left(1 - h\frac{\varphi''}{(1 + (\varphi')^2)^{3/2}}\right) du\, dh.$$

$$(5.4.31)$$

First, consider the integral

$$I := \int_{\mathbb{R}^2} W_\epsilon\left(\sqrt{(\tilde{h} - h)^2 + u^2\left(1 + \frac{\tilde{h} + h}{R_0}\right) + O(\epsilon^4)}\right)\frac{D(u)}{h} du\, dh$$

$$= \frac{1}{\epsilon}\int_{\mathbb{R}^2} W_1\left(\sqrt{(\tilde{q} - q)^2 + s^2\left(1 + \epsilon\frac{\tilde{q} + q}{R_0}\right) + O(\epsilon^2)}\right)\frac{D(\epsilon s)}{q} ds\, dq.$$

$$(5.4.32)$$

Let us suppose for simplicity that $W_1(r) = \Phi(r^2)$ for some smooth Φ. Note that this is frequently the case in practice (see e.g. mollifier (5.2.22)). Simple inspection of the passage from (5.4.29) to (5.4.30) shows that $O(\epsilon^2)$ in (5.4.32) is smooth in q and $\frac{\partial}{\partial q}O(\epsilon^2) = O(\epsilon^2)$. Hence, integration by parts in q implies

$$I = \frac{1}{\epsilon}\int_{\mathbb{R}^2} W_1\left(\sqrt{(\tilde{q} - q)^2 + s^2\left(1 + \epsilon\frac{\tilde{q} + q}{R_0}\right)}\right)\frac{D(\epsilon s)}{q} ds\, dq + O(\epsilon). \quad (5.4.33)$$

Similarly to (5.4.10), consider the integral with respect to s:

$$\int_{\mathbb{R}} W_1\left(\sqrt{(\tilde{q} - q)^2 + s^2\left(1 + \epsilon\frac{\tilde{q} + q}{R_0}\right)}\right) D(\epsilon s) ds$$

$$= D(0)\int_{\mathbb{R}} W_1\left(\sqrt{(\tilde{q} - q)^2 + s^2\left(1 + \epsilon\frac{\tilde{q} + q}{R_0}\right)}\right) ds + \chi_\epsilon(\tilde{q}, q)$$

$$= \frac{D(0)}{\sqrt{1 + \epsilon\frac{\tilde{q} + q}{R_0}}} w_1(\tilde{q} - q) + \chi_\epsilon(\tilde{q}, q), \quad (5.4.34)$$

where

$$\chi_\epsilon(\tilde{q}, q) = \int_{\mathbb{R}} W_1\left(\sqrt{(\tilde{q} - q)^2 + s^2\left(1 + \epsilon\frac{\tilde{q} + q}{R_0}\right)}\right)$$

$$\times (D(\epsilon s) - D(0) - \epsilon s D'(0)) ds.$$

$$(5.4.35)$$

Since $\frac{\partial}{\partial q}\chi_\epsilon(\tilde{q}, q) = O(\epsilon^2)$, integrating by parts we conclude

$$\frac{1}{\epsilon} \int_{\mathbb{R}} \frac{\chi_\epsilon(\tilde{q}, q)}{q} dq = O(\epsilon).$$

Therefore, substituting (5.4.34) into (5.4.33), we get

$$I = \frac{D(0)}{\epsilon} \int_{\mathbb{R}} \frac{w_1(\tilde{q} - q)}{q} \left(1 - \epsilon\frac{\tilde{q} + q}{2R_0}\right) dq + O(\epsilon)$$

$$= \frac{D(0)}{\epsilon} \left(1 - \epsilon\frac{\tilde{q}}{2R_0}\right) \psi(\tilde{q}) - \frac{D(0)}{2R_0} + O(\epsilon), \qquad (5.4.36)$$

because $\int_{\mathbb{R}} w_1(\tilde{q} - q) dq = 1$.

Returning to Equation (5.4.31), we see that it remains to estimate the integral

$$J := \int_{\mathbb{R}^2} W_\epsilon \left(\sqrt{(\tilde{h} - h)^2 + u^2\left(1 + \frac{\tilde{h} + h}{R_0}\right) + O(\epsilon^4)} \right)$$

$$\times D(u) \left(-\frac{\varphi''}{(1 + (\varphi')^2)^{3/2}} \right) du\, dh.$$

This is not difficult. Indeed, using the scaling as in Equation (5.4.32) and properties of φ, we derive

$$J := \int_{\mathbb{R}^2} W_\epsilon \left(\sqrt{(\tilde{q} - q)^2 + s^2\left(1 + \epsilon\frac{\tilde{q} + q}{R_0}\right) + O(\epsilon^2)} \right) \frac{D(\epsilon s)}{R_0} ds\, dq + O(\epsilon)$$

$$= \frac{D(0)}{R_0} \int_{\mathbb{R}} w_1(\tilde{q} - q)\left(1 - \epsilon\frac{\tilde{q}}{2R_0}\right) dq + O(\epsilon) = \frac{D(0)}{R_0} + O(\epsilon).$$
$$(5.4.37)$$

Substituting (5.4.36) and (5.4.37) into (5.4.31), we obtain

$$f_{\Lambda\epsilon}(x_0 + \tilde{q}\epsilon n_0) \sim \frac{D(x_0)}{\pi}\frac{\psi(\tilde{q})}{\epsilon}(1 + O(\epsilon)) + \psi_\epsilon(x_0, \tilde{q}) + O(\epsilon), \quad (5.4.38)$$

where $\psi_\epsilon(x_0, \tilde{q}) = D(x_0)/(2R_0)$ is even in $\tilde{q}$. We see that Equations (5.4.5) and (5.4.38) are in agreement. Considering lower order terms of f_Λ, we prove Equation (5.4.5) for the case of the convex S.

Exercise 5.4.3. Following the approach developed in this section, prove (5.4.6).

The proof of Theorem 5.4.1 is complete. $\square$

5.5. Numerical implementation

5.5.1. The first numerical scheme for computing values of jumps

First, let us describe an algorithm which works in the x-domain with the already computed local tomography function $f_{\Lambda\epsilon}(x)$. We assume that values of $f_{\Lambda\epsilon}$ are calculated on a square grid with step size h: $x_{ij} = (x_i^{(1)}, x_j^{(2)}) = (ih, jh)$, $i,j \in \mathbb{Z}$. Let us choose a grid node $x_{i_0 j_0}$ on S. Assuming that h and ϵ are sufficiently small, we can neglect the change of $f_{\Lambda\epsilon}$ in the direction parallel to S. In particular, we assume $D(x_s) = D(x_{i_0 j_0})$ if $|x_s - x_{i_0 j_0}| \leq \epsilon$. Thus we can rewrite (5.4.5) as

$$f_{\Lambda\epsilon}(x) \approx \frac{D(x_{i_0 j_0})}{\pi\epsilon} \psi\left(\frac{x - x_{i_0 j_0}}{\epsilon} \cdot n_0\right) + \psi_\epsilon\left(x_{i_0 j_0}, \frac{x - x_{i_0 j_0}}{\epsilon} \cdot n_0\right), \quad \epsilon \ll 1.$$
(5.5.1)

Fix $n_1, n_2 \in \mathbb{N}$ and consider a $(2n_1 + 1) \times (2n_2 + 1)$ window around $x_{i_0 j_0}$. To use (5.5.1) for finding $D(x_{i_0 j_0})$, first, we need to estimate n_0. This can be easily done by computing partial derivatives

$$n_0 \approx N_0 :=$$

$$\frac{(f_{\Lambda\epsilon}(x_{i_0+1,j_0}) - f_{\Lambda\epsilon}(x_{i_0-1,j_0}), f_{\Lambda\epsilon}(x_{i_0,j_0+1}) - f_{\Lambda\epsilon}(x_{i_0,j_0-1}))}{\sqrt{(f_{\Lambda\epsilon}(x_{i_0+1,j_0}) - f_{\Lambda\epsilon}(x_{i_0-1,j_0}))^2 + (f_{\Lambda\epsilon}(x_{i_0,j_0+1}) - f_{\Lambda\epsilon}(x_{i_0,j_0-1}))^2}}.$$
(5.5.2)

From (5.5.1) we see that it is convenient to find $D(x_{i_0 j_0})$ by solving the minimization problem

$$\sum_{\substack{|i-i_0|\leq n_1 \\ |j-j_0|\leq n_2}} \left(f_{\Lambda\epsilon}(x_{ij}) - \frac{D}{\pi\epsilon}\psi\left(\frac{x_{ij} - x_{i_0 j_0}}{\epsilon} \cdot N_0\right)\right.$$

$$\left. - \psi_\epsilon\left(x_{i_0 j_0}, \frac{x_{ij} - x_{i_0 j_0}}{\epsilon} \cdot N_0\right)\right)^2 \to \min$$
(5.5.3)

with respect to D and an even function ψ_ϵ. Thus, since $\psi(q)$ is odd and $\psi_\epsilon(x_s, q)$ is even in q, we have

$$D(x_{i_0 j_0}) \approx \pi\epsilon \frac{\sum_{\substack{|i-i_0|\leq n_1 \\ |j-j_0|\leq n_2}} f_{\Lambda\epsilon}(x_{ij})\psi\left(\frac{x_{ij}-x_{i_0 j_0}}{\epsilon} \cdot N_0\right)}{\sum_{\substack{|i-i_0|\leq n_1 \\ |j-j_0|\leq n_2}} \psi^2\left(\frac{x_{ij}-x_{i_0 j_0}}{\epsilon} \cdot N_0\right)}.$$
(5.5.4)

Equations (5.4.4), (5.4.5) and (5.4.7) imply that larger values of $f_{\Lambda\epsilon}$ correspond to the side of S with larger values of f. Thus, we came to the following algorithm for estimating values of jumps of f from $f_{\Lambda\epsilon}$:

(1) Estimate vector n_0 by computing its approximation N_0 according to (5.5.2);

(2) Compute the estimate of $D(x_{ij})$ by formula (5.5.4); and

(3) The vector N_0 given by (5.5.2) points from the smaller values of f to the larger values of f.

REMARK 5.5.1. From Item (3) above and the relation

$$D(x_0) = \lim_{t \to 0+} [f(x_0 + tn(x_0)) - f(x_0 - tn(x_0))], \quad x_0 \in S,$$

we see that values of jumps $D(x_{i_0 j_0})$, estimated according to (5.5.4), should always be nonnegative. Therefore, in practice, large negative values of $D(x_{i_0 j_0})$ will indicate that the tomographic data are contaminated by considerable amount of noise.

Clearly, calculation of $D(x_{i_0 j_0})$ from (5.5.3), (5.5.4) is most stable if the function ψ on the interval

$$X := \left[\min_{\substack{|i-i_0| \le n_1 \\ |j-j_0| \le n_2}} \left(\frac{x_{ij} - x_{i_0 j_0}}{\epsilon} \cdot n_0 \right), \ \max_{\substack{|i-i_0| \le n_1 \\ |j-j_0| \le n_2}} \left(\frac{x_{ij} - x_{i_0 j_0}}{\epsilon} \cdot n_0 \right) \right] \quad (5.5.5)$$

differs from a constant as much as possible. The graph of $\psi(q)$ presented in Figure 5.4.1 suggests that the interval X should be close to the interval between q-coordinates of the local minimum and the local maximum of $\psi(q)$, that is we should have $X \approx [-0.31, 0.31]$. From the derivation of the algorithm it follows also that for the reliable and accurate recovery of the jump magnitude at a point $x_s \in S$, the radius of curvature $R(x_s)$ should be much larger than the radius of mollification ϵ, and the grid step size h should be much smaller than ϵ, i.e. $h \ll \epsilon \ll R(x_s)$. The precise relationship between the ratios h/ϵ and $\epsilon/R(x_s)$ and the resulting accuracy of jump magnitude recovery still needs to be investigated. However, the results of numerical experiments presented below indicate that these ratios need not be very small in order to provide good accuracy of the recovery.

The performance of the proposed algorithm was tested on the simulated data. In Figure 5.5.1, we see a phantom used for generating the Radon transform data. The densities are: exterior: 0, ellipse: 1, exterior annulus: 0.8, area between the annulus and the ellipse: 0, three small disks off the center: 1.8, the small disk at the center: 0.1. The radius of the phantom: 0.9, the half-axes of the ellipse: 0.2 and 0.4, the radii of the four small discs: 0.05, the radius of the disc containing the ellipse: 0.5. The Radon transform was computed for 350 angles equispaced on $[0, \pi)$ and 601 projections per each angle. In Figure 5.5.2, we see the density plot of $f_{\Lambda \epsilon}$ computed using the mollifier (5.4.8) with $\epsilon = 9\Delta p$ at the nodes of a square 201×201 grid, where Δp is the discretization step of the p-variable. This value of ϵ means that the discrete convolutions were computed using 19 points per integral. The grid step size was $h = 3\Delta p$, hence $h/\epsilon = 1/3$. The smallest value of the radius of curvature on S equals 0.05, hence $\epsilon/R_{\min} = 0.6$. In Figure 5.5.3, we see the density plot of the function $D(x)$ estimated by formula (5.5.4) with $n_1 = n_2 = 1$, i.e. we used a 3×3 window. The presented results show that the true jump values are accurately estimated. We see that for the given ratio $h/\epsilon = 1/3$ and window size $n_1 = n_2 = 1$ the interval X (see (5.5.5)) was very close to the optimal one: $X = [-1/3, 1/3]$ if n_0 is parallel to one of the coordinate axes and $X = [-\sqrt{2}/3, \sqrt{2}/3] \approx [-0.47, 0.47]$ if n_0 is at $45°$ to one of the axes. The horizontal central cross-section of the true density distribution f is given in Figure 5.5.4. The horizontal central cross-section of Figure 5.5.3 is represented in Figure 5.5.5. A line with peaks is a graph of $D(x)$. Big dots represent positions and amplitudes of jumps of the original density function f (see Figure 5.5.4). We see a good agreement between the dots and maxima of the peaks. Results of the reconstruction from the noisy data are shown in Figures 5.5.6, 5.5.7, and 5.5.8. The exact Radon transform data was corrupted by a Gaussian noise, the standard deviation of which was 3% of the value of the Radon transform at this point.

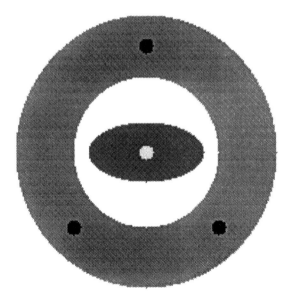

FIGURE 5.5.1. The phantom used for generating the Radon transform data

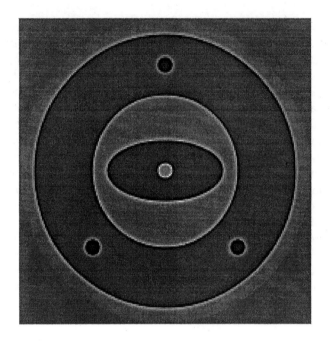

FIGURE 5.5.2. The density plot of $f_{\Lambda\epsilon}$

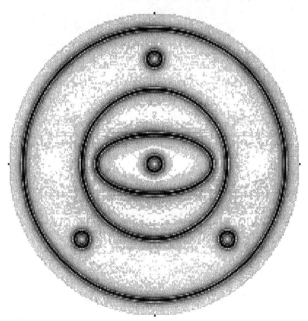

FIGURE 5.5.3. The density plot of the estimated $D(x)$

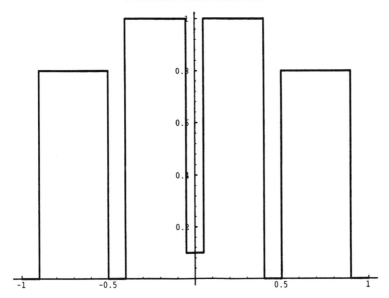

FIGURE 5.5.4. The central horizontal cross-section of the true density distribution f

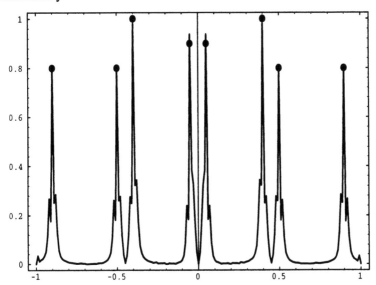

FIGURE 5.5.5. The central horizontal cross-section of the plot of the estimated $D(x)$

5.5.2. The second numerical scheme for computing values of jumps

Now we describe the second algorithm for computing values of jumps of f using the local tomography approach. This algorithm does not

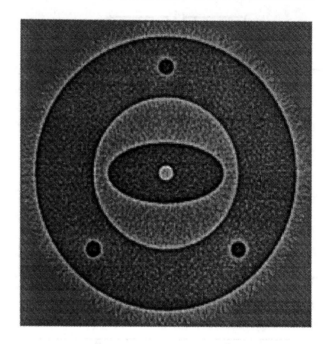

FIGURE 5.5.6.   The density plot of $f_{\Lambda\epsilon}$ computed from the noisy data

Now we describe the second algorithm for computing values of jumps of f using the local tomography approach. This algorithm does not require any calculations in the x-domain and provides values of jumps from the tomographic data, basically, in one step. Recall that to use the algorithm from Section 5.5.1 we have to compute $f_{\Lambda\epsilon}$ first. Neglecting the variation of $f_{\Lambda\epsilon}(x_s + hn_0)$ with respect to x_s, we get $|\nabla f_{\Lambda\epsilon}(x_0)| \approx |\frac{\partial}{\partial h} f_{\Lambda\epsilon}(x_0 + hn_0)|_{h=0}$, $x_0 \in S$. Therefore, using (5.4.6), we obtain an approximate equality:

$$|\nabla f_{\Lambda\epsilon}(x_{i_0 j_0})| \approx \frac{|D(x_{i_0 j_0})|}{\pi\epsilon^2}\psi'(0), \quad x_{i_0 j_0} \in S, \epsilon \ll 1,$$

hence

$$|D(x_{i_0 j_0})| \approx |\nabla f_{\Lambda\epsilon}(x_{i_0 j_0})|\frac{\pi\epsilon^2}{\psi'(0)}. \tag{5.5.6}$$

The gradient $\nabla f_{\Lambda\epsilon}(x_{ij})$ can be calculated using (5.4.3) without considerable increase in the amount of computational work as compared with calculation of $f_{\Lambda\epsilon}$. Indeed, differentiating (5.4.3) with respect to x_1 and x_2, we get

$$\frac{\partial f_{\Lambda\epsilon}(x)}{\partial x_1} = -\frac{1}{4\pi}\int\limits_{S^1}\cos(\theta)\left[\int\limits_{\mathbb{R}} w_\epsilon'''(\Theta \cdot x - p)\hat{f}(\theta,p)dp\right]d\theta, \tag{5.5.7a}$$

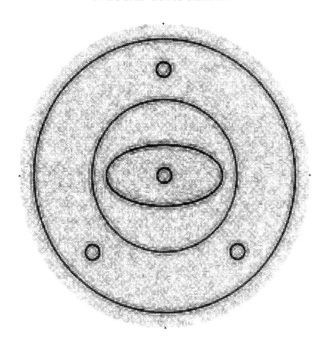

FIGURE 5.5.7. The density plot of the estimated $D(x)$ using $f_{\Lambda\epsilon}$ computed from the noisy data (see Figure 5.5.6)

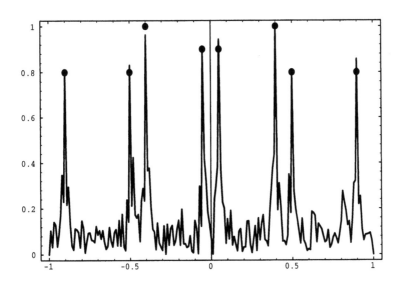

FIGURE 5.5.8. The central horizontal cross-section of the plot of the estimated $D(x)$ from the noisy data (see Figure 5.5.7)

$$\frac{\partial f_{\Lambda\epsilon}(x)}{\partial x_2} = -\frac{1}{4\pi} \int\limits_{S^1} \sin(\theta) \left[\int\limits_{\mathbb{R}} w_\epsilon'''(\Theta \cdot x - p)\hat{f}(\theta, p)dp \right] d\theta, \qquad (5.5.7b)$$

$$|\nabla f_{\Lambda\epsilon}(x)| = \left(\left(\frac{\partial f_{\Lambda\epsilon}(x)}{\partial x_1}\right)^2 + \left(\frac{\partial f_{\Lambda\epsilon}(x)}{\partial x_2}\right)^2 \right)^{1/2}. \qquad (5.5.7c)$$

As we can see, the integrals in brackets in (5.5.7a,b) are identical. Therefore, for each θ, it is convenient to compute

$$g(q, \theta) = \int\limits_{q-\epsilon}^{q+\epsilon} w_\epsilon'''(q - p)\hat{f}(\theta, p)dp \qquad (5.5.8)$$

for all q in the projection of the support of f on the direction Θ, and
then compute

$$\frac{\partial f_{\Lambda\epsilon}(x)}{\partial x_1} = -\frac{1}{4\pi} \int\limits_{S^1} \cos(\theta)g(\Theta \cdot x, \theta)d\theta, \qquad (5.5.9a)$$

$$\frac{\partial f_{\Lambda\epsilon}(x)}{\partial x_2} = -\frac{1}{4\pi} \int\limits_{S^1} \sin(\theta)g(\Theta \cdot x, \theta)d\theta. \qquad (5.5.9b)$$

Thus, we came to the following algorithm for estimating values of jumps
of f:

(1) Compute $\nabla f_{\Lambda\epsilon}(x_{ij})$ using (5.5.8) and (5.5.9);
(2) Compute the estimate of $|D(x_{ij})|$ using (5.5.7c) and (5.5.6); and
(3) The vector $\nabla f_{\Lambda\epsilon}(x_{ij})$ points from the smaller values of f to the
larger values of f.

For a mollifier (5.4.8) we can compute $\psi'(0)$ theoretically. Using (5.4.13),
we get

$$w_1(p) = \frac{m+1}{\pi} \int\limits_{-\sqrt{1-p^2}}^{\sqrt{1-p^2}} (1 - p^2 - t^2)^m dt = c(1 - p^2)^{m+0.5},$$

$$c = \frac{2^{m+1}(m+1)!}{\pi(2m+1)!!}.$$

Differentiation of (5.4.7) with respect to q and substitution of the last
equation yields

$$\psi'(0) = -\int\limits_{-1}^{1} \frac{w_1'(t)}{t}dt = c(2m+1)\int\limits_{-1}^{1}(1-t^2)^{m-0.5}dt = 2(m+1).$$

The performance of the proposed algorithm was tested on the same data as in Section 5.5.1. Values of all parameters are also the same. In Figure 5.5.9, we see the density plot of the function $|D(x)|$ estimated by formula (5.5.1). Again, the presented results show that the true jump values are accurately estimated. The horizontal central cross-section of Figure 5.5.9 is represented in Figure 5.5.10. A line with peaks is a graph of $|D(x)|$. Big dots represent positions and amplitudes of jumps of the original density function f. We see a good agreement between the dots and maxima of the peaks. Results of the reconstruction from the data corrupted by the 3% noise (as in Section 5.5.1) are shown in Figures 5.5.11 and 5.5.12.

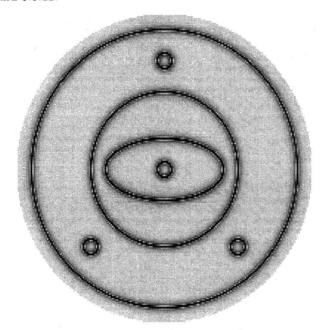

FIGURE 5.5.9.   The density plot of the estimated $|D(x)|$

5.6. Local tomography for the exponential Radon transform

Similarly to the regular local tomography (see (5.4.1)), we define local tomography function for the exponential Radon transform (or, simply, local exponential tomography function) by the formula

$$f_\Lambda^{(\mu)}(x) := -\frac{1}{4\pi} \int\limits_{S^1} e^{-\mu(x \cdot \Theta^\perp)} g_{pp}(\theta, \Theta \cdot x) d\theta, \qquad (5.6.1)$$

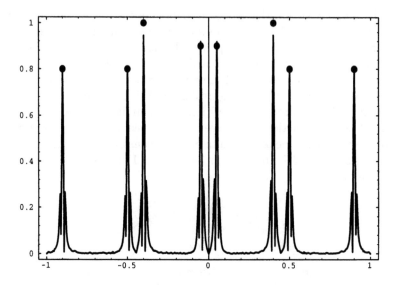

FIGURE 5.5.10.   The central horizontal cross-section of the plot of the
estimated $|D(x)|$

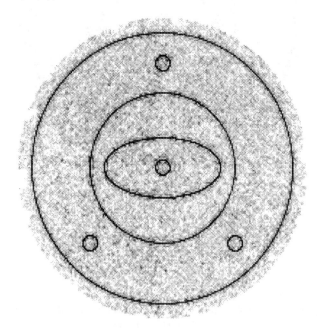

FIGURE 5.5.11.   The density plot of $|D(x)|$ estimated from the noisy
data

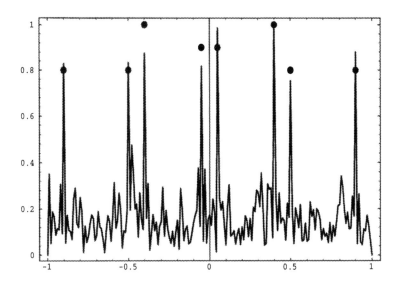

FIGURE 5.5.12.   The central horizontal cross-section of the plot of $|D(x)|$ estimated from the noisy data (see Figure 5.5.11)

where $g = \hat{f}^{(\mu)}$ is the exponential Radon transform of f (see Section 2.8.1, Equation (2.8.3)). To check that (5.6.1) is indeed the generalization of the regular local tomography function f_Λ defined by (5.4.1), let us substitute (2.8.3) into (5.6.1).

$$f_\Lambda^{(\mu)}(x) = -\frac{1}{4\pi} \int_{S^1} e^{-\mu(x\cdot\Theta^\perp)} \frac{\partial^2}{\partial p^2} \left[\int_{\mathbb{R}^2} e^{\mu(y\cdot\Theta^\perp)} f(y)\delta(\Theta\cdot y - p)dy \right]_{p=\Theta\cdot x} d\theta.$$

Using the formula $\delta(t) = (2\pi)^{-1} \int_\mathbb{R} \exp(its)ds$, let us represent the integral in brackets as an oscillatory integral which, as it is known, can be differentiated with respect to a parameter:

$$f_\Lambda^{(\mu)}(x) = \frac{1}{8\pi^2} \int_{S^1} e^{-\mu(x\cdot\Theta^\perp)} \int_\mathbb{R} \int_{\mathbb{R}^2} t^2 e^{-it(x-y)\cdot\Theta} e^{\mu(y\cdot\Theta^\perp)} f(y)dydtd\theta$$

$$= \frac{1}{8\pi^2} \int_{S^1} \int_\mathbb{R} \int_{\mathbb{R}^2} t^2 e^{-it(x-y)\cdot\Theta} e^{-\mu(x-y)\cdot\Theta^\perp} f(y)dydtd\theta.$$

After simple transformations, we get

$$f_\Lambda^{(\mu)}(x) = \frac{1}{4\pi^2} \int_{S^1} \int_0^\infty \int_{\mathbb{R}^2} t^2 \cosh(\mu(x-y)\cdot\Theta^\perp) e^{-it(x-y)\cdot\Theta} f(y)dydtd\theta$$

$$= \frac{1}{4\pi^2} \int_{\mathbb{R}^2} \int_{\mathbb{R}^2} |\xi| \cosh(\mu(x-y)\cdot\xi^\perp/|\xi|) e^{-it(x-y)\cdot\Theta} f(y)dyd\xi,$$
$$\tag{5.6.2}$$

where $t\Theta = \xi = (\xi_1, \xi_2)$ and $\xi^{\perp} = (-\xi_2, \xi_1)$. Denoting

$$a(x, y, \xi) = |\xi| \cosh\left(\mu \frac{(x - y) \cdot \xi^{\perp}}{|\xi|}\right), \qquad (5.6.3)$$

we see that $f_{\Lambda}^{(\mu)}$ is the result of action of a PDO with amplitude (5.6.3) on f.

Lemma 5.6.1. *Let $f(x)$, $x \in \mathbb{R}^2$, be compactly supported. Then one has*

$$WF(f_{\Lambda}^{(\mu)}) = WF(f). \qquad (5.6.4)$$

In particular, this implies that singsupp $f_{\Lambda}^{(\mu)}$ = singsupp f.

Proof. Fix any $\psi \in C^{\infty}(\mathbb{R}^4)$, $\psi(x, y) = 1$ for $|x - y| \leq c_1$ and $\psi(x, y) = 0$ for $|x - y| \geq c_2$ for some $0 < c_1 < c_2$. Let us represent $a(x, y, \xi)$ in the form $a(x, y, \xi) = \psi(x, y)a(x, y, \xi) + (1 - \psi(x, y))a(x, y, \xi)$. Differentiating ψa with respect to x, y, and ξ (note that $a(x, y, \xi)$ is positive homogeneous of degree 1 in ξ), we obtain that the PDO A_{ψ}, corresponding to the symbol ψa, belongs to the class $L_{1,0}^1(\mathbb{R}^2) = L^1(\mathbb{R}^2)$. Clearly, A_{ψ} is properly supported. Moreover, letting $y = x$ in (5.6.3) and using that $\psi(x, x) = 1$, we obtain that the principal symbol of A_{ψ} equals $|\xi|$, hence A_{ψ} is an elliptic PDO of order 1. Since PDO with the amplitude $(1 - \psi)a$ is smoothing and f is compactly supported, we conclude that $WF(f_{\Lambda}^{(\mu)}) = WF(f)$. □

Using notation from the proof of Lemma 5.6.1, recalling that the principal symbol of the PDO A_{ψ} equals $|\xi|$, and using the results from Section 5.9, we obtain that in a neighborhood of S, the discontinuity curve of f, the asymptotic behavior of $f_{\Lambda}^{(\mu)}$ is given by the formula

$$f_{\Lambda}^{(\mu)}(x_s + hn(x_s)) = \frac{D(x_s)}{\pi} \frac{1 + O(h \ln |h|)}{h}, \quad h \to 0,$$

$$D(x_s) = \lim_{t \to 0^+} [f(x_s + tn(x_s)) - f(x_s - tn(x_s))], \qquad (5.6.5)$$

for all $x_s \in S$ in a neighborhood of which S is smooth. Here $n(x_s)$ is the unit vector perpendicular to S at x_s. Formula (5.6.5) is convenient, because, similarly to the case of regular tomography (see (5.4.4)), it explicitly relates $D(x_s)$ with $f_{\Lambda}^{(\mu)}(x)$ for x close to $x_s \in S$. Therefore, it allows one to estimate values of jumps of f from local exponential tomographic reconstruction using techniques described in Sections 5.4 and 5.5. Indeed, comparing (5.6.1) and (2.7.5), we conclude that $f_{\Lambda}^{(\mu)} = -(4\pi)^{-1}T_{-\mu}^*g_{pp}$. Using (2.7.7a), we get

$$f_{\Lambda}^{(\mu)} * W_{\epsilon} = -\frac{1}{4\pi}(T_{-\mu}^*g_{pp}) * W_{\epsilon} = -\frac{1}{4\pi}T_{-\mu}^*(g_{pp} \circledast (T_{\mu}W_{\epsilon}))$$

$$= -\frac{1}{4\pi}\int_{S^1} e^{-\mu(x \cdot \Theta^{\perp})} \int_{\mathbb{R}} w_{\epsilon}''(\Theta \cdot x - p)g(\theta, p)dp\, d\theta, \qquad (5.6.6)$$

where we assumed that W_ϵ is radial, $W_\epsilon(x) := W_\epsilon(|x|)$, and denoted

$$w_\epsilon(p) := (T_\mu W_\epsilon)(p) = \int_{\mathbb{R}^2} e^{\mu(x \cdot \Theta^\perp)} W_\epsilon(x)\delta(\Theta \cdot x - p)dx.$$

The last equation implies

$$w_\epsilon(p) = \begin{cases} 2\int_{|p|}^\epsilon r \frac{\cosh(\mu\sqrt{r^2-p^2})}{\sqrt{r^2-p^2}} W_\epsilon(r)dr, & |p| \leq \epsilon, \\ 0, & |p| \geq \epsilon. \end{cases} \tag{5.6.7}$$

Exercise 5.6.1. Derive Equation (5.6.7).

Comparing (5.6.6) and (5.4.3), we see that the behavior of $f_{\Lambda\epsilon}^{(\mu)} := f_\Lambda^{(\mu)} * W_\epsilon, \epsilon \to 0$, in a neighborhood of S is asymptotically the same as that of $f_{\Lambda\epsilon}$ (see Theorem 5.4.1). Therefore, the numerical schemes described in Section 5.5 can be applied to local exponential tomography as well.

Now let us choose any function $b(\Theta)$ which has the following properties:

$$b(\Theta) \in C^\infty(S^1), \quad \min_{\Theta \in S^1} b(\Theta) > 0, \quad b(\Theta) = b(-\Theta). \tag{5.6.8}$$

Inserting function $b(\Theta)$ into Equation (5.6.1), we obtain the family of local exponential tomography functions $f_b^{(\mu)}$:

$$f_b^{(\mu)}(x) := -\frac{1}{4\pi} \int_{S^1} b(\Theta)e^{-\mu(x \cdot \Theta^\perp)}g_{pp}(\theta, \Theta \cdot x)d\theta,$$

which is a generalization of the family of regular local tomography functions described in Section 5.2.

Exercise 5.6.2. Prove that $WF(f_b^{(\mu)}) = WF(f)$. Show that the principal symbol of the corresponding PDO is given by $|\xi|b(\xi/|\xi|)$.

5.7. Local tomography for the generalized Radon transform

For simplicity, in this section we consider only the case of $\mathbb{R}^2$. However, the results can be generalized to any $\mathbb{R}^n, n = 2k, k \geq 1$.

5.7.1. The first approach
Let us define a local tomography function for the generalized Radon transform (see Section 2.8.3) as follows:

$$f_\Lambda^{(\Phi)}(x) := -\frac{1}{4\pi} \int_{S^1} \frac{\hat{f}_{pp}^{(\Phi)}(\theta, \Theta \cdot x)}{\Phi(x, \Theta, \Theta \cdot x)}d\theta. \tag{5.7.1}$$

To check that $f_\Lambda^{(\Phi)}$ is indeed the generalization of regular local tomography, let us substitute (2.8.19) (with $n = 2$) into (5.7.1).

$$
f_\Lambda^{(\Phi)}(x) = -\frac{1}{4\pi} \int_{S^1} \frac{1}{\Phi(x, \Theta, \Theta \cdot x)}
$$

$$
\times \frac{\partial^2}{\partial p^2} \left[\int_{\mathbb{R}^2} \Phi(y, \Theta, p) f(y) \delta(p - \Theta \cdot y) dy \right]_{p=\Theta \cdot x} d\theta
$$

$$
= -\frac{1}{8\pi^2} \int_{S^1} \frac{1}{\Phi(x, \Theta, \Theta \cdot x)}
$$

$$
\times \frac{\partial^2}{\partial p^2} \left[\int_{\mathbb{R}} \int_{\mathbb{R}^2} \Phi(y, \Theta, p) f(y) e^{-it(p - \Theta \cdot y)} dy\, dt \right]_{p=\Theta \cdot x} d\theta
$$

$$
= \frac{1}{8\pi^2} \int_{S^1} \int_{\mathbb{R}} \int_{\mathbb{R}^2} t^2 \frac{\Phi(y, \Theta, \Theta \cdot x)}{\Phi(x, \Theta, \Theta \cdot x)} f(y) e^{-it\Theta \cdot (x-y)} dy\, dt\, d\theta
$$

$$
+ \frac{1}{4\pi^2} \int_{S^1} \int_{\mathbb{R}} \int_{\mathbb{R}^2} it \frac{\Phi_p(y, \Theta, \Theta \cdot x)}{\Phi(x, \Theta, \Theta \cdot x)} f(y) e^{-it\Theta \cdot (x-y)} dy\, dt\, d\theta
$$

$$
- \frac{1}{8\pi^2} \int_{S^1} \int_{\mathbb{R}} \int_{\mathbb{R}^2} \frac{\Phi_{pp}(y, \Theta, \Theta \cdot x)}{\Phi(x, \Theta, \Theta \cdot x)} f(y) e^{-it\Theta \cdot (x-y)} dy\, dt\, d\theta.
$$

Splitting integration with respect to t into two intervals $(-\infty, 0]$ and $[0, \infty)$, we get

$$
f_\Lambda^{(\Phi)}(x) = \frac{1}{4\pi^2} \int_{S^1} \int_0^\infty \int_{\mathbb{R}} t^2 a_1(x, y, \Theta) f(y) e^{-it\Theta \cdot (x-y)} dy\, dt\, d\theta
$$

$$
+ \frac{i}{4\pi^2} \int_{S^1} \int_0^\infty \int_{\mathbb{R}} t a_2(x, y, \Theta) f(y) e^{-it\Theta \cdot (x-y)} dy\, dt\, d\theta
$$

$$
- \frac{1}{4\pi^2} \int_{S^1} \int_0^\infty \int_{\mathbb{R}} a_3(x, y, \Theta) f(y) e^{-it\Theta \cdot (x-y)} dy\, dt\, d\theta,
\tag{5.7.2}
$$

where

$$
a_1(x, y, \Theta) = \frac{1}{2} \left[\frac{\Phi(y, \Theta, \Theta \cdot x)}{\Phi(x, \Theta, \Theta \cdot x)} + \frac{\Phi(y, -\Theta, -\Theta \cdot x)}{\Phi(x, -\Theta, -\Theta \cdot x)} \right],
$$

$$
a_2(x, y, \Theta) = \frac{\Phi_p(y, \Theta, \Theta \cdot x)}{\Phi(x, \Theta, \Theta \cdot x)} - \frac{\Phi_p(y, -\Theta, -\Theta \cdot x)}{\Phi(x, -\Theta, -\Theta \cdot x)},
$$

$$
a_3(x, y, \Theta) = \frac{1}{2} \left[\frac{\Phi_{pp}(y, \Theta, \Theta \cdot x)}{\Phi(x, \Theta, \Theta \cdot x)} + \frac{\Phi_{pp}(y, -\Theta, -\Theta \cdot x)}{\Phi(x, -\Theta, -\Theta \cdot x)} \right].
\tag{5.7.3}
$$

Equations (5.7.2) and (5.7.3) yield

$$f_\Lambda^{(\Phi)}(x) = \frac{1}{4\pi^2} \int\limits_{\mathbb{R}^2} \int\limits_{\mathbb{R}^2} a(x,y,\xi) f(y) e^{-i\xi\cdot(x-y)} dy\, d\xi, \qquad (5.7.4)$$

where

$$a(x,y,\xi) = |\xi| a_1(x,y,\xi/|\xi|) + i a_2(x,y,\xi/|\xi|) - |\xi|^{-1} a_3(x,y,\xi/|\xi|). \qquad (5.7.5)$$

Equation (5.7.4) shows that $f_\Lambda^{(\Phi)}$ is the result of action of the PDO with amplitude $a(x,y,\xi)$ on f.

Lemma 5.7.1. *Let $f(x)$, $x \in \mathbb{R}^2$, be compactly supported. Then one has*

$$WF(f_\Lambda^{(\Phi)}) = WF(f). \qquad (5.7.6)$$

In particular, this implies that $\operatorname{singsupp} f_\Lambda^{(\Phi)} = \operatorname{singsupp} f$.

Proof. Fix any $\psi \in C^\infty(\mathbb{R}^4)$, $\psi(x,y) = 1$ for $|x - y| \leq c_1$ and $\psi(x,y) = 0$ for $|x - y| \geq c_2$ for some $0 < c_1 < c_2$. Fix also any $\varphi \in C^\infty(\mathbb{R}^2)$ such that $\varphi(\xi) > 0$ for $|\xi| \neq 0$, $\varphi(\xi)$ vanishes with all derivatives at the origin, and $\varphi(\xi) = 1$ for $|\xi| \geq c_2$. Let us represent $a(x,y,\xi)$ in the form

$$\begin{aligned} a(x,y,\xi) = &\psi(x,y)\varphi(\xi)a(x,y,\xi) + (1 - \psi(x,y))a(x,y,\xi) \\ &+ \psi(x,y)(1 - \varphi(\xi))a(x,y,\xi). \end{aligned}$$

Differentiating $\psi\varphi a$ with respect to x, y, and ξ, we obtain that the PDO A_1, corresponding to the symbol $\psi\varphi a$, belongs to the class $L_{1,0}^1(\mathbb{R}^2) = L^1(\mathbb{R}^2)$ (see Section 14.3.3 for the definitions of classes of PDO). Clearly, A_1 is properly supported. Moreover, letting $y = x$ in (5.7.5) and using that $\psi(x,x) = 1$, we obtain that the principal symbol of A_1 equals $|\xi|$, hence A_1 is an elliptic PDO of order 1. Since PDO with the amplitude $(1 - \psi)a + \psi(1 - \varphi)a$ is smoothing and f is compactly supported, we conclude that $WF(f_\Lambda^{(\Phi)}) = WF(f)$. The second assertion of Lemma 5.7.1 is now obvious. $\square$

As it has been established, the principal symbol of the PDO A_1 equals $|\xi|$. Therefore, Equation (5.4.2) yields that the principal terms of the asymptotics of $f_\Lambda^{(\Phi)}(x)$ and of $f_\Lambda(x)$ as $x \to S$ are the same, and we have similarly to (5.4.4):

$$f_\Lambda^{(\Phi)}(x_s + hn(x_s)) = \frac{D(x_s)}{\pi} \frac{1 + O(h \ln |h|)}{h}, \qquad h \to 0. \qquad (5.7.7)$$

Equation (5.7.7) shows that if $f_\Lambda^{(\Phi)}$ is computed, we can recover values of jumps of f from $f_\Lambda^{(\Phi)}$.

5.7.2. The second approach

Let us define the second local tomography function as follows:

$$\tilde{f}_\Lambda^{(\Phi)}(x) := -\frac{1}{4\pi} \int_{S^1} \hat{f}_{pp}^{(\Phi)}(\theta, \Theta \cdot x) d\theta. \qquad (5.7.8)$$

Equation (5.7.8) has advantage over (5.7.1) in that it can be used even if the weight function $\Phi(x, \Theta, p)$ is not known. To investigate properties of $\tilde{f}_\Lambda^{(\Phi)}$, let us substitute (2.8.19) into (5.7.8). Similarly to (5.7.4) and (5.7.5), we obtain

$$\tilde{f}_\Lambda^{(\Phi)}(x) = \frac{1}{4\pi^2} \int_{\mathbb{R}^2} \int_{\mathbb{R}^2} \tilde{a}(x, y, \xi) f(y) e^{-i\xi \cdot (x-y)} dy \, d\xi, \qquad (5.7.9)$$

$$\tilde{a}(x, y, \xi) = |\xi| \tilde{a}_1(x, y, \xi/|\xi|) + i\tilde{a}_2(x, y, \xi/|\xi|) - |\xi|^{-1} \tilde{a}_3(x, y, \xi/|\xi|), \qquad (5.7.10)$$

$$\begin{aligned}
\tilde{a}_1(x, y, \Theta) &= 0.5\big[\Phi(y, \Theta, \Theta \cdot x) + \Phi(y, -\Theta, -\Theta \cdot x)\big], \\
\tilde{a}_2(x, y, \Theta) &= \Phi_p(y, \Theta, \Theta \cdot x) - \Phi_p(y, -\Theta, -\Theta \cdot x), \\
\tilde{a}_3(x, y, \Theta) &= 0.5\big[\Phi_{pp}(y, \Theta, \Theta \cdot x) + \Phi_{pp}(y, -\Theta, -\Theta \cdot x)\big].
\end{aligned}$$
$$(5.7.11)$$

Choosing functions ψ and φ as in the proof of Lemma 5.7.1, we see that the PDO $\tilde{A}_1$ with the amplitude $\psi(x, y)\varphi(\xi)\tilde{a}(x, y, \xi)$ is classical, properly supported PDO of order 1. The principal symbol of $\tilde{A}_1$ equals $|\xi|\tilde{a}_1(x, x, \xi/|\xi|)$. Fix an arbitrary compact $K \subset \mathbb{R}^2$. Since Φ is smooth and strictly positive, we have

$$0 < \min_{x \in K, \Theta \in S^1} \tilde{a}_1(x, x, \Theta) \le \max_{x \in K, \Theta \in S^1} \tilde{a}_1(x, x, \Theta) < \infty.$$

Therefore, $\tilde{A}_1$ is an elliptic PDO. Since $A - \tilde{A}_1$ is smoothing, we prove

Lemma 5.7.2. *Let* $f(x)$, $x \in \mathbb{R}^2$, *be compactly supported. Then one has*

$$WF(\tilde{f}_\Lambda^{(\Phi)}) = WF(f).$$

In particular, this implies that singsupp $\tilde{f}_\Lambda^{(\Phi)} =$ singsupp f.

Using the results obtained in Section 5.9.3, we get that the analogue of (5.7.7) is

$$\tilde{f}_\Lambda^{(\Phi)}(x_s + hn(x_s)) = \frac{D(x_s)}{\pi} \frac{\tilde{a}_1(x_s, x_s, n(x_s))}{h}(1 + O(h \ln |h|)), \quad h \to 0. \qquad (5.7.12)$$

Comparing (5.7.7) and (5.7.12), we see that if Φ is known, it is probably better to use the local tomography function $f_\Lambda^{(\Phi)}$ instead of $\tilde{f}_\Lambda^{(\Phi)}$, because jumps of f are imaged in $f_\Lambda^{(\Phi)}$ with intensities directly proportional to magnitudes of jumps. In particular, larger jumps will have higher intensities than smaller jumps, and one will be able to recover qualitative information from images of $f_\Lambda^{(\Phi)}$ without further processing. On the contrary, raw images of $\tilde{f}_\Lambda^{(\Phi)}$ can mask differences between sharp variations of the original function f, especially if the weight function Φ changes considerably inside the support of f. However, if Φ is known, one can compute $\tilde{a}_1(x_s, x_s, n(x_s))$ using (5.7.11) and renormalize values of $\tilde{f}_\Lambda^{(\Phi)}$, so that jumps are imaged with intensities proportional to their magnitudes.

If the weight function Φ is not known, Lemma 5.7.2 ensures that we still can recover locations of discontinuities of f. However, in this case $\tilde{a}_1(x_s, x_s, n(x_s))$ is not known either, and (5.7.12) shows that, in general, we cannot find values of jumps of f.

5.7.3. Remarks on numerical implementation

In Section 2.1.8, in case of the regular Radon transform R, we derived the following useful formula

$$W * (R^*g) = R^*(RW \circledast g), \qquad (5.7.13)$$

where R^* is the operator adjoint to R, and '$\circledast$' denotes convolution in only the p variable. Using this relation, one has a very convenient way for regularizing local tomography formula (5.4.1) (see Equation (5.4.3)). Moreover, using (5.7.13) for practical calculations, one has the following two advantages:

(1) The filter RW, with which g is convolved, is a function on $\mathbb{R}^1$ (if W is radial), and can be computed only once before the tomographic data processing;

(2) The regularization of any derivatives, which appear in g (e.g., $g = \hat{f}_{pp}$ in regular local tomography), by convolving with RW has a very precise effect on the result R^*g. Namely, R^*g is convolved with W. Since the asymptotic behavior of R^*g in a neighborhood of S is known, this allows one to construct simple and efficient algorithms for finding locations and values of jumps of a function using local tomography.

Unfortunately, there is no analogue of Equation (5.7.13) for the generalized Radon transform. More precisely, there is no simple and efficient formula of the type (5.7.13) with R^* replaced by $(R^{(\Phi)})^*$, where $(R^{(\Phi)})^*$ is the operator adjoint to $R^{(\Phi)}$:

$$[(R^{(\Phi)})^*g](x) = \int_{S^1} \Phi(x, \Theta, \Theta \cdot x) g(\theta, \Theta \cdot x) d\theta.$$

To solve this problem, we propose an approximate analogue of (5.7.13), such that the two above advantages still hold. Define a function

$$g_\Lambda^{(\Phi)}(x',x) := -\frac{1}{4\pi}\int_{S^1}\frac{\hat{f}_{pp}^{(\Phi)}(\theta,\Theta\cdot x)}{\Phi(x',\Theta,\Theta\cdot x')}d\theta, \quad x',x\in\mathbb{R}^2. \qquad (5.7.14)$$

Let $x'\in\mathbb{R}^2$ be fixed. Similarly to (5.7.1) – (5.7.3), we see that $g_\Lambda^{(\Phi)}(x',x)$, as a function of x, is a result of action of an elliptic PDO on f. Therefore, $g_\Lambda^{(\Phi)}(x',x)$ is C^∞ in x for $x\notin S$. This implies that

$$g_\Lambda^{(\Phi)}(x',x)\to f_\Lambda^{(\Phi)}(x'), \quad x\to x'\notin S, \qquad (5.7.15)$$

because $g_\Lambda^{(\Phi)}(x',x') = f_\Lambda^{(\Phi)}(x')$. Clearly, one has

$$\int_{\mathbb{R}^2}W_\epsilon(x'-x)g_\Lambda^{(\Phi)}(x',x)dx$$

$$= -\frac{1}{4\pi}\int_{S^1}\frac{1}{\Phi(x',\Theta,\Theta\cdot x')}\int_{\mathbb{R}}w_\epsilon(\Theta\cdot x'-p)\hat{f}_{pp}^{(\Phi)}(\theta,p)dpd\theta$$

$$= -\frac{1}{4\pi}\int_{S^1}\frac{1}{\Phi(x',\Theta,\Theta\cdot x')}\int_{\mathbb{R}}w_\epsilon''(\Theta\cdot x'-p)\hat{f}^{(\Phi)}(\theta,p)dpd\theta, \qquad (5.7.16)$$

where W_ϵ is assumed to be radial and $w_\epsilon = RW_\epsilon$ is the regular Radon transform of W_ϵ. The importance of Equation (5.7.16) becomes obvious if we choose W_ϵ to be a sequence of mollifiers with radius of support ϵ, $\epsilon\to 0$. In this case, (5.7.15) implies

$$\int_{\mathbb{R}^2}W_\epsilon(x'-x)g_\Lambda^{(\Phi)}(x',x)dx\to f_\Lambda^{(\Phi)}(x'), \quad x'\notin S, \epsilon\to 0. \qquad (5.7.17)$$

From (5.7.17) it follows that in practice we can compute an approximation to $f_\Lambda^{(\Phi)}$ using the right-hand side of Equation (5.7.16). Clearly, the numerical scheme based on such an approximation is as efficient as in the regular local tomography case.

Now let us consider what happens in a neighborhood of S. Similarly to (5.7.3), (5.7.4), (5.7.7), and (5.7.12), we obtain an asymptotic formula for $g_\Lambda^{(\Phi)}(x',x)$, $x\to S$:

$$g_\Lambda^{(\Phi)}(x',x_s+hn(x_s)) = \frac{D(x_s)}{\pi}\frac{G(x',x_s)}{h}(1+O(h\ln|h|)), \quad h\to 0, \qquad (5.7.18)$$

$$G(x', x_s) = \frac{1}{2}\left[\frac{\Phi(x_s, n(x_s), n(x_s) \cdot x_s)}{\Phi(x', n(x_s), n(x_s) \cdot x')} + \frac{\Phi(x_s, -n(x_s), -n(x_s) \cdot x_s)}{\Phi(x', -n(x_s), -n(x_s) \cdot x')}\right]$$

$$= 1 + A(x', x_s), \quad A \in C^\infty,$$

$$|A(x', x_s)| \le O(|x' - x_s|) \text{ as } x' \to x_s. \tag{5.7.19}$$

We have omitted $n(x_s)$ from the arguments of G and A, because $n(x_s)$ is a function of x_s.

Comparing (5.4.4) and (5.7.7) with (5.7.18), we see that in Equation (5.7.18) there is an additional factor $G(x', x_s)$. Since, in practice, we compute mollified versions of functions f_Λ and $g_\Lambda^{(\Phi)}$ (e.g., according to (5.7.16)), it is important to compare the behavior of $W_\epsilon * f_\Lambda$ and $W_\epsilon * g_\Lambda^{(\Phi)}$ in a neighborhood of S. For simplicity, we shall compare mollifications of only the principal terms of the asymptotic expansions of the regular local tomography function f_Λ and of the function $g_\Lambda^{(\Phi)}$. Let W_ϵ be a sequence of sufficiently smooth mollifiers with the same properties as in Section 5.4 (see below (5.4.2)). Let us assume that ϵ is sufficiently small, so that we can neglect the curvature of S in ϵ-neighborhood of x_s. Define $x' = x_s' + h'n_0$, $x_s' \in S$, where n_0 is a unit vector perpendicular to S at x_s' and $n_0 \approx n(x_s)$. Choose a coordinate system such that its origin coincides with x_s', and a unit vector along the first axis is n_0. In this coordinate system we have $x' = (x_s', h')$, $|A(x', x_s)| = |A(x_s', h', x_s)| \le O\left[((x_s' - x_s)^2 + (h')^2)^{1/2}\right]$. Since now we are interested in what happens in a neighborhood of S, and the radius of mollification ϵ is a natural choice for a scaling factor, we put below $h' = q'\epsilon$, $|q'| < \infty$, $\epsilon \to 0$. For regular local tomography, we have using the scaling property of W_ϵ and Equation (5.4.4):

$$\int W_\epsilon(x' - x) f_\Lambda(x) dx$$

$$\sim \iint W_\epsilon(x_s' - x_s, h' - h) \frac{D(x_s)}{\pi h} dx_s\, dh$$

$$= \frac{D(x_s')}{\pi} \int \frac{1}{h}\left[\int W_\epsilon(x_s' - x_s, h' - h) dx_s\right] dh$$

$$+ \int \frac{D(x_s) - D(x_s')}{\pi} \int \frac{W_\epsilon(x_s' - x_s, h' - h)}{h} dh\, dx_s$$

$$= \frac{D(x_s')}{\pi\epsilon} \int \frac{w_1(q' - q)}{q} dq + \int O(\epsilon^2) \int \frac{W_1(q' - q, t' - t)}{\epsilon t} dt\, dq$$

$$= \frac{1}{\epsilon}\frac{D(x_s')}{\pi} \int \frac{w_1(q' - q)}{q} dq + O(\epsilon), \quad x' = x_s' + (q'\epsilon)n_0, \ \epsilon \to 0, \tag{5.7.20}$$

where we have used that under our assumptions $W_\epsilon(\cdot, \cdot)$ is an even function of its arguments and denoted $w_1 = RW_1$. In (5.7.20) and in (5.7.21) below, we assume that integrals are taken over bounded domains where corresponding mollifiers: W_ϵ, W_1, or w_1 are positive. Similarly, substituting (5.7.18) and (5.7.19) into the left-hand side of (5.7.16), we obtain:

$$\int W_\epsilon(x'-x)g_\Lambda^{(\Phi)}(x',x)dx$$

$$\sim \iint W_\epsilon(x'_s - x_s, h' - h)\frac{D(x_s)}{\pi}\frac{1+A(x'_s,h',x_s)}{h}dx_s\,dh$$

$$= \iint W_1(t'-t, q'-q)\frac{D(\epsilon t)}{\pi}\frac{1+A(\epsilon t',\epsilon q',\epsilon t)}{\epsilon q}dt\,dq$$

$$= \frac{1}{\epsilon}\int\frac{1}{q}\left[\int W_1(t'-t,q'-q)\frac{D(\epsilon t)}{\pi}dt\right]dq$$

$$+ \int\frac{D(\epsilon t)A(\epsilon t',\epsilon q',\epsilon t)}{\pi\epsilon}\left[\int\int\frac{W_1(t'-t,q'-q)}{q}dq\right]dt$$

$$= \frac{1}{\epsilon}\frac{D(x'_s)}{\pi}\int\frac{w_1(q'-q)}{q}dq + O(\epsilon)$$

$$+ \int O(1)\int\frac{W_1(t'-t,q'-q)}{q}dq\,dt$$

$$= \frac{1}{\epsilon}\frac{D(x'_s)}{\pi}\int\frac{w_1(q'-q)}{q}dq + O(1), \quad x' = x'_s + (q'\epsilon)n_0,\ \epsilon \to 0. \tag{5.7.21}$$

Comparing (5.7.20) and (5.7.21), we see that the leading terms in the expansion as $\epsilon \to 0$ of the mollified regular local tomography function is the same as that of the local tomography function for the generalized Radon transform.

Now let us consider regularization of the second local tomography function defined by (5.7.8). Applying (5.7.13), we get

$$(W * \tilde{f}_\Lambda^{(\Phi)})(x) = -\frac{1}{4\pi}\int_{S^1}\int_\mathbb{R} w''_\epsilon(\Theta \cdot x' - p)\hat{f}^{(\Phi)}(\theta,p)dpd\theta. \tag{5.7.22}$$

Suppose that Φ is known. Since Equation (5.7.22) is exact, it is tempting to use it in place of (5.7.16) with subsequent dividing of $(W * \tilde{f}_\Lambda^{(\Phi)})(x_s)$ by $\tilde{a}_1(x_s, x_s, n(x_s))$. However, this approach appears to have the same order of accuracy as the one based on Equation (5.7.16) (see (5.7.21)). Indeed, let $x' = x'_s + (q'\epsilon)n_0$ be a point in a neighborhood of S. Since n_0 is not known, we can estimate it as a unit vector in the direction of the gradient of $W * \tilde{f}_\Lambda^{(\Phi)}$ at x'. Let such an estimate be denoted $n_0(x')$. Similarly to (5.7.20), we have using (5.7.12):

$$(W * \tilde{f}_\Lambda^{(\Phi)})(x'_s+(q'\epsilon)n_0) \sim \frac{1}{\epsilon}\frac{D(x'_s)\tilde{a}_1(x'_s, x'_s, n_0)}{\pi}\int\frac{w_1(q'-q)}{q}dq + O(\epsilon). \tag{5.7.23}$$

Dividing both sides of (5.7.23) by $\tilde{a}_1(x', x', n_0(x'))$ yields

$$\frac{(W * \tilde{f}_\Lambda^{(\Phi)})(x')}{\tilde{a}_1(x', x', n_0(x'))} \sim \frac{1}{\epsilon} \frac{D(x'_s)\tilde{a}_1(x'_s, x'_s, n_0)}{\pi(\tilde{a}_1(x'_s, x'_s, n_0) + O(\epsilon))} \int \frac{w_1(q' - q)}{q} dq + O(\epsilon)$$

$$= \frac{1}{\epsilon} \frac{D(x'_s)}{\pi} \int \frac{w_1(q' - q)}{q} dq + O(1),$$

$$x' = x'_s + (q'\epsilon)n_0, \ \epsilon \to 0,$$

which coincides with the right-hand side of (5.7.21).

5.8. Local tomography for the limited-angle data

Let $b(\Theta)$, $\Theta \in S^1$, be a piecewise-smooth even function: $b(\Theta) = b(-\Theta)$. Consider the family of local tomography functions ψ, defined by (5.2.4):

$$\psi(x) := -\frac{1}{4\pi} \int\limits_0^{2\pi} b(\Theta)\hat{f}_{pp}(\theta, \Theta \cdot x)d\theta. \qquad (5.8.1)$$

Suppose the Radon transform $\hat{f}(\theta, p)$ is given for $\theta \in [\theta_1, \theta_2]$ and $p \in \mathbb{R}$. Since $\hat{f}(\theta, p)$ is even: $\hat{f}(\theta + \pi, p) = \hat{f}(\theta, -p)$, we may assume that $\hat{f}(\theta, p)$ is known for $\theta \in [\theta_1, \theta_2]$, $\theta \in [\theta_1 + \pi, \theta_2 + \pi]$ and $p \in \mathbb{R}$. Denote $\Omega :=$ $\{\Theta \in S^1 : \theta \in [\theta_1, \theta_2] \text{ or } \theta \in [\theta_1 + \pi, \theta_2 + \pi]\}$. Putting $b(\Theta) = 0, \Theta \in \Omega$, in (5.8.1), we obtain the local tomography function which uses only the known data. From (5.8.1) one easily gets using the Fourier slice theorem (see Section 5.2.1):

$$\psi = \mathcal{F}^{-1}(b(\xi_0)|\xi|\hat{f}(\xi)), \quad \hat{f} = \mathcal{F}f, \ \xi_0 = \frac{\xi}{|\xi|}. \qquad (5.8.2)$$

Let us suppose for simplicity that $b(\Theta) = 1, \Theta \in \Omega$. Note that in this case $b(\Theta)$ is discontinuous. In Section 5.2, the function b was taken from the class $C^\infty(S^1)$. Equation (5.8.2) shows that ψ is the result of action on f of a PDO with the symbol, which is discontinuous along the boundary of a cone. Therefore, the theory developed in Section 5.9.4 is directly applicable to the analysis of singularities of ψ. The results from Section 5.9.4 imply that the singular support of ψ consists of

(1) corner points of S;
(2) points $x_s \in S$, where S is smooth and $n(x_s) \in \Omega$; and
(3) the lines which are tangent to S and which are perpendicular to vectors Θ_1 or Θ_2, $\Theta_k = (\cos\theta_k, \sin\theta_k), k = 1, 2.$

Pick any x_s such that $n(x_s)$ is strictly inside Ω. Then, using Equation (5.9.15), we get

$$\psi(x_s + hn(x_s)) \sim \frac{D(x_s)}{\pi}h^{-1}, \ h \to 0, \ x_s \in S. \qquad (5.8.3)$$

Now pick any line $L := \{x : \Theta \cdot x = p\}$, where $\Theta = \Theta_1$ or $\Theta = \Theta_2$, which is tangent to S. Take, for example, $\Theta = \Theta_1$ and let y_0 be a point of contact. Fix any $x_0 \in L, x_0 \neq y_0$. Since the angular interval of available data is symmetrical, we may always assume that Θ_1 points in the same direction as $n(y_0)$ (that is from the center of curvature towards y_0). Thus $\Theta_1 = n(y_0)$ and, according to Theorem 5.9.4, Equation (5.9.47) yields with $B_0(x, y, \Theta) = b(\Theta) \equiv 1, \Theta \in \Omega$, and $\gamma = 1$:

$$\psi(x_0 + hn(y_0)) \sim ie^{-i\frac{3\pi}{4}} \frac{\sqrt{R_0}D(y_0)}{(2\pi)^{1.5}(x_0 - y_0) \cdot n^\perp(y_0)} \int_0^\infty \psi_{-0.5}(t, x)e^{-ith} dt$$

$$+ie^{i\frac{3\pi}{4}} \frac{\sqrt{R_0}D(y_0)}{(2\pi)^{1.5}(x_0 - y_0) \cdot (-n^\perp(y_0))} \int_0^\infty \psi_{-0.5}(t, x)e^{ith} dt,$$

where $n^\perp(y_0)$ is a unit vector perpendicular to $n(y_0)$ such that $n^\perp(y_0)$ is obtained by rotating $n(y_0)$ 90 degrees counter clockwise. The first and the second terms on the right-hand side of the last equation correspond to the contributions from the discontinuities of $b(\xi_0)$ at $\xi_0 = n(y_0)$ and $\xi_0 = -n(y_0)$, respectively. In the second term we had to replace h by $-h$ so that the point under consideration $x_0 + hn(y_0)$ does not change when we replace $n(y_0)$ by $-n(y_0)$. After simple transformations, we get

$$\psi(x_0 + hn(y_0)) \sim \frac{2\sqrt{R_0}D(y_0)}{(2\pi)^{1.5}(x_0 - y_0) \cdot n^\perp(y_0)} \mathrm{Re}\left[ie^{-i\frac{3\pi}{4}} \int_0^\infty t_{-0.5}e^{-ith} dt\right]$$

$$= \frac{2\sqrt{R_0}D(y_0)}{(2\pi)^{1.5}(x_0 - y_0) \cdot n^\perp(y_0)} \mathrm{Re}\left[-i\Gamma(1/2)(h - i0)^{-1/2}\right]$$

$$= \frac{\sqrt{2R_0}D(y_0)}{2\pi(x_0 - y_0) \cdot n^\perp(y_0)} \mathrm{Re}(-ih_+^{-1/2} + h_-^{-1/2})$$

$$= \frac{\sqrt{2R_0}D(y_0)}{2\pi(x_0 - y_0) \cdot n^\perp(y_0)} h_-^{-1/2}, \quad h \to 0. \quad (5.8.4)$$

From (5.8.4) we see that the leading singular term of $\psi(x)$ as $x \to x_0 \in L, x_0 \neq y_0$, is on the same side of L as S in a neighborhood of y_0. In (5.8.4) we took into account the contribution of the leading term of $\psi_{-0.5}$ as $t \to \infty$. The second term of the expansion of $\psi_{-0.5}$ is $O(t^{-1.5})$ as $t \to \infty$. Since the function $\int_0^\infty O(t^{-1.5})e^{-ith} dt$ is continuous at $h = 0$, together with (5.8.4) this implies that there exists a limit of $\psi(x)$ as x approaches $x_0 \in L, x_0 \neq y_0$ from the side of L opposite to the location of S in a neighborhood of y_0.

Exercise 5.8.1. Analyze the singularities of ψ in a neighborhood of the line $L = \{x : \Theta \cdot x = p\}$, where $\Theta = \Theta_2$.

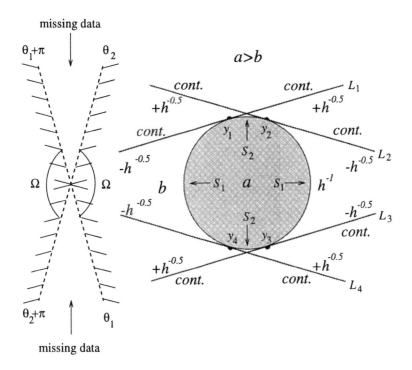

FIGURE 5.8.1. Schematic behavior of the local tomography function ψ in a neighborhood of its singular support in case of the limited angle data. Ω - angular interval of available data, S_1 - pieces of boundary of the phantom which are in a singular support of ψ, S_2 - pieces of boundary of the phantom which are not in a singular support of ψ. One has singsupp $\psi = S_1 \cup L_1 \cup L_2 \cup L_3 \cup L_4$.

Equations (5.8.3) and (5.8.4) are illustrated by Figure 5.8.1, where the behavior of ψ in a neighborhood of singsupp ψ is sketched. The shaded disc represents a phantom, which is more dense than the surrounding medium. According to (5.8.3), $\psi(x) \sim \text{const} h^{-1}$ as $h = \text{dist}(x, S_1) \to 0$, where S_1 is a piece of S which is in singsupp ψ. Now let us consider, for example, the line L_1 (see Figure 5.8.1). The function ψ is continuous as x approaches L_1 from the side opposite to S. In Figure 5.8.1 this is denoted by '*cont.*'. Equation (5.8.4) implies that $\psi(x)$ is proportional to $h^{-0.5}$ as $h = \text{dist}(x, L_1) \to 0$ if x approaches L_1 from the side of S. Moreover, since the disc is more dense than the surrounding medium, the coefficient of proportionality is positive to the right of the point of contact y_1, and it is negative to the left of y_1. In Figure 5.8.1 this is denoted by $+h^{-0.5}$ and $-h^{-0.5}$, respectively.

In Figure 5.8.2 we see the density plot of $\psi(x)$ computed for the same phantom as in Figure 5.8.1. The intervals of missing data are $[80°, 100°]$

and $[260°, 280°]$. The vertical cross-section of Figure 5.8.2 along the black line is shown in Figure 5.8.3. Let us note that Figures 5.8.1 and 5.8.3 are in complete agreement.

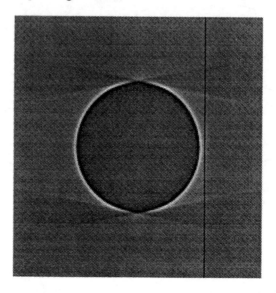

FIGURE 5.8.2. Density plot of the local tomography function computed from the limited angle data. The phantom consists of one disk.

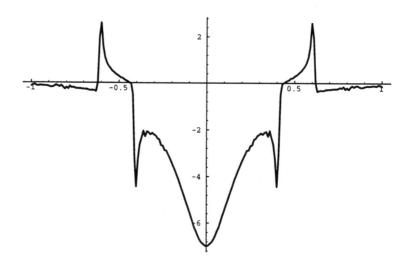

FIGURE 5.8.3. Vertical cross-section of the local tomography function.

5.9. Asymptotics of pseudodifferential operators, acting on a piecewise-smooth function f, near the singular support of f

5.9.1. The case of a convex boundary

In this section we consider the action of a PDO $\mathcal{B}$ on a piecewise-smooth function f:

$$(\mathcal{B}f)(x) = \frac{1}{(2\pi)^2} \int_{\mathbb{R}^2} B(\xi)\tilde{f}(\xi)e^{-i\xi \cdot x}d\xi, \quad B \in C^\infty(\mathbb{R}^2), \ \tilde{f} = \mathcal{F}f.$$

The case of more general pseudodifferential operators is considered in Section 5.9.3. Let us suppose that the amplitude B satisfies the following asymptotic condition

$$B(\xi) = B(t\Theta) = \psi_\gamma(t, \Theta)b(\Theta), \ t = |\xi|, \ \Theta = \xi/|\xi|,$$
$$b \in C^\infty(S^1), \ \min_{\Theta \in S^1} |b(\Theta)| > 0, \tag{5.9.1}$$

for some functions ψ_γ and b. In (5.9.1) and everywhere in Section 5.9, by $\psi_\gamma(t, \zeta)$ we denote any C^∞ function with properties:

(i) $\psi_\gamma(t, \zeta) = t^\gamma + O(t^{\gamma-1}), \ t \to \infty$;
(ii) $O(t^{\gamma-1})$ is uniform with respect to other parameters ζ, ψ_γ depends on; and
(iii)

$$\frac{\partial^k}{\partial t^k}\frac{\partial^\alpha}{\partial\zeta^\alpha}\psi_\gamma(t, \zeta) = O(t^{\gamma-k}), \ k = 0, 1, \ldots,$$

for any multiindex α.

In particular, $\zeta = \Theta$ in (5.9.1). Also, wherever variables $\Theta \in S^1$ and θ, $0 \le \theta \le 2\pi$, are used in the same equation, they are assumed to be related by the formula $\Theta = (\sin\theta, \cos\theta)$.

Let us pick any $x_0 \in S$ such that S is smooth in a neighborhood of x_0, and S has positive finite radius of curvature R_0 at x_0. We use the following coordinate system: the origin coincides with the center of curvature of S at x_0, and the x_1-axis is perpendicular to S at x_0. Clearly, one has $x_0 = (R_0, 0)$. Let $w(x)$ be a C_0^∞ cut-off function with a sufficiently small support supp w, and such that $x_0 \in$ supp w and $w(x) = 1$ in an open set U, $x_0 \in U \subset$ supp w. Denoting $f_w(x) = f(x)w(x)$, we have $\mathcal{B}f - \mathcal{B}f_w \in C^\infty(U)$. Using the Fourier slice theorem (see Equation (2.1.20)), we can write

$$(\mathcal{B}f_w)(x) = \frac{1}{(2\pi)^2} \int_0^\infty \int_0^{2\pi} B(t\Theta)\tilde{f}_w(t\Theta)e^{-it\Theta \cdot x} d\theta \, tdt$$

$$= \frac{1}{(2\pi)^2} \int_0^\infty \int_0^{2\pi} B(t\Theta) \int_{-\infty}^\infty \hat{f}_w(\theta, p)e^{ipt}dp \, e^{-it\Theta \cdot x} d\theta \, tdt, \tag{5.9.2}$$

where $\tilde{f}_w(t\theta)$ and $\hat{f}_w(\theta,p)$ are the Fourier and Radon transforms of $f_w(x)$, respectively. Since $\hat{f}_w(\theta,p)$ is even: $\hat{f}_w(\theta,p) = \hat{f}_w(\theta+\pi,-p)$, Equation (5.9.2) becomes

$$(\mathcal{B}f_w)(x) = \frac{1}{(2\pi)^2} \int\limits_0^\infty \int\limits_{-\pi/2}^{\pi/2} B(t\Theta) \int\limits_{-\infty}^\infty \hat{f}_w(\theta,p)e^{ipt} dp\, e^{-it\Theta\cdot x} d\theta\, tdt$$

$$+ \frac{1}{(2\pi)^2} \int\limits_0^\infty \int\limits_{-\pi/2}^{\pi/2} B(-t\Theta) \int\limits_{-\infty}^\infty \hat{f}_w(\theta,p)e^{-ipt} dp\, e^{it\Theta\cdot x} d\theta\, tdt. \tag{5.9.3}$$

Let the first integral in (5.9.3) be denoted $B_1(x)$. We introduce some notation. Let $[-\theta_1,\theta_2] \subset [-\pi/2,\pi/2]$, $\theta_1,\theta_2 > 0$, be the interval such that $\forall\theta$, $\theta \in [-\theta_1,\theta_2]$, the line $\Theta\cdot x = p$ is tangent to $S \cap \text{supp}\, w$ for some p, $p := p_0(\theta)$. Since the radius of curvature R_0 is finite and $\text{supp}\, w$ can be made arbitrarily small, we may assume that S does not have inflection points inside $\text{supp}\, w$. The lines tangent to $S \cap \text{supp}\, w$ at the end points are defined as limits of neighboring tangent lines. The Radon transform $\hat{f}_w(\theta,p)$ is a C_0^∞ function in p for $\theta \in [-\pi/2,\pi/2] \setminus [-\theta_1,\theta_2]$ (see Section 4.2). Therefore, we can write

$$B_1(x) - \frac{1}{(2\pi)^2} \int\limits_0^\infty \int\limits_{-\theta_1}^{\theta_2} B(t\Theta) \int\limits_{-\infty}^\infty \hat{f}_w(\theta,p)e^{ipt} dp\, e^{-it\Theta\cdot x} d\theta\, tdt$$

$$= \frac{1}{(2\pi)^2} \int\limits_0^\infty \left(\int\limits_{-\pi/2}^{-\theta_1} + \int\limits_{\theta_2}^{\pi/2} \right) B(t\Theta) \int\limits_{-\infty}^\infty \hat{f}_w(\theta,p)e^{ipt} dp\, e^{-it\Theta\cdot x} d\theta\, tdt$$

$$\in C^\infty(\mathbb{R}^2), \tag{5.9.4}$$

because

$$\left| \int\limits_{-\infty}^\infty \hat{f}_w(\theta,p)e^{ipt} dp \right| = o(t^{-N}), \quad t \to \infty, \quad \forall N > 0,$$

$$\theta \in [-\pi/2,\pi/2] \setminus [-\theta_1,\theta_2].$$

Let $\tilde{x}(\theta)$, $\theta \in [-\theta_1,\theta_2]$, be the point of contact between S and the line $\Theta\cdot x = p_0(\theta)$ (see Figure 5.9.1), $R(\theta)$ be the radius of curvature of S at $\tilde{x}(\theta)$, and $D_w(\theta)$ be the jump of $f_w(x)$ across S at $x = \tilde{x}(\theta) \in S$. To be more precise, $D_w(\theta) = \lim_{t\to0+} \left[f_w(\tilde{x}(\theta) + tn(\theta)) - f_w(\tilde{x}(\theta) - tn(\theta)) \right]$, where $n(\theta)$ is a vector pointing at $\tilde{x}(\theta)$ away from the origin.

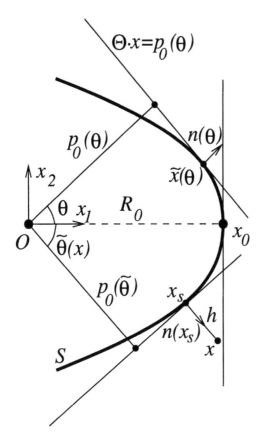

FIGURE 5.9.1.   Illustration of the notation in case of the convex curve S

According to the results obtained in Section 4.2, the function $\hat{f}_w(\theta, p)$ can be represented as $\hat{f}_w(\theta, p) = -\varphi_1(\theta, p)2\sqrt{2R(\theta)}D_w(\theta)(p-p_0(\theta))_-^{0.5} + \varphi_2(\theta, p)$ for each $\theta \in [-\theta_1, \theta_2]$, where $(p - p_0)_- = \max(-(p - p_0), 0)$, φ_1, φ_2 are C_0^∞ in the p and θ variables, and $\varphi_1(p_0(\theta), \theta) = 1$. Using this, we obtain for the integral on the left-hand side of (5.9.4):

$$B_2(x) := \frac{1}{(2\pi)^2} \int_0^\infty \int_{-\theta_1}^{\theta_2} B(t\Theta) \int_{-\infty}^\infty \hat{f}_w(\theta,p)e^{ipt}dp\, e^{-it\Theta\cdot x}d\theta\, tdt$$

$$= \frac{-1}{(2\pi)^2} \int_0^\infty \int_{-\theta_1}^{\theta_2} B(t\Theta)2\sqrt{2R(\theta)}D_w(\theta)$$

$$\times \left\{ \int_{-\infty}^\infty \varphi_1(\theta,p)(p-p_0(\theta))_-^{0.5}e^{ipt}dp \right\}e^{-it\Theta\cdot x}d\theta\, tdt$$

$$+ \frac{1}{(2\pi)^2} \int_0^\infty \int_{-\theta_1}^{\theta_2} B(t\Theta) \int_{-\infty}^\infty \varphi_2(\theta,p)e^{ipt}dp\, e^{-it\Theta\cdot x}d\theta\, tdt. \tag{5.9.5}$$

Since φ_2 is C^∞ in p and θ, and φ_2 is compactly supported in p, the last integral in (5.9.5) is a C^∞-function in x. Applying Lemma 14.5.5 to the integral in braces, we get

$$\int_{-\infty}^\infty \varphi_1(\theta,p)(p-p_0(\theta))_-^{0.5}e^{ipt}dp = \Gamma(1.5)e^{-i3\pi/4}e^{ip_0(\theta)t}\psi_{-1.5}(t,\theta), \tag{5.9.6a}$$

$$\psi_{-1.5}(t,\theta) \sim \sum_{k=0}^\infty a_k(\theta)t^{-(k+1.5)},$$

$$a_k(\theta) = (-1)^k \frac{\partial^k \varphi_1(\theta,p)}{\partial p^k}\bigg|_{p=p_0(\theta)} \frac{\Gamma(k+1.5)}{k!\Gamma(1.5)}e^{-i\pi k/2}, \tag{5.9.6b}$$

where '$\sim$' denotes an asymptotic expansion as $t \to \infty$ (see Section 14.5.1). Clearly, the integral in (5.9.6a) is a C^∞ function of t and θ, therefore $\psi_{-1.5}(t,\theta)$ is also C^∞ in t and θ. Using the equality $\varphi_1(p_0(\theta),\theta) = 1$, we check that $a_0 \equiv 1$, hence $\psi_{-1.5}(t,\theta) = t^{-1.5}(1+O(1/t))$ and $O(1/t)$ is uniform in θ for $\theta \in [-\theta_1,\theta_2]$. Thus the function $\psi_{-1.5}$ in (5.9.6) has properties (i) and (ii) (see below (5.9.1)). Differentiating on both sides of (5.9.6a) with respect to θ and using that the asymptotic expansion in (5.9.6b) can be differentiated with respect to t, we verify property (iii).

Equations (5.9.5) and (5.9.6a) yield

$$B_2(x) + \frac{1}{(2\pi)^2} \int_0^\infty \int_{-\theta_1}^{\theta_2} B(t\Theta)2\sqrt{2R(\theta)}D_w(\theta)\Gamma(1.5)e^{-3i\pi/4}e^{ip_0(\theta)t} \times$$

$$\times\, \psi_{-1.5}(t,\theta)e^{-it\Theta\cdot x}d\theta\, tdt \in C^\infty(\mathbb{R}^2). \tag{5.9.7}$$

Using asymptotics (5.9.1) of the symbol $B(t\Theta)$ of the PDO $\mathcal{B}$: $B(t\Theta) = \psi_\gamma(t,\theta)b(\Theta)$, the property $\psi_{\gamma_1+\gamma_2}(t,\cdot) = \psi_{\gamma_1}(t,\cdot)\psi_{\gamma_2}(t,\cdot)$, and denoting $g(\theta) = 2\sqrt{2R(\theta)}D_w(\theta)b(\Theta)$, the integral over θ in (5.9.7) can be written as

$$G(t,x) = \int_{-\theta_1}^{\theta_2} \psi_{\gamma-1.5}(t,\theta)g(\theta)e^{it(p_0(\theta)-\Theta\cdot x)}d\theta. \tag{5.9.8}$$

To compute the asymptotics of $G(t,x)$ as $t \to \infty$, we need to find stationary points of the function $a(\theta,x) := p_0(\theta) - \Theta \cdot x$ in θ. One has:

Lemma 5.9.1.

(1) $p_0 \in C^\infty([-\theta_1,\theta_2])$ if the interval $[-\theta_1,\theta_2]$ is sufficiently small;
(2) $p_0(0) = R_0$, where R_0 is the radius of curvature of S at x_0;
(3) $p_0'(0) = p_0''(0) = 0$, $p_0'(\pi) = p_0''(\pi) = 0$; and
(4) $a_{\theta\theta}(0,x_0) \neq 0$.

Proof. See Section 5.9.5.1.

The stationary point of $a(\theta,x)$ is found from the equation

$$a_\theta(\theta,x) := \frac{\partial a(\theta,x)}{\partial\theta} = p_0'(\theta) + x_1\sin\theta - x_2\cos\theta = 0, \quad x = (x_1,x_2). \tag{5.9.9}$$

Let $x = x_0$, that is $x_1 = R_0$, $x_2 = 0$. Using assertion (3) of Lemma 5.9.1, we get $a_\theta(0,x_0) = 0$. Since $a_{\theta\theta}(0,x_0) \neq 0$, then, if the interval $[-\theta_1,\theta_2]$ is sufficiently small (this interval shrinks to zero as the support of w shrinks to zero), there are no stationary points of $a(\theta,x_0)$, except $\theta = 0$, inside the interval $[-\theta_1,\theta_2]$. Differentiating (5.9.9) with respect to x_1 and x_2 and assuming that the stationary point θ_0 is a function of x: $\theta_0 = \theta(x)$, we find $\partial\theta(x)/\partial x_1|_{x=x_0} = 0$ and $\partial\theta(x)/\partial x_2|_{x=x_0} = 1/R_0$. Thus the implicit function theorem implies that $\theta(x)$ is a C^∞ function in a neighborhood of $x = x_0$. Note that $\theta(x_0) = 0$. According to the choice of the cut-off function, $D_w(\theta)$ and, hence, $g(\theta)$ both vanish with all derivatives at $\theta = -\theta_1$ and $\theta = \theta_2$. Using property (iii) of the function $\psi_{\gamma-1.5}$ from (5.9.8), Remark 14.5.1, and Theorem 14.5.2, we get

$$G(t,x) = \sqrt{\frac{2\pi}{|S(x)|}}\psi_{\gamma-2}(t,x)g(\theta(x))e^{\frac{i\pi}{4}\operatorname{sgn}(S(x))}e^{ita(\theta(x),x)}, \tag{5.9.10}$$

where $S(x) = \partial^2 a(\theta,x)/\partial\theta^2|_{\theta=\theta(x)}$. In what follows we assume that x is sufficiently close to x_0. Note that if assumption (5.9.1) holds, then $g(\theta(x)) \neq 0$. From (5.9.8) we see that $G(t,x)$ is C^∞ in t and x. Therefore, $\psi_{\gamma-2}(t,x)$ is also C^∞ in t and x. Now let us check that the function $\psi_{\gamma-2}(t,x)$ has properties (i) – (iii). Since the function $\psi_{\gamma-1.5}$ from

(5.9.8) has properties (i) – (iii), Theorem 14.5.2 immediately implies that $\psi_{\gamma-2}(t,x)$ has properties (i) and (ii). From (5.9.10) we have

$$\psi_{\gamma-2}(t,x) = A(x)G(t,x)e^{-ita(\theta(x),x)}$$

$$= A(x) \int_{-\theta_1}^{\theta_2} \psi_{\gamma-1.5}(t,\theta)g(\theta)e^{it(a(\theta,x)-a(\theta(x),x))}\, d\theta,$$

where $A(x)$ is C^∞ in x. Differentiating the last equality with respect to t and x, using that $a(\theta,x) - a(\theta(x),x) = (\theta - \theta(x))^2\psi(\theta,x)$ for some smooth $\psi(\theta,x)$ such that $\psi(\theta(x),x) \neq 0$, and appealing to Remark 14.5.1 and Theorem 14.5.2 one more time, we verify property (iii). Here the subscript 'x' denotes the first order partial derivative with respect to either x_1 or x_2.

For a given x, find x_s such that $|x_s - x| = \min_{y\in S}|y - x|$ and consider the line $\tilde{\Theta}\cdot y = p_0(\tilde{\theta})$ tangent to S at the point x_s (see Figure 5.9.1). Clearly, $\tilde{\theta} = \tilde{\theta}(x)$.

Lemma 5.9.2. *One has:*

(1) $\theta(x) = \tilde{\theta}(x)$ *and* $\Theta(x) = n(x_s)$, *where* $\Theta(x) = (\cos\theta(x),\sin\theta(x))$ *and* $n(x_s)$ *is a unit vector normal to S at x_s pointing away from the center of curvature;*

(2) $R(\theta(x)) = R(x_s)$ *and* $D(\theta(x)) = D(x_s)$, *where $R(x_s)$ and $D(x_s)$ are the radius of curvature of S at x_s and the jump of f across S at x_s, respectively;*

(3) $a(\theta(x),x) = -n(x_s)\cdot(x - x_s)$; *and*

(4) $S(x) = R(x_s) + n(x_s)\cdot(x - x_s)$.

Proof. See Section 5.9.5.2.

Now let x be represented in the form (see Figure 5.9.1)

$$x = x_s + hn(x_s), \quad x_s \in S. \tag{5.9.11}$$

Lemma 5.9.2 and Equation (5.9.10) yield

$$G(t,x) = g(\theta(x_s))\sqrt{\frac{2\pi}{R(x_s) + h}}e^{i\pi/4}\psi_{\gamma-2}(t,x)e^{-ith}$$

for x in a neighborhood of $x = x_0$. Denoting the integral in (5.9.7) by

$B_3(x)$, we obtain from (5.9.8) and the last equation

$$B_3(x) = -\frac{\Gamma(1.5)}{(2\pi)^2}e^{-3i\pi/4}\int_0^\infty G(t,x)t\,dt$$

$$= -\frac{\Gamma(1.5)\sqrt{2\pi}}{(2\pi)^2}e^{-i\pi/2}\frac{g(\theta(x_s))}{\sqrt{R(x_s)+h}}\int_0^\infty \psi_{\gamma-1}(t,x)e^{-ith}\,dt$$

$$= \frac{b(n(x_s))D_w(x_s)}{2\pi}\sqrt{\frac{R(x_s)}{R(x_s)+h}}\,i\int_0^\infty \psi_{\gamma-1}(t,x)e^{-ith}\,dt. \tag{5.9.12}$$

The second integral in (5.9.3) can be estimated similarly, and we get

$$(\mathcal{B}f_w)(x) - \frac{D_w(x_s)}{2\pi}\sqrt{\frac{R(x_s)}{R(x_s)+h}}\,i\bigg\{b(n(x_s))\int_0^\infty \psi_{\gamma-1}(t,x)e^{-ith}\,dt$$

$$- b(-n(x_s))\int_0^\infty \psi_{\gamma-1}(t,x)e^{ith}\,dt\bigg\} \in C^\infty(\mathbb{R}^2).$$

According to the choice of the cut-off function, the jumps of f and f_w across $x \in S$ in a neighborhood of x_0 are the same, therefore we obtain from the last equation the following result.

Theorem 5.9.1. *Fix any $x_0 \in S$ such that S is smooth in a neighborhood of x_0 and the radius of curvature R_0 of S at x_0 is finite, $0 < R_0 < \infty$. Let U be a sufficiently small neighborhood of x_0. Consider a point $x = x_s + hn(x_s) \in U$, $x_s \in S$, $h \in \mathbb{R}$, where $n(x_s)$ is a unit vector normal to S at x_s and pointing away from the center of curvature. Let $R(x_s)$ and $D(x_s)$ be the radius of curvature of S at the point x_s and the jump of f across S at x_s, respectively, $D(x_s) = \lim_{t\to0+}\left[f(x_s+tn(x_s))-f(x_s-tn(x_s))\right]$. Suppose property (5.9.1) holds. Then one has*

$$(\mathcal{B}f)(x_s+hn(x_s))$$

$$= \frac{D(x_s)}{2\pi}\sqrt{\frac{R(x_s)}{R(x_s)+h}}\,i\bigg\{b(n(x_s))\int_0^\infty \psi_{\gamma-1}(t,x)e^{-ith}\,dt$$

$$- b(-n(x_s))\int_0^\infty \psi_{\gamma-1}(t,x)e^{ith}\,dt\bigg\} + \eta(x_s + hn(x_s)). \tag{5.9.13}$$

In (5.9.13), $\eta \in C^\infty(U)$ and $\psi_{\gamma-1}$ denotes a function with properties (i) – (iii), where $\zeta = x \in U$ (see below (5.9.1)).

REMARK 5.9.1. Using (5.9.6) and considering all the terms in expansion (5.9.10), one can obtain explicit formulas for the functions $\psi_{\gamma-1}$ in

(5.9.13), thus obtaining a full asymptotics in smoothness of $(\mathcal{B}f)(x)$ as $h \to 0$.

Using Equations (5.9.3), (5.9.12), and Theorem 5.9.1, one proves

Corollary 5.9.1. *Let $B(t\Theta)$ be even, $B(t\Theta) = B(-t\Theta)$. Then one has*

$$(\mathcal{B}f)(x_s + hn(x_s))$$

$$= \frac{D(x_s)}{\pi} \sqrt{\frac{R(x_s)}{R(x_s) + h}}\, b(n(x_s)) Im\left\{ \int_0^\infty \psi_{\gamma-1}(t, x) e^{ith}\, dt \right\}$$

$$+ \eta(x_s + hn(x_s)). \qquad (5.9.14)$$

Let $\sim$ denote the leading singular term in the h variable only. Using formula (14.2.13) for the Fourier transform of x_+^λ:

$$\int_{-\infty}^\infty x_+^\lambda e^{ixh}\, dx = ie^{i\lambda(\pi/2)}\Gamma(\lambda+1)(h+i0)^{-(\lambda+1)}, \quad \lambda \neq -1, -2, \ldots,$$

we obtain from (5.9.14):

Corollary 5.9.2. *Let $\gamma > 0$ and $B(t\Theta)$ be even. Then one has*

$$(\mathcal{B}f)(x_s + hn(x_s)) \sim \frac{D(x_s)b(n(x_s))}{\pi} Im\left\{ e^{i\gamma\pi/2}\Gamma(\gamma)(h+i0)^{-\gamma} \right\},$$

$$h \to 0.$$

In particular,

$$(\mathcal{B}f)(x_s + hn(x_s)) \sim \frac{D(x_s)b(n(x_s))}{\pi} h^{-1}, \ h \to 0, \quad \text{if } \gamma = 1, \quad (5.9.15)$$

and

$$(\mathcal{B}f)(x_s + hn(x_s)) \sim D(x_s)b(n(x_s))\delta'(h), \ h \to 0, \quad \text{if } \gamma = 2.$$

5.9.2. The case of a flat boundary

Consider now the case when S is a line segment in a neighborhood of $x_0 \in S$. Let us choose the coordinate system with the origin at x_0 such that the local equation of S becomes $S = \{(x_1, x_2) : x_1 = 0\}$, that is the x_1-axis is perpendicular to S at the origin. Fix a sufficiently small $a > 0$ such that S is still a line segment inside the square $U := \{(x_1, x_2) : |x_1| \leq a, |x_2| \leq a\}$, and choose a cut-off function $w \in C_0^\infty(U)$ such that $x_0 \in \operatorname{supp} w$ and $w(x) = 1$ in an open set U_1, $x_0 \in U_1 \subset \operatorname{supp} w$.

Clearly, the asymptotics of $(\mathcal{B}f)(x)$ and of $(\mathcal{B}f_w)(x)$ as $x \to 0$ are the same (recall that $f_w = f \cdot w$). The expression for $(\mathcal{B}f_w)(x)$ becomes

$$(\mathcal{B}f_w)(x) = \frac{1}{(2\pi)^2} \int\!\!\!\int_{-\infty}^{\infty} B(\xi_1, \xi_2) e^{-i(\xi_1 x_1 + \xi_2 x_2)}$$

$$\times \int\!\!\!\int_{-a}^{a} f_w(y_1, y_2) e^{i(\xi_1 y_1 + \xi_2 y_2)} dy_1\, dy_2 d\xi_1\, d\xi_2,$$

where $B(\xi_1, \xi_2)$ is a symbol of the PDO $\mathcal{B}$. Clearly, the Fourier transform $\tilde{f}_w(\xi_1, \xi_2)$ of $f_w(y_1, y_2)$ decays faster than any power of $t = (\xi_1^2 + \xi_2^2)^{1/2}$ as $t \to \infty$ outside the cone $K_\delta := \{(\xi_1, \xi_2) : |\xi_2/\xi_1| \leq \delta\}$, where $\delta > 0$ is fixed. Thus we can integrate in the last equation only over the cone K_δ:

$$(\mathcal{B}_1 f_w)(x) = \frac{1}{(2\pi)^2} \int_{-\infty}^{\infty} \int_{-\delta|\xi_1|}^{\delta|\xi_1|} B(\xi_1, \xi_2) e^{-i(\xi_1 x_1 + \xi_2 x_2)}$$

$$\times \int\!\!\!\int_{-a}^{a} f_w(y_1, y_2) e^{i(\xi_1 y_1 + \xi_2 y_2)} dy_1\, dy_2 d\xi_2\, d\xi_1,$$
$$\tag{5.9.16}$$

and $(\mathcal{B}_1 f_w - \mathcal{B}f_w) \in C^\infty(\mathbb{R}^2)$. Note that asymptotics (5.9.1) can be written in a slightly different way

$$B(\xi_1, \xi_2) = \psi_\gamma(t, \theta) b(\Theta) = \psi_\gamma[(\xi_1^2 + \xi_2^2)^{1/2}, \arctan(\frac{\xi_2}{\xi_1})] b(\arctan \frac{\xi_2}{\xi_1})$$

$$= \psi_\gamma(\xi_1, \frac{\xi_2}{\xi_1}) \tilde{b}_+(\frac{\xi_2}{\xi_1}), \quad \xi_1 \to +\infty, \ (\xi_1, \xi_2) \in K_\delta,$$

where $b(\theta) := b(\Theta)$ and $\tilde{b}_+(0) = b(0)$. Similarly,

$$B(\xi_1, \xi_2) = \psi_\gamma(|\xi_1|, \frac{\xi_2}{|\xi_1|}) \tilde{b}_-(\frac{\xi_2}{|\xi_1|}), \quad \xi_1 \to -\infty, \ (\xi_1, \xi_2) \in K_\delta,$$

where $\tilde{b}_-(0) = b(\pi)$. Similarly to (5.9.4), Equation (5.9.16) becomes

$$
(\mathcal{B}_1 f_w)(x) = \frac{1}{(2\pi)^2} \int\limits_0^\infty \int\limits_{-\delta|\xi_1|}^{\delta|\xi_1|} \psi_\gamma(\xi_1, \frac{\xi_2}{\xi_1}) \tilde{b}_+(\frac{\xi_2}{\xi_1}) e^{-i(\xi_1 x_1 + \xi_2 x_2)}
$$

$$
\times \int\!\!\int\limits_{-a}^{a} f_w(y_1, y_2) e^{i(\xi_1 y_1 + \xi_2 y_2)} dy_1 dy_2 d\xi_2 d\xi_1
$$

$$
+ \frac{1}{(2\pi)^2} \int\limits_0^\infty \int\limits_{-\delta|\xi_1|}^{\delta|\xi_1|} \psi_\gamma(\xi_1, \frac{\xi_2}{\xi_1}) \tilde{b}_-(-\frac{\xi_2}{\xi_1}) e^{i(\xi_1 x_1 + \xi_2 x_2)}
$$

$$
\times \int\!\!\int\limits_{-a}^{a} f_w(y_1, y_2) e^{-i(\xi_1 y_1 + \xi_2 y_2)} dy_1 dy_2 d\xi_2 d\xi_1. \tag{5.9.17}
$$

Let us consider the first integral in (5.9.17) separately. Fix any y_2, $|y_2| < a$. Since $f_w(y_1, y_2)$ is discontinuous at $y_1 = 0$ and vanishes with all derivatives at $|y_1| = a$, one has

$$
\int\limits_{-a}^{a} f_w(y_1, y_2) e^{i\xi_1 y_1} dy_1 = i D_w(y_2) \psi_{-1}(\xi_1, y_2), \quad \xi_1 \to +\infty, \tag{5.9.18}
$$

where $D_w(y_2) := \lim_{y_1 \to 0^+} \big(f_w(y_1, y_2) - f_w(-y_1, y_2)\big)$ is the jump of f_w across S at $y = (0, y_2)$. As a simple exercise the reader can check that the function $\psi_{-1}(\xi_1, y_2)$ is C^∞ and that it has properties (i) - (iii). Substitution of (5.9.18) into the first integral in (5.9.17) yields

$$
(\mathcal{B}_1^+ f_w)(x)
$$

$$
= \frac{i}{(2\pi)^2} \int\limits_0^\infty \int\limits_{-\delta|\xi_1|}^{\delta|\xi_1|} \psi_\gamma(\xi_1, \frac{\xi_2}{\xi_1}) \tilde{b}_+(\frac{\xi_2}{\xi_1}) e^{-i(\xi_1 x_1 + \xi_2 x_2)}
$$

$$
\times \int\limits_{-a}^{a} D_w(y_2) \psi_{-1}(\xi_1, y_2) e^{i\xi_2 y_2} dy_2 d\xi_2 d\xi_1 = \tag{5.9.19}
$$

$$
\frac{i}{(2\pi)^2} \int\limits_0^\infty e^{-i\xi_1 x_1} \left[\int\limits_{-a}^{a} \int\limits_{-\delta}^{\delta} (\psi_\gamma(\xi_1, s, y_2) D_w(y_2) \tilde{b}_+(s)) e^{i\xi_1(y_2 - x_2)s} ds dy_2 \right] d\xi_1,
$$

where we have made the change of variables $s = \xi_2/\xi_1$. The asymptotics of the double integral in brackets in (5.9.19) can be computed using the two-dimensional stationary phase method (see Equations (14.5.34) - (14.5.37) in Section 14.5.3)

$$\int\limits_{-a}^{a}\int\limits_{-\delta}^{\delta}\left(\psi_{\gamma}(\xi_1,s,y_2)D_w(y_2)\tilde{b}_+(s)\right)e^{i\xi_1(y_2-x_2)s}\,ds\,dy_2$$

$$= 2\pi D_w(x_2)\tilde{b}_+(0)\psi_{\gamma-1}(\xi_1,x_2),\ \ |x_2| < a,\ \xi_1 \to +\infty.$$

Again, it is clear that the function $\psi_{\gamma-1}(\xi_1,x_2)$ has all the required properties. Since $\tilde{b}_+(0) = b(0)$, Equation (5.9.19) implies

$$(\mathcal{B}_1^+ f_w)(x) = D_w(x_2)b(0)\frac{i}{2\pi}\int\limits_{0}^{\infty}\psi_{\gamma-1}(\xi_1,x_2)e^{-i\xi_1 x_1}\,d\xi_1. \qquad (5.9.20)$$

Comparing (5.9.20) with (5.9.12), we see that (5.9.20) is obtained from (5.9.12) by taking $x_1 = h$ and formally letting $R(x_s) \to \infty$. Thus, Theorem 5.9.1 and Corollaries 5.9.1, 5.9.2 are valid for the case of a locally flat S, that is $R(x_s) = \infty$, $x_s \in U \cap S$.

5.9.3. Further generalizations

The main goal of this section is to generalize Theorem 5.9.1 to PDO from the class $L_{10}^{\gamma}(\mathbb{R}^2)$ (see Section 14.3.3). We shall assume that a PDO $\mathcal{B} \in L_{10}^{\gamma}(\mathbb{R}^2)$ is properly supported. Let $B(x,y,\xi)$ be the amplitude of $\mathcal{B}$, and $\sigma_B(x,\xi)$ be its symbol. Since $B(x,y,\xi) \in S_{10}^{\gamma}(\mathbb{R}^6)$, Equations (14.3.1) and (14.3.18) – (14.3.20) imply that

$$\sigma_B(x,\xi) - B(x,x,\xi) \in S_{10}^{\gamma-1}(\mathbb{R}^4),$$

where $B(x,x,\xi)$ is the principal symbol of the PDO $\mathcal{B}$. Suppose that $B(x,x,\xi)$ has the asymptotics:

$$B(x,x,t\Theta) = \psi_{\gamma}(t,x,\theta)b(x,\Theta),\ t \to \infty,\ b(x,\Theta) \in C^{\infty}(\mathbb{R}^2 \times S^1).$$
$$(5.9.21)$$

Let $x_0 \in S$ be chosen so that

$$b(x,\Theta) \neq 0\ \forall(x,\Theta) \in V \qquad (5.9.22)$$

for some open set V, $(x_0,n(x_0)) \in V \subset \mathbb{R}^2 \times S^1$. Then

$$\sigma_B(x,t\Theta) = B(x,x,t\Theta)(1 + O(1/t)) = \psi_{\gamma}(t,x,\theta)b(x,\Theta),\ t \to \infty.$$

Thus we obtain the following generalization of Theorem 5.9.1:

Theorem 5.9.2. *Let us consider the PDO $B \in L_{10}^{\gamma}(\mathbb{R}^2)$:*

$$(\mathcal{B}f)(x) = \frac{1}{(2\pi)^2} \int\limits_{\mathbb{R}^2} \int\limits_{\mathbb{R}^2} B(x, y, \xi) f(y) e^{i\xi \cdot (x-y)} \, dy \, d\xi,$$

such that its principal symbol satisfies condition (5.9.21). Fix $x_0 \in S$ such that, in addition to the assumptions from Theorem 5.9.1, relation (5.9.22) holds. Then one has

$$(\mathcal{B}f)(x) = \frac{D(x_s)}{2\pi} \sqrt{\frac{R(x_s)}{R(x_s) + h}} \, i \left\{ b(x, n(x_s)) \int\limits_0^{\infty} \psi_{\gamma-1}(t, x) e^{-ith} \, dt \right.$$

$$\left. -b(x, -n(x_s)) \int\limits_0^{\infty} \psi_{\gamma-1}(t, x) e^{ith} \, dt \right\} + \eta(x),$$

$$x = x_s + hn(x_s), \tag{5.9.23}$$

where we have used the notation from Theorem 5.9.1.

Exercise 5.9.1. Prove Theorem 5.9.2 following the proof of Theorem 5.9.1.

Corollary 5.9.3. *Let $b(x, \Theta)$ be even in Θ: $b(x, \Theta) = b(x, -\Theta)$. Then one has*

$$(\mathcal{B}f)(x) = \frac{D(x_s)}{\pi} \sqrt{\frac{R(x_s)}{R(x_s) + h}} \, b(x, n(x_s)) \left[Im \left\{ \int\limits_0^{\infty} \psi_{\gamma-1}(t, x_s) e^{ith} \, dt \right. \right.$$

$$\left. \left. + \int\limits_0^{\infty} O(t^{\gamma-2}) e^{ith} \, dt \right] + \eta(x), \ x = x_s + hn(x_s). \right.$$

$$\tag{5.9.24}$$

REMARK 5.9.2. Equation (5.9.24) is more complicated than its analogue (5.9.14), because in Corollary 5.9.1 we assumed that the amplitude $B(\xi)$ of the PDO B is even, whereas in Corollary 5.9.3 we assumed that only the asymptotics of the principal symbol of B is even. However, the dominant behavior of $(\mathcal{B}f)(x_s + hn(x_s))$ as $h \to 0$ is determined by the first term in brackets on the right-hand side of (5.9.24), which coincides with a similar term in (5.9.14).

Corollary 5.9.4. *Let $\gamma > 0$ and $b(x, \Theta)$ be even in Θ. Then one has*

$$(\mathcal{B}f)(x_s + hn(x_s)) \sim \frac{D(x_s) b(x_s, n(x_s))}{\pi} Im \left\{ e^{i\gamma\pi/2} \Gamma(\gamma)(h + i0)^{-\gamma} \right\},$$

$$h \to 0.$$

In particular,

$$(\mathcal{B}f)(x_s + hn(x_s)) \sim \frac{D(x_s)b(x_s, n(x_s))}{\pi} h^{-1}, \quad h \to 0, \quad \text{if } \gamma = 1,$$

(5.9.25)

and

$$(\mathcal{B}f)(x_s + hn(x_s)) \sim D(x_s)b(x_s, n(x_s))\delta'(h), \quad h \to 0, \quad \text{if } \gamma = 2.$$

Exercise 5.9.2. Check that Theorem 5.9.2 and Corollaries 5.9.3, 5.9.4 are also valid for the case of a locally flat S, and we have to formally let $R(x_s) \to \infty$ in the corresponding equations.

For future reference, we shall need a more precise version of Equation (5.9.25). Suppose that the amplitude $B(x, y, \xi)$ of the PDO $\mathcal{B}$ can be represented in the form

$$B(x, y, t\Theta) = B_0(x, y, \Theta)t^\gamma + B_1(x, y, \Theta)t^{\gamma-1} + O(t^{\gamma-2}), \quad t \to \infty.$$

Then, using Equations (5.9.21) and (14.3.18), we get for the symbol σ_B:

$$\sigma_B(x, t\Theta) = b_0(x, \Theta)t^\gamma + b_1(x, \Theta)t^{\gamma-1} + O(t^{\gamma-2}), \quad t \to \infty,$$

where $b_0(x, \Theta) = b(x, \Theta)$ (see Equation (5.9.21)). Thus, from (5.9.6) and (5.9.10) we see that the terms $\psi_{\gamma-1}(t, x)$ and $O(t^{\gamma-2})$ on the right-hand side of Equation (5.9.24) can be written as follows

$$\psi_{\gamma-1}(t, x) = t^{\gamma-1} + a_1(x)t^{\gamma-2} + O(t^{\gamma-3}),$$
$$O(t^{\gamma-2}) = a_2(x)t^{\gamma-2} + O(t^{\gamma-3})$$

for some smooth a_1 and a_2. Combining the similar terms, we obtain for positive γ:

$$(\mathcal{B}f)(x) = \frac{D(x_s)}{\pi}\sqrt{\frac{R(x_s)}{R(x_s) + h}}\, b(x, n(x_s))\left[\mathrm{Im}\left\{\int_0^\infty t^{\gamma-1}e^{ith}\,dt\right\}\right.$$
$$\left. + \int_1^\infty (a_3(x)t^{\gamma-2} + O(t^{\gamma-3}))e^{ith}\,dt\right] + \tilde{\eta}(x),$$
$$x = x_s + hn(x_s), \quad \gamma > 0,$$

where $\tilde{\eta} - \eta \in C^\infty(\mathbb{R}^2)$. Let $\gamma = 1$. Substitution of the well-known formulas

$$\int_1^\infty \frac{\cos(th)}{t} dt = -\ln|h| - c + O(h^2),$$

$$\int_1^\infty \frac{\sin(th)}{t} dt = \frac{\pi}{2}\operatorname{sgn}(h) - h + O(h^3),$$

$$\int_1^\infty \frac{\cos(th)}{t^2} dt = 1 + \frac{\pi}{2}|h| + O(h^2),$$

where $c = 0.57\ldots$ is the Euler constant, implies

$$(\mathcal{B}f)(x) = \frac{D(x_s)}{\pi}\sqrt{\frac{R(x_s)}{R(x_s) + h}}\, b(x, n(x_s)) \left[\frac{1}{h} + c_1(x)\ln|h|\right.$$

$$\left. + c_2(x)\operatorname{sgn}(h) + c_3(x, h)\right] + \tilde{\eta}(x), \quad x = x_s + hn(x_s),$$

where $c_i, i = 1, 2, 3$, are smooth functions of x in a neighborhood of x_0, and $c_3(x, h)$ satisfies the conditions:

$$c_3(x, h) = O(1), \quad \frac{\partial}{\partial h}c_3(x, h) = O(\ln|h|), \quad h \to 0.$$

Finally, including $\tilde{\eta}$ in c_3 and taking into account that

$$\sqrt{\frac{R(x_s)}{R(x_s) + h}}\, b(x_s + hn(x_s), n(x_s)) = b(x_s, n(x_s)) + O(h),$$

$$c_i(x) = c_i(x_s) + O(h), i = 1, 2, \quad c_3(x, h) = c_3(x_s, h) + O(h),$$

we prove

Theorem 5.9.3. *Let* $\mathcal{B} \in L_{10}^1(\mathbb{R}^2)$ *be a PDO with the amplitude* $B(x, y, \xi)$, *which satisfies the condition*

$$B(x, y, t\Theta) = B_0(x, y, \Theta)t + B_1(x, y, \Theta) + O(t^{-1}), \quad t \to \infty. \quad (5.9.26)$$

Let us denote $b(x, \Theta) := B_0(x, x, \Theta)$ *and suppose that* $b(x, \Theta)$ *is even in* Θ. *Fix* $x_0 \in S$ *such that* S *is smooth in a neighborhood of* x_0. *Then one has*

$$(\mathcal{B}f)(x_s + hn(x_s))$$
$$= \frac{D(x_s)b(x_s, n(x_s))}{\pi}\left[\frac{1}{h} + c_1(x_s)\ln|h| + c_2(x_s)\operatorname{sgn}(h) + c_3(x_s, h)\right]$$
$$\tag{5.9.27}$$

and

$$\frac{\partial}{\partial h}(\mathcal{B}f)(x_s + hn(x_s))$$

$$= \frac{D(x_s)b(x_s, n(x_s))}{\pi}\left[-\frac{1}{h^2} + \frac{c_1(x_s)}{h} + 2c_2(x_s)\delta(h) + O(\ln|h|)\right],$$

$$(5.9.28)$$

where $c_i, i = 1, 2, 3$, are smooth functions of $x_s \in S$ in a neighborhood of x_0, and $c_3(x_s, h)$ satisfies the conditions:

$$c_3(x_s, h) = O(1), \quad \frac{\partial}{\partial h}c_3(x_s, h) = O(\ln|h|), \quad h \to 0. \qquad (5.9.29)$$

5.9.4. Asymptotics of PDO, symbols of which have discontinuities on a conical surface

Let us fix the angular interval $\Omega \subsetneq S^1$, and suppose for simplicity that Ω is symmetrical: if $\Theta \in \Omega$, then $-\Theta \in \Omega$. Define the cone $K :=$ $\{\lambda\Theta : \lambda \in \mathbb{R}, \Theta \in \Omega\}$. In this section we study the truncated operator

$$(\mathcal{B}_K f)(x) = \frac{1}{(2\pi)^2}\int\limits_K\int\limits_{\mathbb{R}^2} B(x, y, \xi)f(y)e^{-i\xi\cdot(x-y)}dy\,d\xi. \qquad (5.9.30)$$

Clearly, the operator $\mathcal{B}_K$ is no longer a PDO. Fix any function $\chi \in C^\infty(\mathbb{R}^2)$ such that

(1) $\chi(\xi) = 0, \xi \notin K$, and $\chi(\xi) > 0$ in the interior of K;
(2) χ vanishes identically in a neighborhood of a boundary of K;
(3) $\chi(\xi)$ depends only on $\xi/|\xi|$ if $|\xi| \geq 1$; and
(4) $\chi(\Theta) \equiv 1$ for $\Theta \in \Omega' \subset \Omega$, where $|\Omega \setminus \Omega'| < \epsilon$ for a sufficiently small $\epsilon > 0$.

Let $\mathcal{B}_\chi$ and $\mathcal{B}_{1-\chi}$ be the operators corresponding to the amplitudes $B_\chi(x, y, \xi) := \chi(\xi)B(x, y, \xi)$ and $B_{1-\chi}(x, y, \xi) := (1_K - \chi(\xi))B(x, y, \xi)$, respectively, where 1_K is the characteristic function of the cone K. Thus,

$$B_\chi + B_{1-\chi} = \begin{cases} B, & \xi \in K, \\ 0, & \xi \notin K. \end{cases}$$

If $\mathcal{B} \in L_{10}^\gamma(\mathbb{R}^2)$, then, clearly, $\mathcal{B}_\chi \in L_{10}^\gamma(\mathbb{R}^2)$ too, and singsupp $\mathcal{B}_\chi f$ can be easily found:

$$x \in \text{singsupp}\,\mathcal{B}_\chi f := S_\Omega \text{ if and only if the second component of the}$$

$$\text{intersection } WF(f) \cap (x, K) \text{ is nonempty.}$$

$$(5.9.31)$$

Let $n(x)$ denote a unit vector normal to S at a point $x \in S$, where S is smooth, and pointing away from the center of curvature of S at x. Then condition (5.9.31) means that $n(x) \in \Omega$ implies $x \in S_\Omega$. If x is a corner point, then x belongs to S_Ω too. Clearly, $S_\Omega \subset S$. Theorem 5.9.2 gives precise characterization of $(\mathcal{B}_\chi f)(x)$ in a neighborhood of S_Ω.

Now let us investigate the behavior of $\mathcal{B}_{1-\chi} f$ near its singular support. Let $\Theta_0 := (\cos\theta_0, \sin\theta_0)$ be one of the boundary points of Ω. Without loss of generality we may assume that in a neighborhood of θ_0 the available data correspond to $\theta \geq \theta_0$. Using partition of unity and linearity of the problem, we conclude that it is sufficient to consider an amplitude $B_{1-\chi}(x, y, t\Theta)$ which is zero for $\theta \notin [\theta_0, \theta_0 + \Delta)$, where $\Delta > 0$ is a sufficiently small number:

$$(\mathcal{B}_{1-\chi} f)(x) = \frac{1}{(2\pi)^2} \int_0^\infty \int_{\theta_0}^{\theta_0+\Delta} \int_{\mathbb{R}^2} B_{1-\chi}(x, y, t\Theta) f(y) e^{-it\Theta \cdot (x-y)} dy \, d\theta \, t dt,$$

(5.9.32a)

$$
\begin{aligned}
B_{1-\chi}(x, y, t\Theta) &= B_0(x, y, \Theta)(1 - \chi(\Theta)) t^\gamma + O(t^{\gamma-1}) \\
&= B_0(x, y, \Theta)(1 - \chi(\Theta)) \psi_\gamma(t, x, y, \Theta), \quad t \to \infty.
\end{aligned}
$$

(5.9.32b)

Here $1 - \chi(\Theta) = 1$ at $\theta = \theta_0$, and $1 - \chi(\Theta)$ vanishes with all derivatives at $\theta = \theta_0 + \Delta$.

Fix any Θ such that $\theta_0 \leq \theta \leq \theta_0 + \Delta$, and let $y_j(\theta) \in S$, $j = 1, \ldots, N$, be a finite set of points of contact between a family of lines $\Theta \cdot x = p, p \in \mathbb{R}$, and S. Denote also $p_j(\theta) = \Theta \cdot y_j(\theta), S_j := \{ y_j(\theta) : \theta \in [\theta_0, \theta_0 + \Delta] \}, j = 1, \ldots, N$. Fix a sufficiently small $\epsilon > 0$ and find any function $\mu(p) \in C^\infty(\mathbb{R})$ such that μ vanishes identically inside all intervals $[p_j(\theta) - \epsilon, p_j(\theta) + \epsilon]$ and $1 - \mu$ vanishes identically outside of all intervals $[p_j(\theta) - 2\epsilon, p_j(\theta) + 2\epsilon]$. Then

$$
\begin{aligned}
I(x, t, \Theta) &:= \int_{\mathbb{R}^2} B_{1-\chi}(x, y, t\Theta) f(y) e^{it\Theta \cdot y} dy \\
&= \int_{\mathbb{R}} \left(\int_{\mathbb{R}} B_{1-\chi}(x, p\Theta + s\Theta^\perp, t\Theta) f(p\Theta + s\Theta^\perp) ds \right) (1 - \mu(p)) e^{itp} dp \\
&\quad + \int_{\mathbb{R}} \left(\int_{\mathbb{R}} B_{1-\chi}(x, p\Theta + s\Theta^\perp, t\Theta) f(p\Theta + s\Theta^\perp) ds \right) \mu(p) e^{itp} dp,
\end{aligned}
$$

(5.9.33)

where $\Theta^\perp = (-\sin\theta, \cos\theta)$. Clearly, the last integral on the right-hand side of Equation (5.9.33) is $O(t^{-\infty})$, that is it decays faster than any power of t as $t \to \infty$. This can be verified integrating by parts with respect to p, using the properties of an amplitude from the class S_{10}^{γ} (Equation (14.3.1) with $\rho = 1$ and $\delta = 0$), and the properties of μ. Using results from Section 4.5, we obtain the following formula:

$$\int_{\mathbb{R}} B_{1-\chi}(x, p\Theta + s\Theta^{\perp}, t\Theta) f(p\Theta + s\Theta^{\perp}) ds$$

$$= -\varphi_j^{(1)}(p, x, t, \Theta) 2\sqrt{2R_j(\theta)} B_{1-\chi}(x, y_j(\theta), t\Theta) D(y_j(\theta))(p - p_j(\theta))_{\mp}^{1/2}$$

$$+ \varphi_j^{(2)}(p, x, t, \Theta), \qquad (5.9.34)$$

when p is in a neighborhood of $p_j(\theta)$ and $B_{1-\chi}(x, y_j(\theta), t\Theta) \neq 0$. Here $R_j(\theta)$ is the radius of curvature of S at $y_j(\theta)$, $\varphi_j^{(l)}, l = 1, 2$, are C^{∞} functions of their arguments, and $\varphi_j^{(1)}(p_j(\theta), x, t, \Theta) = 1$. In (5.9.34) we used the notation $(p - p_j)_{\mp} := \max(\mp(p - p_j), 0)$, and the sign, '$-$' or '$+$', is chosen according to the location of the center of curvature of S at $y_j(\theta) \in S$. More precisely, if $O_j(\theta)$ is the center of the circle which is tangent to S at $y_j(\theta)$, then '$-$' is chosen if $(y_j(\theta) - O_j(\theta)) \cdot \Theta > 0$, and '$+$' is chosen if $(y_j(\theta) - O_j(\theta)) \cdot \Theta < 0$.

Let us represent f as $f = f_1 + f_2$, such that for x in $O(\sqrt{\epsilon})$ - neighborhoods of all $y_j(\theta), j = 1, \ldots, N$, we have:

(1) $f_1(x)$ has a jump across S and is smooth otherwise;

(2) $f_1(x)$ identically equals zero on the side of S which is opposite to $O_j(\theta)$ (see Figure 5.9.2); and

(3) f_2 is smooth across S.

Then, clearly,

$$\varphi_j^{(1)}(p, x, t, \Theta) 2\sqrt{2R_j(\theta)} B_{1-\chi}(x, y_j(\theta), t\Theta) D(y_j(\theta))$$

$$= -(p - p_j(\theta))_{\mp}^{-1/2} \int_{\mathbb{R}} B_{1-\chi}(x, p\Theta + s\Theta^{\perp}, t\Theta) f_1(p\Theta + s\Theta^{\perp}) ds. \qquad (5.9.35)$$

In Section 5.9.5.3 below it is proved that

$$\frac{\partial^k}{\partial p^k} \varphi_j^{(1)}(p, x, t, \Theta) = O(1), \ t \to \infty, \ k = 0, 1, 2, \ldots,$$

$$|p - p_j(\theta)| \leq 2\epsilon, \ j = 1, \ldots, N. \qquad (5.9.36a)$$

Differentiating on both sides of (5.9.35) with respect to x, θ, and t, and using (14.3.1), we establish

$$\frac{\partial^k}{\partial t^k} \frac{\partial^{k_1}}{\partial x_1^{k_1}} \frac{\partial^{k_2}}{\partial x_2^{k_2}} \frac{\partial^{k_3}}{\partial \theta^{k_3}} \varphi_j^{(1)}(p, x, t, \Theta) = O(t^{-k}), \ t \to \infty,$$

$$k, k_1, k_2, k_3 = 0, 1, 2, \ldots, \ |p - p_j(\theta)| \leq 2\epsilon, \ j = 1, \ldots, N. \qquad (5.9.36b)$$

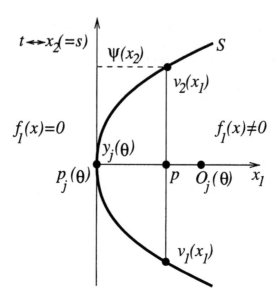

FIGURE 5.9.2. Illustration of the notation in case of the convex curve S

For the function $\varphi_j^{(2)}$ we get

$$\varphi_j^{(2)}(p, x, t, \Theta) = \int_{\mathbb{R}} B_{1-\chi}(x, p\Theta + s\Theta^\perp, t\Theta) f_2(p\Theta + s\Theta^\perp) ds,$$

$$|p - p_j(\theta)| \leq 2\epsilon, \, j = 1, \ldots, N.$$

Since f_2 is smooth, similarly to (5.9.33), Equations (14.3.1) and integration by parts yield:

$$\int_{\mathbb{R}} \varphi_j^{(2)}(p, x, t, \Theta)(1 - \mu(p)) e^{itp} dp = O(t^{-\infty}), \, t \to \infty.$$

Substituting (5.9.34) and the last equation into (5.9.33), we find

$$I(x, t, \Theta) = -\sum_{j=1}^{N} \int_{\mathbb{R}} \varphi_j^{(1)}(p, x, t, \Theta) 2\sqrt{2R_j(\theta)} B_{1-\chi}(x, y_j(\theta), t\Theta) D(y_j(\theta))$$

$$\times (p - p_j(\theta))_{\mp}^{1/2}(1 - \mu(p)) e^{itp} dp + O(t^{-\infty}).$$

$$(5.9.37)$$

Using Equations (5.9.32b), (5.9.36a), and applying Remark 14.5.1 and

Lemma 14.5.5, we compute

$$I(x,t,\Theta) = -\sum_{j=1}^{N} 2\sqrt{2R_j(\theta)}B_0(x, y_j(\theta), \Theta)(1-\chi(\Theta))D(y_j(\theta))$$

$$\times e^{\mp i\frac{3\pi}{4}}\Gamma\left(\frac{3}{2}\right)\psi_{\gamma-1.5}(t,x,\theta)e^{ip_j(\theta)t} + O(t^{-\infty}).$$

It is important to check that the function $\psi_{\gamma-1.5}$ on the right-hand side of the last equation has properties (i)-(iii) formulated below (5.9.1) with $\zeta = (x,\theta)$. Indeed, properties (i) and (ii) follow directly from Lemma 14.5.5. Property (iii) can be checked differentiating (5.9.37) and the last equation with respect to x, θ, and t and using Equations (5.9.36b) and (14.3.1). Substitution into (5.9.32a) now yields:

$$(\mathcal{B}_{1-\chi}f)(x) = \frac{1}{(2\pi)^2}\int_0^\infty \int_{\theta_0}^{\theta_0+\Delta} I(x,t,\Theta)e^{-it\Theta\cdot x}\,d\theta\,tdt$$

$$= \frac{-1}{(2\pi)^2}\sum_{j=1}^{N} e^{\mp i\frac{3\pi}{4}}\Gamma\left(\frac{3}{2}\right)\int_0^\infty \int_{\theta_0}^{\theta_0+\Delta} 2\sqrt{2R_j(\theta)}B_0(x, y_j(\theta), \Theta)(1-\chi(\Theta))$$

$$\times D(y_j(\theta))\psi_{\gamma-1.5}(t,x,\theta)e^{it(p_j(\theta)-\Theta\cdot x)}\,d\theta\,tdt + \eta_1(x),$$
$$(5.9.38)$$

where $\eta_1 \in C^\infty(\mathbb{R}^2)$. Consider the integral with respect to θ separately:

$$J_j(x,t) := \int_{\theta_0}^{\theta_0+\Delta} 2\sqrt{2R_j(\theta)}B_0(x, y_j(\theta), \Theta)(1-\chi(\Theta))D(y_j(\theta))$$

$$\times \psi_{\gamma-1.5}(t,x,\theta)e^{it(p_j(\theta)-\Theta\cdot x)}\,d\theta.$$
$$(5.9.39)$$

The integrand in (5.9.39) is $C^\infty([\theta_0, \theta_0+\Delta])$ and vanishes with all derivatives at $\theta = \theta_0+\Delta$. Therefore, the asymptotic of $J_j(x,t)$ as $t \to \infty$ comes from a neighborhood of a point θ_0 and from neighborhoods of stationary points of the phase $p_j(\theta) - \Theta \cdot x$. Equations (5.9.38) and (5.9.39) yield that the singular support of $\mathcal{B}_{1-\chi}f$, besides corner points of S, consists of x such that

(I) $p_j(\theta_0) = \Theta_0 \cdot x$; and

(II) there exists $\tilde{\theta}$ (which depends on x) with the properties:

(a) $\tilde{\theta} \in [\theta_0, \theta_0 + \Delta)$,

(b) $p_j(\tilde{\theta}) = \tilde{\Theta} \cdot x$, and

(c) $\frac{\partial}{\partial\theta}(p_j(\theta) - \Theta \cdot x)|_{\theta=\tilde{\theta}} = 0$.

Lemma 5.9.2 implies that conditions (b) and (c) are satisfied only for points on S. Moreover, for a fixed $x \in S$, the stationary point $\tilde{\theta}$ corresponds to the vector $\tilde{\Theta}$, which is normal to S at x. Together with condition (a) this implies that the singular points defined in Item (II) fill in the segment S_j of the curve S. However, if $\tilde{\theta} \in (\theta_0, \theta_0 + \Delta)$, that is $\tilde{\theta} \neq \theta_0$ and $x \in S_j$, $x \neq y_j(\theta_0)$, the behavior of $\mathcal{B}_{1-\chi}f$ in a neighborhood of such x is basically the same as the one described in Theorem 5.9.2. Thus, we have to consider the behavior of $\mathcal{B}_{1-\chi}f$ only in a neighborhood of the line $L_j := \{y : p_j(\theta_0) = \Theta_0 \cdot y\}$ and distinguish two cases $x \neq y_j(\theta_0)$ and $x = y_j(\theta_0)$.

Let us consider the first case. Fix any line L_m and any $x_0 \in L_m$, $x_0 \neq y_m(\theta_0)$. Since x_0 is away from S_m and Δ can be made arbitrarily small, we may assume without loss of generality that there are no stationary points of the phase $p_j(\theta) - \Theta \cdot x$ inside the interval $[\theta_0, \theta_0 + \Delta)$. Integration by parts in Equation (5.9.39) yields

$$
J_m(x, t) = 2\sqrt{2R_m(\theta_0)} B_0(x, y_m(\theta_0), \Theta_0) D(y_m(\theta_0)) \psi_{\gamma - 2.5}(t, x)
$$

$$
\times \frac{i}{\frac{\partial}{\partial \theta}(p_m(\theta) - \Theta \cdot x)|_{\theta = \theta_0}} e^{it(p_m(\theta_0) - \Theta_0 \cdot x)}, \tag{5.9.40}
$$

where we have used the fact that $\chi(\Theta_0) = 0$. It is easy to check that the function $\psi_{\gamma - 2.5}$ on the right-hand side of (5.9.40) satisfies properties (i) - (iii) formulated below (5.9.1) with $\zeta = x$. Let us introduce the following coordinate system (see Figure 5.9.3): the origin coincides with the center of curvature of S at $y_m(\theta_0)$, and the x_1-axis points from the center of curvature towards $y_m(\theta_0)$. In the new coordinate system, we have $\theta_0 = 0$ or $\theta_0 = \pi$, $y_m(\theta_0) = (R_m(\theta_0), 0)$, and $p_m(\theta_0) = \pm R_m(\theta_0)$. Moreover,

$$
\frac{\partial}{\partial \theta}(p_m(\theta) - \Theta \cdot x)|_{\theta = \theta_0} = p'_m(\theta_0) - \Theta_0^\perp \cdot x = -\Theta_0^\perp \cdot x,
$$

$$
\Theta_0^\perp := (-\sin \theta_0, \cos \theta_0), \tag{5.9.41}
$$

where we have used assertion (3) of Lemma 5.9.1, which states that $p'_m(\theta_0) = 0$ when $\theta_0 = 0$ or π. Let us represent x in the form: $x = x_0 + \Theta_0 h$ with $h \to 0$. Recall that $x_0 \in L_m$, $x_0 \neq y_m(\theta_0)$, and $y_m(\theta_0) = \pm R_m(\theta_0)\Theta_0$. Substitution of (5.9.41) into (5.9.40) yields

$$
J_m(x, t) = 2\sqrt{2R_m(\theta_0)} B_0(x, y_m(\theta_0), \Theta_0) D(y_m(\theta_0)) \psi_{\gamma - 2.5}(t, x)
$$

$$
\times \frac{-i}{(x_0 - y_m(\theta_0)) \cdot \Theta_0^\perp} e^{-ith}. \tag{5.9.42}
$$

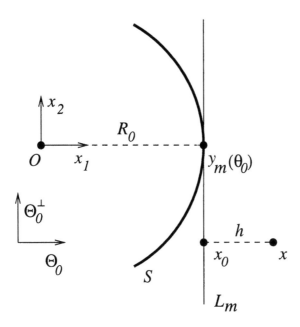

FIGURE 5.9.3.   Illustration of the notation used for finding asymptotics of PDO, symbols of which have discontinuities on a conical surface

Using (5.9.39) and (5.9.42) in (5.9.38), we get

$$
(\mathcal{B}_{1-\chi}f)(x) = \frac{ie^{\mp i\frac{3\pi}{4}}}{(2\pi)^{1.5}} \frac{\sqrt{R_m(\theta_0)}B_0(x, y_m(\theta_0), \Theta_0)D(y_m(\theta_0))}{(x_0 - y_m(\theta_0)) \cdot \Theta_0^{\perp}}
$$

$$
\times \int_0^{\infty} \psi_{\gamma-1.5}(t, x)e^{-ith}\,dt + \eta(x), \quad x = x_0 + \Theta_0 h,
\tag{5.9.43}
$$

where η is C^{∞} in a neighborhood of x_0. This function is of the form $\eta = \eta_1$ (cf. (5.9.38)) + a contribution of the terms with $j \neq m$ from the summation in (5.9.38).

Now let us consider the case $x = y_m(\theta_0) + \Theta_0 h$, $h \to 0$. For such a choice of x, Equation (5.9.39) yields

$$
J_m(x, t) = \int_{\theta_0}^{\theta_0+\Delta} 2\sqrt{2R_m(\theta)}B_0(x, y_m(\theta), \Theta)(1 - \chi(\Theta))D(y_m(\theta))
$$

$$
\times \psi_{\gamma-1.5}(t, x, \theta)e^{it(p_m(\theta) - |x|\cos\theta)}\,d\theta.
$$

Since $p_m'(\theta_0) = p_m''(\theta_0) = 0$ when $\theta_0 = 0$ or π, we see that the phase $a(\theta) := p_m(\theta) - |x|\cos\theta$ is strictly increasing on the interval $[\theta_0, \theta_0 + \Delta)$

if $\theta_0 = 0$, and it is strictly decreasing there if $\theta_0 = \pi$, provided Δ is sufficiently small. Moreover, $a'(\theta_0) = 0$, $a''(\theta_0) = \pm|x| \neq 0$. Hence, the stationary phase method implies

$$J_m(x, t)$$

$$= 2\sqrt{\pi} e^{\pm i\frac{\pi}{4}} B_0(x, y_m(\theta_0), \Theta_0) D(y_m(\theta_0)) \sqrt{\frac{R_m(\theta_0)}{R_m(\theta_0) \pm h}} \, t^{\gamma-2} e^{-ith}$$

$$+ O(t^{\gamma-2.5}), \quad x = y_m(\theta_0) + \Theta_0 h, \ t \geq 1.$$

Finally, substituting into (5.9.38), we derive

$$(\mathcal{B}_{1-\chi} f)(x)$$

$$= \frac{\pm i}{4\pi} B_0(x, y_m(\theta_0), \Theta_0) D(y_m(\theta_0)) \sqrt{\frac{R_m(\theta_0)}{R_m(\theta_0) \pm h}} \int_1^\infty t^{\gamma-1}(t, x) e^{-ith} \, dt$$

$$+ \int_0^\infty O(t^{\gamma-1.5}) e^{-ith} \, dt + \eta(x), \quad x = y_m(\theta_0) + \Theta_0 h. \tag{5.9.44}$$

Collecting Equations (5.9.43) and (5.9.44) and absorbing $\eta(x)$ in the integral $\int_0^\infty O(t^q) e^{-ith} \, dt$, $q \in \mathbb{R}$, for x in a neighborhood of x_0, that is for small h, we prove

Theorem 5.9.4. *Let us consider the action of the operator $\mathcal{B}_{1-\chi}$ on f:*

$$(\mathcal{B}_{1-\chi} f)(x) = \frac{1}{(2\pi)^2} \int_0^\infty \int_{\theta_0}^{\theta_0+\Delta} \int_{\mathbb{R}^2} B(x, y, t\Theta) D(y) e^{-it\Theta \cdot (x-y)} \, dy \, d\theta \, t \, dt,$$

$$\tag{5.9.45}$$

where $B \in C^\infty(\mathbb{R}^6)$ satisfies the conditions

$$B(x, y, t\Theta) = B_0(x, y, \theta) \psi_\gamma(t, x, y, \Theta);$$

$$B_0(x, y, \theta_0) \neq 0; \quad \frac{\partial^k}{\partial \theta^k} B_0(x, y, \theta)\big|_{\theta=\theta_0+\Delta} \equiv 0, \ k = 0, 1, 2, \ldots. \tag{5.9.46}$$

Fix any line L_0 perpendicular to Θ_0 which is tangent to S, and let y_0 be a point of contact. Let R_0 be the radius of curvature of S at y_0, $0 < R_0 < \infty$. Fix any $x_0 \in L_0$. Then, if $x_0 \neq y_0$, we have

$$(\mathcal{B}_{1-\chi} f)(x) = i e^{\mp i\frac{3\pi}{4}} \frac{\sqrt{R_0} B_0(x, y_0, \Theta_0) D(y_0)}{(2\pi)^{1.5}(x_0 - y_0) \cdot \Theta_0^\perp} \int_0^\infty \psi_{\gamma-1.5}(t, x) e^{-ith} \, dt,$$

$$x = x_0 + \Theta_0 h. \tag{5.9.47}$$

If $x_0 = y_0$, we have

$$(\mathcal{B}_{1-\chi}f)(x) = \frac{1}{4\pi}B_0(x,y_0,\Theta_0)D(y_0)\sqrt{\frac{R_0}{R_0 \pm h}}(\pm i)\int_1^\infty t^{\gamma-1}e^{-ith}dt$$

$$+ \int_0^\infty O(t^{\gamma-1.5})e^{-ith}dt, \quad x = y_0 + \Theta_0 h. \tag{5.9.48}$$

Let O be the center of curvature of S at y_0. Then the top expressions from $\pm$ and $\mp$ are chosen if $(y_0 - O)\cdot\Theta_0 > 0$, and the bottom expressions are chosen if $(y_0 - O)\cdot\Theta_0 < 0$.

REMARK 5.9.3. Similarly to the proof of Theorem 5.9.4, it is easy to show that in case of the operator given by

$$(\mathcal{B}_{1-\chi}f)(x) = \frac{1}{(2\pi)^2}\int_0^\infty\int_{\theta_0-\Delta}^{\theta_0}\int_{\mathbb{R}^2} B(x,y,t\Theta)D(y)e^{-it\Theta\cdot(x-y)}dy\,d\theta\,tdt,$$

where $B \in C^\infty(\mathbb{R}^6)$ satisfies the conditions

$$B(x,y,t\Theta) = B_0(x,y,\theta)\psi_\gamma(t,x,y,\Theta);$$

$$B_0(x,y,\theta_0) \neq 0; \quad \frac{\partial^k}{\partial\theta^k}B_0(x,y,\theta)\Big|_{\theta=\theta_0-\Delta} \equiv 0, \ k = 0,1,2,\ldots,$$

the analogue of Equation (5.9.47) becomes

$$(\mathcal{B}_{1-\chi}f)(x) = -ie^{\mp i\frac{3\pi}{4}}\frac{\sqrt{R_0}B_0(x,y_0,\Theta_0)D(y_0)}{(2\pi)^{1.5}(x_0 - y_0)\cdot\Theta_0^\perp}\int_0^\infty \psi_{\gamma-1.5}(t,x)e^{-ith}dt,$$

$$x = x_0 + \Theta_0 h, \tag{5.9.49}$$

and the form of Equation (5.9.48) remains unchanged.

5.9.5. Proof of the auxiliary results

5.9.5.1. Proof of Lemma 5.9.1.

1. Let $\rho = \rho(\varphi)$ be the equation of S in cylindrical coordinates in a neighborhood of x_0. Since the family of lines $\Theta \cdot x = p_0(\theta)$ is the envelope of S, we have

$$p_0(\theta) = \rho(\varphi)\cos(\varphi - \theta) \tag{5.9.50}$$

and

$$\frac{\rho'(\varphi)}{\rho(\varphi)} = \tan(\varphi - \theta). \tag{5.9.51}$$

Since the curve S is smooth in a neighborhood of x_0, the function $\rho(\varphi)$ is C^∞ in a neighborhood of $\varphi = 0$. Assuming $\varphi = \varphi(\theta)$, differentiation of (5.9.51) with respect to θ yields

$$\frac{\rho''\rho - (\rho')^2}{\rho^2}\varphi' = \frac{\varphi' - 1}{\cos^2(\varphi - \theta)}. \tag{5.9.52}$$

According to the choice of the coordinate system, $\varphi(0) = 0$ and $\rho(0) = R_0 > 0$; hence, (5.9.51) yields $\rho'(0) = 0$. Putting $\theta = 0$ and $\rho'(0) = 0$ in (5.9.52) and solving the resulting expression for $\varphi'(0)$, we find $\varphi'(0) = \rho(0)/(\rho(0) - \rho''(0))$. Let $R(\varphi)$ be the radius of curvature of S at a point $(\rho(\varphi), \varphi)$. One has [KK]:

$$R(\varphi) = \frac{(\rho^2 + (\rho')^2)^{3/2}}{\rho^2 + 2(\rho')^2 - \rho\rho''}. \tag{5.9.53}$$

According to the choice of the coordinate system, $R_0 = R(0) = \rho(0)$. Using this and putting $\theta = 0$ and $\rho'(0) = 0$ in (5.9.53), we find that $\rho''(0) = 0$. Thus $\rho(0) \neq \rho''(0)$ and, therefore, $\varphi'(0) = 1$. By the implicit function theorem, Equation (5.9.51) can be solved for φ, in a neighborhood of the point $(\varphi = 0, \theta = 0)$. Moreover, this implies that the function $\varphi(\theta)$ is smooth in a neighborhood of 0, and so Equation (5.9.50) implies that $p_0(\theta)$ is also smooth in a neighborhood of 0.

2. Statement 2 follows from the choice of the coordinate system.

3. Differentiating (5.9.50) with respect to θ and using the chain rule, we get

$$p_0'(\theta) = \rho'\varphi'\cos(\varphi - \theta) - \rho\sin(\varphi - \theta)(\varphi' - 1),$$

where $\rho' = d\rho(\varphi)/d\varphi$ and $\varphi' = d\varphi(\theta)/d\theta$. Since $\varphi(0) = 0$, $\rho(0) = R_0 > 0$, and $\rho'(0) = 0$, Equation (5.9.52) implies $p_0'(0) = 0$. Differentiating the last equation with respect to θ, we obtain

$$p_0''(\theta) = \rho''(\varphi')^2\cos(\varphi - \theta) + \rho'\varphi''\cos(\varphi - \theta) - 2\rho'\varphi'\sin(\varphi - \theta)(\varphi' - 1)$$
$$- \rho\cos(\varphi - \theta)(\varphi' - 1)^2 - \rho\sin(\varphi - \theta)\varphi''.$$

Since $\varphi(0) = 0$ and $\rho'(0) = 0$, we get

$$p_0''(0) = \rho''(0)(\varphi'(0))^2 - \rho(0)(\varphi'(0) - 1)^2.$$

In the proof of statement 1 it was shown that $\rho''(0) = 0$ and $\varphi'(0) = 1$. Thus the above equation implies $p_0''(0) = 0$. The proof of the assertion $p_0'(\pi) = p_0''(\pi) = 0$ is analogous.

4. According to the choice of the coordinate system,

$$a(\theta, x_0) = p_0(\theta) - R_0\cos\theta.$$

Differentiating with respect to θ twice and using that $p_0''(0) = 0$, we prove statement 4. $\square$

5.9.5.2. Proof of Lemma 5.9.2.

1. Let us fix x in a neighborhood of $x_0 \in S$. Equation (5.9.9) can be written as

$$p_0'(\theta) = \Theta^\perp \cdot x, \qquad (5.9.54)$$

where $\Theta^\perp = (-\sin\theta, \cos\theta)$. Consider the line $y \cdot \Theta = p_0(\theta)$ with $\theta = \theta(x)$, which is tangent to S at some point, and find projection of x onto this line, which will be denoted $\tilde{x}_s$. Clearly,

$$p_0'(\theta) = \Theta^\perp \cdot \tilde{x}_s, \qquad (5.9.55)$$

because by the construction $(x - \tilde{x}_s) \parallel \Theta$. Let $\rho(\varphi)$ be the equation of S in polar coordinates. Since the line $y \cdot \Theta = p_0(\theta)$ is tangent to S, Equations (5.9.50) and (5.9.51) hold for some $\varphi = \varphi(\theta)$. Let $\tilde{x}_s = r(\cos\psi, \sin\psi)$. Since $\tilde{x}_s$ belongs to this line,

$$p_0(\theta) = r\cos(\psi - \theta). \qquad (5.9.56)$$

Equation (5.9.55) yields

$$p_0'(\theta) = r\sin(\psi - \theta). \qquad (5.9.57)$$

Since $\varphi = \varphi(\theta)$ in (5.9.50) and (5.9.51), differentiating (5.9.50) with respect to θ, and using (5.9.51), we get

$$p_0'(\theta) = \rho(\varphi)\sin(\varphi - \theta). \qquad (5.9.58)$$

Comparing (5.9.50), (5.9.58) with (5.9.56), (5.9.57) implies that $r = \rho(\varphi)$ and $\psi = \varphi$ for some φ, thus $\tilde{x}_s \in S$ and, by the definition, $\tilde{x}_s = x_s$. This immediately proves the first assertion of Lemma 5.9.2.

2. This assertion follows immediately from the previous one and from the definitions of $R(\theta(x))$ and $D(\theta(x))$ (see below (5.9.4)).

3. Let x_s be the projection of x onto S. By the first assertion, the line $p_0(\theta(x)) = \Theta(x) \cdot y$ contains x_s and is tangent to S at x_s. Therefore, $n(x_s) = \Theta(x)$ and we have

$$a(\theta(x), x) = p_0(\theta(x)) - \Theta(x) \cdot x = p_0(\theta(x)) - \Theta(x) \cdot (x_s + x - x_s)$$
$$= 0 - \Theta(x) \cdot (x - x_s) = -n(x_s) \cdot (x - x_s).$$

4. Since $S(x) = \partial^2 a(\theta, x)/\partial\theta^2 |_{\theta = \theta(x)}$, we compute

$$S(x) = p_0''(\theta(x)) + \Theta(x) \cdot x = \left[p_0''(\theta(x)) + \Theta(x) \cdot x_s\right] + n(x_s) \cdot (x - x_s). \qquad (5.9.59)$$

Since $x_s \in S$, then $x_s = \rho(\varphi)(\sin\varphi, \cos\varphi)$ for some φ. Omitting the dependence on x, we have

$$p_0''(\theta) + \Theta \cdot x_s = p_0''(\theta) + \rho(\varphi)\cos(\varphi - \theta). \qquad (5.9.60)$$

By the first assertion, the line $p_0(\theta) = \Theta \cdot y$ is tangent to S at x_s. Therefore, Equations (5.9.50), (5.9.51), and hence (5.9.52) can be used. Solving (5.9.52) for φ' and using (5.9.53), we obtain

$$
\begin{aligned}
\varphi' &= \left(1 + \frac{\cos^2(\varphi - \theta)}{\rho^2}((\rho')^2 - \rho\rho'')\right)^{-1} \\
&= \left(1 + \frac{\cos^2(\varphi - \theta)}{\rho^2}\left(\frac{(\rho^2 + (\rho')^2)^{3/2}}{R(\varphi)} - \rho^2 - (\rho')^2\right)\right)^{-1} \\
&= \left\{\left[\sin^2(\varphi - \theta) - \cos^2(\varphi - \theta)\frac{(\rho')^2}{\rho^2}\right] \right. \\
&\qquad\qquad \left. + \frac{\cos^2(\varphi - \theta)}{R\rho^2}(\rho^2 + (\rho')^2)^{3/2}\right\}^{-1} \\
&= R(\varphi)\frac{\cos(\varphi - \theta)}{\rho}, \qquad\qquad\qquad (5.9.61)
\end{aligned}
$$

where we have used (5.9.51). Differentiation of (5.9.58) with respect to θ yields

$$
\begin{aligned}
p_0''(\theta) &= \rho'\varphi'\sin(\varphi - \theta) + \rho\cos(\varphi - \theta)(\varphi' - 1) \\
&= \varphi'(\rho'\sin(\varphi - \theta) + \rho\cos(\varphi - \theta)) - \rho\cos(\varphi - \theta).
\end{aligned}
$$

Using (5.9.51), (5.9.61), and (5.9.60), we get

$$p_0''(\theta) + \Theta \cdot x_s = R(\varphi)\frac{\cos(\varphi - \theta)}{\rho}\left(\frac{\rho\sin^2(\varphi - \theta)}{\cos(\varphi - \theta)} + \rho\cos(\varphi - \theta)\right) = R(\varphi),$$

which, together with (5.9.59) proves the desired assertion. $\square$

5.9.5.3. *Proof of (5.9.36a).* Let us denote $B_{1-\chi}(x, p\Theta + s\Theta^\perp, t\Theta)$ $\cdot f_1(p\Theta + s\Theta^\perp) = G(s, p)$, where t and other variables are treated as parameters. From (5.9.35) and Figure 5.9.2 it follows that we have to study the integral

$$\omega_1(p) = \frac{1}{(p - p_j(\theta))_\mp^{1/2}}\int_{v_1((p-p_j(\theta))_\mp)}^{v_2((p-p_j(\theta))_\mp)} G(s, p)ds, \quad |p - p_j(\theta)| \le 2\epsilon,$$

for some functions v_1 and v_2. Below we will always assume that $|p - p_j(\theta)| \le 2\epsilon$, where ϵ is sufficiently small. Fix the coordinate system such that the origin coincides with $y_j(\theta)$ and the x_1-axis points in the same direction as Θ. Let $x_1 = \psi(x_2)$ be the local equation of S in a neighborhood of $y_j(\theta)$ (see Figure 5.9.2). Without loss of generality we may assume that $p_j(\theta) = 0$. For brevity, we will consider only the case $(y_j(\theta) - O_j(\theta)) \cdot \Theta < 0$. The case $(y_j(\theta) - O_j(\theta)) \cdot \Theta > 0$ can be considered analogously. Thus $x_1 = (p - p_j(\theta))_+ = p$ and $x_2 = s$. Clearly, $\psi(0) = \psi'(0) = 0$ and, under the assumption that the radius of curvature of S at $y_j(\theta)$ is positive, we get $\psi''(0) \neq 0$. According to the Morse lemma (see Section 14.5.4), there exists a diffeomorphism $s = \eta(t)$ such that the local equation of S becomes

$$p = \psi(s) = \psi(\eta(t)) = t^2.$$

Solving for t, we get $t = \pm\sqrt{p}$ and

$$s = \eta(t) = \eta(\pm\sqrt{p}) = v_{1,2}(p).$$

By the assumption, S is analytic near $y_j(\theta)$; therefore, η is also analytic and we have

$$v_{1,2}(p) = \sum_{m=0}^{\infty} \frac{\eta^{(m)}(0)}{m!}(\pm\sqrt{p})^m = \eta_1(p) \pm \sqrt{p}\,\eta_2(p),$$

where η_1 and η_2 are analytic. This implies

$$\omega_1(p) = \frac{1}{\sqrt{p}} \int_{\eta_1(p)-\sqrt{p}\,\eta_2(p)}^{\eta_1(p)+\sqrt{p}\,\eta_2(p)} G(s,p)ds = \int_{-\eta_2(p)}^{\eta_2(p)} G(\eta_1(p) + s\sqrt{p}, p)ds.$$

$$(5.9.62)$$

Below, by $G(s,p)$ we denote any function, such that its arbitrary order derivatives with respect to s and p do not grow faster than $O(t^\gamma)$ as $t \to \infty$, where γ is the same as in (5.9.21) and (5.9.32b). Thus we have

$$\omega_1(p) = \int_{-\eta_2(p)}^{\eta_2(p)} G(s\sqrt{p}, p)ds. \qquad (5.9.63)$$

Let ω_1 denote any function which can be represented in the form

$$\omega_1(p) = \int_{-\eta_2(p)}^{\eta_2(p)} A(s^2)G(s\sqrt{p}, p)ds, \qquad (5.9.64)$$

where A is some smooth function and G is the G-type function. Clearly, ω_1, defined in (5.9.63), is also of the type (5.9.64). Differentiating (5.9.64) with respect to p, we obtain

$$\omega_1'(p) = [G(\eta_2(p)\sqrt{p}, p) + G(-\eta_2(p)\sqrt{p}, p)]\eta_2'(p)A(\eta_2^2(p))$$

$$+ \int_{-\eta_2(p)}^{\eta_2(p)} A(s^2)\frac{\partial}{\partial q}G(s\sqrt{p}, q)\Big|_{q=p} ds$$

$$+ \int_{-\eta_2(p)}^{\eta_2(p)} A(s^2)\frac{s}{2\sqrt{p}}\frac{\partial}{\partial q}G(q, p)\Big|_{q=s\sqrt{p}} ds$$

$$= [G(\eta_2(p)\sqrt{p}, p) + G(-\eta_2(p)\sqrt{p}, p)] + \omega_1(p)$$

$$+ \int_{-\eta_2(p)}^{\eta_2(p)} A(s^2)\frac{s}{\sqrt{p}}G(s\sqrt{p}, p)ds, \qquad (5.9.65)$$

where we have used that $G(s, p)\eta_2'(p)A(\eta_2^2(p))$ and derivatives of G are also the G-type functions. In (5.9.65) and below we use the convention: if two G-type functions are in brackets, then these two functions are the same. Let ω_2 denote any function which can be represented in the form

$$\omega_2(p) = [G(\eta_2(p)\sqrt{p}, p) + G(-\eta_2(p)\sqrt{p}, p)]. \qquad (5.9.66)$$

Using this definition, we get from (5.9.65):

$$\omega_1'(p) = \omega_2(p) + \omega_1(p) + \int_{-\eta_2(p)}^{\eta_2(p)} A(s^2)\frac{s}{\sqrt{p}}G(s\sqrt{p}, p)ds. \qquad (5.9.67)$$

Taking into account our definition of the G-type functions, we derive

$$\omega_2'(p) = \frac{[G(\eta_2(p)\sqrt{p}, p) - G(-\eta_2(p)\sqrt{p}, p)]}{\sqrt{p}} + \omega_2(p)$$

$$= \frac{1}{\sqrt{p}} \int_{-\eta_2(p)\sqrt{p}}^{\eta_2(p)\sqrt{p}} G(s, p)ds + \omega_2(p)$$

$$= \int_{-\eta_2(p)}^{\eta_2(p)} G(s\sqrt{p}, p)ds + \omega_2(p) = \omega_1(p) + \omega_2(p). \qquad (5.9.68)$$

For the integral on the right-hand side of (5.9.67), we get

$$\int\limits_{-\eta_2(p)}^{\eta_2(p)} A(s^2)\frac{s}{\sqrt{p}}G(s\sqrt{p},p)ds$$

$$= \int\limits_{0}^{\eta_2(p)} A(s^2)\frac{s}{\sqrt{p}}[G(s\sqrt{p},p) - G(-s\sqrt{p},p)]ds$$

$$= \int\limits_{0}^{\eta_2(p)} A(s^2)\frac{s}{\sqrt{p}} \int\limits_{-s\sqrt{p}}^{s\sqrt{p}} G(u,p)du\,ds = \int\limits_{0}^{\eta_2(p)} A(s^2)s \int\limits_{-s}^{s} G(u\sqrt{p},p)du\,ds$$

$$= \left[\int\limits_{0}^{\eta_2(p)} G(u\sqrt{p},p) \int\limits_{u}^{\eta_2(p)} A(s^2)sds\,du + \int\limits_{-\eta_2(p)}^{0} G(u\sqrt{p},p) \int\limits_{-u}^{\eta_2(p)} A(s^2)sds\,du \right]$$

$$= \int\limits_{-\eta_2(p)}^{\eta_2(p)} G(u\sqrt{p},p)\frac{1}{2} \int\limits_{u^2}^{\eta_2^2(p)} A(s)ds\,du$$

$$= \int\limits_{-\eta_2(p)}^{\eta_2(p)} G(u\sqrt{p},p)(B(\eta_2^2(p)) - B(u^2))du = \omega_1^{(1)} + \omega_1^{(2)},$$

where $\omega_1^{(1)}$ and $\omega_1^{(2)}$ are two different functions of the type (5.9.64). Therefore, Equations (5.9.67) and (5.9.68) yield

$$\omega_1' = \omega_2 + \omega_1 + \omega_1^{(1)} + \omega_1^{(2)}, \quad \omega_2' = \omega_1 + \omega_2. \qquad (5.9.69)$$

Definitions (5.9.64) and (5.9.66) imply that the ω_1- and ω_2-type functions do not grow faster than t^γ as $t \to \infty$. Hence Equations (5.9.69) imply that any order derivatives of ω_1 do not grow faster than t^γ as $t \to \infty$. Since in (5.9.35) we have explicitly factored out the term $B_{1-\chi}(x, y_j(\theta), t\Theta) = O(t^\gamma)$, this proves Equations (5.9.36a).

CHAPTER 6

PSEUDOLOCAL TOMOGRAPHY

6.1. Introduction

Local tomography described in Chapter 5 is attractive because

(a) It does not require collecting and processing of the tomographic data of the whole object if one is interested only in finding discontinuities inside a certain part of the object – the reconstruction at a point requires the knowledge of integrals of the density function f along lines close to that point; and

(b) One does not compute the initial density function f, but Λf, where Λ is the square root of $-\Delta$ and Δ is the Laplace operator. Therefore, since Λ is a pseudolocal pseudodifferential elliptic operator, the location of singularities of f and Λf are the same, but they are sharper in Λf than in f.

The main purpose of this chapter is to introduce the alternative concept: pseudolocal tomography. The pseudolocal tomography function, on one hand, has locality properties and, on the other hand, preserves sizes of discontinuities of the original density function f and of its derivatives. Moreover, images of discontinuities of f computed from the pseudolocal tomography function are also sharper than those in standard (global) tomography.

Throughout the entire chapter, we assume that the original density f is a compactly supported, piecewise-continuous bounded function. Additional (local) assumptions on f are stated where necessary. As usual, S denotes the curve across which f or its derivative of some order is discontinuous. We always assume that there exist one-sided limits of $f(x)$ and of its derivatives (up to a certain order) as x approaches S. The discontinuity curve S is assumed to be piecewise-smooth. The class of functions we use includes most, if not all, densities considered in practical tomography.

Let $\rho > 0$ be fixed. The following pseudolocal tomography function is proposed

$$f_\rho(x) := \frac{1}{4\pi^2} \int\limits_{S^1} \int\limits_{\Theta \cdot x - \rho}^{\Theta \cdot x + \rho} \frac{\hat{f}_p(\theta, p)}{\Theta \cdot x - p} \, dp \, d\theta, \quad x \in \mathbb{R}^2, \tag{6.1.1}$$

In Equation (6.1.1) and everywhere below in this chapter the variables Θ and θ are related by the equation $\Theta = (\cos\theta, \sin\theta)$. One sees that the formula for f_ρ is obtained from the standard inversion formula (2.2.19) by keeping only the interval of length 2ρ centered at the singularity of the Cauchy kernel. We call formula (6.1.1) pseudolocal because at a point x, $f_\rho(x)$ is computed using $\hat{f}(\theta, p)$ for (θ, p) satisfying $|\Theta \cdot x - p| \leq \rho$, that is we use integrals of f along lines passing at a distance not exceeding ρ from the point x. Local tomography formulas from Section 5.2 require the knowledge of the second derivative $\hat{f}_{pp}(\theta, p)$ only for (θ, p) satisfying $\Theta \cdot x = p$.

The chapter is organized as follows. In Section 6.2 we introduce the function f_ρ and prove that discontinuities of f and f_ρ are precisely the same, that is, $f_\rho^c := f - f_\rho \in C(\mathbb{R}^2)$. In Section 6.3 we investigate the convergence $f_\rho^c \to f$ as $\rho \to 0$ in three cases:

a) on compact sets not intersecting S (the discontinuity curve of f),
b) at the points of S, and
c) in a neighborhood of S.

In particular, we establish the existence of a layer of width $O(\rho)$ around S inside which f_ρ^c does not converge to f in the sup-norm. Note that removing the interval $[\Theta \cdot x - \rho, \Theta \cdot x + \rho]$ where the Cauchy kernel is singular can be considered as a possible method of regularizing the inversion formula. Therefore, convergence $f_\rho^c \to f$ can be considered as convergence of a regularized convolution and backprojection algorithm to the original density function. In Section 6.4, more results on functions f_ρ^c, f_ρ and on convergence $f_\rho^c \to f$ are presented. In Section 6.5 we discuss numerical implementation of pseudolocal tomography and present results of its testing on synthetic tomographic data. In Section 6.7, the generalization of pseudolocal tomography to the exponential Radon transform is developed.

6.2. Definition of a pseudolocal tomography function. Basic property

Using Radon's inversion formula (2.2.26), let us fix $\rho > 0$ and introduce the following two functions

$$f_\rho(x) := \frac{-1}{\pi} \int_0^\rho \frac{F_q(x, q)}{q} dq, \qquad (6.2.1)$$

and

$$f_\rho^c(x) := \frac{-1}{\pi} \int_\rho^\infty \frac{F_q(x, q)}{q} dq = f(x) - f_\rho(x), \qquad (6.2.2)$$

where $F(x,q)$ is defined in (2.2.22):

$$F(x,q) := \frac{1}{2\pi} \int_0^{2\pi} \hat{f}(\theta, q + \Theta \cdot x)d\theta, \tag{6.2.3}$$

and $F_q(x,q) = \partial F(x,q)/\partial q$.

Exercise 6.2.1. Show that Equations (6.2.1) and (6.2.2) are equivalent to the following equations:

$$f_\rho(x) = \frac{1}{4\pi^2} \int_{S^1} \int_{\Theta \cdot x - \rho}^{\Theta \cdot x + \rho} \frac{\hat{f}_p(\theta, p)}{\Theta \cdot x - p} dp d\theta, \tag{6.2.1'}$$

$$f_\rho^c(x) = \frac{1}{4\pi^2} \int_{S^1} \left(\int_{-\infty}^{\Theta \cdot x - \rho} + \int_{\Theta \cdot x + \rho}^{\infty} \right) \frac{\hat{f}_p(\theta, p)}{\Theta \cdot x - p} dp d\theta. \tag{6.2.2'}$$

One has

Lemma 6.1.1. *Let $\rho > 0$ be fixed. Then the function $f_\rho^c(x)$ defined in (6.2.2) is continuous.*

Proof. Integrating by parts in (6.2.2), we get

$$f_\rho^c(x) = \frac{-1}{\pi} \left[\frac{F(x,q)}{q} \Big|_\rho^\infty + \int_\rho^\infty \frac{F(x,q)}{q^2} dq \right] = \frac{1}{\pi} \left[\frac{F(x,\rho)}{\rho} - \int_\rho^\infty \frac{F(x,q)}{q^2} dq \right]. \tag{6.2.4}$$

Actually, $F(x,q)$ has compact support as a function of q, so that the integrals above are taken over compact sets. Thus, in order to prove continuity of $f_\rho^c(x)$, it is sufficient to prove that $F(x,q)$ is continuous with respect to x. The Radon transform $\hat{f}(\theta, p)$ is discontinuous with respect to p only for such pairs (θ_0, p_0), that the lines $\Theta_0 \cdot x = p_0$ intersect the discontinuity curve of $f(x)$ over intervals of positive lengths (see Chapter 4). Since the number of such pairs (θ_0, p_0) is at most countable and $\sup_{\Theta \in S^1, p \in \mathbb{R}} |\hat{f}(\theta, p)| < \infty$, Equation (6.2.3) implies continuity of $F(x,q)$ in x. □

Exercise 6.2.2. Let S be a piecewise-smooth curve without self intersections. Prove that the number of lines intersecting S over intervals of positive length is at most countable.

Since $f = f_\rho + f_\rho^c$ and the function f_ρ^c is continuous, we conclude that

(1) the location of discontinuities of $f_\rho(x)$ is the same as that of $f(x)$, and

(2) at each discontinuity point ξ of f_ρ, the jump $f_\rho(\xi + 0) - f_\rho(\xi - 0)$ is precisely the same as that of $f(x)$ at $x = \xi$.

Thus discontinuities of f can be recovered knowing the function f_ρ. Note that $f_\rho(x)$ is computed using $\hat{f}(\theta, p)$ for (θ, p) satisfying $|\Theta \cdot x - p| \leq \rho$. Therefore, we call the function f_ρ pseudolocal. Local tomography formulas from Section 5.1 require the knowledge of the second derivative $\hat{f}_{pp}(\theta, p)$ only for (θ, p) satisfying $\Theta \cdot x = p$.

From Remark 2.2.2 and Equation (6.2.2) we get $f(x) = \lim_{\rho \to 0} f_\rho^c(x)$. Since $f_\rho = f - f_\rho^c$, or, more precisely, $f_\rho = \lim_{\epsilon \to 0} f_\epsilon^c - f_\rho^c$, we need to investigate the convergence $f_\rho^c \to f$, $\rho \to 0$, in order to better understand properties of f_ρ.

6.3. Investigation of the convergence $f_\rho^c(x) \to f(x)$ as $\rho \to 0$

Let us introduce the following function

$$\bar{f}(r, x) := \frac{1}{2\pi} \int_{S^1} f(x + r\theta) d\theta, \qquad (6.3.1)$$

and define

$$\bar{f}(0, x) = \lim_{r \to 0} \bar{f}(r, x). \qquad (6.3.2)$$

Exercise 6.3.1. Prove the identity

$$\int_{S^1} \delta(\Theta \cdot x - q) d\theta = \begin{cases} 2(|x|^2 - q^2)^{-1/2}, & |x| > |q|, \\ 0, & |x| < |q|, \end{cases} \quad x \in \mathbb{R}^2, \ q \in \mathbb{R}.$$

Exercise 6.3.2. Prove the formula

$$F(x, q) = 2 \int_{r > |q|} \bar{f}(r, x) \left[1 - (q/r)^2\right]^{-1/2} dr.$$

Hint. Use definition (6.2.3), definition of the Radon transform, and Exercise 6.3.1.

Using Exercise 6.3.2 in Equation (6.2.4) and changing the order of integration in the second integral, we get

$$f_\rho^c(x) = \frac{2}{\pi} \left(\frac{1}{\rho} \int_\rho^\infty \frac{\bar{f}(r, x)}{[1 - (\rho/r)^2]^{1/2}} dr - \int_\rho^\infty \bar{f}(r, x) \int_\rho^r \frac{dq}{[1 - (q/r)^2]^{1/2} q^2} dr \right)$$

$$= \frac{2}{\pi} \left(\frac{1}{\rho} \int_\rho^\infty \frac{\bar{f}(r, x)}{[1 - (\rho/r)^2]^{1/2}} dr - \frac{1}{\rho} \int_\rho^\infty \bar{f}(r, x)[1 - (\rho/r)^2]^{1/2} dr \right)$$

$$= \rho \frac{2}{\pi} \int_\rho^\infty \frac{\bar{f}(r, x)}{[1 - (\rho/r)^2]^{1/2}} \frac{dr}{r^2} = \bar{f}(0, x) + \rho \frac{2}{\pi} \int_\rho^\infty \frac{\bar{f}(r, x) - \bar{f}(0, x)}{[1 - (\rho/r)^2]^{1/2}} \frac{dr}{r^2}. \quad (6.3.3)$$

One has

Theorem 6.3.1. *Suppose $f \in C^2(U)$ for some open set U, $U \subset \mathbb{R}^2$. Then*

$$|f_\rho^c(x) - f(x)| = O(\rho) \quad \text{as } \rho \to 0, \ x \in U. \qquad (6.3.4)$$

Moreover, the convergence in (6.3.4) is uniform on all compact subsets of U. If $x_0 \in S$ is fixed and there exists an open neighborhood V of x_0 such that S is smooth in V and f is piecewise-C^2 in V, then

$$\left| f_\rho^c(x_0) - \frac{f_+(x_0) + f_-(x_0)}{2} \right| = O(\rho |\ln \rho|), \ \rho \to 0, \qquad (6.3.5)$$

where $f_\pm(x_0)$ are the limiting values of $f(x)$ as x approaches x_0 from different sides of S along any path nonintersecting S.

REMARK 6.3.1. Theorem 6.4.3 in Section 6.4 is a generalization of Theorem 6.3.1 which relaxes the smoothness assumptions on f.

Proof. First, we prove (6.3.4). Let $f \in C^2(U)$. Choose any $x_0 \in U$ and find $R > 0$ such that $B(x_0, R) \subset U$, where $B(x_0, R) := \{x \in \mathbb{R}^2 : |x - x_0| \le R\}$. Then $\bar{f}(0, x_0) = f(x_0)$ and

$$f(x_0 + r\Theta) = f(x_0) + \nabla f(x_0) \cdot \Theta r + \frac{1}{2} r^2 \frac{\partial^2 f(x_0 + t\Theta)}{\partial t^2}\bigg|_{t=t(r,\theta)},$$

$$\Theta \in S^1, \ r < R, \qquad (6.3.6)$$

for some function $t(r, \theta)$, $0 < t(r, \theta) < r$. Substitution of (6.3.6) into (6.3.1) yields

$$|\bar{f}(r, x_0) - f(x_0)| = \frac{r^2}{2} \frac{1}{2\pi} \left| \int_{S^1} \frac{\partial^2 f(x_0 + t\Theta)}{\partial t^2} d\theta \right| \le r^2 M_2(r, x_0), \quad (6.3.7)$$

where $M_2(r, x_0) = \max_{x,i,j} |\partial^2 f(x)/(\partial x_i \partial x_j)|$, the maximum is taken over x such that $|x - x_0| < r$ and $i, j = 1, 2$. Equations (6.3.3) and (6.3.7) imply for $\rho < R$:

$$|f_\rho^c(x_0) - f(x_0)| \le \frac{2\rho}{\pi} \int_\rho^R \frac{M_2(r, x_0)}{[1 - (\rho/r)^2]^{1/2}} dr$$

$$+ \frac{2\rho}{\pi} \int_R^\infty \frac{|\bar{f}(r, x_0) - \bar{f}(0, x_0)|}{[1 - (\rho/r)^2]^{1/2}} \frac{dr}{r^2}$$

$$\le \frac{2\rho}{\pi} (R^2 - \rho^2)^{1/2} M_2(R, x_0) + M_0(x_0) \frac{2}{\pi} \arcsin(\rho/R)$$

$$\le \frac{2\rho}{\pi} \left[R M_2(R, x_0) + \frac{M_0(x_0)}{R} \right] + O(\rho^3/R^3), \quad \rho \to 0, \qquad (6.3.8)$$

where $M_0(x_0) := \sup_{x \in \mathbb{R}^2} |f(x) - f(x_0)|$. To prove that f_ρ^c converges to f uniformly on any compact set K, $K \subset U$, let us fix δ, $0 < \delta <$ $\text{dist}(K, \partial U)$, where ∂U is the boundary of U. Equation (6.3.8) implies

$$\sup_{x \in K} |f_\rho^c(x) - f(x)| \leq \frac{2\rho}{\pi} \left(\text{diam}(U) \max_{\xi \in K} M_2(\delta, \xi) \right.$$

$$+ \left. \frac{\max_{\xi \in K} M_0(\xi)}{\delta} \right) + O(\rho^3/\delta^3), \quad \rho \to 0,$$

which proves the desired assertion.

Now let us pick $x_0 \in S$. Let R_0, $0 < R_0 \leq \infty$, be the radius of curvature of S at x_0. Under the assumptions of the theorem, there exists $R > 0$ such that

(a) S divides $B(x_0, R)$ into two sets $B_+(x_0, R)$ and $B_-(x_0, R)$ with $f \in C^2(B_+(x_0, R))$, $f \in C^2(B_-(x_0, R))$;
(b) S is smooth inside $B(x_0, R)$; and
(c) intersection of S with any circle centered at x_0 and with radius r, $0 < r \leq R$, contains exactly two points.

Let us denote

$$f_\pm(x_0) = \lim_{x \to x_0, x \in B_\pm(x_0, R)} f(x), \quad \nabla f_\pm(x_0) = \lim_{x \to x_0, x \in B_\pm(x_0, R)} \nabla f(x),$$

$$(6.3.9)$$

where the limits are taken along the paths not intersecting S. Clearly,

$$\bar{f}(0, x_0) = (f_-(x_0) + f_+(x_0))/2. \qquad (6.3.10)$$

To estimate the difference $\bar{f}(r, x_0) - \bar{f}(0, x_0)$ for $r < R$, we use the notation depicted in Figure 6.3.1. Let Q_1 and Q_2 be two points on S such that $|Q_1 - x_0| = |Q_2 - x_0| = r$. Suppose first that $R_0 < \infty$. By Ω_+ we denote the smaller of the two angles $\angle Q_1 x_0 Q_2$, and $\Omega_- = S^1 \setminus \Omega_+$.

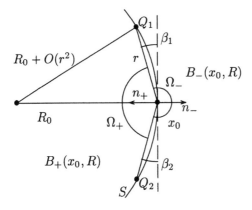

FIGURE 6.3.1. Figure illustrating the proof of Theorem 6.3.1.

Since $\beta_1 = r/(2R_0) + O(r^2)$ and $\beta_2 = r/(2R_0) + O(r^2)$, we get

$$\bar{f}(r, x_0) - \bar{f}(0, x_0)$$

$$= \frac{1}{2\pi} \int_{\Omega_+} (f_+(x_0) + \nabla f_+(x_0) \cdot \Theta r + O(r^2))d\theta$$

$$+ \frac{1}{2\pi} \int_{\Omega_-} (f_-(x_0) + \nabla f_-(x_0) \cdot \Theta r + O(r^2))d\theta - \frac{f_+(x_0) + f_-(x_0)}{2}$$

$$= f_+(x_0)\left(\frac{\pi - r/R_0 + O(r^2)}{2\pi} - 1/2\right)$$

$$+ f_-(x_0)\left(\frac{\pi + r/R_0 + O(r^2)}{2\pi} - 1/2\right)$$

$$+ \frac{r}{2\pi}\left(\int_{\Omega_+} \nabla f_+(x_0) \cdot \Theta d\theta + \int_{\Omega_-} \nabla f_-(x_0) \cdot \Theta d\theta\right) + O(r^2)$$

$$= -\frac{D(x_0)}{2\pi R_0}r + \frac{r}{2\pi}\left(\int_{\Omega_+ + \beta_1 + \beta_2} \nabla f_+(x_0) \cdot \Theta d\theta \right.$$

$$\left. + \int_{\Omega_- - \beta_1 - \beta_2} \nabla f_-(x_0) \cdot \Theta d\theta\right) + O(r^2)$$

$$= -\frac{D(x_0)}{2\pi R_0}r + \frac{r}{\pi}(f_{n_+}(x_0) + f_{n_-}(x_0)) + O(r^2) = Ar + O(r^2),$$

$$\tag{6.3.11}$$

where $f_{n_\pm}(x_0) = \nabla f_\pm(x_0) \cdot n_\pm$, n_+ and $n_- = -n_+$ are unit vectors perpendicular to S at x_0 and pointing inside B_+ and B_-, respectively, $D(x_0) = f_+(x_0) - f_-(x_0)$, and

$$A := \frac{1}{2\pi}\left(-\frac{D(x_0)}{R_0} + 2(f_{n_+}(x_0) + f_{n_-}(x_0))\right). \tag{6.3.12}$$

If $R_0 = \infty$ (the curvature of S at x_0 equals zero), then for an arbitrary choice of $\Omega_+ = \angle Q_1 x_0 Q_2$ and $\Omega_- = S^1 \setminus \Omega_+$ we get $\beta_1 = O(r^2)$, $\beta_2 = O(r^2)$, and the value of the constant A in (6.3.11) is given by $A = (f_{n_+}(x_0) + f_{n_-}(x_0))/\pi$. Clearly, this formula can be obtained from (6.3.12) by putting $R_0 = \infty$. Substitution of (6.3.11) into (6.3.3) yields

$$|f_\rho^c(x_0) - \bar{f}(0, x_0)| \leq \frac{\rho}{\pi} \int_\rho^R \frac{|Ar + O(r^2)|}{[1 - (\rho/r)^2]^{1/2}} dr$$

$$+ \frac{2\rho}{\pi} \int_R^\infty \frac{|\bar{f}(r, x_0) - \bar{f}(0, x_0)|}{[1 - (\rho/r)^2]^{1/2}} \frac{dr}{r^2}$$

$$\leq \frac{2}{\pi} |A| \rho \ln(2R/\rho) + O(\rho). \qquad (6.3.13)$$

In particular, if $R_0 = \infty$ and $f_{n_+}(x_0) = f_{n_-}(x_0) = 0$ (for instance, if $f(x)$ is constant in $B_+(x_0, R)$ and $B_-(x_0, R)$), then $A = 0$ and $|f_\rho^c(x_0) - \bar{f}(0, x_0)| = O(\rho)$. Equation (6.3.13) together with (6.3.10) proves (6.3.5). $\square$

Equation (6.3.5) asserts pointwise convergence of f_ρ^c at the discontinuity curve S. Convergence (or, rather, "nonconvergence") in a neighborhood of S is investigated in the following theorem.

Theorem 6.3.2. Let $x_0 \in S$, S is smooth in some neighborhood of x_0 and f is piecewise-C^2 there. Let n_0 be a unit vectors normal to S at x_0. Then for an arbitrary fixed γ, $\gamma \neq 0$, one has

$$\lim_{\rho \to 0} \left[f(x_0 + \gamma \rho n_0) - f_\rho^c(x_0 + \gamma \rho n_0) \right] = D(x_0) \psi(\gamma), \qquad (6.3.14)$$

where $D(x_0) := \lim_{t \to 0^+} \left(f(x_0 + t n_0) - f(x_0 - t n_0) \right)$ and

$$\psi(\gamma) := \frac{2}{\pi^2} \int_0^{\min(1, 1/\gamma)} \frac{\arccos(\gamma t)}{(1 - t^2)^{1/2}} dt, \quad \gamma > 0; \quad \psi(\gamma) = -\psi(|\gamma|), \ \gamma < 0.$$

The function $\psi(\gamma)$ is strictly positive, monotonically decreasing with $\lim_{\gamma \to 0} \psi(\gamma) = 1/2$ and $\psi(\gamma) = 2/(\pi^2 \gamma) + O(\gamma^{-3})$, $\gamma \to \infty$.

Proof. Under the assumptions of the theorem, there exists $R > 0$ such that conditions (a), (b), and (c) in the proof of Theorem 6.3.1 are satisfied (see above (6.3.9)). Suppose, for example, that n_0 points into B_+. Let us consider a point $x = x_0 + (\gamma \rho) n_0 \in B_+(x_0, R)$ (see Figure 6.3.2), where γ, $0 < \gamma < \infty$, is fixed.

Suppose $R_0 < \infty$. First, consider the case $r > \gamma \rho$. Using the cosine theorem for the triangle OQx, we get

$$(R_0 + O(r^2))^2 = (R_0 - \gamma \rho)^2 + r^2 + 2(R_0 - \gamma \rho) r \cos \alpha,$$

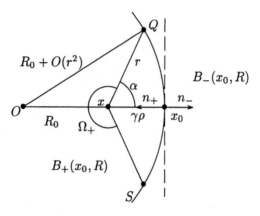

FIGURE 6.3.2. Figure illustrating the proof of Theorem 6.3.2.

and so,

$$\cos\alpha = \frac{2R_0\gamma\rho + O(r^2)}{2(R_0 - \gamma\rho)r} = \frac{\gamma\rho}{r} + O(r), \quad \gamma\rho < r < R. \qquad (6.3.15)$$

Since $f \in C^2(B_+(x_0, R))$ and $x \in B_+(x_0, R)$, we get that $\bar{f}(0, x) = f(x)$ and

$$\bar{f}(r, x) - \bar{f}(0, x) = O(r^2), \quad r < \gamma\rho. \qquad (6.3.16)$$

Exercise 6.3.3. Prove inequality

$$|\arccos(a + b) - \arccos(b)| \le 2a^{1/2}, \quad a, b \ge 0, \ a + b \le 1.$$

Exercise 6.3.3 and Equation (6.3.15) imply that $\alpha = \arccos(\gamma\rho/r) + O(r^{1/2})$ if $r > \gamma\rho$. Using this, we obtain from (6.3.1)

$$\bar{f}(r, x) - \bar{f}(0, x) = \frac{1}{2\pi} \int_{\Omega_+} [f_+(x_0) + O(r)]d\theta + \frac{1}{2\pi} \int_{\Omega_-} [f_-(x_0) + O(r)]d\theta$$

$$- (f_+(x_0) + O(\gamma\rho))$$

$$= f_+(x_0)\frac{2\pi - 2\alpha}{2\pi} + f_-(x_0)\frac{2\alpha}{2\pi} - f_+(x_0) + O(r)$$

$$= - D(x_0)\frac{\arccos(\gamma\rho/r)}{\pi} + O(r^{1/2}), \quad r > \gamma\rho.$$

$$(6.3.17)$$

Substitution of (6.3.16) and (6.3.17) into (6.3.3) yields

$$f(x) - f_\rho^c(x) = \frac{2\rho}{\pi} \int\limits_{\rho}^{\max(\gamma\rho,\rho)} \frac{O(r^2)}{[1-(\rho/r)^2]^{1/2}} \frac{dr}{r^2}$$

$$+ \frac{2\rho}{\pi} \int\limits_{\max(\gamma\rho,\rho)}^{R} \frac{D(x_0)\frac{\arccos(\gamma\rho/r)}{\pi} + O(r^{1/2})}{[1-(\rho/r)^2]^{1/2}} \frac{dr}{r^2}$$

$$- \frac{2\rho}{\pi} \int\limits_{R}^{\infty} \frac{\bar{f}(r,x) - \bar{f}(0,x)}{[1-(\rho/r)^2]^{1/2}} \frac{dr}{r^2}$$

$$= O(\rho^2) + \frac{2D(x_0)}{\pi^2} \rho \int\limits_{\max(\gamma\rho,\rho)}^{R} \frac{\arccos(\gamma\rho/r)}{[1-(\rho/r)^2]^{1/2}} \frac{dr}{r^2}$$

$$+ O(\rho^{1/2}) + O(\rho)$$

$$= \frac{2D(x_0)}{\pi^2} \int\limits_{0}^{\min(1,1/\gamma)} \frac{\arccos(\gamma t)}{(1-t^2)^{1/2}} dt + O(\rho^{1/2}), \quad \rho \to 0.$$

If $R_0 = \infty$, then the analogue of (6.3.15) becomes $\cos\alpha = \gamma\rho/r + O(r^2)$, and in the last equation we should replace $O(\rho^{1/2})$ by $O(\rho)$. Thus, in both cases $0 < R_0 < \infty$ and $R_0 = \infty$, we obtain

$$\lim_{\rho\to0} f_\rho(x_0 + \gamma\rho n_0) = \lim_{\rho\to0} \left[f(x_0 + \gamma\rho n_0) - f_\rho^c(x_0 + \gamma\rho n_0) \right]$$

$$= \frac{2D(x_0)}{\pi^2} \int\limits_{0}^{\min(1,1/\gamma)} \frac{\arccos(\gamma t)}{(1-t^2)^{1/2}} dt. \tag{6.3.18}$$

By considering a point x on the other side of x_0, $x = x_0 + (\gamma\rho)n_0 \in B_-(x_0, R)$, $\gamma < 0$, we can obtain similarly to (6.3.15) that $\cos\alpha = -\frac{|\gamma|\rho}{r} + O(r)$, $r > |\gamma|\rho$. Similarly to (6.3.17), this yields

$$\bar{f}(r,x) - \bar{f}(0,x) = -D(x_0)\frac{\arccos(|\gamma|\rho/r)}{\pi} + O(r^{1/2}), \quad r > |\gamma|\rho,$$
$$x = x_0 + (\gamma\rho)n_0 \in B_-(x_0, R), \quad \gamma < 0.$$

Thus, Equation (6.3.18) can be generalized as follows

$$\lim_{\rho\to0} \left[f(x_0 + \gamma\rho n_0) - f_\rho^c(x_0 + \gamma\rho n_0) \right] = D(x_0)\psi(\gamma), \tag{6.3.19}$$

where

$$\psi(\gamma) := \frac{2}{\pi^2} \int_0^{\min(1,1/\gamma)} \frac{\arccos(\gamma t)}{(1-t^2)^{1/2}} dt, \ \gamma > 0; \quad \psi(\gamma) := -\psi(|\gamma|), \ \gamma < 0.$$

(6.3.20)

Properties of $\psi(\gamma)$ formulated in Theorem 6.3.2 can be easily established by noting that $\arccos(\gamma t)$ is a decreasing function of γ on $(0, 1/t)$ if $t > 0$ is fixed and that

$$\lim_{\gamma \to 0+} \psi(\gamma) = \frac{2}{\pi^2} \int_0^1 \frac{\pi/2}{(1-t^2)^{1/2}} dt = 1/2,$$

$$\psi(\gamma) = \frac{2}{\pi^2} \frac{1}{\gamma} \int_0^1 \frac{\arccos t}{[1-(t/\gamma)^2]^{1/2}} dt = \frac{2}{\pi^2} \frac{1}{\gamma} \left(\int_0^1 \arccos t \, dt + O(\gamma^{-2}) \right)$$

$$= \frac{2}{\pi^2} \gamma^{-1} + O(\gamma^{-3}), \ \gamma \to +\infty.$$

Theorem 6.3.2 is proved. □

Summarizing the results of this section, we conclude that f_ρ^c converges to f uniformly on compact subsets of $\mathbb{R}^2$ not intersecting S, converges pointwise to the average $(f_+(x_0) + f_-(x_0))/2$ at $x_0 \in S$ where S is smooth, and there is a boundary layer of width $O(\rho)$ in which the difference $|f - f_\rho^c|$ monotonically increases (with the bound $|D(x_0)|/2$) as x approaches x_0. Let us describe the last point in more detail. Neglecting the terms of order $\rho^{1/2}$, we have from (6.3.19):

$$f(x_0 + \gamma \rho n_0) - f_\rho^c(x_0 + \gamma \rho n_0) \approx D(x_0)\psi(\gamma).$$

We see that as γ changes within compact intervals $0 < a \le \gamma \le b < \infty$, the point $x(\gamma) := x_0 + \gamma \rho n_0$ moves within a layer of width $O(\rho)$. Moreover, according to the properties of $\psi(\gamma)$ formulated in Theorem 6.3.2, the modulus of the difference $|f(x(\gamma)) - f_\rho^c(x(\gamma))|$ increases monotonically (neglecting the terms of order $\rho^{1/2}$) as $\gamma \to 0$ (or, as $x(\gamma) \to x_0$). Since $f_\rho(x) = \lim_{\epsilon \to 0} f_\epsilon^c(x) - f_\rho^c(x)$, we get that:

(a) $f_\rho(x)$ converges to zero as $\rho \to 0$ on compact subsets of $\mathbb{R}^2$ outside S and at the points of S;

(b) for a sufficiently small fixed $\rho > 0$, $f_\rho(x)$ approaches the limit $D(x_0)/2 + O(\rho^{1/2})$ or $-D(x_0)/2 + O(\rho^{1/2})$ as $x \to x_0 \in S$ from the corresponding side of S.

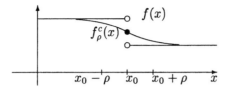

FIGURE 6.3.3. Schematic graphs of functions f_ρ^c, f, and f_ρ.

Thus, at the cross-section perpendicular to S at x_0 the graphs of f_ρ^c, f, and f_ρ can be schematically represented as in Figure 6.3.3.

6.4. More results on functions f_ρ^c, f_ρ, and on convergence $f_\rho^c \to f$

In this section we will prove the pseudolocality property of the function f_ρ and broaden the range of applicability of the pseudolocal tomography formulas. First, let us formulate basic results. We suppose that $\rho > 0$ is fixed. By definition, the 0-th order derivative of any function denotes this function itself. For any set U, $U \subset \mathbb{R}^2$, we define $U_\rho := \{x \in \mathbb{R}^2 : \text{dist}(x, U) \le \rho\}$. Also we define $C(x_0, \rho) := \{x \in \mathbb{R}^2 : |x - x_0| = \rho\}$, $S(x_0, a, b) := \{x \in \mathbb{R}^2 : a \le |x - x_0| \le b\}$ for $0 \le a \le b$.

Theorem 6.4.1. *Let us fix any $x_0 \in \mathbb{R}^2$ such that $C(x_0, \rho)$ is not tangent to S. In the case $C(x_0, \rho) \cap S \ne \varnothing$ we assume that there exists $k \ge 0$ such that*

(i) *the set $C(x_0, \rho) \cap S$ consists of finitely many points;*
(ii) *S is C^k in a neighborhood of $C(x_0, \rho) \cap S$;*
(iii) *f is C^k in a neighborhood of any x, $x \in C(x_0, \rho)$ and $x \notin S$.*

In the case $C(x_0, \rho) \cap S = \varnothing$ we assume that f is C^k in a neighborhood of $C(x_0, \rho)$. Then, in the above two cases, f_ρ^c is C^k in some neighborhood of x_0.

Corollary. "The pseudolocality property". *Suppose there exists an open set U, $U \subset \mathbb{R}^2$, such that $f \in C^k(U_\rho)$. Then $f_\rho \in C^k(U)$. In particular, $f \in C^\infty(U_\rho) \Rightarrow f_\rho \in C^\infty(U)$.*

Theorem 6.4.2. "The preservation of discontinuities property". *Let us fix an open set U, $U \subset \mathbb{R}^2$. Suppose there exists $k \ge 0$ such that*

(a) *$f \in C^{k-1}(U_\rho)$ if $k \ge 1$, if $k = 0$ this assumption is dropped;*

(b) *k-th order derivatives of f exist in U_ρ and some (or all of them) are discontinuous across S; and*

(c) *S is piecewise-smooth inside U_ρ.*

Then $f_\rho^c = f - f_\rho \in C^k(U)$.

REMARK 6.4.1. Lemma 6.1.1 is a particular case of Theorem 6.4.2 with $k = 0$.

REMARK 6.4.2. Loosely speaking, Theorems 6.4.1 and 6.4.2 mean that the function f_ρ (or its derivatives) has precisely the same location and size of discontinuities inside any open set U as those of the original function f (or its derivatives) regardless of the behavior of f outside U_ρ.

We will use some lemmas in the proof of Theorems 6.4.1 and 6.4.2.

Lemma 6.4.1. *Suppose there exist $x_0 \in \mathbb{R}^2$ and ϵ, $0 < \epsilon \le \rho$, such that $f(x) \equiv 0$ on $S(x_0, \rho - \epsilon, \rho + \epsilon)$. Then $f_\rho^c \in C^\infty(B^o(x_0, \epsilon))$, where B^o denotes the interior of the ball.*

Proof. From (6.3.3) and (6.3.1) we get

$$f_\rho^c(x) = \rho \frac{2}{\pi} \int_\rho^\infty \frac{\bar{f}(r, x)}{[1 - (\rho/r)^2]^{1/2}} \frac{dr}{r^2}$$

$$= \frac{\rho}{\pi^2} \int_\rho^\infty \int_{S^1} (1 - (\rho/r)^2)^{-1/2} r^{-3} f(x + r\Theta) d\theta r dr$$

$$= \int_{|y|>\rho} \psi(y) f(x + y) dy = \int_{|y-x|>\rho} \psi(y - x) f(y) dy, \qquad (6.4.1)$$

where

$$\psi(y) := \frac{\rho}{\pi^2} (1 - (\rho/|y|)^2)^{-1/2} |y|^{-3}. \qquad (6.4.2)$$

Clearly,

$$\psi(y) \in C^\infty(\{y \in \mathbb{R}^2 : |y| > \rho\}). \qquad (6.4.3)$$

Since $C(x, \rho) \subset S(x_0, \rho - \epsilon, \rho + \epsilon)$ for any x, $|x - x_0| < \epsilon$, we have $f \equiv 0$ on any $C(x, \rho)$, $|x - x_0| < \epsilon$, and, in view of (6.4.3), we can differentiate (6.4.1) any number of times. □

Corollary to Lemma 6.4.1. *Suppose there exists an open set U, $U \subset \mathbb{R}^2$, such that $f \equiv 0$ on U_ρ. Then $f_\rho^c \in C^\infty(U)$.*

Let us denote the dependence of f_ρ^c on the original function f by $f_\rho^c[f]$, and let D^k be any derivative of order $k \ge 0$, $D^k := \partial^{k_1+k_2}/(\partial^{k_1} x_1 \partial^{k_2} x_2)$, $x := (x_1, x_2)$, $k_1 + k_2 = k$. Recall that $D^0 f := f$. The following simple lemma follows easily from the third equality in (6.4.1).

Lemma 6.4.2. *One has*

$$f_\rho^c[\alpha_1 f_1 + \alpha_2 f_2] = \alpha_1 f_\rho^c[f_1] + \alpha_2 f_\rho^c[f_2], \quad \alpha_1, \alpha_2 \in \mathbb{R}, \quad (6.4.4)$$

$$W * f_\rho^c[f] = f_\rho^c[W * f], \quad W \in L_1(\mathbb{R}^2), \quad (6.4.5)$$

$$D^k f_\rho^c[f] = f_\rho^c[D^k f], \quad k \geq 0, \quad (6.4.6)$$

where the last equality holds if the function on the right-hand side of (6.4.6) exists.

Proof of Theorem 6.4.1. First, let us consider the case $C(x_0, \rho) \cap S \neq \emptyset$. Then, by assumptions (i)-(iii), there exists ϵ, $0 < \epsilon \leq \rho$, such that

(a) all sets $C(x, r) \cap S$ have the same (finite) number of points for any x, $|x - x_0| < \epsilon$, and r, $\rho \leq r \leq \rho + 2\epsilon$;

(b) no circle $C(x, r)$ for any x, $|x - x_0| < \epsilon$, and r, $\rho \leq r \leq \rho + 2\epsilon$ is tangent to S;

(c) S is C^k in a neighborhood of $S(x_0, \rho - \epsilon, \rho + \epsilon) \cap S$;

(d) f is C^k in a neighborhood of any x, $x \in S(x_0, \rho - \epsilon, \rho + \epsilon)$ and $x \notin S$.

Let us fix any δ, $0 < \delta < \epsilon$, and find $\varphi(x) \in C_0^\infty(S(x_0, \rho - \epsilon, \rho + \epsilon))$ such that $\varphi \equiv 1$ on $S(x_0, \rho - (\epsilon - \delta), \rho + (\epsilon - \delta))$. Lemma 6.4.1 implies that $f_\rho^c[(1 - \varphi)f] \in C^\infty(B^o(x_0, \epsilon - \delta))$. By Lemma 6.4.2 it remains to be proved that $f_\rho^c[\varphi f]$ is C^k in a neighborhood of x_0. Since

$$f_\rho^c[\varphi f](x) = \rho \frac{2}{\pi} \int\limits_\rho^{\rho + 2\epsilon} \frac{\overline{(\varphi f)}(r, x)}{[1 - (\rho/r)^2]^{1/2}} \frac{dr}{r^2}, \quad |x - x_0| \leq \epsilon, \quad (6.4.7)$$

it is sufficient to prove that k-th order derivatives of $\overline{(\varphi f)}(r, x)$ with respect to x are continuous both with respect to r and x for $(r, x) \in D$, where $D := \{(r, x) : \rho \leq r \leq \rho + 2\epsilon, |x - x_0| \leq \epsilon\}$. Using (a) and (d), let us find domains $U_i \subset \mathbb{R}^2$, $i = 1, \ldots, m$, for some $m < \infty$ such that $\cup_{i=1}^m U_i = S(x_0, \rho - \epsilon, \rho + \epsilon)$ and $\varphi f \in C^k(U_i)$, $i = 1, \ldots, m$. From (6.3.1) we have

$$\overline{(\varphi f)}(r, x) = \frac{1}{2\pi} \sum_{i=1}^m \int\limits_{\theta \in S^1, x + r\theta \in U_i} (\varphi f)(x + r\Theta) d\theta. \quad (6.4.8)$$

The boundary of each set U_i consists only of pieces of $C(x_0, \rho - \epsilon)$, $C(x_0, \rho + \epsilon)$, and S. Since circles $C(x, r)$ are transversal to S for $(r, x) \in D$ and S is C^k in a neighborhood of $S(x_0, \rho - \epsilon, \rho + \epsilon) \cap S$ (see (b) and (c) above), the implicit function theorem implies that coordinates of the points from the set $C(x, r) \cap S$ are C^k functions in r and x for $(r, x) \in D$.

Also $(\varphi f)(x) \equiv 0$ on $C(x_0, \rho - \epsilon)$ and $C(x_0, \rho + \epsilon)$, hence (6.4.8) implies that $\overline{(\varphi f)}$ is C^k in r and x for $(r, x) \in D$, and this proves the assertion of the theorem in the case $C(x_0, \rho) \cap S \neq \emptyset$.

In the case $C(x_0, \rho) \cap S = \emptyset$ we find ϵ, $0 < \epsilon \leq \rho$, such that $f \in C^k(S(x_0, \rho - \epsilon, \rho + \epsilon))$ and choose $\varphi(x)$ as above. Lemma 6.4.1 implies that $f_\rho^c[(1 - \varphi)f] \in C^\infty(B^o(x_0, \epsilon - \delta))$. Clearly, $f_\rho^c[\varphi f] \in C^k(\mathbb{R}^2)$ and the proof of Theorem 6.4.1 is complete. □

Proof of Theorem 6.4.2. Let $D^k f$, $k \geq 0$, denote any k-th order derivative of f which is discontinuous across S inside U_ρ. We fix any open set V, $V \subset U$, and find $\varphi \in C_0^\infty(U_\rho)$ such that $\varphi \equiv 1$ on V_ρ. Using Lemma 6.4.2, we have

$$D^k f_\rho^c[f] = D^k f_\rho^c[(1 - \varphi)f] + f_\rho^c[D^k(\varphi f)].$$

By Corollary to Lemma 6.4.1, $D^k f_\rho^c[(1 - \varphi)f] \in C^\infty(V)$. By the assumptions of the theorem, $D^k(\varphi f)$ exists, is compactly supported and piecewise-continuous. Therefore, Lemma 6.1.1 implies that $f_\rho^c[D^k(\varphi f)] \in C(\mathbb{R}^2)$. Thus $D^k f_\rho^c[f] \in C(V)$. Since $U \setminus V$ can be made arbitrarily small, this proves Theorem 6.4.2. □

Finally, we will present a generalization of Theorem 6.3.1 on convergence $f_\rho^c \to f$ at the points where f is k times continuously differentiable, $k \geq 0$.

Theorem 6.4.3. *Suppose there exist $x_0 \in \mathbb{R}^2$ and $R > 0$ such that $f \in C^{k_0}(B(x_0, R))$ for some $k_0 \geq 0$. Let D^k denote any k-th order derivative, $k \geq 0$. Then*

$$|(D^k f_\rho^c)(x) - (D^k f)(x)| = o(1) \quad \text{if} \quad k = k_0, \qquad (6.4.9)$$

$$|(D^k f_\rho^c)(x) - (D^k f)(x)| = o(\rho |\log \rho|) \quad \text{if} \quad k = k_0 - 1, \qquad (6.4.10)$$

$$|(D^k f_\rho^c)(x) - (D^k f)(x)| = O(\rho) \quad \text{if} \quad 0 \leq k \leq k_0 - 2 \qquad (6.4.11)$$

for $x \in B(x_0, R)$. Moreover, the convergence in (6.4.9) - (6.4.11) is uniform on any ball $B(x_0, R')$, $0 < R' < R$.

Proof. First, let us prove (6.4.9). Let us fix any R', $0 < R' < R$, and denote $\delta := R - R'$. Then (6.3.3) yields with $\rho < \delta$ and $k = k_0$:

$$(D^k f_\rho^c)(x) - (D^k f)(x) = \rho \frac{2}{\pi} \int_\rho^\delta \frac{D^k[\bar{f}(r, x) - f(x)]}{[1 - (\rho/r)^2]^{1/2}} \frac{dr}{r^2}$$

$$+ D^k \left[\rho \frac{2}{\pi} \int_\delta^\infty \frac{\bar{f}(r, x)}{[1 - (\rho/r)^2]^{1/2}} \frac{dr}{r^2} \right]$$

$$- (D^k f)(x) \rho \frac{2}{\pi} \int_\delta^\infty \frac{1}{[1 - (\rho/r)^2]^{1/2}} \frac{dr}{r^2}. \qquad (6.4.12)$$

where D^k denotes derivative with respect to x. Let us consider each integral on the right-hand side of (6.4.12) separately. Since $D^{k_0} f \in C(B(x_0, R))$, we have $D^{k_0}[\bar{f}(r, x) - f(x)] = o(1)$ as $r \to 0$ uniformly for $x \in B(x_0, R')$. Therefore,

$$C_1(x) = \rho \frac{2}{\pi} \int_\rho^\delta \frac{o(1)}{[1 - (\rho/r)^2]^{1/2}} \frac{dr}{r^2} = o(1)$$

as $\rho \to 0$ uniformly on $B(x_0, R')$.   (6.4.13)

Transforming the second integral in (6.4.12) similarly to (6.4.1), we obtain

$$C_2(x) = D^{k_0} \left[\int_{|y-x|\geq\delta} \psi(y - x) f(y) dy \right] = O(\rho) \quad \text{uniformly on } B(x_0, R').$$

(6.4.14)

The last equality in (6.4.14) holds because all the derivatives of f at the set $\{y \in \mathbb{R}^2 : y - x = \delta, x \in B(x_0, R')\}$ up to the order $k_0 - 1$ exist and because $(D^k \psi)(x) = O(\rho)$ as $\rho \to 0$ $(\rho < \delta)$ uniformly in x, $|x| \geq \delta$. Since $\max_{x\in B(x_0,R')} |(D^{k_0} f)(x)| < \infty$, the last integral in (6.4.12) is of order $O(\rho)$. This together with (6.4.12) - (6.4.14) proves (6.4.9).

Now suppose $k = k_0 - 1$. In this case, similarly to (6.3.6) and (6.3.7), we get $D^k[\bar{f}(r, x) - f(x)] = o(r)$ uniformly on $B(x_0, R')$ and the first integral in (6.4.12) becomes

$$C_1(x) = \rho \frac{2}{\pi} \int_\rho^\delta \frac{o(r)}{[1 - (\rho/r)^2]^{1/2}} \frac{dr}{r^2} = \rho \frac{2}{\pi} \int_\rho^\delta \frac{o(1)}{\sqrt{r^2 - \rho^2}} dr = o(\rho |\log \rho|).$$

(6.4.15)

As above, $C_2(x) = O(\rho)$ and $C_3(x) = O(\rho)$, $x \in B(x_0, R')$, and (6.4.10) is proved.

Similarly to (6.3.6) - (6.3.8), it is easy to establish that $C_1(x) = O(\rho)$ if $0 \leq k \leq k_0 - 2$. Again, $C_2(x) = O(\rho)$ and $C_3(x) = O(\rho)$, and (6.4.11) is proved. $\square$

6.5. A family of pseudolocal tomography functions

6.5.1. Definition of a family. Basic property

Let $\rho > 0$ be fixed. We define a family of pseudolocal tomography (PLT) functions:

$$f_{\sigma\rho}(x) := \frac{1}{4\pi^2} \int_{S^1} \int_{\Theta\cdot x - \rho}^{\Theta\cdot x + \rho} \frac{\sigma_\rho(\Theta \cdot x - p)}{\Theta \cdot x - p} \hat{f}_p(\theta, p) dp \, d\theta,$$

(6.5.1)

where $\sigma_\rho(p)$ satisfies the properties

  (1) $\sigma_\rho(p)$ is real-valued and even;
  (2) $\sigma_\rho(p)$ is piecewise-continuously differentiable and there are at most finitely many points at which σ_ρ is discontinuous; and
  (3)

$$\sigma_\rho(p) = \sigma_1(p/\rho), \quad |\sigma_\rho(p) - 1| \le O(p), \; p \to 0. \tag{6.5.2}$$

REMARK 6.5.1. Taking $\sigma_\rho(p) \equiv 1$ in (6.5.1), we get the PLT function defined in (6.2.1')

Theorem 6.5.1. *The difference $f - f_{\sigma\rho}$ is continuous, hence functions f and $f_{\sigma\rho}$ have the same discontinuity curve S and the same values of jumps across S.*

Proof. Let us extend $\sigma_\rho(p)$ by 0 for $|p| > \rho$. Then, using inversion formula (2.2.19), we get

$$f_{\sigma\rho}(x) - f(x) = \frac{1}{4\pi^2} \int\limits_{S^1} \int\limits_{\mathbb{R}} \frac{\sigma_\rho(\Theta \cdot x - p) - 1}{\Theta \cdot x - p} \hat{f}_p(\theta, p) dp \, d\theta$$

$$= \int\limits_{S^1} \int\limits_{\mathbb{R}} \psi(\Theta \cdot x - p) \hat{f}_p(\theta, p) dp d\theta,$$

$$\psi(p) := \frac{\sigma_\rho(p) - 1}{4\pi^2 p}. \tag{6.5.3}$$

Let $p_i, i = 1, \ldots, I$, be the points where $\psi(p)$ is discontinuous, and ψ_i be the corresponding jumps of ψ: $\psi_i := \psi(p_i + 0) - \psi(p_i - 0)$. Under assumption (6.5.2), $|\psi_i| < \infty, i = 1, \ldots, I$, so (6.5.3) yields

$$f_{\sigma\rho}(x) - f(x) = \sum_{i=1}^{I} \psi_i \int\limits_{S^1} \hat{f}(\theta, \Theta \cdot x - p_i) d\theta$$

$$+ \sum_{i=1}^{I-1} \int\limits_{S^1} \int\limits_{p_i}^{p_{i+1}} \psi'(\Theta \cdot x - p) \hat{f}(\theta, p) dp d\theta.$$

The Radon transform $\hat{f}(\theta, p)$ is discontinuous with respect to p only for such pairs (θ_0, p_0), that the lines $\Theta_0 \cdot x = p_0$ intersect the discontinuity curve of $f(x)$ over intervals of positive lengths (see Chapter 4). Since the number of such pairs (θ_0, p_0) is at most countable and $\sup_{\theta \in S^1, p \in \mathbb{R}} |\hat{f}(\theta, p)| < \infty$, the last equation implies continuity of the difference $f_{\sigma\rho}(x) - f(x)$ in x. □

6.5.2. Relation between pseudolocal and local tomography functions

An alternative approach for obtaining functions with discontinuities identical (in location and in size) to these of f is based on Equation (5.4.2). Indeed, let $M(x)$ be a function, the Fourier transform of which has an asymptotic $\tilde{M}(\xi) = |\xi|^{-1}(1 + \psi(\xi))$, $\psi(\xi) = o(1)$, $|\xi| \to \infty$. Then $M * f_\Lambda = \mathcal{F}^{-1}\{|\xi|\tilde{M}(\xi)\hat{f}(\xi)\}$, and the difference $M * f_\Lambda - f = \mathcal{F}^{-1}\{\psi(\xi)\hat{f}(\xi)\}$ is continuous if $\psi(\xi)$ decreases sufficiently fast as $|\xi| \to \infty$. Here $*$ denotes convolution in $\mathbb{R}^2$. Similarly to (5.4.3), we have

$$(M * f_\Lambda)(x) = -\frac{1}{4\pi} \int_{S^1} \int_{\mathbb{R}} \hat{M}''(\Theta \cdot x - p)\hat{f}(\theta, p)\,dp\,d\theta, \quad \hat{M} = RM. \quad (6.5.4)$$

In (6.5.4), for simplicity, we assumed that $M(x)$ is radial. Thus the problem reduces to finding compactly supported $M(x)$ with the desired spectral asymptotics. As an example of $M(x)$ we can take the function whose Radon transform is

$$\hat{M}(p) = -\frac{1}{\pi}\eta(p) \ln|p|, \quad (6.5.5)$$

where

$$\eta(0) = 1, \quad \eta(p) = \eta(-p), \quad \eta(p) = 0 \quad \text{for} \quad |p| \geq d, \quad \eta(p) \in C^2(\mathbb{R}). \quad (6.5.6)$$

Using the Erdelyi lemma (see Section 14.5.3) and the formula $\tilde{M}(\xi) = \int_{\mathbb{R}} \hat{M}(p)e^{ip|\xi|}\,dp$, one can check that

$$\tilde{M}(\xi) = |\xi|^{-1}(1 + O(|\xi|^{-2})), \quad |\xi| \to \infty. \quad (6.5.7)$$

Theorem 6.5.2. *The difference $f - M * f_\Lambda$ is continuous, hence the function $M * f_\Lambda$ has the same discontinuity curve S and the same values of jumps across S as those of f.*

Proof. Let us denote $f_1 = M * f_\Lambda$. It follows from the results in Section 4.7 that if S is a closed smooth curve with nonvanishing curvature then $\tilde{f}(\xi) = O(|\xi|^{-3/2})$, $|\xi| \to \infty$. Therefore, $\widetilde{f_1 - f} = O(|\xi|^{-2})\hat{f} \in L^1(\mathbb{R}^2)$, and the difference $f_1(x) - f(x)$ is continuous. In the general case, when S is non-smooth and may have vanishing curvature, one has $\tilde{f}(\xi) = O(|\xi|^{-\gamma})$ as $|\xi| \to \infty$, where $\gamma \geq 1$ (the case $\gamma = 1$ takes place when S contains a piece of line segment), and the above argument remains valid. Note that the results in Section 4.7 require S to be a union of finitely many smooth submanifolds. This allows one to have various types of singular points on S in $\mathbb{R}^n$, $n > 2$. In $\mathbb{R}^2$ this assumption means that S is a union of finitely many smooth curves, so that S may have finitely many singular points (corner points or spikes). □

Substituting (6.5.5) into (6.5.4) and integrating by parts, we obtain

$$(M * f_\Lambda)(x)$$

$$= -\frac{1}{4\pi} \int_{S^1} \int_R \hat{M}'(\Theta \cdot x - p) \hat{f}_p(\theta, p) dp d\theta$$

$$= \frac{1}{4\pi^2} \int_{S^1} \int_R \left(\frac{\eta(\Theta \cdot x - p)}{\Theta \cdot x - p} + \eta'(\Theta \cdot x - p) \ln |\Theta \cdot x - p| \right) \hat{f}_p(\theta, p) dp d\theta$$

$$= \frac{1}{4\pi^2} \int_{S^1} \int_R \frac{\sigma(\Theta \cdot x - p)}{\Theta \cdot x - p} \hat{f}_p(\theta, p) dp d\theta, \quad \sigma(p) = \eta(p) + \eta'(p)p \ln |p|.$$

$$(6.5.8)$$

Comparing (6.5.8) and (6.5.1), we see that the function $M * f_\Lambda$ is a PLT function if $\operatorname{supp} \eta = [-\rho, \rho]$. As an example we can take

$$\eta(p) = (1 - (p/\rho)^{2n})^{2m}, \ |p| \le \rho, \quad \eta(p) = 0, \ |p| > \rho, \quad m, n \in \mathbb{N}.$$

Equation (6.5.8) in the forward direction implies that the convolution of the LT function f_Λ with a kernel from a certain class equals to a PLT function. The opposite is also true. Indeed, let us consider (6.5.8) in the opposite direction. Let a family of functions $\sigma_\rho(p)$ satisfying conditions (a) – (c) from Section 6.5.1 be given. Then, defining

$$\hat{M}'_{\sigma\rho}(p) := -\frac{\sigma_\rho(p)}{\pi p}, \ |p| \le \rho, \quad \hat{M}_{\sigma\rho}(p) = 0, \ |p| > \rho, \quad (6.5.9)$$

that is $\hat{M}_{\sigma\rho}(p) = -\int_{-\rho}^p \frac{\sigma_\rho(s)}{\pi s} ds, \ |p| \le \rho$, we obtain $f_{\sigma\rho} = M_{\sigma\rho} * f_\Lambda$.

As one can see, the usage of a PLT function $f_{\sigma\rho}$ for finding values of jumps of f requires investigation of the properties of $f_{\sigma\rho}$ as $x \to S$. An example of such analysis for $\sigma_\rho(p) = 1, \ |p| \le \rho$, is given in Section 6.3. However, this analysis depends heavily on the cut-off function $\sigma_\rho(p)$ and, therefore, it should be done for each function $\sigma_\rho(p)$ separately. The results from Section 4.4 and the relation between PLT and LT functions which we have just established allow us to obtain the behavior of $f_{\sigma\rho}, x \to S$, in much easier way. Indeed, assuming for simplicity that S is flat inside ρ - neighborhood of x_0 for a sufficiently small $\rho > 0$: $n(x_s) = n(x_0) =: n_0, \ |x_s - x_0| \le \rho$, and using the relation $f_{\sigma\rho} = M_{\sigma\rho} * f_\Lambda$, where the Radon transform of M satisfies (6.5.9), we derive

$$f_{\sigma\rho}(x_s + hn_0) = (M_{\sigma\rho} * f_\Lambda)(x_s + hn_0)$$

$$= \int_{\mathbb{R}^2} M_{\sigma\rho}((h - t)n_0 + (x_s - x_r))\left(\frac{D(x_r)}{\pi t} + O(\ln|t|)\right)dx_r dt$$

$$= \frac{D(x_s)}{\pi}\int_{h-\rho}^{h+\rho}\frac{\hat{M}_{\sigma\rho}(h - t)}{t}dt + O(\rho^2) + \int_{h-\rho}^{h+\rho}\hat{M}_{\sigma\rho}(h - t)O(\ln|t|)dt$$

$$= \frac{D(x_s)}{\pi}\int_{h-\rho}^{h+\rho}\hat{M}_{\sigma\rho}(h - t)d(\ln|t|) + O(\rho\ln\rho)$$

$$= -\frac{D(x_s)}{\pi^2}\int_{h-\rho}^{h+\rho}\frac{(\ln|t|)\sigma_\rho(h - t)}{h - t}dt + O(\rho\ln\rho)$$

$$= D(x_s)\psi_{\sigma\rho}(h) + O(\rho\ln\rho), \quad \frac{|h|}{\rho} < \infty, \rho \to 0. \qquad (6.5.10)$$

In the last equation, we have used the result from Section 6.5.3 below and have denoted

$$\psi_{\sigma\rho}(h) = -\frac{1}{\pi^2}\int_{-\rho}^{\rho}\ln|h - t|\frac{\sigma_\rho(t)}{t}dt. \qquad (6.5.11)$$

Substituting $\sigma_\rho(t) = 1$, $|t| \le \rho$, and $\gamma = h/\rho$ into (6.5.10) and (6.5.11), we get

$$\lim_{\rho\to 0}f_{\sigma\rho}(x_s + \gamma\rho n_0) = D(x_s)\frac{-1}{\pi^2}\int_{-1}^{1}\frac{\ln|\gamma - t|}{t}dt.$$

Exercise 6.5.1. Check the following identity

$$\frac{-1}{\pi^2}\int_{-1}^{1}\frac{\ln|\gamma - t|}{t}dt = \psi(\gamma), \quad 0 < |\gamma| < \infty,$$

where $\psi(\gamma)$ is defined in Theorem 6.3.2.

Using Exercise 6.5.1 we see that Theorem 6.3.2 is in complete agreement with (6.5.10) and (6.5.11).

6.5.3. Proof of auxiliary results

In this section we prove the results used in (6.5.10). Since $M_{\sigma\rho}$ is radial, we get

$$
\int_{\mathbf{R}^2} M_{\sigma\rho}((h-t)n_0 + (x_s - x_r)) \frac{D(x_r)}{\pi t} dx_r dt
$$

$$
= \frac{D(x_s)}{\pi} \int_{h-\rho}^{h+\rho} \frac{\hat{M}_{\sigma\rho}(h-t)}{t} dt
$$

$$
+ \int_{\mathbf{R}^2} M_{\sigma\rho}((h-t)n_0 + (x_s - x_r)) \frac{\tilde{D}(x_r)}{\pi t} dx_r dt,
$$

where we have denoted $\tilde{D}(x_r) := D(x_r) - D(x_s) - D'(x_s)(x_r - x_s)$. Clearly, $\tilde{D}(x_r) = O(|x_r - x_s|^2)$ as $x_r \to x_s$. Let us introduce the following coordinate system: the origin coincides with x_s, and the x_1-axis is perpendicular to S at x_s, that is it points in the same direction as n_0. In the new coordinate system, we have

$$
I_\rho := \int_{\mathbf{R}^2} M_{\sigma\rho}((h-t)n_0 + (x_s - x_r)) \frac{\tilde{D}(x_r)}{t} dx_r dt
$$

$$
= \int_{\mathbf{R}^2} M_{\sigma\rho}(h-t, -x_2) \frac{\tilde{D}(x_2)}{t} dx_2 \, dt. \tag{6.5.12}
$$

Using Equations (6.5.2) and (6.6.2) below, we get

$$
M_{\sigma\rho}(x_1, x_2) = \rho^{-1} M_{\sigma 1}(x_1/\rho, x_2/\rho). \tag{6.5.13}
$$

Denoting $h = \gamma\rho$ and changing variables in (6.5.12), we find

$$
I_\rho = \int_{\mathbf{R}^2} M_{\sigma 1}(\gamma - u, -v) \frac{\tilde{D}(\rho v)}{u} dv \, du, \tag{6.5.14}
$$

and we have to show that $I_\rho = O(\rho^2)$ as $\rho \to 0$. Using property (6.5.2) and Equation (6.5.9), we see that the leading singular term of $\hat{M}_{\sigma 1}(p)$ is $\hat{M}'_{\sigma 1}(p) \sim -1/(\pi p), p \to 0$. Since M is radial, we obtain using Equation (2.2.43) (cf. also Equation (6.6.2) below)

$$
M_{\sigma 1}(r) \sim \frac{1}{\pi^2} \int_r^1 \frac{ds}{\sqrt{s^2 - r^2}\, s} = \frac{1}{\pi^2} \frac{\arccos(r)}{r} \sim \frac{1}{2\pi r}, \quad r \to 0.
$$

Substituting into (6.5.14) and using polar coordinates, we conclude that
it is sufficient to verify the property

$$
J_\rho = \int\limits_0^1 \int\limits_{-\pi/2}^{\pi/2} \frac{1}{r} \frac{\tilde{D}(\rho r \cos\alpha)}{\gamma - r\sin\alpha}\, d\alpha\, r dr = O(\rho^2).
$$

Canceling r and defining $s = r\sin\alpha$, we get

$$
J_\rho = \int\limits_0^1 \int\limits_{-r}^{r} \frac{\tilde{D}\left(\rho\sqrt{r^2 - s^2}\right)/\sqrt{r^2 - s^2}}{\gamma - s}\, ds\, dr. \tag{6.5.15}
$$

Since $\tilde{D}(\epsilon) = O(\epsilon^2), \epsilon \to 0$, we have

$$
\frac{\tilde{D}\left(\rho\sqrt{r^2 - s^2}\right)}{\sqrt{r^2 - s^2}} = O\left(\rho^2\sqrt{r^2 - s^2}\right),\ \rho\to 0 \text{ or } s\to r,
$$

$$
\frac{\partial}{\partial s}\left(\frac{\tilde{D}\left(\rho\sqrt{r^2 - s^2}\right)}{\sqrt{r^2 - s^2}}\right) = O\left(\rho^2 \frac{s}{\sqrt{r^2 - s^2}}\right),\ \rho\to 0 \text{ or } s\to r.
$$

Integrating by parts in (6.5.15) with respect to s and substituting the
last equation, we establish the desired assertion:

$$
J_\rho = \int\limits_0^1 \int\limits_{-r}^{r} \ln|\gamma - s|\frac{\partial}{\partial s}\left(\frac{\tilde{D}\left(\rho\sqrt{r^2 - s^2}\right)}{\sqrt{r^2 - s^2}}\right) ds\, dr = O(\rho^2),\ \rho\to 0.
$$

Equation (6.5.13) implies also that $\hat{M}_{\sigma\rho}(t) = \hat{M}_{\sigma 1}(t/\rho)$. As it was
established, $\hat{M}'_{\sigma 1}(p) = O(p^{-1}), p\to 0$. Hence $\hat{M}_{\sigma 1}(p) = O(\ln|p|), p\to 0$.
Using these properties and changing variables, we immediately obtain
the other desired assertion:

$$
\int\limits_{h-\rho}^{h+\rho} \hat{M}_{\sigma\rho}(h - t)O(\ln|t|)dt = O(\rho\ln\rho),\ \rho\to 0.
$$

6.6. Numerical implementation of pseudolocal tomography

As in Chapter 5, we will compute the mollified version of $f_{\sigma\rho}$: $f_{\sigma\rho,\epsilon} := W_\epsilon * f_{\sigma\rho}$. Recall that $f_{\sigma\rho} = M_{\sigma\rho} * f_\Lambda$. Using Equation (5.4.4), we find

$$\frac{\partial}{\partial h} f_{\sigma\rho,\epsilon}(x_s + hn_0)$$

$$= M_{\sigma\rho} * \frac{\partial}{\partial h}(W_\epsilon * f_\Lambda)$$

$$= \int_{\mathbb{R}^2} M_{\sigma\rho}((h-t)n_0 + (x_s - x_r))$$

$$\times \left\{ \frac{D(x_r)}{\pi} \int_{t-\epsilon}^{t+\epsilon} w_\epsilon(t-p)\left[-\frac{1}{p^2} + \frac{c_1(x_r)}{p} + 2c_2(x_r)\delta(p)\right.\right.$$

$$\left.\left. + O(\ln|p|)\right] dp + O(\ln\epsilon) \right\} dx_r \, dt. \tag{6.6.1}$$

Let us assume for simplicity that $\sigma_\rho(p)$, defined in (6.5.2), is positive. Since $M_{\sigma\rho}$ is radial, one easily establishes using Equations (2.2.43) and (6.5.9) that $M_{\sigma\rho}$ is positive too:

$$M_{\sigma\rho}(r) = -\frac{1}{\pi} \int_r^\infty \frac{\hat{M}'_{\sigma\rho}(p)}{\sqrt{p^2 - r^2}} dp$$

$$= \frac{1}{\pi^2} \int_r^\rho \frac{\sigma_\rho(p)}{p\sqrt{p^2 - r^2}} dp \geq 0, \quad 0 < r \leq \rho. \tag{6.6.2}$$

Consider the first term in Equation (6.6.1).

$$I(h) := \int_{\mathbb{R}^2} M_{\sigma\rho}((h-t)n_0 + (x_s - x_r))\frac{D(x_r)}{\pi} \int_{t-\epsilon}^{t+\epsilon} w_\epsilon(t-p)\frac{-1}{p^2} dp \, dx_r \, dt$$

$$= \frac{D(x_s)}{\pi} \int_{\mathbb{R}} \hat{M}_{\sigma\rho}(h-t) \int_{t-\epsilon}^{t+\epsilon} w_\epsilon(t-p)\frac{-1}{p^2} dp dt$$

$$+ \int_{\mathbb{R}} \hat{M}_{\sigma\rho}(h-t)O(\rho^2)\left[\int_{t-\epsilon}^{t+\epsilon} w_\epsilon(t-p)\frac{-1}{p^2} dp\right] dt$$

$$= \frac{D(x_s)}{\pi} \int_{\mathbb{R}} \hat{M}'_{\sigma\rho}(h-t) \int_{t-\epsilon}^{t+\epsilon} w_\epsilon(t-p)\frac{dp}{p} dt$$

$$+ \int_{h-\rho}^{h+\rho} O(\ln |h - t|) O(\rho^2) O(\epsilon^{-2}) dt, \tag{6.6.3}$$

because $\hat{M}_{\sigma\rho}(h) = O(\ln |h|), h \to 0$. Letting $\rho = q\epsilon$ and $h = 0$ in the last equation and using (6.5.9) yields

$$I(0) = D(x_s) \frac{-1}{\pi} \int_{-\epsilon}^{\epsilon} w_\epsilon(h) \int_{h-\rho}^{h+\rho} \frac{\sigma_\rho(h-p)}{(h-p)p} dp + O(\epsilon \ln \epsilon)$$

$$= D(x_s) \frac{-1}{\pi} \int_{-1}^{1} w_1(r) \int_{r-q}^{r+q} \frac{\sigma_q(r-t)}{(r-t)t} dt + O(\epsilon \ln \epsilon). \tag{6.6.4}$$

Other terms on the right-hand side of Equation (6.6.1) can be treated similarly to (6.6.3), (6.6.4). Suppose functions $w_1(r)$ and $\sigma_q(r)$ are chosen so that

$$A_{\sigma q} := \int_{-1}^{1} w_1(r) \int_{r-q}^{r+q} \frac{\sigma_q(r-t)}{(r-t)t} dt dr \neq 0. \tag{6.6.5}$$

In this case, using Equations (6.6.1) and (6.6.4), we obtain after some transformations

$$\frac{\partial}{\partial h} f_{\sigma\rho,\epsilon}(x_s + hn_0)\bigg|_{h=0} = \frac{D(x_r)}{\epsilon\pi^2} A_{\sigma q}(1 + O(\epsilon)), \quad \rho = q\epsilon, \ \epsilon \to 0. \tag{6.6.6}$$

REMARK 6.6.1. Condition (6.6.5) is not very restrictive. It can be verified in the case $w_1(r) \geq 0$, $|r| \leq 1$, $\sigma_q(r)$ is even and nonnegative, and $\sigma_q(r)$ is nonincreasing on $(0, q]$. Indeed, since the inner integral in (6.6.5) is even, it is sufficient to show that

$$\int_{r-q}^{r+q} \frac{\sigma_q(r-t)}{(r-t)t} dt = 2 \int_{0}^{q} \frac{\sigma_q(s)}{r^2 - s^2} ds \geq 0 \text{ for almost all } r > 0.$$

Denoting $\varphi_r(s) := (2r)^{-1} \ln[(r+s)/(r-s)]$ and assuming that $\sigma_q(r)$ is differentiable, we get

$$\int_{0}^{q} \frac{\sigma_q(s)}{r^2 - s^2} ds = \int_{0}^{q} \sigma_q(s) d\varphi_r(s) = \sigma_q(s)\varphi_r(s)\bigg|_{0}^{q} - \int_{0}^{q} \sigma_q'(s)\varphi_r(s) ds.$$

Hence the desired assertion follows immediately because $\varphi_r(0) = 0$, $\varphi_r(s) \geq 0$ for any $s \geq 0, r > 0$, and by the assumption $\sigma_q'(s) \leq 0$.

If σ_q has jumps (by assumption (2) from Section 6.5.1, there are finitely many of them), we let r to be a point of smoothness of σ_q. Then the integral on the right-hand side of the last equation is well-defined and contains also delta-functions with negative coefficients, hence the desired assertion also follows.

Using (6.5.1), we obtain the following formula for computing $f_{\sigma\rho,\epsilon}$:

$$f_{\sigma\rho,\epsilon}(x) = \frac{1}{4\pi^2} \int_{S^1} \int_{\Theta\cdot x - (\rho+\epsilon)}^{\Theta\cdot x + (\rho+\epsilon)} w_{\sigma\rho,\epsilon}(\Theta \cdot x - p)\hat{f}(\theta,p)dpd\theta,$$

$$w_{\sigma\rho,\epsilon}(s) = \int_{s-\rho}^{s+\rho} \frac{\sigma_\rho(s-p)}{s-p} w'_\epsilon(p)dp.$$

As in Section 5.5.2, neglecting the variation of $f_{\sigma\rho,\epsilon}(x_s + hn_s)$ with respect to x_s, we get $|\nabla f_{\sigma\rho,\epsilon}(x_0)| \approx \left|\frac{\partial}{\partial h}f_{\sigma\rho,\epsilon}(x_0 + hn_0)\right|_{h=0}|$, $x_0 \in S$. Thus we came to the algorithm for computing values of jumps of f using PLT:

(0) Before the actual tomographic data processing, compute the constant $A_{\sigma q}$ according to (6.6.5);

(1) Compute $\nabla f_{\sigma\rho,\epsilon}(x)$ at the grid nodes $x = x_{ij}$ using formulas

$$g(s,\theta) = \int_{s-(\rho+\epsilon)}^{s+(\rho+\epsilon)} w'_{\sigma\rho,\epsilon}(s-p)\hat{f}(\theta,p)dp,$$

$$\frac{\partial f_{\sigma\rho,\epsilon}(x)}{\partial x_1} = \frac{1}{4\pi^2} \int_{S^1} \cos(\theta)g(\Theta \cdot x, \theta)d\theta,$$

$$\frac{\partial f_{\sigma\rho,\epsilon}(x)}{\partial x_2} = \frac{1}{4\pi^2} \int_{S^1} \sin(\theta)g(\Theta \cdot x, \theta)d\theta;$$

(2) Compute the estimate of $|D(x_{ij})|$ using (6.6.6):

$$|D(x_{ij})| \approx |\nabla f_{\sigma\rho,\epsilon}(x_{ij})|\frac{\pi^2\epsilon}{A_{\sigma q}};$$

(3) The vector $\nabla f_{\sigma\rho,\epsilon}(x_{ij})$ points from the smaller values of f to the larger values of f.

The performance of the proposed algorithm was tested on the same data as in Section 5.5.1. In Figure 6.6.1, we see the density plot of the estimated $|D(x)|$. In computations we used mollifier W_ϵ (see (5.4.8)) with $\epsilon = 9\Delta p$, and the cut-off function $\sigma_\rho(p) = (1 - (p/\rho)^4)^4$, $\rho = 10\Delta p$. Thus $q = \rho/\epsilon \approx 1.11$. Here Δp is the discretization step in the p variable. The presented results show that the true jump values are accurately estimated. The horizontal central cross-section of Figure 6.6.1 is represented in Figure 6.6.2. A line with peaks is a graph of $|D(x)|$. Big dots represent positions and amplitudes of jumps of the original density function f. We see a good agreement between the dots and maxima of the peaks. Results of the reconstruction from the data corrupted by the 3% noise (as in Section 5.5) are shown in Figures 6.6.3 and 6.6.4.

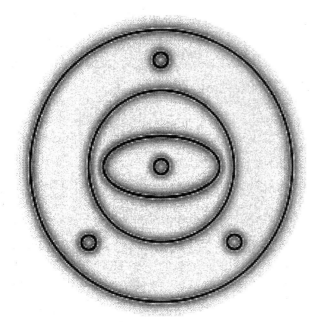

FIGURE 6.6.1. The density plot of the estimated $|D(x)|$

6.7. Pseudolocal tomography for the exponential Radon transform

6.7.1. Definitions. Basic property

Let $g(\theta, p)$ denote the exponential Radon transform of the function $f(x), x \in \mathbb{R}^2$. Similarly to (6.2.1') and (6.2.2'), fix $\rho > 0$ and, using inversion formula (2.8.9), introduce the pseudolocal tomography function

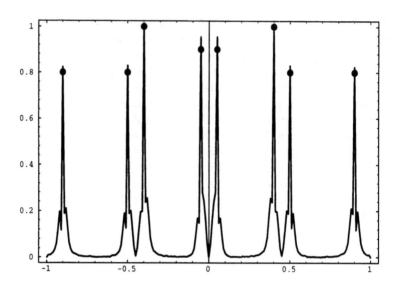

FIGURE 6.6.2.   The central horizontal cross-section of the plot of the estimated $|D(x)|$

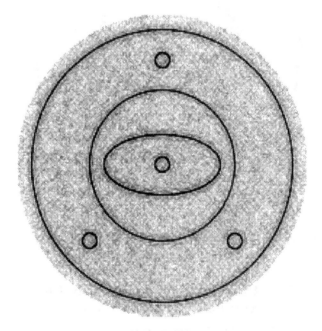

FIGURE 6.6.3.   The density plot of $|D(x)|$ estimated from the noisy data

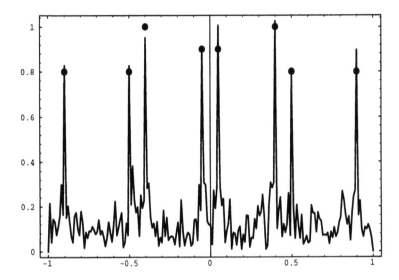

FIGURE 6.6.4. The central horizontal cross-section of the plot of $|D(x)|$ estimated from the noisy data (see Figure 6.6.3)

for the exponential Radon transform f_ρ and its complement f_ρ^c:

$$f_\rho(x) := \frac{1}{4\pi^2} \int_{S^1} e^{-\mu(x\cdot\Theta^\perp)} \int_{\Theta\cdot x-\rho}^{\Theta\cdot x+\rho} \frac{\cos(\mu(\Theta\cdot x-p))}{\Theta\cdot x-p} g_p(\theta,p)dpd\theta, \quad (6.7.1)$$

and

$$f_\rho^c(x) := \frac{1}{4\pi^2} \int_{S^1} e^{-\mu(x\cdot\Theta^\perp)} \int_{|\Theta\cdot x-p|\geq\rho} \frac{\cos(\mu(\Theta\cdot x-p))}{\Theta\cdot x-p} g_p(\theta,p)dp\,d\theta.$$
$$(6.7.2)$$

As usual, integral (6.7.1) is understood in the Cauchy principal value sense.

Lemma 6.7.1. *Let $\rho > 0$ be fixed. Then the function $f_\rho^c(x)$ defined in (6.7.2) is continuous.*

Proof. Changing variables in (6.7.2), we get

$$f_\rho^c(x) = \frac{-1}{\pi} \int_\rho^\infty \frac{\cos(\mu q)}{q} F_q(x,q)dq, \quad (6.7.3)$$

$$F(x,q) = \frac{1}{2\pi} \int_0^{2\pi} e^{-\mu(x\cdot\Theta^\perp)} g(\theta,\Theta\cdot x+q)d\theta, \quad (6.7.4)$$

where $F_q = \partial F/\partial q$, and we have used that $F(x,q)$ is even in q (see (6.7.6) below). Integration by parts in (6.7.3) yields:

$$f_\rho^c(x) = \frac{1}{\pi}\left[\frac{\cos(\mu\rho)}{\rho}F(x,\rho) + \int\limits_\rho^\infty \left(\frac{\cos(\mu q)}{q}\right)' F(x,q)dq\right]. \qquad (6.7.5)$$

Actually, $F(x,q)$ has compact support as a function of q, so the integral above is taken over a compact set. Thus, in order to prove continuity of $f_\rho^c(x)$, it is sufficient to prove that $F(x,q)$ is continuous with respect to x. As a trivial extension of the results obtained in Sections 4.2–4.6, one can show that the exponential Radon transform $g(\theta,p)$ is discontinuous with respect to p only for such pairs (θ_0, p_0), that the lines $\Theta_0 \cdot x = p_0$ intersect the discontinuity curve S of f over intervals of positive lengths. Since the number of such pairs (θ_0, p_0) is at most countable and $\sup_{\Theta \in S^1, p \in \mathbb{R}} |g(\theta,p)| < \infty$, Equation (6.7.4) implies continuity of $F(x,q)$ in x. $\square$

Since $f = f_\rho + f_\rho^c$ and the function f_ρ^c is continuous, we conclude that

(1) the location of discontinuities of $f_\rho(x)$ is the same as that of $f(x)$, and

(2) at each discontinuity point ξ of f_ρ, the jump $f_\rho(\xi+0) - f_\rho(\xi-0)$ is precisely the same as that of $f(x)$ at $x = \xi$.

Thus discontinuities of $f(x)$ can be recovered from the function $f_\rho(x)$. Note that $f_\rho(x)$ is computed using $g(\theta,p)$ for (θ,p) satisfying $|\Theta \cdot x - p| \le \rho$. Therefore, we call the function f_ρ pseudolocal.

From (6.7.2) and (6.7.4) we see that $f(x) = \lim_{\rho \to 0} f_\rho^c(x)$. Since $f_\rho = f - f_\rho^c$, or, more precisely, $f_\rho = \lim_{\epsilon \to 0} f_\epsilon^c - f_\rho^c$, we need to investigate the convergence $f_\rho^c \to f$, $\rho \to 0$, in order to better understand properties of f_ρ.

6.7.2. Some auxiliary results

Substituting (2.8.3) into (6.7.4), we find

$$F(x, q) = \frac{1}{2\pi} \int_0^{2\pi} e^{-\mu(x \cdot \Theta^\perp)} \int_{\mathbb{R}^2} e^{\mu(y \cdot \Theta^\perp)} f(y) \delta(\Theta \cdot y - \Theta \cdot x - q) dy d\theta$$

$$= \frac{1}{2\pi} \int_{\mathbb{R}^2} f(y) \int_0^{2\pi} e^{\mu((y-x) \cdot \Theta^\perp)} \delta(\Theta \cdot (y - x) - q) d\theta \, dy$$

$$= \frac{1}{2\pi} \int_0^\infty \int_0^{2\pi} f(x + r\Phi) \int_0^{2\pi} e^{\mu r \sin(\theta - \phi)} \delta(r \cos(\theta - \phi) - q) d\theta d\phi r dr$$

$$= 2 \int_{|q|}^\infty \frac{\cosh(\mu\sqrt{r^2 - q^2})}{\sqrt{1 - q^2/r^2}} \bar{f}(r, x) dr, \tag{6.7.6}$$

where $\Theta^\perp = (-\sin\theta, \cos\theta)$, $\Phi = (\cos\phi, \sin\phi)$, and we have used definition (6.3.1). Substituting into (6.7.5) and using that $\bar{f}(r, x) = 0$ for $r \geq 2R = B$ if $|x| \leq R$, we obtain

$$f_\rho^c(x) = \frac{2}{\pi} \left\{ \frac{\cos(\mu\rho)}{\rho} \int_\rho^B \frac{\cosh(\mu\sqrt{r^2 - \rho^2})}{\sqrt{r^2 - \rho^2}} \bar{f}(r, x) r dr \right.$$

$$\left. + \int_\rho^B \left(\frac{\cos(\mu q)}{q} \right)' \int_q^B \frac{\cosh(\mu\sqrt{r^2 - q^2})}{\sqrt{r^2 - q^2}} \bar{f}(r, x) r dr dq \right\}$$

$$= K_\rho \bar{f}(0, x) + \int_\rho^B \Psi_\rho(r)(\bar{f}(r, x) - \bar{f}(0, x)) dr, \tag{6.7.7}$$

where

$$\Psi_\rho(r) = \frac{2}{\pi} r \left[\frac{\cos(\mu\rho)}{\rho} \frac{\cosh(\mu\sqrt{r^2 - \rho^2})}{\sqrt{r^2 - \rho^2}} \right.$$

$$\left. + \int_\rho^r \left(\frac{\cos(\mu q)}{q} \right)' \frac{\cosh(\mu\sqrt{r^2 - q^2})}{\sqrt{r^2 - q^2}} dq \right] \tag{6.7.8}$$

and

$$K_\rho = \int_\rho^B \Psi_\rho(r) dr = \frac{2}{\pi} \int_{\rho/B}^1 \frac{\cos(\mu Bq) \cosh(\mu B\sqrt{1 - q^2})}{\sqrt{1 - q^2}} dq. \tag{6.7.9}$$

Let us investigate the behavior of K_ρ and $\Psi_\rho(r)$ as $\rho \to 0$. Using integral (2.5.53.5) from [GR]:

$$\int_0^1 \frac{\cosh(c\sqrt{1-x^2})}{\sqrt{1-x^2}} \cos(bx)dx = \frac{\pi}{2} I_0(\sqrt{c^2 - b^2}), \quad |b| \leq |c|, \qquad (6.7.10)$$

where I_0 is the Bessel function of order zero, we obtain from (6.7.9)

$$K_\rho = 1 - \frac{2}{\pi} \int_0^{\rho/B} \frac{\cos(\mu Bq)\cosh(\mu B\sqrt{1-q^2})}{\sqrt{1-q^2}} dq$$

$$= 1 - \rho \frac{2\cosh(\mu B)}{\pi B} + O(\rho^2), \quad \rho \to 0, \qquad (6.7.11)$$

because $I_0(0) = 1$. Let us represent the function Ψ_ρ as follows

$$\Psi_\rho(r)\frac{\pi}{2r} = \frac{\cos(\mu\rho)\cosh(\mu\sqrt{r^2-\rho^2})}{\rho} - \int_\rho^r \frac{\cosh(\mu\sqrt{r^2-q^2})}{\sqrt{r^2-q^2}} \frac{dq}{q^2}$$

$$- \int_\rho^r \frac{(\mu q)\sin(\mu q) + \cos(\mu q) - 1}{q^2} \frac{\cosh(\mu\sqrt{r^2-q^2})}{\sqrt{r^2-q^2}} dq. \qquad (6.7.12)$$

Lemma 6.7.2. *One has*

$$\int_0^r \frac{(\mu q)\sin(\mu q) + \cos(\mu q) - 1}{q^2} \frac{\cosh(\mu\sqrt{r^2-q^2})}{\sqrt{r^2-q^2}} dq = \frac{\pi}{2} \frac{\mu}{r} I_1(\mu r).$$

$$\qquad (6.7.13)$$

For the proof of Lemma 6.7.2 and of two other lemmas from this section see Section 6.7.1.5. Substitution of (6.7.13) into (6.7.12) yields

$$\Psi_\rho(r)\frac{\pi}{2r} = \left[\frac{1}{\rho} \frac{\cosh(\mu r)}{r} - \int_\rho^r \frac{\cosh(\mu\sqrt{r^2-q^2})}{\sqrt{r^2-q^2}} \frac{dq}{q^2} \right] - \frac{\pi}{2} \frac{\mu}{r} I_1(\mu r)$$

$$+ \left[\int_0^\rho \frac{(\mu q)\sin(\mu q) + \cos(\mu q) - 1}{q^2} \frac{\cosh(\mu\sqrt{r^2-q^2})}{\sqrt{r^2-q^2}} dq \right.$$

$$\left. + \frac{\cos(\mu\rho)\cosh(\mu\sqrt{r^2-\rho^2})}{\rho} - \frac{1}{\rho} \frac{\cosh(\mu r)}{r} \right]. \qquad (6.7.14)$$

The expression in the first brackets can be rewritten as follows

$$\frac{1}{\rho}\frac{\cosh(\mu r)}{r} - \int_{\rho}^{r} \frac{\cosh(\mu\sqrt{r^2 - q^2})}{\sqrt{r^2 - q^2}}\frac{dq}{q^2}$$

$$= \int_{\rho}^{r} \left(\frac{\cosh(\mu r)}{r} - \frac{\cosh(\mu\sqrt{r^2 - q^2})}{\sqrt{r^2 - q^2}} \right)\frac{dq}{q^2} + \frac{\cosh(\mu r)}{r^2}$$

$$= \int_{0}^{r} \left(\frac{\cosh(\mu r)}{r} - \frac{\cosh(\mu\sqrt{r^2 - q^2})}{\sqrt{r^2 - q^2}} \right)\frac{dq}{q^2} + \frac{\cosh(\mu r)}{r^2}$$

$$- \int_{0}^{\rho} \left(\frac{\cosh(\mu r)}{r} - \frac{\cosh(\mu\sqrt{r^2 - q^2})}{\sqrt{r^2 - q^2}} \right)\frac{dq}{q^2}. \tag{6.7.15}$$

Lemma 6.7.3. *One has*

$$\int_{0}^{r} \left(\frac{\cosh(\mu r)}{r} - \frac{\cosh(\mu\sqrt{r^2 - q^2})}{\sqrt{r^2 - q^2}} \right)\frac{dq}{q^2} = -\frac{\cosh(\mu r)}{r^2} + \frac{\pi}{2}\frac{\mu}{r}I_1(\mu r). \tag{6.7.16}$$

Substituting (6.7.15) and (6.7.16) into (6.7.14), we find

$$\Psi_{\rho}(r)\frac{\pi}{2r} = \int_{0}^{\rho} \left[((\mu q)\sin(\mu q) + \cos(\mu q))\frac{\cosh(\mu\sqrt{r^2 - q^2})}{\sqrt{r^2 - q^2}} \right.$$

$$\left. - \frac{\cosh(\mu r)}{r} \right]\frac{dq}{q^2}$$

$$+ \left(\frac{\cos(\mu\rho)}{\rho}\frac{\cosh(\mu\sqrt{r^2 - \rho^2})}{\sqrt{r^2 - \rho^2}} - \frac{1}{\rho}\frac{\cosh(\mu r)}{r} \right). \tag{6.7.17}$$

Estimating the right-hand side of (6.7.17) as $\rho \to 0$, we prove

Lemma 6.7.4. *One has*

$$\Psi_{\rho}(r) = \frac{1}{\pi}(\cosh(\mu r) - (\mu r)\sinh(\mu r) + O(\rho))$$

$$\times \left(\frac{\rho(2 - \mu^2 r^2)}{r\sqrt{r^2 - \rho^2}} + \mu^2 r \arcsin(\rho/r) \right), \quad \rho \to 0, \tag{6.7.18}$$

where $O(\rho)$ is uniform in r, $\rho < r \leq B$.

Using Lemma 6.7.4, one easily gets another useful formula:

$$\Psi_{\rho}(r) = \frac{2}{\pi}(\cosh(\mu r) - (\mu r)\sinh(\mu r) + O(\rho))\frac{\rho}{r\sqrt{r^2 - \rho^2}}, \quad \rho \to 0,$$

$$0 < R_2 \leq r \leq B. \tag{6.7.19}$$

6.7.3. Investigation of the convergence $f_\rho^c(x) \to f(x)$ as $\rho \to 0$

Using results obtained in the previous section, we are now ready to prove

Theorem 6.7.1. *Suppose $f \in C^2(U)$ for some open set U, $U \subset \mathbb{R}^2$. Then*

$$|f_\rho^c(x) - f(x)| = O(\rho) \quad \text{as } \rho \to 0, \; x \in U. \tag{6.7.20}$$

Moreover, the convergence in (6.7.20) is uniform on all compact subsets of U. If $x_0 \in S$ is fixed and there exists an open neighborhood V of x_0 such that S is smooth in V and f is piecewise-C^2 in V, then

$$\left| f_\rho^c(x_0) - \frac{f_+(x_0) + f_-(x_0)}{2} \right| = O(\rho \ln \rho), \; \rho \to 0, \tag{6.7.21}$$

where $f_\pm(x_0)$ are the limiting values of $f(x)$ as x approaches x_0 from different sides of S along any path nonintersecting S.

Proof. First, we prove (6.7.20). Let $f \in C^2(U)$. Choose any $x_0 \in U$ and find $R_1 > 0$ such that $B(x_0, R_1) \subset U$, where $B(x_0, R_1) := \{x \in \mathbb{R}^2 : |x - x_0| \le R_1\}$. Then $\bar{f}(0, x_0) = f(x_0)$ and one has (see (6.3.7))

$$|\bar{f}(r, x_0) - f(x_0)| \le r^2 M_2(r, x_0), \tag{6.7.22}$$

where $M_2(r, x_0) = \max_{x, i, j} |\partial^2 f(x)/(\partial x_i \partial x_j)|$, the maximum is taken over x such that $|x - x_0| < r$ and $i, j = 1, 2$. Equations (6.7.7) and (6.7.11) imply for $\rho < R_1$:

$$|f_\rho^c(x_0) - f(x_0)| \le |f(x_0)| \left(\rho \frac{2 \cosh(\mu B)}{\pi B} + O(\rho^2) \right)$$

$$+ \int_\rho^{R_1} |\Psi_\rho(r)| r^2 M_2(r, x_0) dr + \int_{R_1}^{B} |\Psi_\rho(r)| |\bar{f}(r, x_0) - f(x_0)| dr. \tag{6.7.23}$$

Let us estimate the first integral in (6.7.23). Using (6.7.18), we get

$$\int_{\rho}^{R_1} |\Psi_\rho(r)| r^2 M_2(r, x_0) dr$$

$$= \int_{\rho}^{R_1} \frac{|\cosh(\mu r) - (\mu r)\sinh(\mu r) + O(\rho)|}{\pi}$$

$$\times \left| \frac{\rho(2 - \mu^2 r^2)}{r\sqrt{r^2 - \rho^2}} + \mu^2 r \arcsin(\rho/r) \right| r^2 M_2(r, x_0) dr$$

$$\leq \frac{2M_2(x_0)c(R_1)}{\pi} \left[\rho \int_{\rho}^{R_1} \frac{r dr}{\sqrt{r^2 - \rho^2}} \right.$$

$$\left. + \frac{1}{2} \int_{\rho}^{R_1} (\mu^2 r^3) \left(\frac{\rho}{\sqrt{r^2 - \rho^2}} - \arcsin(\rho/r) \right) dr \right] (1 + O(\rho))$$

$$\leq \frac{2M_2(x_0)c(R_1)}{\pi} \rho R_1 \left(1 + \frac{\mu^2 R_1^2}{6} \right) (1 + O(\rho)), \qquad (6.7.24)$$

where $c(R_1) = \max_{0 \leq r \leq R_1} |\cosh(\mu r) - (\mu r)\sinh(\mu r)|$, and we have used Equations (a) and (b) from the following exercise:

Exercise 6.7.1. Prove the following relations:

$$\arcsin(\rho/r) \leq \rho/\sqrt{\rho^2 - r^2} \qquad (a)$$

$$\int_{\rho}^{R_1} \frac{r^3}{\sqrt{r^2 - \rho^2}} dr \leq \frac{R_1^3}{3} + \rho^2 R_1, \qquad (b)$$

and

$$\int_{\rho}^{R_1} \frac{r^2}{\sqrt{r^2 - \rho^2}} dr \to \frac{R_1^2}{2}, \quad \rho \to 0. \qquad (c)$$

The second integral in (6.7.23) is estimated using (6.7.19):

$$\int_{R_1}^{B} |\Psi_\rho(r)| |\bar{f}(r, x_0) - f(x_0)| dr$$

$$\leq \frac{2\rho M_0(x_0)}{\pi} \int_{R_1}^{B} \frac{|\cosh(\mu r) - (\mu r)\sinh(\mu r) + O(\rho)|}{r^2} (1 + O(\rho)) dr$$

$$< \frac{2\rho M_0(x_0)}{\pi} \frac{c(B)}{R_1} (1 + O(\rho)), \qquad (6.7.25)$$

where $M_0(x_0) := \sup_{x \in \mathbb{R}^2} |f(x) - f(x_0)|$. Substituting (6.7.24) and (6.7.25) into (6.7.23), we obtain

$$|f_\rho^c(x_0) - f(x_0)| \le \frac{2\rho}{\pi} \left[|f(x_0)| \frac{\cosh(\mu B)}{B} + M_2(x_0)c(R_1)R_1 \left(1 + O(R_1^2)\right) \right.$$

$$\left. + M_0(x_0) \frac{c(B)}{R_1} \right] (1 + O(\rho)) = O'(\rho), \quad \rho \to 0.$$
$$(6.7.26)$$

To prove that f_ρ^c converges to f uniformly on any compact set K, $K \subset U$, let us fix δ, $0 < \delta < \text{dist}(K, \partial U)$, where ∂U is the boundary of U. Let us denote the expression in brackets in (6.7.26) by $C(x_0, R_1)$. Clearly,

$$C := \max_{\substack{\xi \in K \\ \delta \le R \le \text{diam}(U)}} C(\xi, R) < \infty.$$

Therefore, $\sup_{x \in K} |f_\rho^c(x) - f(x)| \le \frac{2\rho}{\pi} C(1 + O(\rho))$, which proves the desired assertion.

Now let us pick $x_0 \in S$. Let R_0, $0 < R_0 \le \infty$, be the radius of curvature of S at x_0. Under the assumptions of the theorem, there exists $R_1 > 0$ such that

(a) S divides $B(x_0, R_1)$ into two sets $B_+(x_0, R_1)$ and $B_-(x_0, R_1)$ with $f \in C^2(B_+(x_0, R_1))$, $f \in C^2(B_-(x_0, R_1))$;
(b) S is smooth inside $B(x_0, R_1)$; and
(c) intersection of S with any circle centered at x_0 and with radius r, $0 < r \le R_1$, contains exactly two points.

Using Equations (6.3.11), (6.3.12), and (6.7.11) in (6.7.7), we get

$$|f_\rho^c(x_0) - \bar{f}(0, x_0)| \le |\bar{f}(0, x_0)| \left(\rho \frac{2\cosh(\mu B)}{\pi B} + O(\rho^2) \right)$$

$$+ \int_\rho^{R_1} |\Psi_\rho(r)|(|A|r + O(r^2))dr$$

$$+ \int_{R_1}^B |\Psi_\rho(r)||\bar{f}(r, x_0) - \bar{f}(0, x_0)|dr.$$
$$(6.7.27)$$

Using (6.7.19), we see that the last integral on the right-hand side of (6.7.27) is of order $O(\rho)$, $\rho \to 0$. Let us estimate the first integral in

(6.7.27) using (6.7.18). We have

$$\int\limits_{\rho}^{R_1} |\Psi_\rho(r)|(|A|r + O(r^2))dr$$

$$\leq \int\limits_{\rho}^{R_1} \frac{|\cosh(\mu r) - (\mu r)\sinh(\mu r) + O(\rho)|}{\pi}$$

$$\times \left| \frac{\rho(2 - \mu^2 r^2)}{r\sqrt{r^2 - \rho^2}} + \mu^2 r \arcsin(\rho/r) \right| (|A|r + O(r^2))dr$$

$$\leq \frac{2|A|c(R_1)}{\pi} \left[\rho \int\limits_{\rho}^{R_1} \frac{dr}{\sqrt{r^2 - \rho^2}} \right.$$

$$\left. + \frac{1}{2} \int\limits_{\rho}^{R_1} (\mu^2 r^2) \left(\frac{\rho}{\sqrt{r^2 - \rho^2}} - \arcsin(\rho/r) \right) dr \right] (1 + O(\rho))$$

$$\leq \frac{2|A|c(R_1)}{\pi} \left[\rho \ln(\frac{2R_1}{\rho}) + O(\rho) \right] = O(\rho \ln \rho),$$

which proves the desired assertion. In the last equation we have used formulas (a) and (c) from Exercise 6.7.1. □

Equation (6.7.21) asserts pointwise convergence of f_ρ^c at the discontinuity curve S. Convergence (or, rather, "nonconvergence") in a neighborhood of S is investigated in the following theorem.

Theorem 6.7.2. *Let $x_0 \in S$, S is smooth in some neighborhood of x_0 and f is piecewise-C^2 there. Let n_+ and $n_- = -n_+$ be unit vectors normal to S at x_0 and pointing inside B_+ and B_-, respectively. Then for an arbitrary fixed γ, $0 < |\gamma| < \infty$, one has*

$$\lim_{\rho \to 0} \left[f(x_0 + \gamma\rho n_0) - f_\rho^c(x_0 + \gamma\rho n_0) \right] = D(x_0)\psi(\gamma), \qquad (6.7.28)$$

where $D(x_0)$ and $\psi(\gamma)$ are defined in Theorem 6.3.2.

Proof. Under the assumptions of the theorem, there exists $R_1 > 0$ such that conditions (a), (b), and (c) in the proof of Theorem 6.7.1 are satisfied (see above (6.7.27)). Let us consider a point $x = x_0 + (\gamma\rho)n_0 \in B_+(x_0, R_1)$, where γ, $0 < \gamma < \infty$, is fixed. Suppose $R_0 < \infty$. In this case we have (see (6.3.7))

$$\bar{f}(r, x) - \bar{f}(0, x) = O(r^2), \quad r < \gamma\rho,$$

and

$$\bar{f}(r,x) - \bar{f}(0,x) = -D(x_0)\frac{\arccos(\gamma\rho/r)}{\pi} + O(r^{1/2}), \quad r > \gamma\rho. \quad (6.7.29)$$

Using the above two equations, (6.7.7), and (6.7.11), we get

$$f(x) - f_\rho^c(x) = f(x)O(\rho) + \int_\rho^{\max(\gamma\rho,\rho)} \Psi_\rho(r)O(r^2)dr$$

$$+ \frac{D(x_0)}{\pi} \int_{\max(\gamma\rho,\rho)}^{R_1} \Psi_\rho(r)\arccos(\gamma\rho/r)dr$$

$$+ \int_{\max(\gamma\rho,\rho)}^{R_1} \Psi_\rho(r)O(r^{0.5})dr$$

$$- \int_{R_1}^{B} \Psi_\rho(r)(\bar{f}(r,x) - \bar{f}(0,x))dr. \quad (6.7.30)$$

Equations (6.7.18) and (6.7.19) imply that the first, third, and the fourth integrals in (6.7.30) are of order $O(\rho^2)$, $O(\rho^{0.5})$, and $O(\rho)$, $\rho \to 0$, respectively. Let us compute the limit of the second integral as $\rho \to 0$.

$$\int_{\max(\gamma\rho,\rho)}^{R_1} \Psi_\rho(r)\arccos(\gamma\rho/r)dr = \left[\frac{2\rho}{\pi} \int_{\max(\gamma\rho,\rho)}^{R_1} \frac{\arccos\left(\frac{\gamma\rho}{r}\right)}{r\sqrt{r^2 - \rho^2}}dr\right.$$

$$+ \frac{2\rho}{\pi} \int_{\max(\gamma\rho,\rho)}^{R_1} (\cosh(\mu r) - (\mu r)\sinh(\mu r) - 1 + O(\rho))\frac{\arccos\left(\frac{\gamma\rho}{r}\right)}{r\sqrt{r^2 - \rho^2}}dr$$

$$- \frac{\mu^2}{\pi} \int_{\max(\gamma\rho,\rho)}^{R_1} (\cosh(\mu r) - (\mu r)\sinh(\mu r) + O(\rho))r$$

$$\left. \times \left(\frac{\rho}{\sqrt{r^2 - \rho^2}} - \arcsin(\rho/r)\right)\arccos\left(\frac{\gamma\rho}{r}\right)dr\right]. \quad (6.7.31)$$

Since $\cosh(t) - t\sinh(t) - 1 = O(t^2)$, $t \to 0$, the last two integrals in (6.7.31) are of order $O(\rho)$, $\rho \to 0$. Transforming the first integral on the right-hand side of (6.7.31) as follows

$$\frac{2\rho}{\pi} \int_{\max(\gamma\rho,\rho)}^{R_1} \frac{\arccos(\gamma\rho/r)}{r\sqrt{r^2 - \rho^2}}dr = \frac{2}{\pi} \int_0^{\min(1,1/\gamma)} \frac{\arccos(\gamma t)}{(1 - t^2)^{1/2}}dt + O(\rho),$$

we finally get using (6.7.30)

$$f(x_0 + (\gamma\rho)n_0) - f_\rho^c(x_0 + (\gamma\rho)n_0) = D(x_0)\psi(\gamma) + O(\rho^{0.5}), \quad \rho \to 0,$$
$$0 < \gamma < \infty, \tag{6.7.32}$$

$$\psi(\gamma) = \frac{2}{\pi^2} \int_0^{\min(1,1/\gamma)} \frac{\arccos(\gamma t)}{(1 - t^2)^{1/2}} dt, \quad 0 < \gamma < \infty. \tag{6.7.33}$$

If $R_0 = \infty$, where R_0 is the radius of curvature of S at $x_0 \in S$, then in (6.7.29) we have to replace $O(r^{1/2})$ by $O(r)$, and this leads to a replacement of $O(\rho^{0.5})$ by $O(\rho \ln \rho)$ in (6.7.32). By considering a point x on the other side of x_0, $x = x_0 + (\gamma\rho)n_0 \in B_-(x_0, R_1)$, $\gamma < 0$, we find similarly to (6.7.29) – (6.7.32) that the analogue of (6.7.32) is obtained by defining $\psi(\gamma) = -\psi(|\gamma|)$, $\gamma < 0$. Therefore, Equation (6.7.28) is proved. □

6.7.4. Remarks on numerical implementation

As in Section 6.5.1, we define a family of pseudolocal tomography functions for the exponential Radon transform (PLTE functions):

$$f_{\sigma\rho}(x) := \frac{1}{4\pi^2} \int_{S^1} \int_{\Theta \cdot x - \rho}^{\Theta \cdot x + \rho} \frac{\sigma_\rho(\Theta \cdot x - p)}{\Theta \cdot x - p} g_p(\theta, p) dp \, d\theta, \tag{6.7.44}$$

where $\sigma_\rho(p)$ satisfies the properties

(1) $\sigma_\rho(p)$ is real-valued and even;
(2) $\sigma_\rho(p)$ is piecewise-continuously differentiable and there are at most finitely many points at which σ_ρ is discontinuous; and
(3)

$$\sigma_\rho(p) = \sigma_1(p/\rho), \quad |\sigma_\rho(p) - 1| \leq O(p), \quad p \to 0. \tag{6.7.45}$$

REMARK 6.7.1. Taking $\sigma_\rho(p) = \cos(\mu p)$, $|p| \leq \rho$, in (6.7.44), we get function (6.7.1).

Theorem 6.7.3. *The difference $f - f_{\sigma\rho}$ is continuous, hence functions f and $f_{\sigma\rho}$ have the same discontinuity curve S and the same values of jumps across S.*

Exercise 6.7.2. Prove Theorem 6.7.3 using the proofs of Theorem 6.5.1 and Lemma 6.7.1.

Replacing W_ϵ by M and integrating by parts, we obtain from Equation (5.6.6)

$$(f_\Lambda^{(\mu)} * M)(x) = -\frac{1}{4\pi} \int_{S^1} e^{-\mu(x \cdot \Theta^\perp)} \int_R \hat{M}'(\Theta \cdot x - p) g_p(\theta, p) dp \, d\theta,$$

$$\hat{M} := (T_\mu M)(p) = \int_{R^2} e^{\mu(x \cdot \Theta^\perp)} M(x)\delta(\Theta \cdot x - p) dx$$

$$= \begin{cases} 2 \int_{|p|}^\rho r \dfrac{\cosh(\mu\sqrt{r^2 - p^2})}{\sqrt{r^2 - p^2}} M(r) dr, & |p| \leq \rho, \\ 0, & |p| \geq \rho, \quad (6.7.46) \end{cases}$$

where M is assumed to be radial, $M(x) := M(|x|)$. Let $M := M_{\sigma\rho}$ be chosen so that

$$\hat{M}'_{\sigma\rho}(p) := -\frac{\sigma_\rho(p)}{\pi p}, \; |p| \leq \rho, \quad \hat{M}_{\sigma\rho}(p) = 0, \; |p| > \rho. \qquad (6.7.47)$$

Comparing Equations (6.7.46) and (6.7.44), we obtain the relation between $f_\Lambda^{(\mu)}$ and PLTE functions:

$$f_\Lambda^{(\mu)} * M_{\sigma\rho} = f_{\sigma\rho}. \qquad (6.7.48)$$

Using Equations (6.7.47) and (6.7.48) and following the approach from Section 6.6, one can check that the mollified PLTE function $f_{\sigma\rho,\epsilon} := W_\epsilon * f_{\sigma\rho}$ satisfies Equations (6.6.5) and (6.6.6).

Exercise 6.7.3. Prove the last claim.

Therefore, the numerical scheme for finding values of jumps described in Section 6.6 is applicable in case of the exponential Radon transform as well.

6.7.5. Proofs of Lemmas 6.7.2 – 6.7.4

Proof of Lemma 6.7.2. Changing variable $q = xr$ in (6.7.10) and denoting $c/r = \mu$, $b/r = \eta$, we get

$$\int_0^r \frac{\cosh(\mu\sqrt{r^2 - q^2})}{\sqrt{r^2 - q^2}} \cos(\eta x) dx = \frac{\pi}{2} I_0(r\sqrt{\mu^2 - \eta^2}), \; 0 \leq \eta \leq \mu.$$

$$(6.7.49)$$

Multiplying on both sides by η and integrating with respect to η between 0 and μ, we get (6.7.13). □

Proof of Lemma 6.7.3. Denoting $\mu r = t$ and $q = sr$, we see that one has to check the identity

$$\int_0^1 \left(\cosh(t) - \frac{\cosh(t\sqrt{1-s^2})}{\sqrt{1-s^2}} \right) \frac{ds}{s^2} = -\cosh(t) + \frac{\pi}{2} t I_1(t). \quad (6.7.50)$$

Let the left-hand side of (6.7.50) be denoted $\varphi(t)$. We have

$$\varphi''(t) = \int_0^1 \left(\cosh(t) - \frac{1-s^2}{\sqrt{1-s^2}} \cosh(t\sqrt{1-s^2}) \right) \frac{ds}{s^2}$$

$$= \varphi(t) + \int_0^1 \frac{\cosh(t\sqrt{1-s^2})}{\sqrt{1-s^2}} ds = \varphi(t) + \frac{\pi}{2} I_0(t),$$
$$(6.7.51)$$

where we have used (6.7.49) with $\eta = 0$ and $r = 1$. Using recurrence relation of the Bessel functions, one easily gets that the right-hand side of (6.7.50) also satisfies differential Equation (6.7.51). Moreover, it is easy to check that (6.7.50) holds for $t = 0$. Therefore, (6.7.50) holds for all $t \geq 0$. $\square$

Proof of Lemma 6.7.4. Let us represent the right-hand side of (6.7.17) as follows

$$\Psi_\rho(r)\frac{\pi}{2r} = \left[\int_0^\rho \frac{(\mu q)\sin(\mu q) + \cos(\mu q) - 1}{q^2} \frac{\cosh(\mu\sqrt{r^2-q^2})}{\sqrt{r^2-q^2}} dq \right.$$

$$\left. - \frac{1-\cosh(\mu\rho)}{\rho} \frac{\cosh(\mu\sqrt{r^2-\rho^2})}{\sqrt{r^2-\rho^2}} \right]$$

$$+ \left[\int_0^\rho \left(\frac{\cosh(\mu\sqrt{r^2-q^2})}{\sqrt{r^2-q^2}} - \frac{\cosh(\mu r)}{r} \right) \frac{dq}{q^2} \right]$$

$$+ \left[\left(\frac{\cosh(\mu\sqrt{r^2-\rho^2})}{\sqrt{r^2-\rho^2}} - \frac{\cosh(\mu r)}{r} \right) \frac{1}{\rho} \right].$$
$$(6.7.52)$$

Let the three expressions in brackets in the last equation be denoted C_1, C_2, and C_3, respectively. Since

$$\frac{1-\cos(\mu\rho)}{\rho} = \int_0^\rho \frac{(\mu q)\sin(\mu q) + \cos(\mu q) - 1}{q^2} dq,$$

one has

$$
\begin{aligned}
C_1 &= \int_0^\rho \frac{(\mu q)\sin(\mu q) + \cos(\mu q) - 1}{q^2} \\
&\quad \times \left(\frac{\cosh(\mu\sqrt{r^2 - q^2})}{\sqrt{r^2 - q^2}} - \frac{\cosh(\mu\sqrt{r^2 - \rho^2})}{\sqrt{r^2 - \rho^2}} \right) dq \\
&= \frac{\mu^2}{2} \int_0^\rho (1 + O(q^2)) \\
&\quad \times \left(\frac{\cosh(\mu\sqrt{r^2 - \rho^2}) + \sinh(\mu\xi)\mu(\sqrt{r^2 - q^2} - \sqrt{r^2 - \rho^2})}{\sqrt{r^2 - q^2}} \right. \\
&\quad \left. - \frac{\cosh(\mu\sqrt{r^2 - \rho^2})}{\sqrt{r^2 - \rho^2}} \right) dq \\
&= \frac{\mu^2}{2} \int_0^\rho (1 + O(q^2)) \left[\sinh(\mu\xi)\mu\sqrt{r^2 - \rho^2} - \cosh(\mu\sqrt{r^2 - \rho^2}) \right] \\
&\quad \left[(r^2 - \rho^2)^{-1/2} - (r^2 - q^2)^{-1/2} \right] dq,
\end{aligned}
$$

where ξ is some point, $\sqrt{r^2 - \rho^2} \le \xi \le \sqrt{r^2 - q^2}$. Thus $\sinh(\mu\xi) = \sinh(\mu\sqrt{r^2 - \rho^2}) + O(\rho)$, $r \ge \rho$, and the last equation implies

$$
\begin{aligned}
C_1 &= -\frac{\mu^2}{2} \left[\cosh(\mu r) - (\mu r)\sinh(\mu r) + O(\rho) \right] \\
&\quad \times \left(\frac{\rho}{\sqrt{r^2 - \rho^2}} - \arcsin(\rho/r) \right), \quad \rho \to 0,
\end{aligned}
\tag{6.7.53}
$$

where $O(\rho)$ is uniform in r, $\rho < r \le B$. Estimating C_2, we obtain

$$
\begin{aligned}
C_2 &= \int_0^\rho \left(\frac{\cosh(\mu r) + \sinh(\mu\xi)(\sqrt{r^2 - q^2} - r)}{\sqrt{r^2 - q^2}} - \frac{\cosh(\mu r)}{r} \right) \frac{dq}{q^2} \\
&= \int_0^\rho (\cosh(\mu r) - (\mu r)\sinh(\mu\xi)) \frac{r - \sqrt{r^2 - q^2}}{q^2 r \sqrt{r^2 - q^2}} dq \\
&= \int_0^\rho (\cosh(\mu r) - (\mu r)\sinh(\mu\xi)) \frac{dq}{r(r + \sqrt{r^2 - q^2})\sqrt{r^2 - q^2}} \\
&= \frac{\cosh(\mu r) - (\mu r)\sinh(\mu r) + O(\rho)}{r} \int_0^\rho \frac{dq}{(r + \sqrt{r^2 - q^2})\sqrt{r^2 - q^2}}
\end{aligned}
$$

$$= \frac{\cosh(\mu r) - (\mu r) \sinh(\mu r) + O(\rho)}{r^2} \frac{\rho}{r + \sqrt{r^2 - \rho^2}}, \quad \rho \to 0, \tag{6.7.54}$$

where $O(\rho)$ is uniform in r, $\rho < r \leq B$. In a similar fashion we estimate C_3:

$$C_3 = \left(\cosh(\mu r) - (\mu r) \sinh(\mu \xi)\right) \frac{r - \sqrt{r^2 - \rho^2}}{\rho r \sqrt{r^2 - \rho^2}}$$

$$= \left(\cosh(\mu r) - (\mu r) \sinh(\mu r) + O(\rho)\right) \frac{\rho}{r(r + \sqrt{r^2 - \rho^2}) \sqrt{r^2 - \rho^2}},$$

$$\rho \to 0. \tag{6.7.55}$$

Combining (6.7.53) – (6.7.55) proves Lemma 6.7.4. □

CHAPTER 7

GEOMETRICAL TOMOGRAPHY

7.1. Basic idea

In Chapters 5 and 6 two different methods for finding discontinuity curve S of the original function f from its tomographic data $\hat{f}$ were proposed. Basically, these two methods consist of computing a function different from f, but local maxima or sharp variations of which occur at points on S. Thus, S is then recovered visually from the density plot of a computed function. In this section we propose a different idea for finding S. We want to find a subset $\hat{S}$ of the tomographic data which is in a one-to-one correspondence with the discontinuity surfaces S of the density function $f(x)$, and the map L which sends $\hat{S}$ onto S. If this is done, then the recovery of the discontinuity surfaces S of $f(x)$ is possible by applying the mapping L to $\hat{S}$. There are two steps in the suggested method:

(1) to find $\hat{S}$ from the tomographic data, and
(2) to invert $\hat{S}$ for S.

From the description of the method we see that points of S can be recovered directly.

A theoretical basis of the method has already been developed in Chapter 4. In this section we briefly recall the main facts which are needed for constructing the algorithm. Numerical implementation of the algorithm and results of its testing are described in Section 7.2.

Let $f(x) \in C^\infty(\mathbb{R}^n)$. Consider the function $\psi(x) := f(x)\chi_D(x)$, where $\chi_D(x) = \begin{cases} 1, & x \in D \\ 0, & x \notin D \end{cases}$, $D \subset \mathbb{R}^n$ is a bounded region whose boundary S is a union of smooth hypersurfaces S_j, $j \in J$, J is a finite set of indices. Assume that $f(x) \neq 0$ on S. The case when $\psi(x)$ has several discontinuity surfaces can be reduced to the case when $\psi(x)$ has one discontinuity surface S, and for simplicity we assume that S is a connected piecewise-smooth surface. Define the dual surface $\hat{S}$ as the set of points (α, p), $p \in \mathbb{R}^1$, $\alpha \in \mathbb{R}^n$, of the projective space $\mathbb{RP}_n$, such that the plane $\alpha \cdot x = p$ is tangent to S at some point $\bar{x} \in S$ (see Sections 4.1 - 4.3 for more details). Let $x_n = g(x')$, $x' := (x_1, \ldots, x_{n-1})$

be the equation of S in local coordinates in a neighborhood of a point $\overline{x} \in S$. Define the Legendre transform of g, $Lg := h$, as follows. Let $\beta \in \mathbb{R}^{n-1}$ and consider the system $\nabla_{x'} g = \beta$ $(*)$. If the Hessian of $g(x')$ does not vanish at $x' = \overline{x}'$, then $(*)$ defines $x' = x'(\beta)$. Define $h(\beta) := x'(\beta) \cdot \beta - g(x'(\beta)) := Lg$ and call it the Legendre transform of g. It is proved in Section 4.3 that S and $\hat{S}$ are related by the Legendre transform. More precisely, $\hat{S}$ has the local equation $q = h(\beta)$, where $h(\beta) = Lg$ and $q := -p/\alpha_n$, $\beta := -\alpha'/\alpha_n$, $\alpha_n \neq 0$. The Legendre transform is involutive, so $Lh = g$. Thus, if $\hat{S}$ is known and its equation in local coordinates is $q = h(\beta)$, then the local equation of S is $x_n = g(x')$, where $g = Lh$. Thus, the algorithm for finding S consists of two steps:

(1) One finds $\hat{S}$, the singular support of the Radon transform $\hat{f}(\alpha, p)$ of the function $f(x)$;
(2) Write the local equation of $\hat{S}$ as $q = h(\beta)$ and calculate $Lh := g(x')$. Then the local equation of S, the discontinuity surface of $f(x)$, is $x_n = g(x')$.

It is proved in Chapter 4 that $\hat{S}$ is a singular support of $\hat{f}$. Moreover, in a neighborhood of any $(\overline{\alpha}, \overline{p}) \in \hat{S}$ one has (see Section 4.5)

$$\hat{\psi}(\overline{\alpha}, p) = \begin{cases} \psi(\overline{x})(p - \overline{p})_{\pm}^{(n+m-2)/2} r_1 + r_2, & \text{if} \quad K \quad \text{is even} \\ \psi(\overline{x})(p - \overline{p})^{\frac{n+m-2}{2}} (\ln|p - \overline{p}|)r_1 + r_2, & \text{if} \quad K \quad \text{is odd} \end{cases},$$
$$K := (n + m - 1)I. \tag{7.1.1}$$

Here $r_i := r_i(\alpha, p)$, $i = 1, 2$, are some smooth functions, m is the codimension of the submanifold $S_m \subset S$ on which the point of contact $\overline{x} \in S$ is situated, I is the number of negative eigenvalues of the Hessian matrix of the function $(\overline{\alpha} \cdot x - \overline{p})/|\overline{\alpha}|$ on the submanifold S_m at the point $\overline{x}$. The formula for $r_1(\overline{\alpha}, \overline{p})$ is obtained in Section 4.5, note that $r_1(\overline{\alpha}, \overline{p}) \neq 0$. The sign "−" in (7.1.1) corresponds to the case when I and $n + m$ are both even or both odd, the sign "+" corresponds to all other cases, $p_+ := \max(p, 0)$. Consider for simplicity the case when $n = 2$ and S is a strictly convex smooth curve, so that $m = 1$ and $I = 0$. Then (7.1.1) yields:

$$\hat{\psi}(\overline{\alpha}, p) = \psi(\overline{x})(p - \overline{p})_+^{1/2} r_1 + r_2, \qquad p \to \overline{p},$$

and

$$\frac{\partial^\nu \hat{\psi}}{\partial p^\nu} \sim \psi(\overline{x}) \frac{\partial^\nu \left[(p - \overline{p})_+^{1/2}\right]}{\partial p^\nu} r_1(\overline{\alpha}, \overline{p}), \qquad p \to \overline{p}. \tag{7.1.1'}$$

Therefore, the local maximum of $\left|\frac{\partial^\nu \hat{\psi}}{\partial p^\nu}\right|$ in the p-variable for $\nu \geq 1$ is $\overline{p}$. Thus, $\hat{S}$ can be found by locating the local maxima of $\left|\frac{\partial^\nu \hat{\psi}}{\partial p^\nu}\right|$ in the p-variable.

7.2. Description of the algorithm and numerical experiments

In this section, we consider the two-dimensional tomographic data and develop methods for numerical finding of $\hat{S}$ and for inversion of $\hat{S}$ for S.

1. Suppose one is given the complete data $\hat{f}(\alpha_i, p_j)$, $1 \le i \le N_\alpha$, $1 \le j \le N_p$, where N_α and N_p are the number of angles and the number of projections per angle, respectively. Let ν, n_α, $n_p \ge 1$ be fixed.

Step 1.1. Fix i, $1 \le i \le N_\alpha$, and j, $1 \le j \le N_p$. For each α_k, $i \le k \le i + n_\alpha$, find p_k such that $|\hat{f}^{(\nu)}(\alpha_k, p_k)| = \max_{\ell, j \le \ell \le j + n_p} |\hat{f}^{(\nu)}(\alpha_k, p_\ell)|$.

Suppose that n_α is sufficiently small, so that $\hat{S}$ can be approximated by straight lines on the intervals $[\alpha_i, \alpha_{i+n_\alpha}]$. Suppose also that n_p is sufficiently small, so that for a fixed α, each of the intervals $[p_j, p_{j+n_p}]$ contains at most one point of $\hat{S}$. Let $(\tilde{\alpha}_i, \tilde{p}_j)$, where $\tilde{\alpha}_i := (\alpha_i + \cdots + \alpha_{i+n_\alpha})/n_\alpha$ and $\tilde{p}_j := (p_j + \cdots + p_{j+n_p})/n_p$, be the center of the current $n_\alpha \times n_p$ window. If $(\tilde{\alpha}_i, \tilde{p}_j) \in \hat{S}$, then the points (α_k, p_k), $k = i, \ldots, i + n_\alpha$, determined in Step 1.1 also belong to $\hat{S}$. Using the assumptions about smallness of n_α and n_p, we see that these points lie close to a straight line through $(\tilde{\alpha}_i, \tilde{p}_j)$, i.e. close to a line which locally approximates $\hat{S}$. Let $\theta^{(0)} := (\theta_\alpha^{(0)}, \theta_p^{(0)})$ be the unit vector in the (α, p)-space perpendicular to $\hat{S}$ at $(\tilde{\alpha}_i, \tilde{p}_j)$. Then it can be found by solving the following minimization problem

$$\Phi_{i,j}(\theta) := \sum_{k=i}^{i+n_\alpha} (\theta_\alpha(\alpha_k - \tilde{\alpha}_i) + \theta_p(p_k - \tilde{p}_j))^2 \to \min, \quad \theta_\alpha^2 + \theta_p^2 = 1. \quad (7.2.1)$$

The solution to (7.2.1) is given by

$$\theta_\alpha^{(0)} = \cos\phi, \quad \theta_p^{(0)} = \sin\phi,$$

$$\tan(2\phi) = 2 \frac{\sum_{k=i}^{i+n_\alpha} (\alpha_k - \tilde{\alpha}_i)(p_k - \tilde{p}_j)}{\sum_{k=i}^{i+n_\alpha} (\alpha_k - \tilde{\alpha}_i)^2 - \sum_{k=i}^{i+n_\alpha} (p_k - \tilde{p}_j)^2}, \quad (7.2.2)$$

and the minimal value $\Phi_{i,j}^{(\min)} := \Phi_{i,j}(\theta^{(0)})$ is obtained by substituting (7.2.2) into (7.2.1). Therefore, if $(\tilde{\alpha}_i, \tilde{p}_j) \in \hat{S}$, then $\Phi_{i,j}^{(\min)} \approx 0$. In the opposite case, when $(\tilde{\alpha}_i, \tilde{p}_j) \notin \hat{S}$, the computed value $\Phi_{i,j}^{(\min)}$ will be large (see below). Using this fact, we obtain a decision rule to distinguish between the two cases. Let the threshold $A > 0$ be fixed.

Step 1.2. By substituting (7.2.2) into (7.2.1), compute $\Phi_{i,j}^{(\min)}$. If $\Phi_{i,j}^{(\min)} \le A$, we assume that $(\tilde{\alpha}_i, \tilde{p}_j) \in \hat{S}$. If $\Phi_{i,j}^{(\min)} > A$, we assume that $(\tilde{\alpha}_i, \tilde{p}_j) \notin \hat{S}$.

Repeating Steps 1.1 and 1.2 for all pairs (i, j), $1 \leq i \leq N_\alpha$, $1 \leq j \leq N_p$, we obtain the singularity curve $\hat{S}$. To find the threshold A, let us use a standard approach to statistical hypotheses testing. Fix a window centered at a point $(\tilde{\alpha}_i, \tilde{p}_j)$. We say that the null hypothesis H_0 holds if $(\tilde{\alpha}_i, \tilde{p}_j) \notin \hat{S}$. The alternative H holds if $(\tilde{\alpha}_i, \tilde{p}_j) \in \hat{S}$. Fix ϵ, $0 < \epsilon < 1$, the probability of the first type error: rejecting H_0 when H_0 holds, and determine A from the equation $P\{\Phi_{i,j}^{(min)} > A | H_0\} = \epsilon$. Since the noise is supposed to be independent from the signal (exact tomographic data), we may assume that, if H_0 holds, there is no correlation between positions (α_k, p_k) of local maxima of $|\hat{f}^{(\nu)}|$ within the window. Thus, a convenient way to compute A is by using the Monte-Carlo method: one models random distributions of local maxima (α_k, p_k), $k = i, \ldots, i + n_\alpha$, within the window and computes $\Phi_{i,j}^{(min)}$ for each distribution. A value of the threshold A is chosen so that $\Phi_{i,j}^{(min)} > A$ in $100\epsilon\%$ cases. Clearly, this procedure is shift invariant, i.e. it does not depend on indices i and j, so it can be done only once. Using the argument very close to the one presented in Chapter 12 (e.g., in Section 12.5), one can prove the local consistency of the proposed algorithm. Recall that $\psi(\overline{x}) r_1(\overline{\alpha}, \overline{p}) \neq 0$ in (7.1.1). Let us formulate the result:

Theorem 7.2.1. *Let ϵ, the probability of the first type error, be fixed. Suppose one is given the noisy values of the Radon transform*

$$\hat{g}(\alpha_i, p_j) = \hat{f}(\alpha_i, p_j) + n_{ij}, \quad \alpha_{i+1} = \alpha_i + \Delta\alpha, \quad p_{j+1} = p_j + \Delta p,$$

where n_{ij} are independent and identically distributed random variables with finite first and second moments. Consider an arbitrary point $(\tilde{\alpha}_i, \tilde{p}_j) \in \hat{S}$ and suppose that $\Delta\alpha, \Delta p \to 0$, $n_\alpha, n_p \to \infty$, and $n_\alpha\Delta\alpha$, $n_p\Delta p \to 0$. Then, the probability of accepting H_0 at this point goes to zero as $n_\alpha, n_p \to \infty$.

Exercise 7.2.1. Prove Theorem 7.2.1.

Now we may assume that the singularity curve $\hat{S}$ has been found. Since calculation of derivatives enhances noise, some of the points from $\hat{S}$ will be missing, and some erroneously detected points will appear. Thus, one needs a stable algorithm for the inversion of $\hat{S}$ to obtain its dual curve S, the discontinuity curve of $f(x)$. In order to calculate S, we will compute a function $\Phi(x)$, $x \in \mathbb{R}^2$, such that its maxima correspond to the points of S. Fix an arbitrary point $x \in \mathbb{R}^2$ and consider a circle $B_\delta(x)$ with radius $\delta > 0$ centered at x. First, consider a model in which S is a circle with radius R centered at $(0, 0)$. Taking into account discretization, suppose that all the points $(\tilde{\alpha}_i, \tilde{p}_j) \in \hat{S}$ have been found in Step 1.2. Define the function $\Phi(x)$ as the number of lines

$\tilde{\alpha}_i \cdot \tilde{x} = \tilde{p}_j$, $\tilde{x} \in \mathbb{R}^2$, which pass through $B_\delta(x)$, that is

$$\Phi(x) := \#\{(i,j) : |\tilde{\alpha}_i \cdot x - \tilde{p}_j| \leq \delta, \ (\tilde{\alpha}_i, \tilde{p}_j) \in \hat{S}\}. \tag{7.2.3}$$

Suppose that $\Delta\alpha N_p \Delta p \ll \delta \ll R$. The following properties of $\Phi(x)$ can be established: $\Phi(x) = 0$ if $|x| < R - \delta$,

$$\Phi(x) = \frac{1}{\Delta\alpha} \frac{4\delta}{|x|} (1 + O(|x|^{-2})) \quad \text{as} \quad |x| \to \infty, \tag{7.2.4}$$

and $\Phi(x)$ attains its maximum

$$\Phi(x) = \frac{4}{\Delta\alpha} \sqrt{\frac{\delta}{R}} (1 + O(\delta/R)) \tag{7.2.5}$$

when $|x| = R + \delta$, that is when $B_\delta(x)$ is tangent to S.

Exercise 7.2.2. Prove Equations (7.2.4) and (7.2.5).

One sees that if δ is sufficiently small, the value of the function $\Phi(x)$ is determined only by a small part of the boundary S, namely, by the part of S which can be touched by straight lines passing through $B_\delta(x)$. Thus, in the case of a more general discontinuity curve S, formulas (7.2.4) and (7.2.5) are replaced by

$$\Phi(x) = \frac{1}{\Delta\alpha} \frac{4\delta}{\text{dist}(x, S(x))} (1 + O(\text{dist}^{-2}(x, S(x)))) \tag{7.2.4'}$$

and

$$\Phi(x) = \frac{4}{\Delta\alpha} \sqrt{\frac{\delta}{R(x)}} (1 + O(\delta/R(x))), \tag{7.2.5'}$$

where $\text{dist}(x, S(x))$ denotes the distance between x and the part of S which is touched by lines through $B_\delta(x)$, and $R(x)$ is the radius of the curvature of S at a point of contact between $B_\delta(x)$ and S. Note also that $\Phi(x)$ is additive, i.e. if S is a union of several smooth closed curves then $\Phi(x)$ is a sum of contributions from each of these curves. Now we describe an algorithm of the second step.

Step 2. Let $(\tilde{\alpha}_i, \tilde{p}_j)$ be the points detected at Step 1.2 as belonging to $\hat{S}$. Compute the function $\Phi(x)$ by the formula

$$\Phi(x) := \#\{(i,j) : |\tilde{\alpha}_i \cdot x - \tilde{p}_j| \leq \delta, \ (\tilde{\alpha}_i, \tilde{p}_j) \in \hat{S}\}.$$

Points at which $\Phi(x)$ is large correspond to discontinuity curve S.

Let us estimate the number of operations needed for the algorithm. Steps 1.1 and 1.2 require $O(N_\alpha N_p)$ operations, Step 2 requires $O(NN_\alpha)$ operations, where N is the number of points on the (x_1, x_2) - plane, where the function $\Phi(x)$ is evaluated. Thus the total operation count is $O(N_\alpha(N_p + N))$. Note that the filtered back-projection (FBP) algorithm requires $O(N_\alpha(N_p^2 + N))$ operations or $O(N_\alpha(N_p \log N_p + N))$ if one uses FFT for the calculation of convolutions. Thus the proposed algorithm is more economical than FBP.

In Figure 7.2.1, one sees the phantom used for generating the Radon transform data. The densities are: exterior: 0, ellipse: 1.5, exterior annulus: 1, area between the annulus and the ellipse: 2, small circle inside the ellipse: 1, three small circles inside the annulus: 0.5. The radius of the phantom: 0.9, the half-axes of the ellipse: 0.2 and 0.4, the radii of the four small discs: 0.05, the radius of the disc containing the ellipse: 0.5. The Radon transform was computed for 250 angles (equispaced on $[0, \pi)$) and 301 projections per each angle. In Figure 7.2.2, one sees an intensity plot of $|\partial^2 \hat{f}(\alpha, p)/\partial p^2|$ on the (α, p)-plane, and in Figure 7.2.3 – an image of detected singularities by Steps 1.1 and 1.2. In Figures 7.2.2 and 7.2.3, the horizontal axis is the α - axis, and the vertical one is the p - axis. The second derivative was computed by the three-point scheme from the non-noisy data. The parameters of the algorithm were: $n_\alpha = 7$, $n_p = 7$, the probability of the first type error $\epsilon = 0.01$. In Figure 7.2.4, an intensity plot of the function $\Phi(x)$ computed from the data in Figure 7.2.3 is presented. The radius δ was chosen $\delta = \Delta p$. In Figure 7.2.5, one sees an intensity plot of $\Phi(x)$ computed from the limited angle data. As the initial data we took detected singularities, which are represented in Figure 7.2.3, and deleted angles from the range $[130°, 150°]$. In Figures 7.2.6 and 7.2.7, the further results for the limited-angle data inversion are presented. In Figures 7.2.8 and 7.2.9, the results for the exterior data problem are presented. The intervals of the missing data in p - variable are $[-0.3, 0.3]$ in Figure 7.2.8, and $[-0.4, 0.4]$ in Figure 7.2.9. The full data correspond to the interval $[-0.9, 0.9]$ in p - variable.

At present, geometrical tomography does not allow one to compute values of jumps and, moreover, its numerical realization is much less stable with respect to noise in the data than local tomography or pseudolocal tomography. However, in our opinion, the beauty of the geometrical relationship between singular supports of f and $\hat{f}$ and the fact that this relationship can be used for the recovery of $S = \text{sing supp} f$ (as presented numerical experiments demonstrate) warrants further investigation of geometrical tomography.

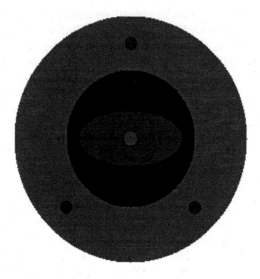

FIGURE 7.2.1. The phantom used for generating the Radon transform data.

FIGURE 7.2.2. An intensity plot of $|\partial^2 \hat{f}(\alpha,p)/\partial p^2|$ on the (α,p)-plane.

FIGURE 7.2.3. An image of detected singularities.

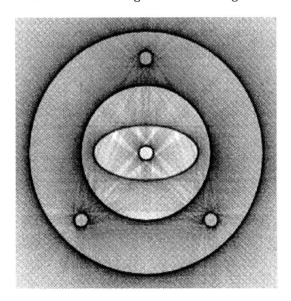

IGURE 7.2.4. An intensity plot of $\Phi(x)$ computed from the data in Figure 7.2.3.

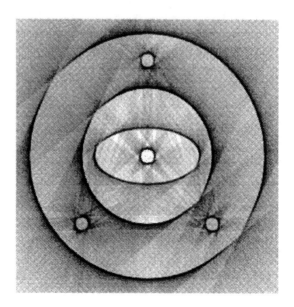

FIGURE 7.2.5. An intensity plot of $\Phi(x)$ computed from the limited angle data. The range of angles $[130°, 150°]$ is missing.

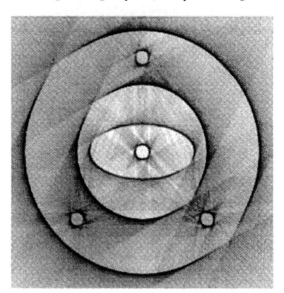

FIGURE 7.2.6. The same as in Figure 7.2.5. The range of angles $[120°, 150°]$ is missing.

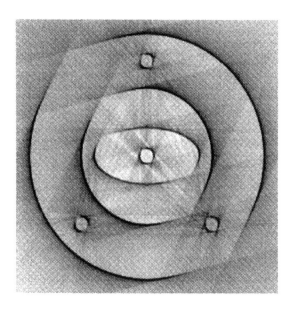

FIGURE 7.2.7. The same as in Figure 7.2.5. The range of angles $[105°, 150°]$ is missing.

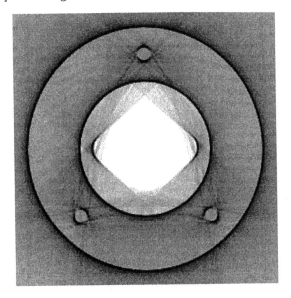

FIGURE 7.2.8. An intensity plot of $\Phi(x)$ for the exterior data problem with missing values of p in the interval $[-0.3, 0.3]$.

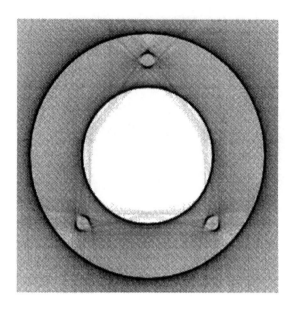

FIGURE 7.2.9. The same as in Figure 7.2.8 with missing values of p in the interval $[-0.4, 0.4]$.

CHAPTER 8

INVERSION OF INCOMPLETE TOMOGRAPHIC DATA

8.1. Inversion of incomplete Fourier transform data

8.1.1. The basic result

In this section a method is given for analytical inversion of incomplete Fourier transform data. Suppose $f \in L^2(B_a)$, $B_a \subset \mathbb{R}^n$, $n \geq 2$, $f(x) = 0$ for $|x| > a$. Suppose we are given the incomplete data

$$\int_{B_a} \exp(i\xi \cdot y)f(y)dy = \tilde{f}(\xi), \quad \xi \in D, \qquad (8.1.1)$$

where D is a proper subdomain of $\mathbb{R}^n$. Multiply (8.1.1) by $(2\pi)^{-n}\tilde{\delta}_N(\xi)$ $\times \exp(-i\xi \cdot x)$, where $\tilde{\delta}_N := \mathcal{F}\delta_N$, δ_N is defined by formulas (8.1.3) below, and integrate over D in ξ to get:

$$f_N(x) := \int_{B_a} f(y)\delta_N(x-y)dy = (2\pi)^{-n}\int_D \tilde{f}(\xi)\tilde{\delta}_N(\xi)\exp(-i\xi \cdot x)d\xi,$$
$$(8.1.2)$$

where N is an integer. Let us choose

$$\delta_N(x) := P_N(|x|^2)(\mathcal{F}^{-1}\tilde{h})(x), \quad P_N(r^2) := \left(\frac{N}{4\pi a_1^2}\right)^{n/2}\left(1 - \frac{r^2}{4a_1^2}\right)^N,$$
$$(8.1.3)$$

where $\tilde{h} \in C_0^\infty(D)$, $(2\pi)^{-n}\int_D \tilde{h}(\xi)d\xi = 1$, and $a_1 > a$. If D is unbounded, we may assume, instead of $\tilde{h} \in C_0^\infty(D)$, that $\tilde{h} \in S(\mathbb{R}^n)$ and $\tilde{h}(\xi) = 0$ for $\xi \notin D$. It turns out that if N grows to infinity, then f_N becomes a good approximation to f. More precisely, one has

Theorem 8.1.1. *The function* $\tilde{\delta}_N = \mathcal{F}\delta_N$ *vanishes outside* D. *The function* δ_N *is a delta-sequence in* B_a *in the following sense: for any* $f \in L^2(B_a)$ *one has*

$$\|f - f_N\| \to 0 \quad as \quad N \to \infty, \qquad (8.1.4)$$

where $|| \cdot || = || \cdot ||_{L^2(B_a)}$ *if* $f \in L^2(B_a)$, $|| \cdot || = || \cdot ||_{C(B_a)}$ *if* $f \in C(B_a)$,
and

$$f_N(x) := \int\limits_{B_a} \delta_N(x-y)f(y)dy = \mathcal{F}^{-1}_{\xi \to x}(\tilde{\delta}_N(\xi)\tilde{f}(\xi)). \qquad (8.1.5)$$

Moreover, if

$$||f||_{C^1(B_a)} := \max_{x \in B_a}|f(x)| + \max_{x \in B_a}|\nabla f(x)| \le m_1, \qquad (8.1.6)$$

where $m_1 = const > 0$, *then*

$$||f - f_N||_{C(B_a)} \le cm_1 N^{-1/2} \quad as \quad N \to \infty, \qquad (8.1.7)$$

where $c > 0$ *does not depend on* f, m_1, *and* N.

First, we state and prove an auxiliary lemma.

Lemma 8.1.1. *Let* $f \in C^1(B_a)$ *and* $||f||_{C^1(B_a)} \le m_1$. *Let* δ_N *be as in Theorem 8.1.1. Then*

$$\int\limits_{|y| \le a} \delta_N(x-y)f(y)dy = f(x) + \epsilon_N(x), \qquad (8.1.8)$$

where

$$||\epsilon_N||_{C(B_a)} \le c_\epsilon m_1 N^{-1/2} \quad as \quad N \to \infty, \qquad (8.1.9)$$

where the constant $c_\epsilon > 0$ *does not depend on* f, m_1, *and* N.

Proof of Lemma 8.1.1. Let us use Lemma 14.5.1: if $g \in C^1([0,1])$, then

$$\int\limits_0^1 \exp(-\lambda t^\alpha)t^{\beta-1}g(t)dt = \frac{1}{\alpha}\left[\lambda^{-\beta/\alpha}\Gamma\left(\frac{\beta}{\alpha}\right)g(0) + O\left(\lambda^{-\frac{\beta+1}{\alpha}}\right)\right],$$

$$\alpha > 0, \beta > 0, \lambda \to +\infty. \qquad (8.1.10)$$

Using (8.1.3), we get

$$\int\limits_{|y| \le a} \delta_N(y)dy = \left(\frac{N}{4\pi a_1^2}\right)^{n/2}|S^{n-1}|\int\limits_0^a \left(1 - \frac{r^2}{4a_1^2}\right)^N r^{n-1}\mu(r)dr$$

$$= \left(\frac{N}{4\pi}\right)^{n/2}|S^{n-1}|\int\limits_0^{a/a_1} \left(1 - \frac{t^2}{4}\right)^N t^{n-1}\mu(a_1 t)dt := I, \qquad (8.1.11)$$

where

$$\mu(r) = |S^{n-1}|^{-1} \int_{S^{n-1}} h(r\alpha)d\alpha, \quad \mu(0) = 1, \quad h := \mathcal{F}^{-1}\tilde{h},$$

and $\left(1 - \frac{t^2}{4}\right)^N = \exp\left\{-\frac{Nt^2}{4}[1 + O(t^2)]\right\}$ as $t \to 0$. Thus $\alpha = 2$, $\beta = n$, $\lambda = \frac{N}{4}$, $g(t) = \mu(a_1 t)$, $g(0) = 1$, and (8.1.10), applied to I, yields:

$$I = \frac{N^{n/2}}{(4\pi)^{1/2}} \frac{|S^{n-1}|}{2} \left[\left(\frac{N}{4}\right)^{-n/2} \Gamma\left(\frac{n}{2}\right) + O(N^{-\frac{n+1}{2}})\right] = 1 + O(N^{-1/2}),$$

$$N \to \infty. \tag{8.1.12}$$

One has

$$\left|\int_{|y|\le a} \delta_N(x-y)f(y)dy - f(x)\right|$$

$$\le \left|\int_{|y|\le a} \delta_N(x-y)(f(y) - f(x))dy\right| + \left|f(x)\left(\int_{|y|\le a} \delta_N(x-y)dy - 1\right)\right|$$

$$:= I_1 + I_2. \tag{8.1.13}$$

Since $\|f\|_{C^1(B_a)} \le m_1$, then $|f(y) - f(x)| \le m_1|y - x|$ if $x, y \in B_a$. Let $c_h = \max_{y\in B_{2a}} |h(y)|$. Using the inequality $P_N(r) \ge 0$, we get

$$I_1 \le c_h m_1 \int_{|y|\le 2a} P_N(|y|^2)|y|dy = c_h m_1 O(N^{-1/2}), \tag{8.1.14}$$

where we have used formula (8.1.10) with $\alpha = 2$, $\beta = n+1$, and $\lambda = N$, to get:

$$\left(\frac{N}{a_1^2}\right)^{n/2} \int_0^{2a} \left(1 - \frac{r^2}{4a_1^2}\right)^N r^n dr = O(N^{-1/2}).$$

Using (8.1.6), we can estimate the integral I_2 in (8.1.13) as follows:

$$I_2 \le m_1 \left(\left|\int_{|y|\le a} P_N(|x-y|^2)dy - 1\right|\right.$$

$$\left. + \left|\int_{|y|\le a} P_N(|x-y|^2)(h(x-y) - 1)dy\right|\right). \tag{8.1.15}$$

Since $h(0) = 1$, we obtain similarly to (8.1.14) that the second integral in (8.1.15) is of order $O(N^{-1/2})$. To estimate the first integral in (8.1.15), suppose that $|x| \leq a$. Then

$$\left| \int_{|y| \leq a} P_N(|x - y|^2)dy - 1 \right|$$

$$\leq \max \left(\left| \int_{|y| \leq a_1 - a} P_N(|y|^2)dy - 1 \right|, \left| \int_{|y| \leq 2a} P_N(|y|^2)dy - 1 \right| \right).$$

Similarly to (8.1.11) and (8.1.12), the two integrals on the right-hand side of the last inequality are of order $O(N^{-1/2})$. This, combined with (8.1.8) and (8.1.13) – (8.1.15), proves (8.1.9). □

Proof of Theorem 8.1.1. The property $\tilde{\delta}_N(\xi) = 0$ for $\xi \notin D$ immediately follows from the relation $\tilde{\delta}_N(\xi) = P_N(-\Delta_\xi)\tilde{h}(\xi)$, because, by the assumption, $\tilde{h}(\xi) = 0$ for $\xi \notin D$.

Estimate (8.1.7) was established in the course of the proof of Lemma 8.1.1.

Assume now that $f \in L^2(B_a)$. We want to prove (8.1.4) with $|| \cdot || = || \cdot ||_{L^2(B_a)}$. Let us use an approximation argument. Pick an arbitrary small $\epsilon > 0$ and an f_ϵ such that $f_\epsilon \in C^1(B_a)$ and $||f - f_\epsilon||_{L^2(B_a)} < \epsilon$. One has, with $|| \cdot || = || \cdot ||_{L^2(B_a)}$,

$$||f_N - f|| \leq ||f - f_\epsilon|| + ||f_\epsilon - f_{\epsilon N}|| + ||f_{\epsilon N} - f_N|| := I_1 + I_2 + I_3.$$

By the assumption, $I_1 < \epsilon$. By Lemma 8.1.1, there exists $N(\epsilon)$ such that $I_2 < \epsilon$ for $N > N(\epsilon)$. Let us prove that $I_3 < \epsilon$ for $N > N_1(\epsilon)$. One has with $\phi_\epsilon := f_\epsilon - f$, $||\phi_\epsilon|| < \epsilon$, and $c_h = \max_{y \in B_{2a}} |h(y)|$:

$$I_3^2 = ||f_{\epsilon N} - f_N||^2 \leq \left\| \int_{|y| \leq a} \delta_N(x - y)\phi_\epsilon(y)dy \right\|^2$$

$$\leq c_h^2 \int_{B_a} \int_{B_a} \int_{B_a} P_N(|x - y|^2)P_N(|x - z|^2)|\phi_\epsilon(y)\phi_\epsilon(z)|dydzdx$$

$$\leq c_h^2 \int_{B_a} \int_{x - B_a} \int_{x - B_a} P_N(|u|^2)P_N(|v|^2)|\phi_\epsilon(x - u)\phi_\epsilon(x - v)|dudvdx$$

$$\leq \frac{c_h^2}{2} \int_{B_a} \int_{x - B_a} \int_{x - B_a} P_N(|u|^2)P_N(|v|^2)[\phi_\epsilon^2(x - u) + \phi_\epsilon^2(x - v)]dudvdx$$

$$\leq c_h^2 \epsilon^2 \left(\int_{B_{2a}} P_N(|u|^2)du \right)^2 \leq c\epsilon^2.$$

Thus, given an arbitrary small $\epsilon > 0$, one can find $N(\epsilon)$ such that

$$\|f_N - f\| < c\epsilon \quad \text{for} \quad N > N(\epsilon).$$

The case when $f \in C(B_a)$ can be considered similarly. Let us estimate, for example, the norm $\|f_{\epsilon N} - f_N\|_{C(B_a)}$, where f_ϵ is such that $\|f_\epsilon - f\|_{C(B_a)} < \epsilon$. One has

$$\|f_{\epsilon N} - f_N\|_{C(B_a)} = \sup_{x \in B_a} \left| \int_{|y| \le a} \delta_N(x - y)\phi_\epsilon(y)dy \right|$$

$$\le c_h\epsilon \int_{B_{2a}} P_N(|y|^2)dy \le c\epsilon.$$

Theorem 8.1.1 is proved. $\quad\square$

Exercise 8.1.1. Suppose D is a ball of radius λ_0 centered at the origin, $D = B_{\lambda_0}$. Show that the function δ_N, defined by

$$\delta_N(x) = P_N(x)\left[h\left(\frac{x}{2N + n + \gamma}\right)\right]^{2N+n+\gamma},$$

$$h(x) = \frac{1}{|B_{\lambda_0}|}\int_{|\xi| \le \lambda_0} \exp(i\xi \cdot x)d\xi,$$

where $|B|$ is the volume of B and $\gamma > 0$ is an integer, is a delta-sequence in B_a in the sense of Theorem 8.1.1.

Hint. Clearly, δ_N is entire and of exponential type λ_0. By the Paley-Wiener-Schwartz theorem, $\tilde{\delta}_N(\xi) = 0$ for $|\xi| \ge \lambda_0$. Using (14.4.48) (with $l = 0$) and the asymptotics of the Bessel functions for large values of the argument (see Section 14.4.2), prove the relations

$$h(x) = n\Gamma\left(\frac{n}{2}\right)2^{\frac{n-2}{2}}(\lambda_0|x|)^{-n/2}J_{n/2}(\lambda_0|x|) = O\left(|x|^{-\frac{n+1}{2}}\right), \quad |x| \to \infty.$$

Therefore,

$$|\delta_N(x)| = O\left(|x|^{2N-(2N+n+\gamma)\frac{n+1}{2}}\right) \quad \text{as} \quad |x| \to \infty.$$

If $\frac{(2N+n+\gamma)(n+1)}{2} - 2N - n > m$, then $(|x| + 1)^m|\delta_N(x)| \in L^1(\mathbb{R}^n)$, so that $h_N \in C^m(\mathbb{R}^n)$. Thus, the parameter γ in the definition of δ_N regulates the rate of decay of $\delta_N(x)$ as $|x| \to \infty$, so that it regulates the smoothness of $\tilde{\delta}_N(\xi)$. Prove also that $[h(x/A)]^A = 1 + O(1/A)$ as $A \to \infty$. This shows that the last factor in the definition of δ_N tends to 1 uniformly on compact subsets of $\mathbb{R}^n$. The rest of the argument is based on the fact that P_N is a polynomial delta-sequence on B_a.

8.1.2. Numerical aspects

Using the theory developed in the previous section, we can construct the following algorithm for the incomplete Fourier transform data inversion.

(1) Compute $\tilde{\delta}_N(\xi) = P_N(-\Delta_\xi)\tilde{h}(\xi)$;
(2) Compute $\tilde{f}_N(\xi) = \tilde{\delta}_N(\xi)\tilde{f}(\xi)$;
(3) Compute inverse Fourier transform $f_N = \mathcal{F}^{-1}\tilde{f}_N$.

Suppose now that the noisy data $\tilde{f}_\delta(\xi)$ are given in (8.1.1) in place of $\tilde{f}(\xi)$, and assume that

$$\sup_{\xi \in D} |\tilde{f}_\delta(\xi) - \tilde{f}(\xi)| < \delta. \qquad (8.1.16)$$

Theorem 8.1.2. *There exists $N(\delta)$ such that $N(\delta) \to \infty$ as $\delta \to 0$ and*

$$\|f_{\delta,N(\delta)} - f\|_{L^2(B_a)} \to 0 \quad as \quad \delta \to 0,$$

where

$$f_{\delta,N(\delta)}(x) := (2\pi)^{-n} \int_D \tilde{f}_\delta(\xi)\tilde{h}_{N(\delta)}(\xi) \exp(-i\xi \cdot x)d\xi.$$

Proof. One has with f_N defined by (8.1.2) and $\|\cdot\| = \|\cdot\|_{L^2(B_a)}$:

$$\|f_{\delta,N(\delta)} - f\| \le \|f_{\delta,N(\delta)} - f_{N(\delta)}\| + \|f_{N(\delta)} - f\| \le \delta a(N(\delta)) + \eta(N(\delta)),$$

where

$$a(N) := (2\pi)^{-n/2} \left(\int_D |\tilde{h}_N(\xi)|^2 d\xi \right)^{1/2} \to \infty \quad as \quad N \to \infty, \quad (8.1.17)$$

$$\eta(N) := \|f_N - f\| \to 0 \quad as \quad N \to \infty. \qquad (8.1.18)$$

For a fixed $\delta > 0$, find $N(\delta)$ such that

$$\epsilon(\delta, N) := \delta a(N) + \eta(N) = \min,$$

where the minimum is taken over all positive integers N. From (8.1.17) and (8.1.18) it follows that there exists $N(\delta)$ such that $\epsilon(\delta) := \epsilon(\delta, N(\delta)) \le \epsilon(\delta, N)$ and $\epsilon(\delta) \to 0$ as $\delta \to 0$. Theorem 8.1.2 is proved. $\square$

REMARK 8.1.2. One can estimate $a(N)$ and, assuming $\|f\|_{C^1(B_a)} \le m_1$, one can estimate $\eta(N)$. This allows one to estimate $\epsilon(\delta)$. For example, if $\|f\|_{C^1(B_a)} \le m_1$, then, by Theorem 8.1.1, $\eta(N) \le cm_1 N^{-1/2}$. The function $a(N)$ does not depend on the data (see (8.1.17)) Theorem 8.1.2 asserts that $f_{\delta,N(\delta)}(x)$ yields a stable approximation to f in the $L^2(B_a)$ norm given the noisy data $\tilde{f}_\delta(\xi)$. The rate $\epsilon(\delta)$ of convergence as $\delta \to 0$ may be very slow.

8.2. Filtered backprojection method for inversion of the limited-angle tomographic data

Let $Rf := \hat{f}$ be the Radon transform of a function $f \in L^2(B_a)$, $f(x) = 0$ for $|x| > a$. Suppose that $\hat{f}(\alpha, p)$ is known for all $|p| \le a$ and $\alpha \in \Omega$, where $\Omega \subsetneq S^{n-1}$ is an open set. Since $\hat{f}(\alpha, p) = 0$ for $|p| > a$ and $\hat{f}(-\alpha, p) = \hat{f}(\alpha, -p)$, we may assume that Ω is symmetrical: if $\alpha \in \Omega$, then $-\alpha \in \Omega$, and that $\hat{f}(\alpha, p)$ is known in the conical set K:

$$K := \{\alpha, p : \ \alpha \in \Omega, \ p \in \mathbb{R}\}. \tag{8.2.1}$$

The problem is: given $\hat{f}(\alpha, p)$ in K, calculate f assuming that $f(x) = 0$ for $|x| > a$, $f \in L^2(B_a)$.

If $\hat{f}(\alpha, p)$ were given for all $p \in \mathbb{R}$ and all $\alpha \in S^{n-1}$, then f could be efficiently recovered using the FBP (filtered backprojection) method, which is based on the formulas (see (2.1.42)):

$$W * f = R^*(w \circledast Rf), \quad W := R^* w. \tag{8.2.2}$$

Here '$*$' is the convolution in $\mathbb{R}^n$, '$\circledast$' is the convolution in the p variable only, and R^* is the backprojection operator. See Section 3.6 for description of the numerical implementation of the FBP algorithm.

Our aim is to derive an analogue of (8.2.2) for the case when the data Rf is given not for all $\alpha \in S^{n-1}$ and $p \in \mathbb{R}$, but for $(\alpha, p) \in K$ only. The basic idea is similar to the idea in Section 8.1. Namely, we choose $W_N(x)$ such that

$$W_N(x) \to \delta(x) \quad \text{on} \quad B_a \quad \text{as } N \to \infty, \tag{8.2.3}$$

and such that (8.2.2) holds with $w = w_N$:

$$W_N = R^* w_N; \quad w_N(\alpha, p) = 0, \ (\alpha, p) \notin K. \tag{8.2.4}$$

Following (8.1.3), define

$$W_N(x) := P_N(|x|^2)(\mathcal{F}^{-1}\tilde{h})(x), \quad \mathcal{F}^{-1}\tilde{h}\Big|_{x=0} = 1, \tag{8.2.5}$$

where $\tilde{h}$ has the same properties as in (8.1.3) (with $D := K$). In particular, $\mathcal{F}^{-1}\tilde{h} \in S(\mathbb{R}^n)$; therefore, $W_N \in S(\mathbb{R}^n)$ for any N. Theorem 8.1.1 asserts that W_N is a delta-sequence in B_a. With Δ_ξ being the Laplacian in the ξ-variable, we have:

$$(\mathcal{F}W_N)(\xi) = P_N(-\Delta_\xi)\tilde{h}(\xi). \tag{8.2.6}$$

From Theorem 2.5.1 (see Equation (2.5.7)) it follows that the solution w_N to the equation in (8.2.4) is given by

$$w_N(\alpha, p) = \frac{1}{2(2\pi)^n} \int\limits_{-\infty}^{\infty} \exp(-i\lambda p)|\lambda|^{n-1} P_N(-\Delta_\xi) \tilde{h}(\xi) \bigg|_{\xi=\lambda\alpha} d\lambda. \quad (8.2.7)$$

It is clear that the function $w_N(\alpha, p)$, defined in (8.2.7), vanishes outside the conical set K. Thus, we obtained the modified FBP algorithm:

(1) Calculate $w_N(\alpha, p)$ by formula (8.2.7);
(2) Calculate

$$f_N := R^*(w_N \circledast Rf). \quad (8.2.8)$$

Our argument proves the following result.

Theorem 8.2.1. *The function $f_N := R^*(w_N \circledast Rf)$ has the property:*

$$f_N(x) = \int\limits_{B_a} W_N(x - y) f(y) dy.$$

Moreover,

$$\|f - f_N\| \to 0 \quad as \quad N \to \infty,$$

where $\|\cdot\| = \|\cdot\|_{L^2(B_a)}$ if $f \in L^2(B_a)$, $\|\cdot\| = \|\cdot\|_{C(B_a)}$ if $f \in C(B_a)$.

We may choose $\tilde{h}(\xi)$, for example, as:

$$\tilde{h}(\xi) = c \exp(-\lambda^2 - \lambda^{-2}) \eta(\alpha) , \xi = \lambda\alpha, \quad c = \text{const} > 0. \quad (8.2.9)$$

where $\eta \in C_0^\infty(\Omega)$ and η is even: $\eta(\alpha) = \eta(-\alpha)$. The constant c in (8.2.9) can be chosen so that the normalizing condition $(2\pi)^{-n} \int_K \tilde{h}(\xi) d\xi = 1$ holds.

REMARK 8.2.1. If $\|f\|_{C^1(B_a)} \le m_1$, Theorem 8.1.1 yields the inequality $\|f - f_N\|_{C(B_a)} \le cm_1 N^{-1/2}$ as $N \to \infty$.

Ill-posedness and, hence, numerical difficulties in practical application of the above method stem from the necessity to calculate the function $P_N(-\Delta_\xi)\tilde{h}(\xi)$. Although $\tilde{h}(\xi)$ can be chosen infinitely smooth, the function $P_N(-\Delta_\xi)\tilde{h}(\xi)$ will oscillate and will take very large values as N grows.

It is possible to look at the results obtained in this section from a different point of view. Theorem 8.2.1 and the discussion after it show that there exists a sequence of functions W_N such that

$$W_N \to \delta \quad \text{on} \quad B_a \quad \text{as} \quad N \to \infty,$$

$$W_N = R^* w_N, \quad w_N(\alpha, p) = 0 \quad \text{for} \quad (\alpha, p) \notin K.$$

For example, one can take W_N in the form (8.2.5). Correspondingly, w_N will be of the form (8.2.7). It is clear that, in fact, there is a relatively large freedom of choice of functions w_N, W_N. Since the problem of the limited angle Radon transform inversion is severely ill-posed, one may use this freedom in order to reduce the ill-posedness. This can be done, for example, by finding the function w_N as a solution to the following optimization problem:

$$\int_{B_{2a} \setminus B_\delta} (R^* w_N)^2 (x) dx < \epsilon, \quad \int_{B_{2a}} (R^* w_N)(x) dx = 1, \qquad (8.2.10)$$

where $\epsilon, \delta > 0$, and

$$w_N(\alpha, p) = 0 \quad \text{for} \quad (\alpha, p) \notin K, \quad w_N(\alpha, p) = w_N(\alpha, -p) = w_N(-\alpha, p).$$

Theorem 8.2.1 guarantees existence of the solution to this problem for any $\epsilon, \delta > 0$. Note that from (8.2.10) it follows that it is sufficient to know the function $w_N(\alpha, p)$ only on the compact $|p| < 2a$, $\alpha \in S^{n-1}$.

8.3. The extrapolation problem

8.3.1. Formulation of the problem

For convenience, in this section we scale the problem so that the parameter a (the radius of support of the unknown function f) equals 1, that is $f \in L^2(B_1)$, $f(x) = 0$ for $|x| \geq 1$. It is frequently the case when the available tomographic data corresponds to $\alpha \in \Omega := \Omega_+ \cup \Omega_-$, where $\Omega_\pm$ are open convex subsets of S^{n-1} and $\Omega_+ = -\Omega_-$. Consider the cones $K_\pm := \{\lambda \alpha : \alpha \in \Omega_\pm, \lambda \geq 0\}$. The Fourier slice theorem implies that given $\hat{f}(\alpha, p)$ for $\alpha \in \Omega_+ \cup \Omega_-$, $|p| \leq a$, we can find $\tilde{f}(\xi)$ for $\xi \in K_+ \cup K_-$. Therefore, it is of interest to extend the values of $\tilde{f}$ from the cones $K_+ \cup K_-$ to the ball $|\xi| \leq \lambda_0$, where $\lambda_0 > 0$ is sufficiently large. If this problem is solved, we may assume that we are given the complete Fourier transform data, and compute f given $\tilde{f}(\xi), |\xi| \leq \lambda_0$, using the usual inverse Fourier transform. Let us consider the following problem.

Problem A. *Let* $G(k) := \int_{-1}^{1} g(x) \exp(ikx) dx$, $g(x) \in L^2([-1, 1])$. *Assume that* $G(k)$ *is known on the set* $E := \{k : k \in \mathbb{C}, k = ib + \sigma, \sigma \geq \lambda_0, \sigma \leq -\lambda_0\}$ *where* $\lambda_0 > 0$ *is a fixed number. Find* $G(k)$ *on the segment* $\ell := \{k : k = ib + \sigma, -\lambda_0 < \sigma < \lambda_0\}$.

This problem arises naturally in the above context. Indeed, if $G(k)$ is known in the angles (the analogues of the cones K_+ and K_-) $|\arg k| <$

ϑ_0, $|\pi - \arg k| < \vartheta_0$, then $G(k)$ is known on the part E of any line $\mathcal{L}$ parallel to the real axis $\mathcal{L} := \{k : k = ib+\sigma, \sigma \in \mathbb{R}\}$, for which $\sigma > |b| \cot \vartheta_0$ and $\sigma < -|b| \cot \vartheta_0$, and we want to find $G(k)$ on the complementary segment $\ell := \{k : k = ib + \sigma, -|b| \cot \vartheta_0 < \sigma < |b| \cot \vartheta_0\}$.

We will outline some methods for solving problem A. Only numerical experiments can demonstrate the relative resolution power of the inversion procedures based on these methods.

8.3.2. The first method of solution

Without loss of generality, we may assume $b > 0$. Map conformally the upper half-plane $\operatorname{Im} k > 0$ onto the unit disk so that the point iq is mapped into the origin:

$$w = \frac{k - iq}{k + iq}, \quad k = -iq\frac{w + 1}{w - 1}.$$

One can choose $0 < q < b$, for example $q = b/2$. The line $\mathcal{L} = \{k : k = ib + x, x \in \mathbb{R}\}$ is mapped onto a curve $\mathcal{L}_w$ in the unit disk, $G(k) = G\left(-iq\frac{w+1}{w-1}\right) := \psi(w)$. The function $\psi(w)$ is analytic in $D_0 := \{w : |w| \le w_0 < 1\}$, where $w_0 > 0$ is an arbitrary number smaller than 1, so that

$$G(k) = \sum_{j=0}^{\infty} c_j w^j, \quad |w| < w_0 < 1,$$

where c_j are the Taylor coefficients of $\psi(w)$ at $w = 0$. Consider the part of $\mathcal{L}'_w := \mathcal{L}_w \setminus \ell_w$ which lies inside D_0. Here ℓ_w is the image of ℓ under the conformal mapping. Fix an arbitrary large integer M and M points w_m on $\mathcal{L}'_w \cap D_0$. Find the coefficients c_j, $0 \le j \le M - 1$, from the following linear system:

$$\sum_{j=0}^{M-1} c_j w_m^j = G\left(-iq\frac{w_m + 1}{w_m - 1}\right), \quad 1 \le m \le M. \qquad (8.3.1)$$

If $w_m \ne w_{m'}$ for $m \ne m'$ one has $\det w_m^j \ne 0$ (the Vandermonde determinant), so that system (8.3.1) is uniquely solvable. If c_j are found then the function $\psi(w)$ can be found approximately on ℓ_w:

$$\psi(w) \simeq \sum_{j=0}^{M-1} c_j w^j, \quad w \in \ell_w,$$

and the function $G(k)$ on ℓ can be found approximately:

$$G(k) \approx \psi\left(\frac{k - iq}{k + iq}\right), \quad k \in \ell.$$

It is of interest to give error estimates for this procedure. This is a problem left for the reader.

8.3.3. The second method of solution

Let us use the classical idea of the Lagrange interpolation formula for entire functions. The interpolation formula is of the form:

$$G(k) = \cosh\{(b^2 - k^2)^{1/2}\} \sum_{j=0}^{\infty} \frac{(-1)^j \pi (j + \frac{1}{2})}{[\pi^2 (j + \frac{1}{2})^2 + b^2]^{1/2}}$$

$$\times \left(\frac{G(a_j)}{a_j - k} + \frac{G(-a_j)}{a_j + k} \right), \quad b > 0, \tag{8.3.2}$$

where $a_j = [(j + \frac{1}{2})^2 a^2 + b^2]^{1/2}$. By choosing $b > 0$ sufficiently large, one can get all the points a_j outside of any finite interval $[-\lambda_0, \lambda_0]$, $\lambda_0 > 0$ is arbitrary fixed.

It is assumed that $G(k)$ in (8.3.2) is of the form given in Problem A. If $k = \sigma + i\lambda$, $\sigma, \lambda \in \mathbb{R}$, then

$$|G(k)| \le \int_{-1}^{1} |g(x)| \exp(-\lambda x) dx$$

$$\le \|g\|_{L^2(-1,1)} \left(\frac{\sinh(2\lambda)}{\lambda} \right)^{1/2} \le \frac{c}{|\lambda|^{1/2}} \|g\| \exp(|\lambda|).$$

From the last inequality it follows that

$$|G(k)| \le \alpha(\lambda)\|g\| \exp(|\lambda|), \quad 0 < \alpha(\lambda) \le \begin{cases} c|\lambda|^{-1/2}, & |\lambda| \ge 1, \\ c, & |\lambda| \le 1. \end{cases}$$

It is also obvious that, with some $\epsilon(\sigma) \to 0$ as $\sigma \to \infty$,

$$|G(k)| \le c\epsilon(\sigma) \exp(|\lambda|) \to 0 \quad \text{as} \quad |\sigma| \to \infty, \quad k = \sigma + i\lambda. \tag{8.3.3}$$

Indeed, if $g \in C^1([-1,1])$, then (8.3.3) follows from an integration by parts with $\epsilon(\sigma) = o\left(\frac{1}{1+|\sigma|}\right)$. If $g \in L^2([-1,1])$ and $\epsilon > 0$ is arbitrary small, then there exists a $g_\epsilon \in C^1([-1,1])$ such that $\|g - g_\epsilon\|_{L^2([-1,1])} < \epsilon$. Thus, for $|\sigma| \ge 1$,

$$|G(k)| \le \left| \int_{-1}^{1} (g - g_\epsilon) \exp(ikx) dx \right| + \left| \int_{-1}^{1} g_\epsilon(x) \exp(ikx) dx \right|$$

$$\le c \exp(|\lambda|)[\epsilon + c(\epsilon)\sigma^{-1}]. \tag{8.3.4}$$

Formula (8.3.3) follows from (8.3.4). Note that $c(\epsilon) \to \infty$ when $\epsilon \to 0$ if $g(x) \notin C^1([-1,1])$.

Lemma 8.3.1. *Let $G(k) = \int_{-1}^{1} g(x)\exp(ikx)dx$, $g \in L^2([-1,1])$. Then (8.3.2) holds.*

Proof. Consider the rectangular contour C_{pd} which consists of the lines $\text{Im}z = \pm p\pi$, where p is an integer, and $\text{Re}z = \pm d$, where $d > 0$ is a large number to be taken to infinity. It follows from (8.3.3) that

$$\left| \int_{\text{Re}z=\pm d} G(z)\frac{dz}{(z-k)\cosh\{(b^2-z^2)^{1/2}\}} \right| \leq O\left(\frac{\epsilon(d)p}{d}\right)$$

$$\text{as} \quad d \to \infty, \quad |d - \pi(j+0.5)| \geq 0.1$$

One estimates the integrals

$$\left| \int_{|\text{Im}z|=p\pi} G(z)\frac{dz}{(z-k)\cosh\{(b^2-z^2)^{1/2}\}} \right| \leq O\left(\frac{d}{p}\right).$$

Here one uses the estimate $|G/\cosh\{(b^2-z^2)^{1/2}\}| \leq c$ for $z = \pm ip\pi + \sigma$, $\sigma p^{-1} \to 0$ as $p \to \infty$. Choose $d \to \infty$ and $p \to \infty$ such that $\epsilon(d)d^{-1}p + dp^{-1} \to 0$. This is clearly possible: for example, one can take $p = \epsilon^{-1/2}(d)d$. Then $\epsilon(d)d^{-1}p + dp^{-1} = 0(\epsilon^{1/2}(d)) \to 0$ as $d \to \infty$. Note that $\cosh\{(b^2-z^2)^{1/2}\}$ has simple zeros $a_j = \pm[b^2 + (j+\frac{1}{2})^2\pi^2]^{1/2}$, $j = 0, 1, 2, \ldots$.

Choosing $d \to \infty$ and $p = \epsilon^{-1/2}(d)d \to \infty$, one has

$$0 = \lim_{d\to\infty} \frac{1}{2\pi i} \int_{C_{pd}} \frac{G(z)dz}{(z-k)\cosh\{(b^2-z^2)^{1/2}\}}$$

$$= \sum \text{Res} \frac{G(z)}{(z-k)\cosh\{(b^2-z^2)^{1/2}\}}$$

$$= \frac{G(k)}{\cosh\{(b^2-k^2)^{1/2}\}} - \sum_{j=0}^{\infty} \left\{ \frac{G(a_j)}{a_j-k}\frac{(-1)^j\pi(j+\frac{1}{2})}{[\pi^2(j+\frac{1}{2})^2+b^2]^{1/2}} \right.$$

$$\left. + \frac{G(-a_j)}{a_j+k}\frac{(-1)^j\pi(j+\frac{1}{2})}{[\pi^2(j+\frac{1}{2})^2+b^2]^{1/2}} \right\}. \quad (8.3.5)$$

Formula (8.3.5) reduces to (8.3.2). Lemma 8.3.1 is proved. $\square$

8.3.4. The third method of solution

Let us use an integral equation used in the theory of antenna synthesis. Let $G(k) := b(k)$ if $k \in \ell$, $G(k) := a(k)$ if $k \in E$. By introducing the new variable $\kappa := k - ib$, one can reduce the problem to the case when $E_\kappa \subset \mathbb{R}_\kappa$ and $\ell_\kappa \subset \mathbb{R}_\kappa$, where $\mathbb{R}_\kappa$ is the real line on the κ complex plane. So, assume that $\ell_\kappa = [-\lambda_0, \lambda_0]$, $E_\kappa = \{\kappa, \text{Im}\kappa = 0, |\kappa| > \lambda_0\}$.

$$
\tilde{g}(\kappa) := \int_{-1}^{1} g(x)\exp(i\kappa x)dx = \begin{cases} A(\kappa), & |\kappa| > \lambda_0, \; \text{Im}\kappa = 0 \\ B(\kappa), & -\lambda_0 \le \kappa \le \lambda_0, \end{cases}
$$

$$
g \in L^2([-1,1]), \tag{8.3.6}
$$

the function $A(\kappa)$ is known, the function $B(\kappa)$ is to be found. Denote $\eta(x) := \begin{cases} 1, & \text{if } x \in [-1,1], \\ 0, & \text{if } x \notin [-1,1]. \end{cases}$ The left-hand side of (8.3.6) can be written as

$$
\int_{-\infty}^{\infty} g(x)\eta(x)\exp(i\kappa x)dx = \frac{1}{2\pi}\int_{-\infty}^{\infty} \tilde{\eta}(\kappa - \lambda)\tilde{g}(\lambda)d\lambda
$$

$$
= \int_{-\infty}^{\infty} \frac{\sin(\kappa - \lambda)}{\pi(\kappa - \lambda)}\tilde{g}(\lambda)d\lambda.
$$

Using (8.3.6) and the last equation, one gets

$$
\int_{-\lambda_0}^{\lambda_0} \frac{\sin(\lambda - \mu)}{\pi(\lambda - \mu)}B(\mu)d\mu + \int_{E_\kappa} \frac{\sin(\lambda - \mu)}{\pi(\lambda - \mu)}A(\mu)d\mu
$$

$$
= \begin{cases} B(\lambda), & -\lambda_0 \le \lambda \le \lambda_0, \\ A(\lambda), & |\lambda| > \lambda_0, \; \text{Im}\lambda = 0. \end{cases}
$$

This yields the integral equation for $B(\lambda)$:

$$
QB := \int_{-\lambda_0}^{\lambda_0} \frac{\sin(\lambda - \mu)}{\pi(\lambda - \mu)}B(\mu)d\mu = B(\lambda) - \phi(\lambda), \quad -\lambda_0 \le \lambda \le \lambda_0, \tag{8.3.7}
$$

where

$$
\phi(\lambda) := \int_{|\mu| > \lambda_0} \frac{\sin(\lambda - \mu)}{\pi(\lambda - \mu)}A(\mu)d\mu, \quad -\lambda_0 \le \lambda \le \lambda_0. \tag{8.3.8}
$$

Let us write (8.3.7) as

$$
B = QB + \phi(\lambda), \quad -\lambda_0 \le \lambda \le \lambda_0. \tag{8.3.9}
$$

This is a Fredholm type equation of the second kind. The operator Q is compact and self-adjoint in the Hilbert space $L^2([-\lambda_0, \lambda_0])$. The norm of Q is less than 1. It is equal to the first eigenvalue of Q. Let us recall that, for $\lambda_0 = 1$,

$$Q\psi_j = \lambda_j\psi_j, \quad 1 > \lambda_1(\lambda_0) > \lambda_2(\lambda_0) > \cdots > 0,$$

where ψ_j are prolate spheroidal functions. These functions and the eigenvalues λ_j are tabulated and well studied. The functions $\psi_j(\lambda)$ have an interesting and important double orthogonality property: they form an orthonormal basis in $L^2([-\lambda_0, \lambda_0])$ and they are orthogonal in $L^2(-\infty, \infty)$ as well. Since $0 < \lambda_1(\lambda_0) < 1$, Equation (8.3.9) can be uniquely solved by interations and convergence is guaranteed at the rate of the geometrical series with the denominator $\lambda_1(\lambda_0)$. Let us summarize the numerical procedure:

(1) Given $G(k)$ for $k \in E$, one puts $\kappa = k - ib$ and reduces Problem A to finding $G_1(\kappa) := G(\kappa + ib)$ on the set $|\kappa| \le \lambda_0$, $\mathrm{Im}\kappa = 0$, from the knowledge of $G_1(\kappa)$ on the set $|\kappa| > \lambda_0$, $\mathrm{Im}\kappa = 0$.

(2) Put $G_1(\kappa) := A(\kappa)$ for $|\kappa| > \lambda_0$, $\mathrm{Im}\kappa = 0$ and $G_1(\kappa) := B(\kappa)$ for $|\kappa| \le \lambda_0$, $\mathrm{Im}\kappa = 0$. Calculate $\phi(\lambda)$, $-\lambda_0 \le \lambda \le \lambda_0$, by formula (8.3.8) given $A(\mu)$.

(3) If $\phi(\lambda)$, $-\lambda_0 \le \lambda \le \lambda_0$, is calculated, solve Equation (8.3.9) by iterations:

$$B_{n+1}(\lambda) = QB_n(\lambda) + \phi(\lambda), \quad B_0(\lambda) = \phi(\lambda). \qquad (8.3.10)$$

This iterative process converges to the unique solution of Equation (8.3.9) at the rate of geometric series with the denominator $\lambda_1(\lambda_0)$.

In conclusion let us discuss briefly the resolution ability of the method. Put $\mu = \lambda_0\nu$, $\lambda = \lambda_0 t$, and $B(\mu) := \beta(\nu)$ in (8.3.7) to get

$$Q\beta = \int\limits_{-1}^{1} \frac{\sin \lambda_0(t-\nu)}{\pi(t-\nu)}\beta(\nu)d\nu. \qquad (8.3.11)$$

Thus the operator in Equation (8.3.9) is $Q = Q(\lambda_0)$, the eigenvalues of $Q(\lambda_0)$ are $\lambda_j(\lambda_0)$, $j = 1, 2, \ldots,$ $1 > \lambda_1(\lambda_0) > \lambda_2(\lambda_0) > \cdots > 0$. If $\lambda_0 \to \infty$ then $\lambda_1(\lambda_0) \to 1$, and Equation (8.3.10) yields very slow convergence. Let us give the values (with 5 digits) of $\lambda_1(\lambda_0)$ for some values of λ_0 : $\lambda_1(0.5) = 0.30969$, $\lambda_1(1) = 0.57258$, $\lambda_1(2) = 0.88056$, $\lambda_1(4) = 0.95589$, $\lambda_1(6) = 0.99990$, $\lambda_1(8) = 1.0000$. One can see that for $\lambda_0 \ge 6$ the iterative procedure (8.3.10) converges very slowly. This is another sign of the severe ill-posedness of the problem of analytic continuation of the limited angle data in the case when the angle ϑ_0 is small.

8.4. The Davison-Grünbaum algorithm

In this section we describe one more FBP-type algorithm for inversion of the limited-angle Radon transform data. Suppose that the unknown function f is from the class $L^2(B_{1/2})$ and vanishes for $|x| \geq 1/2$. Here $B_{1/2} \subset \mathbb{R}^n$ is the ball with radius $1/2$ centered at the origin. The problem is to recover f given its Radon transform $\hat{f}(\alpha_j, p)$ for J distinct (mod 2π) directions $\alpha_j, j = 1, \ldots, J$. Let us look for an approximation f_{DG} to f in the form

$$f_{DG}(x) := \sum_{j=1}^{J} \int_{\mathbb{R}} \varphi_j(\alpha_j \cdot x - p) \hat{f}(\alpha_j, p) dp. \qquad (8.4.1)$$

Substituting the definition of the Radon transform, we find

$$f_{DG}(x) = \sum_{j=1}^{J} \int_{\mathbb{R}} \varphi_j(\alpha_j \cdot x - p) \int_{\mathbb{R}^2} f(y) \delta(\alpha_j \cdot y - p) dy\, dp$$

$$= \int_{\mathbb{R}^2} f(y) \left[\sum_{j=1}^{J} \varphi_j(\alpha_j \cdot (x - y)) \right] dy.$$

Therefore,

$$f_{DG} = \Phi * f, \quad \Phi(x) = \sum_{j=1}^{J} \varphi_j(\alpha_j \cdot x). \qquad (8.4.2)$$

The idea of the algorithm is now clear: we choose the functions $\varphi_j, j = 1, \ldots, J$, so that Φ is close to the δ-function. More precisely, we fix a mollifier W_ϵ and find the functions φ_j such that

$$\|W_\epsilon - \Phi\|_{L^2(B_1)} \to \min. \qquad (8.4.3)$$

The error of the algorithm can be estimated using the Cauchy-Schwarz inequality

$$|f_{DG}(x) - (W_\epsilon * f)(x)| = |[(\Phi - W_\epsilon) * f](x)|$$

$$= \left| \int_{B_{1/2}} (\Phi - W_\epsilon)(x - y) f(y) dy \right|$$

$$\leq \left(\int_{B_{1/2}} (\Phi - W_\epsilon)^2 (x - y) dy \right)^{1/2} \|f\|_{L^2(B_{1/2})}.$$

First, we take the supremum over $x \in B_{1/2}$ on the right-hand side of the last inequality and take into account that a ball with radius $1/2$

centered at any point $x \in B_{1/2}$ is contained inside B_1. Second, we take the supremum over $x \in B_{1/2}$ on the left-hand side of the last inequality. This yields

$$\sup_{x \in B_{1/2}} |f_{DG}(x) - (W_\epsilon * f)(x)| \le \|W_\epsilon - \Phi\|_{L^2(B_1)} \|f\|_{L^2(B_{1/2})}.$$

Hence, if the quantity in (8.4.3) is small, the function f_{DG} will be close to $W_\epsilon * f$, the mollified version of f.

To solve problem (8.4.3), we use the results from Section 2.3. Lemma 2.3.1 asserts that the operator $R_\alpha : L^2(B_1) \to L^2([-1,1], \omega^{1-n})$, $\omega(p) = \sqrt{1 - p^2}$ is continuous. The adjoint operator to R_α is $(\tilde{R}_\alpha^* g)(x) = \omega^{1-n}(\alpha \cdot x) g(\alpha \cdot x)$. Denoting

$$R_J = (R_{\alpha_1}, \ldots, R_{\alpha_J})^T, \quad U = \omega^{n-1}(\varphi_1, \ldots, \varphi_J)^T,$$

where the superscript T denotes transposition, we get $\Phi = \tilde{R}_J^* U$, and Equation (8.4.3) is equivalent to

$$\|W_\epsilon - \tilde{R}_J^* U\|_{L^2(B_1)} \to \min.$$

The solution to this problem satisfies the system of normal equations

$$R_J \tilde{R}_J^* U = R_J W_\epsilon. \qquad (8.4.4)$$

To simplify (8.4.4), we use (2.3.8):

$$R_{\alpha_1} \tilde{R}_{\alpha_2}^* h_m = \psi_m(\alpha_1 \cdot \alpha_2) h_m, \quad h_m = \omega^{n-1} C_m^{(\frac{n}{2})}. \qquad (8.4.5)$$

Using properties of the Gegenbauer polynomials (see Section 14.4.3) and Equation (2.3.11), we represent the functions φ_j and the vector-function $R_J W_\epsilon$ as

$$(\omega^{n-1} \varphi_j)(p) = \sum_{m=0}^{\infty} a_{jm} h_m(p),$$

$$(R_J W_\epsilon)(p) = \sum_{m=0}^{\infty} h_m(p)(b_m(\alpha_1), \ldots, b_m(\alpha_J))^T;$$

$$(8.4.6)$$

$$b_m(\alpha_k) = \frac{1}{\|h_m\|} \int_{-1}^{1} \omega^{1-n}(p) h_m(p)(R_J W_\epsilon)(\alpha_k, p) dp$$

$$= \frac{1}{\|h_m\|} \int_{-1}^{1} C_m^{(\frac{n}{2})}(p)(R_J W_\epsilon)(\alpha_k, p) dp;$$

$$||h_m|| = \int_{-1}^{1} \omega^{1-n}(p)h_m^2(p)dp = \frac{\pi 2^{1-n}\Gamma(m+n)}{m!(m+\frac{n}{2})\Gamma^2(n/2)}.$$

Substituting (8.4.6) into (8.4.4) and using (8.4.5), we get that system (8.4.4) is equivalent to

$$\sum_{j=1}^{J} \psi_m(\alpha_k \cdot \alpha_j)a_{jm} = b_m(\alpha_k), \ 1 \le k \le J; \ m = 0,1,2,\ldots. \qquad (8.4.7)$$

Having solved systems (8.4.7) numerically, we find φ_j:

$$\varphi_j = \sum_{m=0}^{\infty} a_{jm}C_m^{(\frac{n}{2})}, \ 1 \le j \le J, \qquad (8.4.8)$$

which are then used in (8.4.1).

CHAPTER 9

INVERSION OF CONE-BEAM DATA

9.1. Inversion of the complete cone-beam data

Let f be a compactly supported C^1-function on $\mathbb{R}^n$. The cone-beam transform of f is defined by the formula

$$(Df)(x,\alpha) := g(x,\alpha) := \int_0^\infty f(x+t\alpha)dt, \ x \in \mathbb{R}^n, \ \alpha \in S^{n-1}. \quad (9.1.1)$$

Let L be a C^1-curve located outside the support of f. In view of the obvious relation between cone-beam and X-ray transforms

$$(Xf)(x,\alpha) = g(x,\alpha) + g(x,-\alpha), \quad (9.1.2)$$

we see that Theorem 8.1.1 implies that the knowledge of $g(x,\alpha)$, $x \in L, \alpha \in S^{n-1}$, determines f uniquely. However, depending on whether the curve L satisfies a certain condition or not, the inversion of the cone-beam data can be either mildly or severely ill-posed. Suppose that L satisfies the following condition: every plane intersecting $\mathrm{supp} f$ intersects also L. This condition can be stated as follows:

for any $(\alpha,p) \in S^{n-1} \times \mathbb{R}$ such that the plane $y \cdot \alpha = p$ intersects

$\mathrm{supp} f$, there exists $x \in L$, $x = x(\alpha,p)$, with the property

$$x(\alpha,p) \cdot \alpha = p. \quad (9.1.3)$$

If (9.1.3) holds, we say that the cone-beam data are complete, and in this case inversion procedures are only mildly ill-posed. The following result will be used in the sequel.

Lemma 9.1.1. *Let h be a distribution on $\mathbb{R}$. Suppose that h is positively homogeneous of degree $1-n$: $h(ts) = t^{1-n}h(s)$, $t > 0$. Then one has*

$$\int_{\mathbb{R}} \hat{f}(\alpha,p)h(p-\alpha \cdot x)dp = \int_{S^{n-1}} g(x,\beta)h(\alpha \cdot \beta)d\beta. \quad (9.1.4)$$

Proof. Substituting (9.1.1) into (9.1.4) and using that h is positively homogeneous of degree $1 - n$, we get

$$\int_{S^{n-1}} g(x, \beta) h(\alpha \cdot \beta) d\beta$$

$$= \int_{S^{n-1}} \int_0^{\infty} f(x + t\beta) t^{1-n} h(\alpha \cdot \beta) t^{n-1} dt \, d\beta$$

$$= \int_{\mathbb{R}^n} f(x + y) h(\alpha \cdot y) dy = \int_{\mathbb{R}} h(s) \int_{\alpha^{\perp}} f((\alpha \cdot x + s)\alpha + y^{\perp}) dy^{\perp} ds$$

$$= \int_{\mathbb{R}} h(s) \hat{f}(\alpha, s + \alpha \cdot x) ds,$$

and Equation (9.1.4) immediately follows. $\square$

Let us consider the practically important case $n = 3$. Using (9.1.4) and the Radon transform inversion formula (2.2.20), we will derive several different inversion formulas for the cone-beam transform under the assumption that condition (9.1.3) holds.

Let us take $h = \delta'$, which is a homogeneous of order -2 distribution. Thus, Lemma 9.1.1 is applicable in the case $n = 3$. Without loss of generality, we can assume that $\alpha = (0, 0, 1)$. Denoting $\beta = (\sin \beta_1 \cos \beta_2, \sin \beta_1 \sin \beta_2, \cos \beta_1)$, we get from (9.1.4):

$$-\hat{f}_p(\alpha, \alpha \cdot x) = \int_{S^{n-1}} g(x, \beta) \delta'(\alpha \cdot \beta) d\beta$$

$$= \int_0^{2\pi} \int_0^{\pi} g(x, \beta_1, \beta_2) \delta'(\cos \beta_1) \sin \beta_1 d\beta_1 \, d\beta_2$$

$$= \int_0^{2\pi} \int_{-1}^1 g(x, \arccos t, \beta_2) \delta'(t) dt \, d\beta_2$$

$$= \int_0^{2\pi} \frac{\partial}{\partial \beta_1} g(x, \beta_1, \beta_2) \Big|_{\beta_1 = \pi/2} d\beta_2$$

$$= -\int_{S^2 \cap \alpha^{\perp}} \mathcal{D}_{\alpha} g(x, \beta) d\beta,$$

where $\mathcal{D}_{\alpha}$ denotes the partial derivative of $g(x, \beta)$ with respect to the second argument along the direction α, and $S^2 \cap \alpha^{\perp}$ denotes equatorial

circle of the unit sphere which belongs to a plane perpendicular to α:
$S^2 \cap \alpha^\perp := \{\beta \in S^2 : \beta \cdot \alpha = 0\}$. Let p be chosen so that the plane
$\alpha \cdot y = p$ intersects supp$f$. Substituting $p = \alpha \cdot x$ and $x = x(\alpha, p)$ into
the last equation (this is possible because we assume that L satisfies
condition (9.1.3)), we get

$$\hat{f}_p(\alpha, p) = \int_{S^2 \cap \alpha^\perp} \mathcal{D}_\alpha g(x(\alpha, p), \beta) d\beta. \qquad (9.1.5)$$

Clearly, if the plane $\alpha \cdot y = p$ does not intersect supp$f$, then $\hat{f}(\alpha, p) = 0$.
Using Equations (9.1.5) and (2.2.20), we obtain an inversion formula:

$$f(x) = -\frac{1}{8\pi^2} \int_{S^2} \frac{\partial}{\partial p} \left[\int_{S^2 \cap \alpha^\perp} \mathcal{D}_\alpha g(x(\alpha, p), \beta) d\beta \right]\bigg|_{p = \alpha \cdot x} d\alpha. \qquad (9.1.6)$$

Now let a distribution h be defined as

$$h(p) := \frac{1}{2\pi} \int_{-\infty}^{\infty} |\sigma| e^{i\sigma p} d\sigma = -\frac{1}{\pi p^2}.$$

One easily checks that this is a homogeneous of degree -2 function and
that $h * g = -\mathcal{H}g'$, where $\mathcal{H}$ is the Hilbert transform and $g \in \mathcal{S}(\mathbb{R})$.
Therefore, Equation (9.1.4) yields

$$\frac{1}{\pi} \int_{-\infty}^{\infty} \frac{\hat{f}_p(\alpha, p)}{\alpha \cdot x - p} dp = \frac{1}{2\pi} \int_{S^2} g(x, \beta) \int_{-\infty}^{\infty} |\sigma| e^{i\sigma(\alpha \cdot \beta)} d\sigma \, d\beta$$

$$= \frac{1}{2\pi} \int_{S^2} \int_{0}^{\infty} [g(x, \beta) + g(x, -\beta)] e^{i\alpha \cdot (\sigma\beta)} \sigma d\sigma \, d\beta$$

$$= \frac{1}{2\pi} \int_{\mathbb{R}^3} (Xf)(x, \xi) e^{i\alpha \cdot \xi} d\xi = \frac{1}{2\pi} (\widetilde{Xf})(x, \alpha), \qquad (9.1.7)$$

where we have used Equation (9.1.2) and the fact that the cone-beam
transform can be extended from S^2 to $\mathbb{R}^3$ according to

$$g(x, y) = \frac{1}{|y|} g(x, y^0), \ y^0 = y/|y|, \ y \in \mathbb{R}^3 \ |y| > 0.$$

In (9.1.7), $\widetilde{Xf}$ denotes the Fourier transform of $(Xf)(x, y)$ with respect
to the second argument.

Substituting $h(s) = -1/s^2$ into (9.1.4) and integrating by parts, we get

$$\frac{1}{\pi} \int_{-\infty}^{\infty} \frac{\hat{f}_p(\alpha, p)}{\alpha \cdot x - p} dp = -\frac{1}{\pi} \int_{S^2} \frac{g(x, \beta)}{(\alpha \cdot \beta)^2} d\beta. \qquad (9.1.8)$$

Let the right-hand sides of Equations (9.1.7) and (9.1.8) be denoted $G(x, \alpha)$. Then

$$\frac{1}{\pi} \int_{-\infty}^{\infty} \frac{\hat{f}_p(\alpha, p)}{\alpha \cdot x - p} dp = G(x, \alpha).$$

Let $[a(\alpha), b(\alpha)]$ be the projection of the support of f onto direction α. Then the last equation is equivalent to the following one:

$$\frac{1}{\pi} \int_{a(\alpha)}^{b(\alpha)} \frac{\hat{f}_p(\alpha, p)}{\alpha \cdot x - p} dp = G(x(\alpha, t), \alpha), \quad a(\alpha) \le t \le b(\alpha). \qquad (9.1.9)$$

Condition (9.1.3) ensures existence of $x(\alpha, t)$ for $t \in [a(\alpha), b(\alpha)]$. Equation (9.1.9) is a singular integral equation with the Cauchy kernel, for which an inversion formula is known:

$$\hat{f}_p(\alpha, p) = \frac{\sqrt{(b(\alpha) - p)(p - a(\alpha))}}{\pi} \int_{a(\alpha)}^{b(\alpha)} \frac{\tilde{G}(t, \alpha)}{(t - p)\sqrt{(b(\alpha) - t)(t - a(\alpha))}} dt,$$

$$p \in [a(\alpha), b(\alpha)], \qquad (9.1.10)$$

where we have denoted $\tilde{G}(t, \alpha) = G(x(\alpha, t), \alpha)$. We have used the formula for a solution bounded on both ends because, as it follows from the results obtained in Theorem 4.3.1, the derivative of the three-dimensional Radon transform $\hat{f}_p(\alpha, p)$ is continuous with respect to p on the interval $[a(\alpha), b(\alpha)]$ and, therefore, is bounded as $p \to a(\alpha)$ and as $p \to b(\alpha)$. Note that we can use formula (9.1.10) only if

$$\int_{a(\alpha)}^{b(\alpha)} \frac{\tilde{G}(t, \alpha)}{\sqrt{(b(\alpha) - t)(t - a(\alpha))}} dt = 0.$$

This equation can be easily verified: dividing both sides of (9.1.9) by $\sqrt{(b(\alpha) - t)(t - a(\alpha))}$ and integrating with respect to t over the interval

$[a(\alpha), b(\alpha)]$, we obtain

$$\int\limits_{a(\alpha)}^{b(\alpha)} \frac{\tilde{G}(t,\alpha)}{\sqrt{(b(\alpha) - t)(t - a(\alpha))}} \, dt$$

$$= \int\limits_{a(\alpha)}^{b(\alpha)} \frac{1}{\sqrt{(b(\alpha) - t)(t - a(\alpha))}} \int\limits_{a(\alpha)}^{b(\alpha)} \frac{\hat{f}_p(\alpha, p)}{t - p} \, dp \, dt.$$

Changing the order of integration in the second integral in the last equation (this can be done according to the known result about the change of order of integration in a composition of a regular and a singular integrals, which can be applied because $\hat{f}_p(\alpha, p)$ is continuous with respect to p on $[a(\alpha), b(\alpha)]$) and using the well-known identity

$$\int\limits_a^b \frac{1}{(t - p)\sqrt{(b - p)(p - a)}} \, dp = 0, \ a < t < b,$$

one verifies the desired assertion.

Finally, substituting (9.1.10) into (2.2.20), we obtain two more *cone-beam inversion formulas*:

$$f(x) = -\frac{1}{8\pi^3} \int\limits_{S^2} \frac{\partial}{\partial p} \left[\sqrt{(b(\alpha) - p)(p - a(\alpha))} \right.$$

$$\left. \times \int\limits_{a(\alpha)}^{b(\alpha)} \frac{\tilde{G}(t,\alpha)}{(t - p)\sqrt{(b(\alpha) - t)(t - a(\alpha))}} dt \right] \Bigg|_{p=\alpha \cdot x} d\alpha \ , \quad (9.1.11)$$

where

$$\tilde{G}(t, \alpha) = \frac{1}{2\pi} (\widetilde{Xf})(x(\alpha, t), \alpha) \qquad (9.1.12a)$$

or

$$\tilde{G}(t, \alpha) = -\frac{1}{\pi} \int\limits_{S^2} \frac{g(x(\alpha, t), \beta)}{(\alpha \cdot \beta)^2} d\beta. \qquad (9.1.12b)$$

Recall that $\widetilde{Xf}$ denotes the Fourier transform of $(Xf)(x, y)$ with respect to the second argument.

9.2. Inversion of incomplete cone-beam data

In Section 9.1 we assumed that the cone-beam data $g(x, \beta)$, $x \in L, \beta \in S^{n-1}$, are known and the curve L satisfies condition (9.1.3).

Suppose now that condition (9.1.3) is violated and we know $g(x, \beta)$ for $x \in L, \beta \in \Omega$, where Ω is a proper subset of S^{n-1}. The problem we study is this: what are the conditions on L and Ω under which the cone-beam data determine f uniquely, and what are the inversion formulas that allow one to calculate f? Suppose that $f(x) = 0$ for $|x| \geq a$ and denote $D = \operatorname{supp} f$.

We state the results for $n = 3$. The simpler case $n = 2$ is discussed at the end of this section.

Let $l_\alpha := \{y : y = t\alpha, t \in \mathbb{R}, \alpha \in S^2\}$ be a straight line, π_α be the orthogonal projection onto l_α, $\pi_\alpha(D) = [a(\alpha), b(\alpha)]$, $a(\alpha) < b(\alpha)$. Let $D_\epsilon := \{x : \rho(x, D) < \epsilon\}$, $\rho(x, D)$ is the distance from x to D. Our basic assumption is:

there exists $\alpha_0 \in S^2$ and a neighborhood D_ϵ of D such that

$$\pi_{\alpha_0}(D_\epsilon) \subset \pi_{\alpha_0}(L). \tag{9.2.1}$$

In other words, assumption (9.2.1) means that there exists a direction α_0 such that the projection of L onto α_0 strictly contains the projection of D onto α_0.

Theorem 9.2.1. *Fix any α such that the projection of L onto α contains the projection of D onto α. Then one has:*

$$\hat{f}(\alpha, p) = \begin{cases} 0, & p < a(\alpha) \text{ or } p > b(\alpha), \\ \int\limits_{a(\alpha)}^{p} \int\limits_{S^2 \cap \alpha^\perp} \mathcal{D}_\alpha g(x(\alpha, \lambda), \beta) d\beta d\lambda, & a(\alpha) \leq p \leq b(\beta), \end{cases}$$

$$\tag{9.2.2}$$

where the notation is the same as in Equation (9.1.5).

Proof. Let $[a(\alpha), b(\alpha)]$ be the projection of the support of f onto direction α. Clearly, $\hat{f}(\alpha, p) = 0$ for $p \notin [a(\alpha), b(\alpha)]$. Since $f \in C^1(\mathbb{R}^3)$, the function $\hat{f}(\alpha, p)$ is continuous with respect to p, so $\hat{f}(\alpha, a(\alpha)) = 0$. Using the assumption of the theorem, we can integrate Equation (9.1.5) in p, which yields (9.2.2). Theorem 9.2.1 is proved. $\square$

Theorem 9.2.2. *Suppose condition (9.2.1) holds. Let $\Omega \subset S^2$ be an arbitrary small open set containing $S^2 \cap \alpha_0^\perp$:*

$$S^2 \cap \alpha_0^\perp \subset \Omega \subset S^2.$$

Then the data $g(x, \beta)$, $x \in L, \beta \in \Omega$, determine f uniquely.

Proof. Let α_0 be the direction from condition (9.2.1), i.e. $\pi_{\alpha_0}(D_\epsilon) \subset \pi_{\alpha_0}(L)$. By continuity, there exists a sufficiently small neighborhood ω of α_0 such that

(1) $\pi_\alpha(D_\epsilon) \subset \pi_\alpha(L) \; \forall \alpha \in \omega$, and
(2) $S^2 \cap \alpha^\perp \subset \Omega, \; \forall \alpha \in \omega$.

Let $K = \lambda\omega$ be the cone in $\mathbb{R}^3$: $K = \{\xi \in \mathbb{R}^3 : \xi = \lambda\alpha, \lambda \in \mathbb{R}, \alpha \in \omega\}$. The data $g(x,\beta)$, $x \in L$, $\beta \in \mathcal{N}$, determine uniquely $\hat{f}(\alpha,p)$ for $p\alpha \in K$ by formula (9.2.2). If $\hat{f}(\alpha,p)$ is known in K, then $\tilde{f}(\xi)$ is known in K as well (by the Fourier slice theorem $F_{p\to r}\hat{f}(\alpha,p) = \tilde{f}(r\alpha)$). Since $f(x)$ is compactly supported, $\tilde{f}(\xi)$ is an entire function, and the knowledge of $\tilde{f}(\xi)$ in K defines $\tilde{f}(\xi)$ uniquely in $\mathbb{R}^n$. Thus $f(x)$ is uniquely defined, and Theorem 9.2.2 is proved. $\square$

Theorem 9.2.3. *Under the assumptions of Theorem 9.2.2, the Fourier transform of f inside the cone $K = \lambda\omega$ can be computed as follows:*

$$\tilde{f}(t\alpha) = \frac{i}{t} \int_{a(\alpha)}^{b(\alpha)} \exp(ipt) \int_{S^2\cap\alpha^\perp} \mathcal{D}_\alpha g(x(\alpha,p),\beta)d\beta\, dp, \ t \neq 0, \alpha \in \omega.$$

$$(9.2.3)$$

Proof. Recall that $F_{p\to r}\hat{f}(\alpha,p) = \tilde{f}(r\alpha)$. Multiply (9.2.2) by e^{ipt} and integrate in p, take into account that $\hat{f}(\alpha,p) = 0$ for $p \leq a(\alpha)$ and $p \geq b(\alpha)$, integrate by parts in p and get (9.2.3). Theorem 9.2.3 is proved. $\square$

Formulas (9.2.2) and (9.2.3) allow one to calculate $\hat{f}(\alpha,p)$, $p \in \mathbb{R}$, $\alpha \in \omega$, and $\tilde{f}(\xi)$, $\xi \in K$, given the data $g(x,\beta)$, $x \in L$, $\beta \in \mathcal{N}$. We assume that ω is a set in S^2 which defines a convex cone. Calculation by formulas (9.2.2) and (9.2.3) involve two one-dimensional integrations over compact regions.

Given $\hat{f}(\alpha,p)$ for $p \in \mathbb{R}$ and $\alpha \in \omega$, one can calculate $f(x)$ analytically with an arbitrary accuracy. The corresponding formulas are given in Proposition 9.2.1 below and are similar to the ones proved in Section 8.2.

Pick any function $h(\lambda,\alpha)$, $\lambda \in \mathbb{R}$, $\alpha \in S^2$, with the properties

$$h(\lambda,\alpha) = h(\lambda,-\alpha) = h(-\lambda,\alpha), \tag{9.2.4}$$

$$h(\lambda,\alpha) = 0, \quad \lambda\alpha \notin K, \tag{9.2.5}$$

$$\mathcal{F}^{-1}h\big|_{x=0} = 1, \tag{9.2.6}$$

$$\left|\mathcal{F}^{-1}h\right| \leq c_m(1+|x|^2)^{-m} \quad \forall m > 0, \quad c_m = \text{const}. \tag{9.2.7}$$

For example, the desired function h can be constructed as follows. First note that (9.2.7) holds if $h \in C_0^\infty(K)$. Pick any function $\eta_1(\lambda)$ such that

$$\eta_1 \in C_0^\infty(\mathbb{R}), \quad \eta_1(-\lambda) = \eta_1(\lambda) \geq 0,$$

and $\eta_1(\lambda)$ vanishes with all derivatives at $\lambda = 0$. Take

$$\eta_2 \in C_0^\infty(\omega), \quad \eta_2(-\alpha) = \eta_2(\alpha) \geq 0,$$

and let

$$h(\lambda, \alpha) = c\eta_1(\lambda)\eta_2(\alpha),$$

where $c = \text{const} > 0$ is chosen so that (9.2.6) holds. Then conditions (9.2.4)–(9.2.7) are met.

Let

$$P_N(|x|^2) := \left(\frac{N}{4\pi a_1^2}\right)^{n/2}\left(1 - \frac{|x|^2}{4a_1^2}\right)^N, \quad n = 3, \qquad (9.2.8)$$

where $a_1 > a > 0$ is a number such that $f(x) = 0$ for $|x| > a$. Let Δ be the Laplacian in $\mathbb{R}^3$. Define

$$w_N(p, \alpha) := \frac{1}{2}(2\pi)^{-3}\int_{-\infty}^{\infty} \exp(i\lambda p)|\lambda|^2 P_N(-\Delta)h(\lambda, \alpha)d\lambda \qquad (9.2.9)$$

and

$$f_N(x) := \int_{S^2}\int_{-\infty}^{\infty} w_N(\alpha \cdot x - s, \alpha)\hat{f}(\alpha, s)ds\, d\alpha. \qquad (9.2.10)$$

From (9.2.5) and (9.2.9) we see that $w_N(p, \alpha) = 0$ for $\alpha \notin \omega$, so formula (9.2.10) uses only the data $\hat{f}(\alpha, p)$ for $\alpha \in \omega$. The following result is similar to Theorem 8.2.1:

Proposition 9.2.1. *If* $f \in C^1(\mathbb{R}^3)$ *and* $f(x) = 0$ *for* $|x| > a$, *then*

$$\sup_{x\in B_a} |f_N(x) - f(x)| \le cN^{-1/2}, \quad N \to \infty, \qquad (9.2.11)$$

where $c = \text{const} > 0$ *depends on* $\|f\|_{C^1(B_a)}$.

REMARK 9.2.1. If $f \in L^2(B_a)$ and $f(x) = 0$ for $|x| > a$, then $\|f_N(x) - f(x)\|_{L^2(B_a)} \to 0$ as $N \to \infty$.

Formulas (9.2.2) and (9.2.10) give a method for the analytical inversion of the incomplete cone-beam data with an arbitrary accuracy, provided the data are exact. If the data are noisy, then a stable inversion can be obtained by the method described in Section 8.1.2.

In the case $n = 2$, the data $g(x, \beta)$, $\beta \in S^1$, $x \in L$, give directly $\hat{f}(\beta^\perp, p)$. By continuity, a slight rotation of β leads to the direction α for which the basic condition (9.2.1) (with α in place of β) holds. Thus, one gets from the cone-beam data the Radon transform $\hat{f}(\alpha^\perp, p)$ data for which formulas (9.2.8) – (9.2.11) hold (with $n = 2$).

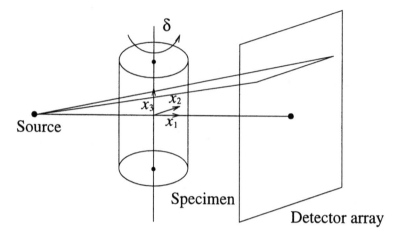

FIGURE 9.3.1. Figure illustrating the protocol, in which the solid cone of X-rays emanates from a point-source, illuminates a specimen, and projects on a detector array.

9.3. An exact reconstruction algorithm for the cone-beam circle geometry

9.3.1. Reconstruction algorithm

Let us consider the following protocol. The solid cone of X-rays emanates from a point-source, illuminates a specimen, and projects on a detector array. Schematically, the system is represented in Figure 9.3.1.

In this section we develop an exact inversion algorithm for the above protocol, which is convenient for numerical implementation, because the three-dimensional problem is reduced to a set of one-dimensional equations. However, since assumption (9.1.3) is violated, solving these one-dimensional equations is severely ill-posed.

Let us remark that since the specimen is rotated around its axis, the given scanning scheme is equivalent to the popular scheme in which sources are located on a circle and the specimen remains fixed. Thus the proposed algorithm can also be used for the reconstruction in the standard circle geometry.

Let us fix the following coordinate system (see Figure 9.3.1). The x_1-axis passes through the X-ray source and is perpendicular to both the body's axis of symmetry and the plane where detectors are located. The axis of symmetry of the body coincides with the x_3-axis, and the x_2-axis is parallel to the plane with detectors. Thus, the source is located at the point $x_0 = (-a, 0, 0)$, $a > 0$. We assume that the body is compactly supported and it is contained within a cylinder $C := \{(x_1, x_2, x_3) : x_1^2 + x_2^2 \leq R, -H_1 \leq x_3 \leq H_2\}$, where $0 < R < a$. Since the body is allowed to be rotated around the x_3-axis, the following data are available:

$$\int_0^\infty f(R_\delta(x_0 + \tau\alpha))d\tau = g_1(\alpha, \delta), \quad \alpha = (\alpha_1, \alpha_2, \alpha_3) \in S^2,$$

$$\alpha_1 > 0, \ 0 \leq \delta < 2\pi, \tag{9.3.1}$$

where S^2 is the unit sphere in $\mathbb{R}^3$, and R_δ is the rotation matrix:

$$R_\delta x := \begin{pmatrix} x_1 \cos\delta + x_2 \sin\delta \\ -x_1 \sin\delta + x_2 \cos\delta \\ x_3 \end{pmatrix}. \tag{9.3.2}$$

Since we assume $\alpha_1 > 0$, it is sufficient to integrate from 0 to $+\infty$ in τ in (9.3.1). Let $\alpha_1 = \cos\phi\cos\gamma$, $\alpha_2 = \sin\phi\cos\gamma$, $\alpha_3 = \sin\gamma$. Substituting (9.3.2) into (9.3.1) we obtain

$$\int_0^\infty f((-a + \tau\cos\phi\cos\gamma)\cos\delta + \tau\sin\phi\cos\gamma\sin\delta,$$

$$(a - \tau\cos\phi\cos\gamma)\sin\delta + \tau\sin\phi\cos\gamma\cos\delta, \tau\sin\gamma)d\tau = g_2(\phi, \delta, \gamma),$$
$$|\phi| < \pi/2, \ 0 \leq \delta < 2\pi, \ |\gamma| < \pi/2.$$

The change of variables $t = \tau\cos\gamma$ yields

$$\int_0^\infty f(-a\cos\delta + t\cos(\phi - \delta), a\sin\delta + t\sin(\phi - \delta), t\tan\gamma)dt$$

$$= g_2(\phi, \delta, \gamma)\cos\gamma. \tag{9.3.3}$$

In Section 9.3.2, a formula is given which relates the fan-beam data to the standard Radon transform. Following an approach from Section 9.3.2 (see below), let us fix any $\theta, p : \theta \in [0, 2\pi), |p| \leq R$, and find $\phi(\theta, p)$, $\delta(\theta, p)$ such that $\sin\phi = -p/a$, $\theta = \phi - \pi/2 - \delta \pmod{2\pi}$, $|\phi| < \pi/2$, $\delta \in [0, 2\pi)$. Simple transformations show that Equation (9.3.3) can be rewritten as

$$\int_0^\infty f(p\cos\theta - (t - \sqrt{a^2 - p^2})\sin\theta, p\sin\theta + (t - \sqrt{a^2 - p^2})\cos\theta, tz)dt$$

$$= g_3(\theta, p, z), \tag{9.3.4}$$

where $g_3(\theta, p, z) = g_2(\phi(\theta, p), \delta(\theta, p), \arctan z)(z^2 + 1)^{-1/2}$. Let us fix any $s > 0$, multiply Equation (9.3.4) by z^{s-1} and integrate with respect to

z over $[0, +\infty)$. Since $f(x_1, x_2, x_3) = 0$ for $x_3 > H_2$, actual integration with respect to z will be over finite interval $[0, \arctan(H_2/(a - R))]$. Changing the order of integration, we get

$$\int_0^\infty \tilde{f}(p \cos\theta - (t - \sqrt{a^2 - p^2}) \sin\theta,$$

$$p \sin\theta + (t - \sqrt{a^2 - p^2}) \cos\theta, s) t^{-s} dt = \tilde{g}(\theta, p, s),$$

$$(9.3.5)$$

where $\tilde{f}$ and $\tilde{g}$ are the Mellin transforms with respect to the third variable of f and g, respectively:

$$\tilde{f}(x_1, x_2, s) = \int_0^\infty f(x_1, x_2, z) z^{s-1} dz,$$

$$\tilde{g}(\theta, p, s) = \int_0^\infty g_3(\theta, p, z) z^{s-1} dz. \qquad (9.3.6)$$

Define $\hat{\alpha} = (\cos\theta, \sin\theta)$. Changing variables $\tau = t - \sqrt{a^2 - p^2}$ in (9.3.5) and transforming the resulting integral to the two-dimensional one, we find

$$\int_{\mathbb{R}^2} \tilde{f}(\hat{x}, s)[(\hat{x} \cdot \hat{\alpha}^\perp) + \sqrt{a^2 - (\hat{x} \cdot \hat{\alpha})^2}]^{-s} \delta(p - \hat{x} \cdot \hat{\alpha}) d\hat{x} = \tilde{g}(\theta, p, s), \quad (9.3.7)$$

where $\delta(\cdot)$ is the delta-function, $\hat{x} = (x_1, x_2)$, and $\hat{\alpha}^\perp = (-\sin\theta, \cos\theta)$. Note that we extended integration in (9.3.7) over $\mathbb{R}^2$, because $\tilde{f}(\hat{x}, s) = 0$ for $|\hat{x}| > R$, and the term in brackets in (9.3.7) is strictly positive for $|\hat{x}| \leq R$. Using the polar coordinate system in (9.3.7) and representing functions $\tilde{f}$ and $\tilde{g}$ as

$$\tilde{f}(\hat{x}, s) = \frac{1}{2\pi} \sum_{n=-\infty}^\infty f_{sn}(r) e^{-in\phi}, \quad \hat{x} = r(\cos\phi, \sin\phi), \qquad (9.3.8a)$$

$$\tilde{g}(\theta, p, s) = \frac{1}{2\pi} \sum_{n=-\infty}^\infty g_{sn}(p) e^{-in\theta}, \quad g_{sn}(p) = \int_0^{2\pi} \tilde{g}(\theta, p, s) e^{in\theta} d\theta,$$

$$(9.3.8b)$$

we obtain

$$\int_0^R \int_0^{2\pi} \sum_{n=-\infty}^\infty f_{sn}(r) e^{-in\phi} \left[r \sin(\phi - \theta) + \sqrt{a^2 - r^2 \cos^2(\phi - \theta)} \right]^{-s}$$

$$\cdot \delta(p - r\cos(\phi - \theta)) d\phi dr = \sum_{n=-\infty}^\infty g_{sn}(p) e^{-in\theta}.$$

Changing variables $\varphi = \phi - \theta$ in the inner integral, we find that the last equation separates into a set of one-dimensional equations

$$\int_p^R G_{sn}(p, r) f_{sn}(r) dr = g_{sn}(p), \quad 0 \le p \le R, \qquad (9.3.9)$$

where

$$G_{sn}(p, r) = r \int_0^{2\pi} e^{-in\varphi} \left[r \sin \varphi + \sqrt{a^2 - r^2 \cos^2 \varphi} \right]^{-s} \delta(p - r \cos \varphi) d\varphi$$

$$= \begin{cases} \dfrac{e^{-in \, \text{arccos}(p/r)}}{\sqrt{1-(p/r)^2}} \left[\sqrt{a^2 - p^2} + \sqrt{r^2 - p^2} \right]^{-s} \\ + \dfrac{e^{in \, \text{arccos}(p/r)}}{\sqrt{1-(p/r)^2}} \left[\sqrt{a^2 - p^2} - \sqrt{r^2 - p^2} \right]^{-s}, \quad 0 \le p < r, \\ 0, \hspace{6.5cm} p \ge r. \end{cases} \qquad (9.3.10)$$

As a result, we came to the following reconstruction algorithm, which is symbolically represented by a diagram:

$$g_3 \xrightarrow{(9.3.6)} \tilde{g} \xrightarrow{(9.3.8b)} g_{sn} \xrightarrow{(9.3.9)} f_{sn} \xrightarrow{(9.3.8a)} \tilde{f} \xrightarrow{(9.3.6)} f(x_1, x_2, x_3),$$

$$0 \le x_3 \le H_2, \qquad (9.3.11)$$

where the numbers above arrows denote the formulas used in the corresponding step. Clearly, the most ill-posed step in the above algorithm is $\tilde{f}(\hat{x}, s) \xrightarrow{(9.3.6)} f(x_1, x_2, x_3)$, $0 \le x_3 \le H_2$, which consists of inverting the Mellin transform. This is a manifestation of the fact that the given scanning geometry is incomplete.

Multiplying Equation (9.3.4) by $(-z)^{s-1}$, $s > 0$, and integrating over $(-\infty, 0]$, one obtains similarly to (9.3.5) - (9.3.10) the analogue of the algorithm (9.3.11) for the recovery of $f(x)$, $-H_1 \le x_3 \le 0$.

9.3.2. Geometry of the fan-beam data

Let us give an elementary procedure for rebinning the data in the plane fan-beam geometry to obtain the classical parallel beam geometry. Let D be a bounded domain on the plane, d be the diameter of D, the origin is assumed to be inside D. Let the source be at the point x_0 outside D, and let $|x_0| > d$. The last condition can be relaxed but we do not go into detail. The fan-beam data are the integrals $\int_{-\infty}^{\infty} f(x_0 + \tau\Theta) d\tau := g(\Theta)$, x_0 is fixed, where f is a piecewise-continuous function with support in D, $\Theta = (\cos \phi, \sin \phi)$, $0 \le \phi < 2\pi$. Fix a coordinate system in which $x_0 = (-a, 0)$. Assume that the domain D can be rotated so that the following data are observed

$$\int_{-\infty}^{\infty} f(R_\delta(x_0 + \tau\Theta)) d\tau = g(\phi, \delta),$$

where R_δ is the rotation matrix:

$$R_\delta x := \begin{pmatrix} x_1 \cos \delta + x_2 \sin \delta \\ -x_1 \sin \delta + x_2 \cos \delta \end{pmatrix}.$$

Thus

$$\int_{-\infty}^{\infty} f((-a + \tau \cos \phi) \cos \delta + \tau \sin \phi \sin \delta,$$

$$(a - \tau \cos \phi) \sin \delta + \tau \sin \phi \cos \delta) d\tau = g(\phi, \delta).$$

This can be written as

$$\int_{-\infty}^{\infty} f(-a \cos \delta + \tau \cos(\phi - \delta), a \sin \delta + \tau \sin(\phi - \delta)) d\tau = g(\phi, \delta). \quad (9.3.33)$$

Fix any θ, p : $\theta \in [0, 2\pi)$, $|p| \le a$, and find ϕ, δ such that $\sin \phi = p/a$, $\theta = \phi + \pi/2 - \delta$ (mod 2π). The Radon transform data are

$$\hat{f}(\theta, p) = \int_{-\infty}^{\infty} f(p \cos \theta + \tau \sin \theta, p \sin \theta - \tau \cos \theta) d\tau.$$

Since

$$\cos \delta = \cos(\phi + \pi/2 - \theta) = -\sin(\phi - \theta) = -(p/a) \cos \theta + (\cos \phi) \sin \theta,$$
$$\sin \delta = \sin(\phi + \pi/2 - \theta) = \cos(\phi - \theta) = (p/a) \sin \theta + (\cos \phi) \cos \theta,$$
$$\cos(\phi - \delta) = \cos(\theta - \pi/2) = \sin \theta, \ \sin(\phi - \delta) = \sin(\theta - \pi/2) = -\cos \theta,$$

substitution into (9.3.33) yields

$$g(\phi, \phi + \pi/2 - \theta)$$

$$= \int_{-\infty}^{\infty} f(p \cos \theta + (\tau - a \cos \phi) \sin \theta, p \sin \theta - (\tau - a \cos \phi) \cos \theta) d\tau$$

$$= \hat{f}(\theta, p),$$

where $\sin \phi = p/a$. Note that if $|p| > a$ then the line $\ell_{\theta p} := \{x : (\cos \theta, \sin \theta) \cdot x = p\}$ does not intersect D because of the assumption $a > d$, so $\hat{f}(\theta, p) = 0$ for $|p| > a$. Let us summarize the result.

Lemma 9.3.1. *If $|p| \le a$ and $\theta \in [0, 2\pi)$, then $\hat{f}(\theta, p) = g(\phi, \phi + \frac{\pi}{2} - \theta)$, $\phi = \arcsin(p/a)$, where the angle ϕ can be chosen so that $\phi \in [0, 2\pi)$ and $\delta = \phi + \frac{\pi}{2} - \theta \in [0, 2\pi)$.*

9.4. γ-ray tomography

9.4.1. Brief description of three different protocols

In Sections 9.1, 9.2, and 9.3 we considered one of the most popular protocols of cone-beam tomography: a source of x-rays moves along a curve in $\mathbb{R}^3$, and for each position of a source one collects the data $g(x, \beta)$ for all $\beta \in S^2$ or $\beta \in \Omega$, where Ω is a proper subset of S^2.

In this chapter three more problems are investigated. These problems deal with various data collecting protocols in the case when a source of radiation is fixed, and rays emanating from the source are situated on a surface of a cone with fixed opening. Clearly, the apex of the cone coincides with the source. Note that the above situation takes place in the case of γ-ray tomography, which is one of the few tomographies that can image explosives. Thus the results obtained in this section can be used, in particular, for tomographic inspection of luggage at airports.

In fact, radiation from the source emanates in all directions. However, the energy of γ-rays is angle-dependent, and the γ-rays which are sensitive to explosives are located on a cone with the opening slightly less then $180°$ degrees.

In each of the problems a three-parametric family of straight lines is given and the integrals over these lines of an unknown compactly supported function $f(x)$, $x := (x_1, x_2, x_3)$, are given. From these data $f(x)$ is to be recovered. Let us formulate the problems.

Problem 9.4.1. Let K be a conical beam (see Figure 9.4.1) described by the equation

$$a^2 x_1^2 = x_2^2 + x_3^2, \quad a = const > 0. \tag{9.4.1}$$

The opening angle of this cone is 2ω, $\tan \omega = a$. The rays which form the conical surface can be written parametrically as

$$x_1 = a^{-1} t, \quad x_2 = t \cos \phi, \quad x_3 = t \sin \phi. \tag{9.4.2}$$

The axis of symmetry of the cone is x_1-axis. Fix a ray, that is, fix $\phi = \phi_0$, $0 \le \phi_0 < 2\pi$. Let $x_0 \in D$,

$$x_0 = (a^{-1} t_0, \ t_0 \cos \phi_0, \ t_0 \sin \phi_0).$$

If the body D is translated along x_2-axis by $-\eta$ and along x_3-axis by $-\xi$, then its density is $f(x_1, x_2 + \eta, x_3 + \xi)$. Let

$$x' = x - x_0.$$

The x'-coordinate system has the origin at the point x_0 and the axes parallel to the axes of the x-coordinate system. Consider a rotation of

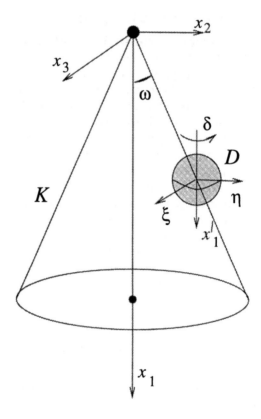

FIGURE 9.4.1. Figure illustrating Problems 9.4.1 and 9.4.2.

the body D about the x_1'-axis by an angle δ. The x'-coordinates are transformed by this rotation according to the formula

$$x_\delta' = \begin{pmatrix} 1 & 0 & 0 \\ 0 & \cos\delta & \sin\delta \\ 0 & -\sin\delta & \cos\delta \end{pmatrix} \begin{pmatrix} x_1' \\ x_2' \\ x_3' \end{pmatrix} = \begin{pmatrix} x_1' \\ x_2'\cos\delta + x_3'\sin\delta \\ -x_2'\sin\delta + x_3'\cos\delta \end{pmatrix}$$

$$= \begin{pmatrix} a^{-1}\tau \\ \tau\cos(\phi_0 - \delta) \\ \tau\sin(\phi_0 - \delta) \end{pmatrix}.$$

Here $\tau = t - t_0$ and x_0 is on the ray (9.4.2) with $\phi = \phi_0$ and $t = t_0$. Thus $x_1' = a^{-1}\tau$, $x_2' = \tau\cos\phi_0$, $x_3' = \tau\sin\phi_0$.

The density of the body D in the original coordinate system is

$$f(x_{01} + a^{-1}\tau, x_{02} + \eta + \tau\cos(\phi_0 - \delta), x_{03} + \xi + \tau\sin(\phi_0 - \delta)).$$

Define

$$\phi_0 - \delta := \psi,$$

then the data are the line integrals

$$g_1(\psi, \eta, \xi) := \int_{-\infty}^{\infty} f(x_{01} + a^{-1}\tau, x_{02} + \eta + \tau \cos \psi, x_{03} + \xi + \tau \sin \psi) d\tau.$$

(9.4.3)

Problem 9.4.1 consists of finding $f(x)$ *given data* $g_1(\psi, \eta, \xi)$ *for all* δ, η, ξ:
$0 \leq \delta < 2\pi$, $-\infty < \eta, \xi < \infty$.

Since $f(x_0 + x'_\delta) = 0$ for $|x_0 + x'_\delta| > R$ ($f(x)$ is compactly supported), the data g_1 vanish outside a finite region on the η, ξ plane. Therefore, practically one needs the data only for a finite region on the plane η, ξ.

REMARK 9.4.1. The above problem allows one to process simultaneously several bodies using one cone beam. Namely, if the diameter of the body is $2R$, the radius of the circle C_0 (by which the cone K intersects the plane $x_1 = x_{01}$) is ax_{01}, then the number of bodies that can be simultaneously tested in one cone-beam is $N \approx \frac{2\pi a x_{01}}{4R} = \frac{\pi a x_{01}}{2R}$.

We will prove that the data in Problem 9.4.1 determine $f(x)$ uniquely and give an analytical inversion formula.

Problem 9.4.2 consists of finding $f(x)$ *given the data*

$$g_2(\phi, \eta, \xi) = \int_{-\infty}^{\infty} f(x_{01} + a^{-1}t, x_{02} + \eta + t\cos\phi, x_{03} + \xi + t\sin\phi) dt. \quad (9.4.4)$$

Formula (9.4.4) is similar to (9.4.3) with $\psi = \phi$, that is, $\delta = 0$ and $\phi_0 = \phi$, where now ϕ is variable. Geometrically these data correspond to the case when the body D is rotated along the circle C_0 and translated along x_2 and x_3 axes by $-\eta$ and $-\xi$, respectively. We will prove that the data g_2 determine $f(x)$ uniquely and give an inversion formula.

Problem 9.4.3. The body D, which is modeled by a cylinder, has the axis of symmetry S perpendicular to K at the point of intersection of S and the cone. The body can be both translated and rotated along its axis S, and one measures integrals of f along all lines on the cone's surface passing through the apex and the body (see Figure 9.4.2 in Section 9.4.3 below). For such a protocol, we prove that the data specify f uniquely and describe two different approaches for the reconstruction of f. The density f is a function of three variables, no symmetry is assumed for f.

In Section 9.4.2, we obtain basic results concerning problems 9.4.1 and 9.4.2. In Section 9.4.3, a detailed discussion of Problem 9.4.3 is given. The results are summarized in theorems at the end of each section. In Section 9.4.4, a general lemma is proved. This lemma gives a symmetry condition sufficient for a separation of variables in tomography.

9.4.2. Uniqueness results and inversion formulas for Problems 9.4.1 and 9.4.2

Problem 9.4.1. Take the Fourier transform of (9.4.3) with respect to the variables η and ξ to get:

$$\check{g}_1(\psi, \lambda_2, \lambda_3) := \int\!\!\!\int_{-\infty}^{\infty} g_1(\psi, \eta, \xi) \exp\{i(\lambda_2\eta + \lambda_3\xi)\} d\eta d\xi$$

$$= \int_{-\infty}^{\infty} \check{f}(x_{01} + a^{-1}t, \lambda_2, \lambda_3) \exp\{-i\lambda_2(t\cos\psi + x_{02})$$

$$- i\lambda_3(t\sin\psi + x_{03})\} dt$$

$$= a\tilde{f}(-\hat{\lambda} \cdot \Psi a, \lambda_2, \lambda_3) \exp\left(ix_{01}\hat{\lambda} \cdot \Psi a - i\hat{\lambda} \cdot \hat{x}_0\right),$$

$$(9.4.5)$$

where

$$\lambda := (\lambda_1, \lambda_2, \lambda_3), \ \hat{\lambda} := (\lambda_2, \lambda_3), \ \hat{x}_0 := (x_{02}, x_{03}),$$

$$\Psi := (\cos\psi, \sin\psi), \ \tilde{f}(\lambda) := \int_{\mathbb{R}^3} f(x) \exp(i\lambda \cdot x) dx.$$

Formula (9.4.5) defines $\tilde{f}(\lambda)$ for all λ_2, λ_3 and $|\lambda_1| \leq a|\hat{\lambda}|$, $\hat{\lambda} = (\lambda_2, \lambda_3)$. This is a solid cone in the space $(\lambda_1, \lambda_2, \lambda_3)$ with the opening angle 2ω, where ω is the same angle as in Figure 9.4.1, $\tan\omega = a$. If $\omega \leq 0.44\pi$ (that is, $80° < \omega < 90°$), or $a > 5.6$ then the knowledge of the Fourier transform in the cone $\mathcal{K}_a := \{\lambda : \lambda \in \mathbb{R}^3, |\lambda_1| \leq a|\hat{\lambda}|\}$ allows one to find $f(x)$ numerically in a relatively stable way. Methods for doing this are given in Section 8.1.

We give a formula for finding $f(x)$:

$$f_N(x) = \frac{1}{(2\pi)^3} \int_{\mathcal{K}_a} \exp(-i\lambda \cdot x) h_N(\lambda)\tilde{f}(\lambda) d\lambda. \qquad (9.4.6)$$

Here

$$\|f_N - f\|_{L^2} \xrightarrow[N\to\infty]{} 0,$$

$$h_N(\lambda) = P_N(-\Delta_\lambda)\tilde{X}(\lambda),$$

$$P_N(-\Delta_\lambda) = \left(\frac{N}{4\pi R_1^2}\right)^{3/2} \left(1 + \frac{\Delta_\lambda}{4R_1^2}\right)^N, \ \Delta_\lambda = \sum_{j=1}^{3} \frac{\partial^2}{\partial\lambda_j}, \ R_1 > R,$$

$$\tilde{X}(\lambda) \in C_0^\infty(\mathcal{K}_a), \ \frac{1}{(2\pi)^3} \int_{\mathbb{R}^3} \tilde{X}(\lambda) d\lambda = 1, \ \tilde{X}(\lambda) = 0 \text{ if } \lambda \notin \mathcal{K}_a.$$

$$(9.4.7)$$

Problem 9.4.2. This problem is analytically identical to Problem 9.4.1 and is solved similarly. Taking the Fourier transform with respect to η and ξ, yields

$$\breve{g}_2(\phi, \hat{\lambda}) = \int\limits_{-\infty}^{\infty} \tilde{f}(x_{01} + a^{-1}t, \hat{\lambda}) \exp\left(-it\Phi \cdot \hat{\lambda} - i\hat{\lambda} \cdot \hat{x}_0\right) dt,$$

$$\Phi := (\cos\phi, \sin\phi).$$

Thus

$$\tilde{f}(-\hat{\lambda} \cdot \Phi a, \hat{\lambda}) = a^{-1} \breve{g}_2(\phi, \hat{\lambda}) \exp\left(i\hat{\lambda} \cdot \hat{x}_0 - ix_{01}a\hat{\lambda} \cdot \Phi\right). \qquad (9.4.8)$$

Formula (9.4.8) defines $\tilde{f}(\lambda)$ for all $\hat{\lambda} \in \mathbb{R}^2$ and $|\lambda_1| \le a|\hat{\lambda}|$. Thus, the Fourier transform $\tilde{f}(\lambda)$ is defined by the data in the cone $\mathcal{K}_a$. Therefore, the density f is recovered as in Problem 9.4.1.

Summarizing the results, we have

Theorem 9.4.1. *Problem 9.4.1 has at most one solution which can be obtained by formulas (9.4.5), (9.4.6), and (9.4.7).*

Theorem 9.4.2. *Problem 9.4.2 has at most one solution which can be obtained by formulas (9.4.8), (9.4.6), and (9.4.7).*

9.4.3. Investigation of Problem 9.4.3

Let us consider the cone K defined by (9.4.1) (see Figure 9.4.2) and assume that the body, which is modeled by a cylinder, has the axis of rotation S perpendicular to K at the point of intersection of the axis S and the cone. The source of γ-rays is situated at the apex of the cone, the body can be both translated along and rotated around the axis S, and one measures integrals of $f(x)$ along all lines on the cone's surface passing through the apex and the body. For such a protocol, we prove two results:

(1) uniqueness: the data specify f uniquely, and
(2) the inversion procedure can be reduced to a problem of solving infinitely many one-dimensional integral equations.

These results are obtained using two different approaches. In this section we use the following coordinate system: $x_3 \, (= S)$ is the axis of the cylinder directed along the normal to the cone's surface, the origin is situated at the intersection of the x_3-axis and the cone, the x_1-axis passes through the apex of the cone and belongs to the cone's surface, the x_2-axis lies in the plane tangent to the cone. Let the distance between the apex and the origin be d. We assume that the radius of the cylinder R is sufficiently small. Using this, fix $\delta > 0$ such that the disk $x_1^2 + x_2^2 \le (R + \delta)^2$ is situated strictly inside the projection of the cone on the (x_1, x_2) plane.

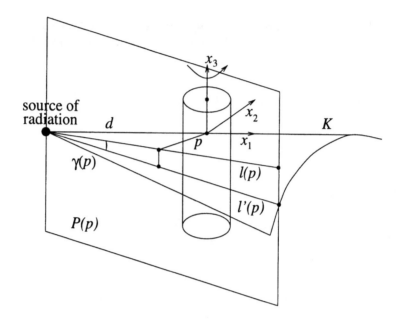

FIGURE 9.4.2. Figure illustrating Problem 9.4.3.

The first approach.

1. *Uniqueness.* Let (r, θ) be the cylindrical coordinate system in the (x_1, x_2) plane. Fix any p, $|p| \leq R + \delta$. Let us consider the plane $P(p)$ passing through the apex, parallel to the x_3-axis, such that $dist(P(p), x_3) = |p|$ and the points at the intersection of $P(p)$ and the cylinder have positive x_2-coordinates if $p > 0$ and negative x_2-coordinates if $p < 0$. Clearly there exists a unique plane $P(p)$ satisfying these conditions. Consider the line

$$\ell(p) : x_1 = \tau e_1 - d, \ \ x_2 = \tau e_2, \ \ x_3 = 0, \ \ \ell(p) \subset P(p),$$
$$e_1^2 + e_2^2 = 1, \ \ e_1 \geq 0, \ \ \tau \in \mathbb{R}.$$

One has $e_1 = e_1(p)$, $e_2 = e_2(p)$, where e_1 and e_2 are smooth functions of p, $|p| < R + \delta$. In fact, $e_1 = \sqrt{1 - (p/d)^2}$, $e_2 = p/d$. Similarly, for the line $\ell'(p) \subset P(p) \cap K$, which is the projection of $\ell(p)$ onto the cone's surface along the x_3-axis, we get

$$\ell'(p) : x_1 = \tau e_1' - d, \ \ x_2 = \tau e_2', \ \ x_3 = \tau e_3', \ \ \ell'(p) \subset P(p) \cap K,$$
$$(e_1')^2 + (e_2')^2 + (e_3')^2 = 1, \ \ e_1' \geq 0, \ \ e_3' \leq 0, \ \ \tau \in \mathbb{R},$$

where $e_1' = e_1(p)\gamma(p)$, $e_2' = e_2(p)\gamma(p)$, $e_3' = -\sqrt{1 - \gamma(p)^2}$, and $\gamma(p)$ is the cosine of the angle between lines $\ell(p)$ and $\ell'(p)$. We do not give explicit formula for $\gamma(p)$ because this is not necessary for the proof of the uniqueness result. Since the body can be arbitrarily rotated around and translated along the x_3-axis, simple reparametrization of the available rays in the (x_1, x_2) plane from the fan-beam to the parallel beam geometry (see Section 9.3.2) shows that the following data are known

$$\int_{-\infty}^{+\infty} f(\Theta^\perp \tau \gamma(p) + \Theta p, \tau e_3'(p) + h)d\tau = g_3(\theta, p, h), \tag{9.4.9}$$

$$\Theta = (\cos\theta, \sin\theta), \quad \Theta^\perp = (-\sin\theta, \cos\theta), \quad 0 \le \theta < 2\pi,$$
$$|p| \le R, \quad -\infty < h < +\infty.$$

Here $\tau e_3'(p)$ is the distance (vertical drop) between the point $(x_1, x_2, 0) \in \ell(p)$, $x_1 = \tau e_1(p)\gamma(p) - d$, $x_2 = \tau e_2(p)\gamma(p)$, and the corresponding point lying directly underneath it on $\ell'(p)$. Clearly, we have the usual property $g_3(\theta, -p, h) = g_3(\theta + \pi, p, h)$. Changing variables in (9.4.9), we get

$$\int_{-\infty}^{+\infty} f(\Theta^\perp \tau + \Theta p, z(\tau, p) + h)d\tau = g_3(\theta, p, h)\gamma(p), \tag{9.4.10}$$

where $z(\tau, p) = \tau e_3'(p)/\gamma(p)$. One can check that $\gamma(p)$ and $z(\tau, p)$ are smooth functions of $\tau \in \mathbb{R}$ and p, $|p| \le R + \delta$.

We wish to prove that (9.4.10) with $g_3 = 0$ implies $f = 0$. Suppose that the body is bounded along x_3-axis, i.e., $f(x_1, x_2, x_3) = 0$ for $|x_3| > Z$ for some $Z > 0$. Taking the Fourier transform with respect to h, we obtain from (9.4.10):

$$\int_{-\infty}^{+\infty} F(\Theta^\perp \tau + \Theta p, \lambda)e^{-i\lambda z(\tau,p)}d\tau = 0, \quad F(\cdot, \lambda) := \int_{-\infty}^{+\infty} f(\cdot, x_3)e^{ix_3\lambda}dx_3, \tag{9.4.11}$$

where we put $g_3 \equiv 0$. Since f is compactly supported, $F(\cdot, \lambda)$ is an entire function of λ, so it suffices to show that $F(\cdot, \lambda) \equiv 0$ for $\lambda \in [-\lambda_0, \lambda_0]$, where $\lambda_0 > 0$ is an arbitrary small fixed number. In what follows, the frequency λ will be fixed so we will drop it from arguments when it does not cause confusion. Equation (9.4.11) can be rewritten as follows

$$\hat{F}(\Theta, p) + \int_{\mathbb{R}^2} F(\hat{x})\epsilon(\hat{x} \cdot \Theta^\perp, \hat{x} \cdot \Theta, \lambda)\delta(p - \Theta \cdot \hat{x})d\hat{x} = 0, \tag{9.4.12}$$

$$\epsilon(\tau, p, \lambda) = \exp(-i\lambda z(\tau, p)) - 1, \tag{9.4.13}$$

where $\hat{x} := (x_1, x_2)$ and $\hat{F}(\Theta, p)$ is the Radon transform of $F(\hat{x})$. Since $f(\hat{x}, x_3) = 0$ for $|\hat{x}| > R$, Equation (9.4.12) will not change if we replace $\epsilon(\hat{x} \cdot \Theta^{\perp}, \hat{x} \cdot \Theta, \lambda)$ by $\epsilon_\chi(\hat{x} \cdot \Theta^{\perp}, \hat{x} \cdot \Theta, \lambda) = \epsilon(\hat{x} \cdot \Theta^{\perp}, \hat{x} \cdot \Theta, \lambda)\chi(|\hat{x}|^2)$, where $\chi \in C_0^\infty(\mathbb{R})$, $\chi(t) = 1$ for $|t| \leq R$ and $\chi(t) = 0$ for $|t| \geq R + \delta$. Since the cut-off function is rotationally symmetric, it does not depend on the choice of Θ. Taking the Fourier transform of (9.4.12) with respect to p, using the Fourier slice theorem, and replacing ϵ by ϵ_χ, we get

$$\tilde{F}(\nu\Theta) + \int_{\mathbb{R}^2} F(\hat{x})\epsilon_\chi(\hat{x} \cdot \Theta^{\perp}, \hat{x} \cdot \Theta, \lambda)e^{i\nu\Theta \cdot \hat{x}}d\hat{x} = 0,$$

$$\tilde{F}(\nu\Theta) := \int_{\mathbb{R}^2} F(\hat{x})e^{i\nu\Theta \cdot \hat{x}}d\hat{x}. \qquad (9.4.14a)$$

Denoting $\xi = \nu\Theta$, $\xi_0 = \Theta$, $\xi_0^{\perp} = \Theta^{\perp}$, we get

$$\tilde{F}(\xi) + \int_{\mathbb{R}^2} F(\hat{x})\epsilon_\chi(\hat{x} \cdot \xi_0^{\perp}, \hat{x} \cdot \xi_0, \lambda)e^{i\xi \cdot \hat{x}}d\hat{x} = 0.$$

Let $\epsilon_\chi(\tau, p, \lambda) = (2\pi)^{-2}\int_{\mathbb{R}^2}\tilde{\epsilon}_\chi(\mu, \theta, \lambda)e^{-i(\tau\mu + p\theta)}d\mu d\theta$. Then Equation (9.4.14a) yields

$$0 = \tilde{F}(\xi) + \int_{\mathbb{R}^2} \frac{1}{(2\pi)^2}\int_{\mathbb{R}^2}\tilde{F}(\eta)e^{-i\eta\hat{x}}d\eta$$

$$\times \frac{1}{(2\pi)^2}\int_{\mathbb{R}}\int_{\mathbb{R}}\tilde{\epsilon}_\chi(\mu, \theta, \lambda)e^{-i((\hat{x} \cdot \xi_0^{\perp})\mu + (\hat{x} \cdot \xi_0)\theta)}d\mu d\theta e^{i\xi \cdot \hat{x}}d\hat{x}$$

$$= \tilde{F}(\xi) + \frac{1}{(2\pi)^2}\int_{\mathbb{R}^2}\tilde{F}(\eta)\int_{\mathbb{R}}\int_{\mathbb{R}}\tilde{\epsilon}_\chi(\mu, \theta, \lambda)$$

$$\times \left[\frac{1}{(2\pi)^2}\int_{\mathbb{R}^2} e^{i\hat{x} \cdot (\xi - \eta - \mu\xi_0^{\perp} - \theta\xi_0)}d\hat{x} \right] d\mu d\theta d\eta$$

$$= \tilde{F}(\xi) + \frac{1}{(2\pi)^2}\int_{\mathbb{R}}\int_{\mathbb{R}}\tilde{\epsilon}_\chi(\mu, \theta, \lambda)\int_{\mathbb{R}^2}\tilde{F}(\eta)\delta(\xi - \eta - \mu\xi_0^{\perp} - \theta\xi_0)d\eta d\mu d\theta$$

$$= \tilde{F}(\xi) + \frac{1}{(2\pi)^2}\int_{\mathbb{R}}\int_{\mathbb{R}}\tilde{\epsilon}_\chi(\mu, \theta, \lambda)\tilde{F}(\xi - \mu\xi_0^{\perp} - \theta\xi_0)d\mu d\theta. \qquad (9.4.14b)$$

The last equation implies

$$\sup_{\xi \in \mathbb{R}^2}|\tilde{F}(\xi)| \leq \sup_{\xi \in \mathbb{R}^2}|\tilde{F}(\xi)|(2\pi)^{-2}\int_{\mathbb{R}}\int_{\mathbb{R}}|\tilde{\epsilon}_\chi(\mu, \theta, \lambda)|d\mu d\theta. \qquad (9.4.15)$$

Thus the uniqueness follows if we find λ_0 such that

$$\int_R \int_R |\bar{\epsilon}_\chi(\mu, \theta, \lambda)| d\mu d\theta < (2\pi)^2 \text{ for all } \lambda, \ |\lambda| \leq \lambda_0. \qquad (9.4.16)$$

One has

$$\int_R \int_R |\bar{\epsilon}_\chi(\mu, \theta, \lambda)| d\mu d\theta$$

$$\leq \left(\int_R \int_R |\bar{\epsilon}_\chi(\mu, \theta, \lambda)(1 + \mu^2 + \theta^2)|^2 d\mu d\theta \right)^{1/2}$$

$$\times \left(\int_R \int_R (1 + \mu^2 + \theta^2)^{-2} d\mu d\theta \right)^{1/2}$$

$$\leq (3\pi)^{1/2} \left(\int_R \int_R |\bar{\epsilon}_\chi(\mu, \theta, \lambda)|^2 d\mu d\theta + \int_R \int_R |\mu^2 \bar{\epsilon}_\chi(\mu, \theta, \lambda)|^2 d\mu d\theta \right.$$

$$\left. + \int_R \int_R |\theta^2 \bar{\epsilon}_\chi(\mu, \theta, \lambda)|^2 d\mu d\theta \right)^{1/2}$$

$$= (3\pi)^{1/2} (2\pi)^2 \left(\iint_{\tau^2 + p^2 \leq R^2} |\epsilon_\chi(\tau, p, \lambda)|^2 d\tau dp \right.$$

$$+ \iint_{\tau^2 + p^2 \leq R^2} |\frac{\partial^2}{\partial \tau^2} \epsilon_\chi(\tau, p, \lambda)|^2 d\tau dp + \iint_{\tau^2 + p^2 \leq R^2} \left. |\frac{\partial^2}{\partial p^2} \epsilon_\chi(\tau, p, \lambda)|^2 d\tau dp \right)^{1/2}, \qquad (9.4.17)$$

where we have used the Parseval identity. From (9.4.13), smoothness of $z(\tau, p)$ for $\tau^2 + p^2 \leq (R + \delta)^2$, and the choice of $\chi(t)$ it follows that each of the three terms on the right-hand side of (9.4.17) is of order $O(\lambda)$ as $\lambda \to 0$. This and (9.4.15) prove that there exists a λ_0 such that $\tilde{F}(\xi, \lambda) \equiv 0$ for all $\lambda, \ |\lambda| \leq \lambda_0$, for a sufficiently small λ_0. Since $\tilde{F}(\xi, \lambda)$ is an entire function, $f \equiv 0$.

2. Reduction to a series of one-dimensional equations. As in (9.4.11), we get from (9.4.10) for the case when $g_3 \not\equiv 0$:

$$\int_{-\infty}^{+\infty} F(\tau \Theta^\perp + p\Theta, \lambda) e^{-i\lambda z(\tau, p)} d\tau = \tilde{g}_3(\theta, p, \lambda),$$

$$\tilde{g}_3(\theta, p, \lambda) = \gamma(p) \int_{-\infty}^{+\infty} g_3(\theta, p, h) e^{ih\lambda} dh. \qquad (9.4.18)$$

Recall that $\Theta = (\cos\theta, \sin\theta)$. Let $\hat{x} := (x_1, x_2)$, $x_1 = r\cos\phi$, $x_2 = r\sin\phi$. Representing $F(x_1, x_2, \lambda)$ and $\tilde{g}_3(\Theta, p, \lambda)$ as

$$F(x_1, x_2, \lambda) = \sum_{n=-\infty}^{+\infty} \tilde{F}_{n,\lambda}(r)e^{in\phi}, \quad \tilde{g}_3(\theta, p, \lambda) = \sum_{n=-\infty}^{+\infty} \tilde{g}_{n,\lambda}(p)e^{in\theta},$$

$$(9.4.19)$$

we get from (9.4.18)

$$\tilde{g}_3(\theta, p, \lambda)$$

$$= \sum_{n=-\infty}^{+\infty} \tilde{g}_{n,\lambda}(p)e^{in\theta} = \int_{-\infty}^{+\infty} F(\tau\Theta^\perp + p\Theta, \lambda)e^{-i\lambda z(\tau, p)}d\tau$$

$$= \int_{\mathbb{R}^2} F(\hat{x}, \lambda)e^{-i\lambda z(\hat{x}\cdot\Theta^\perp, \hat{x}\cdot\Theta)}\delta(p - \hat{x}\cdot\Theta)d\hat{x}$$

$$= \sum_{n=-\infty}^{+\infty} \int_0^\infty \tilde{F}_{n,\lambda}(r)$$

$$\times \int_0^{2\pi} e^{in\phi}e^{-i\lambda z(r\sin(\phi-\theta), r\cos(\phi-\theta))}\delta(p - r\cos(\phi-\theta))d\phi r dr$$

$$= \sum_{n=-\infty}^{+\infty} \int_0^\infty \tilde{F}_{n,\lambda}(r)e^{in\theta} \int_0^{2\pi} e^{in\phi}e^{-i\lambda z(r\sin\phi, r\cos\phi)}\delta(p - r\cos\phi)d\phi r dr$$

$$= \sum_{n=-\infty}^{+\infty} e^{in\theta} \int_0^\infty G_{n,\lambda}(p, r)\tilde{F}_{n,\lambda}(r)dr, \qquad (9.4.20)$$

where

$$G_{n,\lambda}(p, r) = r\int_0^{2\pi} e^{in\phi}e^{-i\lambda z(r\sin\phi, r\cos\phi)}\delta(p - r\cos\phi)d\phi.$$

Since $z(\tau, p)$ is odd in τ and even in p, the above equation implies

$$G_{n,\lambda}(p, r) = 2r\text{Re}\left\{\int_0^\pi e^{in\phi}e^{-i\lambda z(r\sin\phi, r\cos\phi)}\delta(p - r\cos\phi)d\phi\right\}$$

$$= 2r\text{Re}\left\{\int_0^1 e^{in\arccos(t)}e^{-i\lambda z(r\sqrt{1-t^2}, rt)}\delta(p - rt)\frac{dt}{\sqrt{1-t^2}}\right\}$$

$$= 2\text{Re}\left\{\frac{e^{in\arccos(p/r)}e^{-i\lambda z(\sqrt{r^2-p^2}, p)}}{\sqrt{1-(p/r)^2}}\right\}, \quad |p| < r,$$

and $G_{n,\lambda}(p,r) = 0$, $|p| \geq r$. Taking the real part yields

$$G_{n,\lambda}(p,r) = 2\frac{T_n(p/r)}{\sqrt{1-(p/r)^2}} \cos\left(\lambda z(\sqrt{r^2-p^2},p)\right)$$

$$+ 2U_{n-1}(p/r)\sin\left(\lambda z(\sqrt{r^2-p^2},p)\right), \quad |p| < r,$$

$$G_{n,\lambda}(p,r) = 0, \quad |p| \geq r, \tag{9.4.21}$$

where T_n and U_n are the Chebyshev polynomials of the first and second kind, respectively. From (9.4.20) we obtain equations for $\tilde{F}_{n,\lambda}(r)$:

$$\tilde{g}_{n,\lambda}(p) = \int_p^\infty G_{n,\lambda}(p,r)\tilde{F}_{n,\lambda}(r)dr, \tag{9.4.22}$$

which are to be solved for all $n,\lambda:\ n \in \mathbb{Z}, \lambda \in \mathbb{R}$.

REMARK 9.4.2. The fact that we were able to reduce the three-dimensional problem to a set of one-dimensional equations is a consequence of the special symmetry of the problem. More precisely, we have the following property of the protocol described at the beginning of this section. Let (r,θ) be the cylindrical coordinate system in the (x_1,x_2) plane, and let T be a linear operator mapping the unknown density $f(\theta,r,x_3)$ to the data set $g(\alpha,p,z)$. Then, clearly, one has

$$Tf(\theta+\psi,r,x_3+h) = g(\alpha+\psi,p,z+h), \quad \forall \psi, h:\ 0 \leq \psi < 2\pi,\ h \in \mathbb{R}.$$

Using only this property without the explicit formula for T, one can show that the procedure for solving the equation $Tf = g$ for f can be reduced to a set of one-dimensional equations of the type $\tilde{T}\tilde{f}(r) = \tilde{g}(p)$. The proof of this result in a fairly general setting is given in Section 9.4.4.

REMARK 9.4.3. In the proof of the uniqueness result we have used the smallness of λ. However, for any fixed λ, one can prove the uniqueness using smallness of ϵ in the sense (9.4.16). This smallness can be established if the curvature of the conical surface is sufficiently small, or the parameter R/d is sufficiently small. Here R is the radius of the body and d is the distance from the body to the apex of the cone.

REMARK 9.4.4. The analogue of (9.4.14b) for the case $g_3 \neq 0$ is

$$\tilde{F}(\xi,\lambda) + \frac{1}{(2\pi)^2}\int_\mathbb{R}\int_\mathbb{R}\tilde{\epsilon}_\chi(\mu,\theta,\lambda)\tilde{F}(\xi-\mu\xi_0^\perp-\theta\xi_0,\lambda)d\mu d\theta$$

$$= \int_{-\infty}^{+\infty}\tilde{g}_3(\xi/|\xi|,p,\lambda)e^{i|\xi|p}dp. \tag{9.4.14c}$$

If we find λ_0 such that (9.4.16) holds, then the norm of the operator

$$E(\tilde{F}) := \frac{1}{(2\pi)^2} \int_{\mathbb{R}} \int_{\mathbb{R}} \tilde{\epsilon}_\chi(\mu, \theta, \lambda) \tilde{F}(\xi - \mu\xi_0^\perp - \theta\xi_0, \lambda) d\mu d\theta,$$

$$E : \; C(\mathbb{R}^2) \to C(\mathbb{R}^2),$$

is less then 1, and we can compute $\tilde{F}(\xi, \lambda)$ for $|\lambda| \leq \lambda_0$ from (9.4.14c) using the method of successive iterations (by the contraction mapping theorem).

The second approach.

1. Uniqueness. Taking $\lambda = 0$ in (9.4.11), we get

$$\int_{-\infty}^{+\infty} F(\Theta^\perp \tau + \Theta p, 0) d\tau = 0, \quad \forall \Theta, p: \; \Theta \in S^1, \; p \in \mathbb{R}.$$

Since the Radon transform is injective, we get $F(\hat{x}, 0) \equiv 0$. We now use an induction argument. Suppose we have already proved that

$$F_\lambda^{(k)}(\hat{x}, 0) := \frac{\partial^k F(\hat{x}, \lambda)}{\partial \lambda^k}\bigg|_{\lambda=0} \equiv 0, \quad k = 0, \dots, n-1. \qquad (9.4.23)$$

We wish to prove that $F_\lambda^{(n)}(\hat{x}, 0) \equiv 0$. Differentiating (9.4.11) n times with respect to λ and taking $\lambda = 0$, we get

$$\int_{-\infty}^{+\infty} F_\lambda^{(n)}(\Theta^\perp \tau + \Theta p, 0) d\tau$$

$$+ \sum_{k=0}^{n-1} \binom{n}{k} \int_{-\infty}^{+\infty} F_\lambda^{(k)}(\Theta^\perp \tau + \Theta p, 0)(-iz(\tau, p))^{n-k} d\tau = 0, \qquad (9.4.24)$$

where $\binom{n}{k}$ are binomial coefficients. Using (9.4.23) and injectivity of the Radon transform, we conclude from (9.4.24) that $F_\lambda^{(n)}(\hat{x}, 0) \equiv 0$. Thus, by induction, $F_\lambda^{(k)}(\hat{x}, \lambda) \equiv 0 \; \forall k \geq 0$. Since $F(\hat{x}, \lambda)$ is an entire function in λ, this implies that $F(\hat{x}, \lambda) \equiv 0$, and therefore, $f(x) \equiv 0$.

2. A reconstruction algorithm. Using the analogue of (9.4.24) for the case $g_3 \not\equiv 0$, we obtain

$$\int_{-\infty}^{+\infty} F_\lambda^{(n)}(\Theta^\perp \tau + \Theta p, 0) d\tau = \gamma(p) \frac{\partial^n g_3(\Theta, p, \lambda)}{\partial \lambda^n}\bigg|_{\lambda=0}$$

$$- \sum_{k=0}^{n-1} C_k^n \int_{-\infty}^{+\infty} F_\lambda^{(k)}(\Theta^\perp \tau + \Theta p, 0)(-iz(\tau, p))^{n-k} d\tau =: g_n(\Theta, p). \qquad (9.4.25)$$

If $n = 0$ we assume that there is no summation on the right-hand side of (9.4.25). Applying the inverse two-dimensional Radon transform R^{-1} to both sides of (9.4.25), we get

$$F_\lambda^{(n)}(\hat{x}, 0) = R^{-1} g_n. \tag{9.4.26}$$

Note that each successive derivative $F_\lambda^{(n)}$ is recovered from all previous derivatives $F_\lambda^{(k)}$, $k = 0, \ldots, n-1$. Using (9.4.26) and the absolute convergence of the Taylor series expansions of entire functions, we get

$$f(\hat{x}, x_3) = \frac{1}{2\pi} \int_{-\infty}^{+\infty} \left(\sum_{n=0}^{\infty} \frac{F_\lambda^{(n)}(\hat{x}, 0)}{n!} \lambda^n \right) e^{-i\lambda x_3} d\lambda$$

$$= \frac{1}{2\pi} \int_{-\infty}^{+\infty} \left(\sum_{n=0}^{\infty} \frac{(R^{-1} g_n)(\hat{x})}{n!} \lambda^n \right) e^{-i\lambda x_3} d\lambda. \tag{9.4.27}$$

The results obtained in this section prove

Theorem 9.4.3. *Problem 9.4.3 has at most one solution which can be obtained from (9.4.22), (9.4.21), (9.4.11), (9.4.18), and (9.4.19) or by formulas (9.4.27) and (9.4.25).*

9.4.4. Sufficient condition for a linear operator to be a convolution

Let X be a linear topological space of functions defined on $\mathbb{R}^n$, and $C_0^\infty(\mathbb{R}^n)$ be a dense subset in X. Let $T : X \to Y$ be a linear operator, where Y is a linear space of functions defined on $\mathbb{R}^n$. Assume that

$$TF(x + tu) = G(y + tu), \tag{9.4.28}$$

for all $t \in \mathbb{R}^1$, $x \in \mathbb{R}^n$, $F \in C_0^\infty(\mathbb{R}^n)$, and some fixed $u \in \mathbb{R}^n$, $|u| = 1$, that is

$$\int_{\mathbb{R}^n} T(x, y) F(x + tu) dx = G(y + tu) \quad \forall t \in \mathbb{R}^1, \ \forall x \in \mathbb{R}^n, \ \forall F \in C_0^\infty(\mathbb{R}^n). \tag{9.4.29}$$

Lemma 9.4.1. *If (9.4.29) holds, then*

$$T(x, y) = A(\hat{x}, \hat{y}; r - s)$$

where

$$r := x \cdot u, \quad \hat{x} := x - ru, \quad s := y \cdot u, \quad \hat{y} := y - su,$$

and $x \cdot u$ is the dot product.

Corollary. *Consider the equation*

$$Tf = g. \tag{9.4.30}$$

If (9.4.28) holds, the last equation can be written as

$$\int_{\mathbb{R}} \int_{\mathbb{R}^{n-1}} A(\hat{x}, \hat{y}, r - s) f(\hat{x}, r) d\hat{x} dr = g(\hat{y}, s).$$

Taking the Fourier transform with respect to s yields

$$\int_{\mathbb{R}^{n-1}} \tilde{A}(\hat{x}, \hat{y}; -\lambda) \tilde{f}(\hat{x}, \lambda) d\hat{x} = \tilde{g}(\hat{y}, \lambda), \tag{9.4.31}$$

where

$$\tilde{g}(\hat{y}, \lambda) := \int_{-\infty}^{\infty} e^{i\lambda s} g(\hat{y}, s) ds.$$

Thus, the solution of Equation (9.4.30) in $\mathbb{R}^n$ is reduced to the solution of infinitely many Equations (9.4.31) in $\mathbb{R}^{n-1}$, λ in (9.4.31) is a parameter.

Proof of Lemma 9.4.1. Let $y + tu = \hat{y} + (s+t)u$, $x + tu = \hat{x} + (z+t)u$. Write (9.4.29) as

$$\int_{\mathbb{R}} \int_{\mathbb{R}^{n-1}} T(\hat{x}, \hat{y}; r, s) F(\hat{x}, r + t) d\hat{x} dr = G(\hat{y}, s + t).$$

Let $\eta := s + t$, $\xi := r + t$. Then

$$\int_{\mathbb{R}} \int_{\mathbb{R}^{n-1}} T(\hat{x}, \hat{y}; \xi - t, \eta - t) F(\hat{x}, \xi) d\hat{x} d\xi = G(\hat{y}, \eta), \quad \forall t \in \mathbb{R}^1. \tag{9.4.32}$$

Since (9.4.32) holds for all $t \in \mathbb{R}^1$, one can put $t = \eta$ in (9.4.32) and get

$$\int_{\mathbb{R}} \int_{\mathbb{R}^{n-1}} A(\hat{x}, \hat{y}; \xi - \eta) F(\hat{x}, \xi) d\hat{x} d\xi = G(\hat{y}, \eta),$$

where

$$A(\hat{x}, \hat{y}; \xi) := T(\hat{x}, \hat{y}; \xi, 0), \quad \xi \in \mathbb{R}^1, \quad \hat{x}, \hat{y} \in \mathbb{R}^{n-1}.$$

The proof is complete. $\square$

REMARK 9.4.5. The proof shows that condition (9.4.28) is equivalent to T being a convolution with respect to the argument which is the Cartesian coordinate along the u-axis.

REMARK 9.4.6. It is easy to formulate an analogue of Lemma 9.4.1 for functions periodic in one of the variables and operators T which satisfy a condition similar to (9.4.28) in this variable.

$$TF(p, \theta - \tau) = G(\mu, \phi - \tau) \quad \forall \tau \in [0, 2\pi).$$

CHAPTER 10

RADON TRANSFORM OF DISTRIBUTIONS

10.1. Main definitions

In this section we define the Radon transform on distributions using the duality relation

$$(Rf, \mu) = < f, R^*\mu > . \qquad (10.1.1)$$

Recall that $(\cdot, \cdot)$ and $< \cdot, \cdot >$ denote inner products in $L^2(Z)$ and $L^2(\mathbb{R}^n)$, respectively. The function μ in (10.1.1) is always assumed to be real-valued and even. Since $R^*\mu$ may not decay fast at infinity even for $\mu \in C_0^\infty(Z)$ (see Corollary 3.1.1 and Example 3.3.1), the definition based on formula (10.1.1) makes sense only for distributions f with compact support (or decaying sufficiently fast at infinity) if μ runs through all of $S_e(Z)$, the space of even $S(Z)$ functions. To overcome this difficulty, we will use identity (2.2.11'): $R^*KR = I$, which holds on functions from $S(\mathbb{R}^n)$. Suppose we want to define R on distributions from X', i.e. on linear bounded functionals acting on a space of functions X. It is convenient to take μ from the space KRX. Indeed, if the identity $R^*KR = I$ holds on X, then for μ of the form $\mu = KR\varphi$, we have $R^*\mu = R^*KR\varphi = \varphi \in X$, and the expression $< f, R^*\mu >$ is well defined for any $f \in X'$, while $R^*\mu$ runs through the whole X. Consider now two important examples. Taking $X = S := S(\mathbb{R}^n)$, that is $h \in KRS$, formula (10.1.1) defines the Radon transform on S'. Similarly, if we take $X = D := C_0^\infty(\mathbb{R}^n)$, formula (10.1.1) defines the Radon transform on D'. Since the identity $R^*KR = I$ holds on S and D, we introduce

Definition 10.1.1. Fix any $f \in S'$ (or, $f \in D'$). The Radon transform Rf is a linear bounded functional acting on the space KRS (or, KRD) of test functions according to the formula $(Rf, \mu) := < f, R^*\mu >$, that is the value of the functional Rf on a test function μ is defined to be $< f, R^*\mu >$.

If we take $\mu \in C^\infty(Z)$, then identity (10.1.1) defines R also on $\mathcal{E}'$, where $\mathcal{E} := C^\infty(\mathbb{R}^n)$. Clearly, $R^*\mu \in \mathcal{E}$ if $\mu \in C^\infty(Z)$. Next, we have to show that $R^*C^\infty(Z)$ is dense in $\mathcal{E}$ in the topology of $\mathcal{E}$: given any

$\varphi \in \mathcal{E}$, any compact set $U \subset \mathbb{R}^n$, any integer $N > 0$, and any $\epsilon > 0$, there exists $\mu \in C^\infty(Z)$ such that

$$\sup_{x \in U, |j| \leq N} |D^j \varphi - D^j R^* \mu| \leq \epsilon, \tag{10.1.2}$$

where D^j denotes a partial derivative of order $|j|$. Fix a cut-off function $\psi \in C_0^\infty(\mathbb{R}^n)$ such that $\psi \equiv 1$ on U. According to the results from Section 2.5, there exists a unique even solution $\mu_e \in C^\infty(Z)$ to the equation $R^* \mu = \psi \varphi$. Since $\psi \equiv 1$ on U, then $R^* \mu_e \equiv \varphi$ on U and (10.1.2), clearly, holds.

Definition 10.1.2. Fix any $f \in \mathcal{E}'$. The Radon transform Rf is a linear bounded functional acting on the space of even $C^\infty(Z)$ test functions according to the formula $(Rf, \mu) := <f, R^* \mu>$.

Exercise 10.1.1. It is well-known that $\mathcal{E}' \subset \mathcal{S}'$ (see, e.g. [Hor]). Show that if $f \in \mathcal{S}'$ and f is compactly supported, then Definitions 10.1.1 and 10.1.2 are equivalent.

Hint. Suppose that supp $f \subset B_a$. Fix any $\epsilon > 0$. In both definitions, the action of the functional Rf on μ depends only on values of $R^* \mu$ inside $B_{a+\epsilon}$. The desired assertion follows from the inclusions $C_0^\infty(B_{a+\epsilon}) \subset R^* K R \mathcal{S} = \mathcal{S}$ and $C_0^\infty(B_{a+\epsilon}) \subset R^* C^\infty(Z)$.

Using the Plancherel identity (2.1.43), we introduce

Definition 10.1.3. Fix any $f \in \mathcal{S}'$ (or, $f \in \mathcal{D}'$). The Radon transform Rf is a linear bounded functional acting on the space $K_1 R \mathcal{S}$ (or, $K_1 R \mathcal{D}$) of test functions according to the formula

$$<f, h> := \frac{i^n (n-1)!}{(2\pi)^n} \int_{S^{n-1}} d\alpha \int_{-\infty}^{\infty} \int_{-\infty}^{\infty} dp_1 dp_2 \frac{\hat{f}(\alpha, p_1) \hat{h}(\alpha, p_2)}{(p_1 - p_2 + i0)^n}$$
$$:= (Rf, K_1 Rh), \tag{10.1.3}$$

where the operator K_1 is defined in (2.1.51).

Using Equations (2.1.52) and (2.1.52'), we get

$$<f, h> = (Rf, K_1 Rh) = (Rf, KRh).$$

The above identity means that if h runs through X, then the space of test functions in (10.1.3) is KRX, and Definitions 10.1.1 and 10.1.3 are, in fact, equivalent.

Thus, we have defined the Radon transform on the three main classes of distributions: $\mathcal{S}'$, $\mathcal{D}'$, and $\mathcal{E}'$. The presented argument leads to the following result.

Theorem 10.1.1. *The Radon transform, defined on S', D', and $\mathcal{E}'$, is injective.*

Proof. If $Rf \equiv 0, f \in S'$, then (10.1.1) implies $< f, R^*\mu >= 0$, $\mu = KR\phi$, $\forall \phi \in S$. Since $R^*KR\phi = \phi$, one gets $< f, \phi >= 0$ $\forall \phi \in S$. Thus $f \equiv 0$. A similar argument holds for $f \in D'$ and $f \in \mathcal{E}'$. $\square$

Let us give another definition of R on distributions.

Definition 10.1.4. Fix any $f \in S'$. Then Rf is a linear bounded functional computed by the formula

$$Rf := F^{-1}\mathcal{F}f, \qquad (10.1.4)$$

where (10.1.4) is understood as follows:

$$(Rf, \mu) = \frac{1}{\pi} < \mathcal{F}f, |t|^{1-n}F\mu > \quad \forall \mu \in KRS. \qquad (10.1.4')$$

Recall that if $f \in S$, then identity (10.1.4) is a consequence of the Fourier slice theorem.

REMARK 10.1.1. The new Definition 10.1.4 has the following advantages over the earlier ones:

(1) in many cases it allows us to find an explicit formula for Rf simpler than other definitions,
(2) it is conceptually new since it takes formula (10.1.4) as the basis for the definition,
(3) it allows to give a description of the spaces of test functions which does not use the operators R and K: the test functions μ have to be chosen so that $|t|^{1-n}F\mu$ belongs to the space of test functions corresponding to distributions $\mathcal{F}f$, where f runs through the desired space of distributions. We used in (10.1.4') the test functions from the space the definition of which involves operators R and K because we will establish the equivalnce of the new definition to the earlier ones.

We restrict the discussion to the case when $f \in S'$, but similar arguments can be given for $f \in D'$ as well.

Let us show that Definitions 10.1.4 and 10.1.1 are equivalent for $f \in S'$.

Lemma 10.1.1. *One has $|t|^{1-n}F\mu \in S$ if and only if $\mu \in KRS$.*

Proof. If $\mu \in KRS$, then $R^*\mu \in S$. Since $\mathcal{F} : S \to S$ is an isomorphism, the second equation in (2.2.18') implies that $|t|^{1-n}F\mu \in S$. Conversely, assume $|t|^{1-n}F\mu := \mathcal{F}h \in S$. Then $\mu = F^{-1}|t|^{n-1}\mathcal{F}h, h \in S$, and the identity $KR = \gamma F^{-1}|t|^{n-1}\mathcal{F}$ implies that $\mu \in KRS$. $\square$

Lemma 10.1.1'. *Definitions 10.1.1 and 10.1.4 are equivalent for $f \in S$.*

Proof. Let $\mu \in KRS$ and $f \in S'$. Substituting (10.1.4) into the right-hand side of (10.1.1), using the spherical coordinate system for the inner product in $\mathbb{R}^n$, and understanding the integrals below in the distributional sense, we get

$$
< f, R^*\mu > = \frac{1}{\gamma} < f, \mathcal{F}^{-1}|t|^{1-n}F\mu > = \frac{1}{(2\pi)^n \gamma} < \mathcal{F}f, |t|^{1-n}F\mu >
$$

$$
= \frac{1}{\pi} \int\limits_{S^{n-1}} \int\limits_0^\infty \mathcal{F}f(t\alpha)|t|^{1-n}F\mu(\alpha, t)t^{n-1}\,dt\,d\alpha
$$

$$
= \frac{1}{2\pi} \int\limits_{S^{n-1}} \int\limits_{-\infty}^\infty \mathcal{F}f(t\alpha)F\mu(\alpha, t)\,dt\,d\alpha
$$

$$
:= \frac{1}{2\pi}(\mathcal{F}f, F\mu) = (F^{-1}\mathcal{F}f, \mu). \tag{10.1.5}
$$

Since $\mu \in KRS$ was arbitrary, Lemma 10.1.1 and Equations (10.1.1), (10.1.5) prove that Definitions 10.1.1 and 10.1.4 are, indeed, equivalent. □

As a corollary, we immediately get the following new fact: the Fourier slice theorem holds on distribution space S. This is stated in the following lemma:

Lemma 10.1.2 (The Fourier slice theorem). *Let $f \in S'$. Then*

$$
F\hat{f} = \mathcal{F}f, \tag{10.1.6}
$$

where the equality holds in S'.

In fact, Equation (10.1.6) can be viewed as a definition of F on the set of distributions $\hat{f}$ for $f \in S'$.

Exercise 10.1.2. Prove that the definition of R on the Sobolev spaces $H_0^s(B_a)$, given in Section 3.2, coincides with the definition of R on S' if $f \in H_0^s(B_a)$.

Hint. If μ is a test function, $\mu \in S_t = KRS$, and $f_m \in C_0^\infty(B_a)$, then identity (10.1.1) holds. Let $\|f - f_m\|_{H^s} \to 0$. Then $f_m \to f$ in S', and the limit of $(f_m, R^*\mu), m \to \infty$, exists. Hence, any $f \in H_0^s(B_a)$ satisfies (10.1.1).

10.2. Properties of the test function spaces

It turns out that the spaces of test functions KRS and KRD can be described explicitly. This, in fact, has been already done in Section 3.3. Recall that $S_{sm}(Z)$ denotes the space of $S(Z)$ functions which are odd (even) if n is even (odd) and which satisfy the shifted moment conditions:

$$
\int_{-\infty}^{\infty} g(\alpha, p) p^k \, dp = \begin{cases} 0, & 0 \le k \le n - 2, \\ \mathcal{P}_{k-n+1}(\alpha), & k \ge n - 1, \end{cases} \tag{10.2.1}
$$

where $\mathcal{P}_{k+1-n}(\alpha)$ is a restriction to S^{n-1} of a homogeneous polynomial of degree $k + 1 - n$. Similarly, $\mathcal{D}_{sm}(Z)$ denotes the space of $C_0^\infty(Z)$ functions which are odd (even) if n is even (odd) and which satisfy (10.2.1). In what follows, we will use the notation $S_t := KRS$ and $\mathcal{D}_t := KRD$ for the spaces of test functions corresponding to the cases $X = S$ and $X = \mathcal{D}$, respectively.

Theorem 10.2.1. *If n is odd, then $S_t = S_{sm}(Z)$ and $\mathcal{D}_t = \mathcal{D}_{sm}(Z)$. If n is even, then $S_t = \mathcal{H}S_{sm}(Z)$ and $\mathcal{D}_t = \mathcal{H}\mathcal{D}_{sm}(Z)$.*

Proof. Using the definition of the operator K (see (2.2.16′)), the relation $S_m(Z) = RS$, where $S_m(Z)$ is the space of even $S(Z)$ functions which satisfy moment conditions (3.1.2), applying Theorem 3.1.1 and Lemma 3.3.2, we prove that $S_t = S_{sm}(Z)$. Other assertions can be proved similarly. □

Note that the space S_t of test functions has the following property: there exist even distributions $\eta \in S_t'$ such that

$$
(\eta, \mu) = 0 \quad \forall \mu \in S_t. \tag{10.2.2}
$$

First, consider the case of odd n. Using the identity $KR = \gamma F^{-1}|\lambda|^{n-1} F$ and the relation $S_t = KRS$, we see that all $\mu \in S_t$ can be represented in the form $\tilde{\mu}(\alpha, \lambda) = \lambda^{n-1} \tilde{\varphi}(\lambda\alpha)$, where $\tilde{\mu} = F\mu$ and $\tilde{\varphi} = F\varphi \in S$. Equation (10.2.2) now yields

$$
\int_{S^{n-1}} \int_{-\infty}^{\infty} \tilde{\eta}(\alpha, \lambda) \lambda^{n-1} \tilde{\varphi}(\lambda\alpha) d\lambda \, d\alpha = 0 \quad \forall \tilde{\varphi} \in S, \tag{10.2.3}
$$

where $\tilde{\eta} = F\eta$. The functions of the type $\lambda^{n-1}\tilde{\varphi}(\lambda\alpha)$, $\varphi \in S$, include, in particular, all even functions from $S(Z)$, which have the property $\partial^k g(\alpha, \lambda)/\partial\lambda^k|_{\lambda=0} = 0$ for all $k \ge 0$. Indeed, if we take any such $g(\alpha, \lambda)$ and define $\tilde{\varphi}(\lambda\alpha) := g(\alpha, \lambda)/\lambda^{n-1}$, we easily check using Equation (3.1.8) that $\varphi \in S$. Therefore, we conclude from (10.2.3) that $\tilde{\eta}(\alpha, \lambda)$ is a distribution concentrated at $\lambda = 0$, and $\tilde{\eta}$ can be represented as a finite sum

$$\tilde{\eta}(\alpha, \lambda) = \sum_k b_k(\alpha)\delta^{(k)}(\lambda), \quad b_k(-\alpha) = (-1)^k b_k(\alpha),$$

or

$$\eta(\alpha, p) = \sum_k a_k(\alpha)p^k, \quad a_k(-\alpha) = (-1)^k a_k(\alpha), \tag{10.2.4}$$

where $a_k(\alpha)$ differs from $b_k(\alpha)$ by a constant factor. Take $g = \eta = a_k(\alpha)p^k$ in (10.2.1) to get:

$$0 = (\eta, \mu) = \int\limits_{S^{n-1}} \int\limits_{-\infty}^{\infty} a_k(\alpha)p^k \mu(\alpha, p)dp\,d\alpha = 0, \ 0 \le k \le n-2,$$

$$0 = \int\limits_{S^{n-1}} \int\limits_{-\infty}^{\infty} a_k(\alpha)p^k \mu(\alpha, p)dp\,d\alpha = \int\limits_{S^{n-1}} a_k(\alpha)\mathcal{P}_{k-n+1}(\alpha)da, \ k \ge n-1. \tag{10.2.5}$$

Equations (10.2.4) and (10.2.5) imply that the space of even distributions η, which satisfy (10.2.2), is generated by finite linear combinations of the terms $a_k(\alpha)p^k$, where

(1) $a_k(\alpha)$ is even (odd) if k is even (odd), $k \ge 0$, and
(2) $\int_{S^{n-1}} a_k(\alpha)\mathcal{P}_{k-n+1}(\alpha)da = 0$, $k \ge n-1$, for all homogeneous polynomials of degree $k-n+1$.

Considering similarly the case of even n, we obtain that distributions η, which satisfy (10.2.2), are generated by finite linear combinations of the terms $a_k(\alpha)p^k \mathcal{H}$, where

(1) $a_k(\alpha)$ is even (odd) if k is odd (even), $k \ge 0$,
(2) $\int_{S^{n-1}} a_k(\alpha)\mathcal{P}_{k-n+1}(\alpha)da = 0$, $k \ge n-1$, for all homogeneous polynomials of degree $k-n+1$, and
(3) the action of the distribution $a_k(\alpha)p^k \mathcal{H}$ on a test function $\mu \in \mathcal{S}_t$ is defined by

$$(a_k(\alpha)p^k \mathcal{H}, \mu) := \int\limits_{S^{n-1}} \int\limits_{-\infty}^{\infty} a_k(\alpha)p^k (\mathcal{H}\mu)(\alpha, p)dp\,d\alpha.$$

Since $\mathcal{D}_t \subset \mathcal{S}_t$, the distributions η, which satisfy (10.2.2), annihilate also the functions from $\mathcal{D}_t$.

The existence of non-zero distributions η which vanish on any $\mu \in \mathcal{S}_t$ implies that the Radon transform on distributions is not uniquely defined (see, e.g. Example 10.3.2 below, where such nonuniqueness arises in a natural way).

Exercise 10.2.1. Prove the injectivity of R on S' using only the definition of S_t, i.e. without using the identity $R^*KR = I$.

Hint. Let $f \in S'$ and $Rf = 0$. From (10.1.5), we have

$$(Rf, \mu) = \frac{1}{\pi} < Ff, |t|^{1-n}F\mu >= 0. \tag{10.2.6}$$

First, assume $\mu \in S_t$, and $\int_{-\infty}^{\infty} \mu(\alpha, p)p^k\,dp \equiv 0$, $k \geq 0$. Then $(F\mu)(\alpha, t)$ vanishes with all derivatives at $t = 0$, and $|t|^{1-n}(F\mu)(\alpha, t) \in S$, where $\xi = t\alpha$. Conversely, any $\psi \in S$, such that $(D_\xi^j \psi)(\xi = 0) = 0, j \geq 0$, can be represented in the form $\psi(t\alpha) = |t|^{1-n}(F\mu)(\alpha, t)$, where μ is as above. Therefore, Ff is a distribution concentrated at the origin, and we can write

$$(Ff)(\xi) = \sum_{j=0}^{N} Q_j(D_\xi)\delta(\xi), \ 0 \leq N < \infty, \tag{10.2.7}$$

where Q_j are homogeneous polynomials of degree j, and N is some integer. If $\mu \in S_t$ (cf. (10.2.1) and Section 3.1.1), then $|t|^{1-n}(F\mu)(\alpha, t) = \sum_{k=0}^{N} c_k P_k(t\alpha) + O(t^{N+1}), t \to 0$, where c_k are some constants. Fix any $j_0, j_0 \leq N$, and find any $b \in C_0^\infty([-1, 1])$ such that

$$b(-p) = (-1)^{j_0 + n - 1}b(p), \quad \int_{-1}^{1} b(p)p^k\,dp = \begin{cases} 0, & k \neq j_0 + n - 1, \\ 0 \leq k \leq N + n - 1, \\ 1, & k = j_0 + n - 1. \end{cases}$$

Define $\mu(\alpha, p) = b(p)Q_{j_0}(\alpha)$, if n is odd, and $\mu(\alpha, p) = (Hb)(p)Q_{j_0}(\alpha)$, if n is even. Clearly, $\mu \in S_t$. Substituting such μ and (10.2.7) into (10.2.6) and using that Q_j are homogeneous, we find $0 =< Q_{j_0}(D_\xi)\delta, Q_{j_0} >$; therefore, $Q_{j_0} \equiv 0$. Since j_0 was arbitrary, $f \equiv 0$.

10.3. Examples

Example 10.3.1. Let $f(x) = \delta(x)$, $x \in \mathbb{R}^n$. Then, using definition (10.1.1), we get

$$(Rf, \mu) = \int_{\mathbb{R}^n} \delta(x) \int_{S^{n-1}} \mu(\alpha, \alpha \cdot x)d\alpha\,dx$$

$$= \int_{S^{n-1}} \mu(\alpha, 0)d\alpha = \int_{S^{n-1}} \int_{-\infty}^{\infty} \mu(\alpha, p)\delta(p)dp\,d\alpha,$$

therefore,

$$R\delta(\alpha, p) = \delta(p). \tag{10.3.1}$$

The same result follows from definition (10.1.4). Indeed, $\tilde{f}(\xi) = 1$, and

$$R\delta(\alpha, p) = F^{-1}1 = \frac{1}{2\pi} \int_{-\infty}^{\infty} e^{-ip\lambda} d\lambda = \delta(p).$$

Example 10.3.2. Let us consider $f(x) = 1, x \in \mathbb{R}^n$. Then $\tilde{f} = (2\pi)^n \delta(\xi)$. Suppose, first, that n is odd. Using conditions (10.2.1), the second equation in (2.2.18'), writing the inner product in $\mathbb{R}^n$ using spherical coordinates, and noting that $|\lambda|^{1-n} = \lambda^{1-n}$ for $\lambda > 0$, we get:

$$(Rf, \mu) = \gamma^{-1} < \delta(\xi), \lambda^{1-n} F\mu >$$

$$= \gamma^{-1} < \delta(\xi), \sum_{m=n-1}^{N} \frac{i^m}{m!} P_{m+1-n}(\xi) + O(|\xi|^{N+2-n}) >$$

$$= \gamma^{-1} < \delta(\xi), \frac{i^{n-1}}{(n-1)!} c + O(|\xi|) >= \gamma^{-1} \frac{i^{n-1}}{(n-1)!} c, \tag{10.3.2}$$

where $N \geq n - 1$ is an arbitrary integer, $\gamma := 1/[2(2\pi)^{n-1}]$, and

$$c = \mathcal{P}_0(\alpha) = \int_{-\infty}^{\infty} \mu(\alpha, p) p^{n-1} dp, \quad \xi = \lambda\alpha.$$

Therefore,

$$R1 = \gamma^{-1} \frac{i^{n-1} p^{n-1} a(\alpha)}{(n-1)!}, \quad n = 2k + 1, \tag{10.3.3}$$

where $a(\alpha)$ is an arbitrary function such that

$$\int_{S^{n-1}} a(\alpha) d\alpha = 1. \tag{10.3.4}$$

Now let n be even. Theorem 10.2.1 implies that $\mu = \mathcal{H}g$ for some $g \in S_{sm}(Z)$, that is $F\mu = -isgn\lambda Fg$. Similarly to (10.3.2), we find

$$< R1, \mu > = \gamma^{-1} < \delta(\xi), |\lambda|^{1-n} F\mu >= \gamma^{-1} < \delta(\xi), -i\lambda^{1-n} Fg >$$

$$= (-i)\gamma^{-1} \frac{i^{n-1}}{(n-1)!} \int_{-\infty}^{\infty} g(\alpha, p) p^{n-1} dp.$$

Since $g = -\mathcal{H}\mu$, we get

$$< R1, \mu > = \gamma^{-1} \frac{i^n}{(n-1)!} \int_{-\infty}^{\infty} (\mathcal{H}\mu)(\alpha, p) p^{n-1} dp$$

$$= \gamma^{-1} \frac{i^n}{(n-1)!} \int_{S^{n-1}} \int_{-\infty}^{\infty} a(\alpha)(\mathcal{H}\mu)(\alpha, p) p^{n-1} dp \, d\alpha, \tag{10.3.5}$$

where $a(\alpha)$ is an arbitrary function which satisfies (10.3.4). Therefore,

$$R1 = \gamma^{-1} \frac{i^n p^{n-1} a(\alpha)}{(n-1)!} \mathcal{H}, \quad n = 2k, \tag{10.3.6}$$

where the distribution $R1$ acts on $\mu \in S_t$ according to (10.3.5).

Since the functions in the range of the Radon transform are even, we suppose, in addition to (10.3.4), that $a(\alpha)$ in (10.3.3) and (10.3.6) is even: $a(\alpha) = a(-\alpha)$. If one imposes a priori the requirement that the members of the range of R are homogeneous of degree -1 (see (2.1.5), (2.1.7)) or 0 (see (2.1.1), (2.1.3)), then the corresponding homogeneity requirement should be imposed on $a(\alpha)$. We see that $R1$ is not uniquely defined.

Example 10.3.3. Let $f(x) = a(x_1)\delta(y)$, $y = (x_2, \ldots, x_n)$, where $a(x_1) \in L^1(\mathbb{R}^1)$. Then (10.1.5) yields

$$(Rf, \mu) = \int_{-\infty}^{\infty} \int_{\mathbb{R}^{n-1}} a(x_1)\delta(y) \int_{S^{n-1}} \mu(\alpha, \alpha_1 x_1 + \cdots + \alpha_n x_n) d\alpha \, dy \, dx_1$$

$$= \int_{S^{n-1}} \int_{-\infty}^{\infty} a(x_1)\mu(\alpha, \alpha_1 x_1) dx_1 \, d\alpha$$

$$= \int_{S^{n-1}} \int_{-\infty}^{\infty} \mu(\alpha, p) \int_{-\infty}^{\infty} \delta(p - \alpha_1 x_1) a(x_1) dx_1 \, dp \, d\alpha$$

$$= \int_{S^{n-1}} \int_{-\infty}^{\infty} \mu(\alpha, p) \frac{1}{|\alpha_1|} a(p/\alpha_1) dp \, d\alpha.$$

Therefore,

$$Rf = \begin{cases} \frac{1}{|\alpha_1|} a\left(\frac{p}{\alpha_1}\right), & \alpha_1 \neq 0, \\ \delta(p) \int_{-\infty}^{\infty} a(x_1) dx_1, & \alpha_1 = 0. \end{cases} \tag{10.3.7}$$

Since $\mathcal{F}_{(x_1,y)\to\lambda\alpha} a(x_1)\delta(y) = \tilde{a}(\lambda\alpha_1)$, we derive using definition (10.1.4)

$$Rf = F_{\lambda\to p}^{-1}\tilde{a}(\lambda\alpha_1) = \frac{1}{2\pi}\int_{-\infty}^{\infty}\tilde{a}(\lambda\alpha_1)e^{-i\lambda p}d\lambda$$

$$= \begin{cases} \frac{1}{|\alpha_1|}a\left(\frac{p}{\alpha_1}\right), & \alpha_1 \neq 0, \\ \delta(p)\int_{-\infty}^{\infty}a(x_1)dx_1, & \alpha_1 = 0. \end{cases}$$

This is the same formula (10.3.7), which was obtained differently above.

Example 10.3.4. Let $f(x) = \delta(x_1, x_2, \ldots, x_k)$, $1 \leq k < n$. In this case it is convenient to use definition (10.1.4). We have $\tilde{f}(\xi) = (2\pi)^{(n-k)}\delta(\xi_{k+1}, \ldots, \xi_n)$. Therefore,

$$Rf = (2\pi)^{n-k-1}\int_{-\infty}^{\infty}\delta(\lambda\alpha_{k+1}, \ldots, \lambda\alpha_n)e^{-i\lambda p}d\lambda$$

$$= (2\pi)^{n-k-1}\delta(\alpha_{k+1}, \ldots, \alpha_n)\int_{-\infty}^{\infty}|\lambda|^{-n+k}e^{-i\lambda p}d\lambda. \tag{10.3.8}$$

To compute the last integral, we shall use the following formulas

$$F(|\lambda|^{-2m-1}) = c_0^{(2m+1)}p^{2m} - c_{-1}^{(2m+1)}p^{2m}\ln|p|, \tag{10.3.9}$$

$$F(|\lambda|^{-2m}) = (-1)^m\frac{\pi}{(2m-1)!}|p|^{2m-1}, \tag{10.3.10}$$

where the coefficients $c_q^{(2m+1)}$, $q = -1, 0$, are defined from the expansion:

$$-2\sin\frac{\lambda\pi}{2}\Gamma(\lambda+1) = \frac{c_{-1}^{(n)}}{\lambda+n} + c_0^{(n)} + \ldots, \tag{10.3.11}$$

where

$$c_{-1}^{2m+1} = \frac{2(-1)^m}{(2m)!}, \quad c_0^{2m+1} = \frac{2(-1)^m}{(2m)!}\left[1 + \frac{1}{2} + \cdots + \frac{1}{2m} + \Gamma'(1)\right].$$

Therefore,

$$Rf = (2\pi)^{n-k-1}\frac{\pi i^{n-k}}{(n-k-1)!}|p|^{n-k-1}\delta(\alpha_{k+1}, \ldots, \alpha_n), \quad n-k \text{ is even}, \tag{10.3.12a}$$

$$Rf = (2\pi)^{n-k-1}(c_0^{(n-k)} - c_{-1}^{(n-k)}\ln|p|)|p|^{n-k-1}\delta(\alpha_{k+1}, \ldots, \alpha_n), \quad n-k \text{ is odd}, \tag{10.3.12b}$$

where $f(x) = \delta(x_1, x_2, \ldots, x_k)$ and $0 \leq k < n$. For example, let $n = 2$ and $k = 1$. Then Equation (10.3.12b) yields

$$R\delta(x_1) = (c_0^{(1)} - c_{-1}^{(1)}\ln p)\delta(\alpha_2).$$

Exercise 10.3.1. Calculate $R\delta(x_1)$ and $R\delta(x_1, x_2)$ for $n = 3$.

Hint. Use Equations (10.3.12a,b).

10.4. Range theorem for the Radon transform on $\mathcal{E}'$

Take any $f \in \mathcal{E}'$ and suppose supp $f \subset B_a$, where B_a is the ball of radius a centered at the origin. Fix $W_1 \in C_0^\infty(B_1)$ and define

$$f_\epsilon := W_\epsilon * f, \quad W_\epsilon(x) := \epsilon^{-n} W_1(x/\epsilon). \tag{10.4.1}$$

Clearly, $f_\epsilon \in C_0^\infty(B_{a+\epsilon})$ and $Rf_\epsilon \in C_0^\infty(Z_{a+\epsilon})$. Since

$$(Rf_\epsilon, \mu) = <f_\epsilon, R^*\mu> \to <f, R^*\mu> = (Rf, \mu),$$
$$\epsilon \to 0, \quad \mu \in \mathcal{E}(Z) := C^\infty(Z),$$

we conclude that $Rf_\epsilon \to Rf$ in $\mathcal{E}'(Z)$. This implies that $Rf \in \mathcal{E}'(Z)$, $Rf(\alpha, p) \equiv 0, |p| \geq a$, and Rf is even (that is, the distributions $Rf(\alpha, p)$ and $Rf(-\alpha, -p)$ act identically on $\mathcal{E}(Z)$). Let $\mu(\alpha, p) = p^k \varphi(\alpha)$, $\varphi \in C^\infty(S^{n-1})$. We compute:

$$(Rf, p^k \varphi(\alpha)) = \int_{\mathbb{R}^n} f(x) \int_{S^{n-1}} (\alpha \cdot x)^k \varphi(\alpha) d\alpha \, dx$$

$$= \int_{S^{n-1}} \varphi(\alpha) \int_{\mathbb{R}^n} (\alpha \cdot x)^k f(x) dx \, d\alpha$$

$$= <\varphi(\alpha), \mathcal{P}_k(\alpha)>_{L^2(S^{n-1})}, \tag{10.4.2}$$

where $\mathcal{P}_k$ is a homogeneous polynomial of degree k. Thus, in the sense of (10.4.2), we can write

$$\int_{-\infty}^{\infty} Rf(\alpha, p) p^k dp = \mathcal{P}_k(\alpha), \quad k = 0, 1, \ldots. \tag{10.4.3}$$

Now take any even $g \in \mathcal{E}'(Z)$, such that

$$\int_{-\infty}^{\infty} g(\alpha, p) p^k dp = \mathcal{P}_k(\alpha), \quad k = 0, 1, \ldots. \tag{10.4.4}$$

Suppose $g(\alpha, p) \equiv 0, |p| \geq a$. Fix any even function $w_1 \in C_0^\infty([-1, 1])$ and define

$$g_\epsilon(\alpha, p) = \int_{-\infty}^{\infty} w_\epsilon(p - s) g(\alpha, s) ds, \quad w_\epsilon(x) := \epsilon^{-1} w_1(x/\epsilon). \tag{10.4.5}$$

Clearly, $g_\epsilon \in C_0^\infty([-(a+\epsilon), a+\epsilon])$ for any fixed $\alpha \in S^{n-1}$ and $g_\epsilon(\alpha, p)$ is even. It is easy to check that g_ϵ satisfies moment conditions (10.4.4):

$$
\int_{-\infty}^{\infty} g_\epsilon(\alpha, p) p^k \, dp = \int_{-\infty}^{\infty} \int_{-\infty}^{\infty} w_\epsilon(p-s) g(\alpha, s) p^k \, ds \, dp
$$

$$
= \int_{-\infty}^{\infty} g(\alpha, s) \int_{-\infty}^{\infty} w_\epsilon(p-s) p^k \, dp \, ds
$$

$$
= \int_{-\infty}^{\infty} g(\alpha, s) \int_{-\infty}^{\infty} w_\epsilon(t)(t+s)^k \, dt \, ds. \tag{10.4.6}
$$

The change of order of integration in (10.4.6) is justified (cf. [Hor, Vol. 1, §4.1]). Expanding the expression $(t+s)^k$, using that w_ϵ is even, and taking into account (10.4.4), we get for $\alpha \in S^{n-1}$:

$$
\int_{-\infty}^{\infty} g_\epsilon(\alpha, p) p^k \, dp = \int_{-\infty}^{\infty} g(\alpha, s) \sum_{\substack{j=0 \\ k-j \text{ even}}}^{k} c_j s^j \, ds
$$

$$
= \sum_{\substack{j=0 \\ k-j \text{ even}}}^{k} c_j P_j(\alpha)(\alpha \cdot \alpha)^{\frac{k-j}{2}} = Q_k(\alpha), \quad k = 0, 1, \dots, \tag{10.4.7}
$$

where Q_k is a homogeneous polynomial of degree k. Since polynomials are dense in $C^\infty([-(a+\epsilon), a+\epsilon])$, we conclude from (10.4.5) and (10.4.7) that $g_\epsilon \in C_0^\infty(Z_{a+\epsilon})$. Theorem 3.1.1 asserts that there exists $f_\epsilon \in C_0^\infty(B_{a+\epsilon})$ such that $Rf_\epsilon = g_\epsilon$. Clearly, $Rf_\epsilon \to g$ as $\epsilon \to 0$ in $\mathcal{E}'(Z)$. Since the space $R^*\mathcal{E}(Z)$ is dense in $\mathcal{E}$ (see Section 10.1), we conclude using the duality relation $(Rf_\epsilon, \mu) = (f_\epsilon, R^*\mu)$ that the sequence f_ϵ converges in $\mathcal{E}$. Let $f := \lim_{\epsilon \to 0} f_\epsilon \in \mathcal{E}'$. Clearly, $\operatorname{supp} f \subset B_a$ and $Rf = g$. Thus we generalized Theorem 3.1.1 to the case of distributions from $\mathcal{E}'$.

Theorem 10.4.1. *Let $\mathcal{E}_m'(Z)$ denote the class of even compactly supported distributions which satisfy moment conditions (10.4.4). Then the mapping $R: \mathcal{E} \to \mathcal{E}_m'(Z)$ is an isomorphism. Moreover, if $Rf = g$ and $g(\alpha, p) \equiv 0$ for $|p| \geq a$, then $\operatorname{supp} f \subset B_a$.*

Exercise 10.4.1. Prove that $g = Rf$ for some $f \in \mathcal{E}'$, $\operatorname{supp} f \subset B_a$, if and only if $F_{p \to \lambda} g(\alpha, p)$, as a function of $\lambda\alpha$, is entire and of exponential type a.

Hint. Use Definition 10.1.4 and a well known property of the Fourier transform of compactly supported distributions.

10.5. A definition based on spherical harmonics expansion

In this section we give another definition of R on distributions from $\mathcal{S}'$ and $\mathcal{D}'$. Although this definition is equivalent to the one from Section 10.1, we use here a different technique. We will consider the case $f \in \mathcal{S}'$ in detail, and the case $f \in \mathcal{D}'$ can be treated similarly.

Suppose first that $f \in \mathcal{S}$. Then

$$f(r\beta) = \sum_{l=0}^{\infty} f_l(r)Y_l(\beta), \quad f_l(r) = \int_{S^{n-1}} f(r\beta)Y_l(\beta)d\beta, \ r > 0, \beta \in S^{n-1},$$

$$(10.5.1)$$

where Y_l are the normalized spherical harmonics.

Exercise 10.5.1. Prove that the series in (10.5.1) converges in the topology of $\mathcal{S}$.

Hint. First, check the inclusion $f_L(r\beta) := \sum_{l=0}^{L} f_l(r)Y_l(\beta) \in \mathcal{S}$. Let Δ_β be the angular part of the Laplacian. Then $Y_l(\beta) = -(l(l+n-2))^{-1}\Delta_\beta Y_l(\beta)$. Moreover, it can be shown that $\max_{\beta \in S^{n-1}} |D_\beta^j Y_l(\beta)| = O(l^{j+(n/2)-1}), l \to \infty$, for any multiindex j. Since

$$f_l(r) = \frac{(-1)^k}{[l(l+n-2)]^k} \int_{S^{n-1}} f(r\beta)\Delta_\beta^k Y_l(\beta)d\beta$$

$$= \frac{(-1)^k}{[l(l+n-2)]^k} \int_{S^{n-1}} \Delta_\beta^k f(r\beta)Y_l(\beta)d\beta,$$

and $f \in \mathcal{S}$, we conclude that

$$|f_l(r)| \le c_{km}(1+l)^{-k}(1+r)^{-m}, \ \forall k, m \ge 0,$$

where $c_{km} > 0$ are some constants. Hence the series in (10.5.1) converges absolutely, and it is sufficient to show that $\sum_{l=L}^{\infty} f_l Y_l \to 0$ in $\mathcal{S}$ as $L \to \infty$. Using fast decay of $f_l(r)$ as $l \to \infty$ and $r \to \infty$ and taking into account the above mentioned estimate of the derivatives of spherical harmonics, we prove the desired result.

Similarly, for $\mu \in C^\infty(Z)$, we have

$$\mu(\alpha, p) = \sum_{l=0}^{\infty} \mu_l(p)Y_l(\alpha), \quad \mu_l(p) = \int_{S^{n-1}} \mu(\alpha, p)Y_l(\alpha)d\alpha. \quad (10.5.2)$$

The Funk-Hecke Theorem (Equations (2.1.35) and (2.1.36)) implies:

$$(R^*\mu)(r\beta) = \sum_{l=0}^{\infty} \int_{S^{n-1}} \mu_l(\alpha \cdot r\beta)Y_l(\alpha)d\alpha = \sum_{l=0}^{\infty} \nu_l(r)Y_l(\beta), \quad (10.5.3)$$

where

$$\nu_l(r) = \frac{|S^{n-2}|}{C_l^{(\frac{n-2}{2})}(1)} \int_{-1}^{1} \mu_l(rp) C_l^{(\frac{n-2}{2})}(p)(1-p^2)^{\frac{n-3}{2}} dp. \qquad (10.5.4)$$

Since Y_l are normalized, we obtain from (10.5.1) and (10.5.4), provided the product $f_l(r)\nu_l(r)$ decays sufficiently fast as $r \to \infty$:

$$(Rf, \mu) = <f, R^*\mu> = \sum_{l=0}^{\infty} [f_l, \nu_l], \qquad (10.5.5)$$

where

$$[f_l, \nu_l] := \int_0^{\infty} f_l(r)\nu_l(r)r^{n-1} dr. \qquad (10.5.6)$$

Suppose now that f is a distribution, $f \in S'$. Then (10.5.1) remains valid, but we have to assume that $f_l(r)$ are distributions. To study the action of f_l on test functions, we use the identity

$$<f, w_l(r)Y_l(\beta)> = [f_l, w_l], \qquad (10.5.7)$$

which holds, e.g. if $f(x)$ and $w_l(|x|)Y_l(x/|x|) \in S$. Using Equation (3.1.8), we can easily check that $w_l Y_l \in S$ if and only if

$$w_l(r) = r^l \varphi(r^2) \text{ for some } \varphi \in S(\mathbb{R}_+). \qquad (10.5.8)$$

Here $S(\mathbb{R}_+)$ denotes the set of $C^{\infty}([0, \infty))$ functions which decay with all their derivatives faster than any negative power of r as $r \to \infty$. The set of functions which satisfy (10.5.8) will be denoted $S_l(\mathbb{R}_+)$. Thus, if $f \in S'$ and $w_l \in S_l(\mathbb{R}_+)$, then the left-hand side of (10.5.7) is well defined, and this equation can be used to define the distribution f_l on the test function space $S_l(\mathbb{R}_+)$.

Returning to (10.5.5), we see that the function ν_l should be taken from the space $S_l(\mathbb{R}_+), l = 0, 1, \dots$. Equation (10.5.3) implies that ν_l are coefficients of the spherical harmonics expansion of $R^*\mu$. The inclusion $\nu_l \in S_l(\mathbb{R}_+)$ will take place if $R^*\mu \in S$, that is $\mu \in S_t = KRS$.

First, consider the case of odd n. Theorem 10.2.1 asserts that $\mu \in S_t$ if and only if $\mu \in S(Z)$, μ is even, and μ satisfies shifted moment conditions (10.2.1). From (10.5.2), we derive

$$\int_{-\infty}^{\infty} \mu(\alpha, p)p^k dp = \sum_{l=0}^{\infty} \mu_{lk} Y_l(\alpha) = \begin{cases} 0, & 0 \le k \le n-2, \\ \mathcal{P}_{k-n+1}(\alpha), & k \ge n-1, \end{cases}$$

$$\mu_{lk} := \int_{-\infty}^{\infty} \mu_l(p)p^k dp. \qquad (10.5.9)$$

Recall that, by the assumption, P_{k-n+1} is a homogeneous polynomial, and Y_l is the restriction to S^{n-1} of a harmonic homogeneous polynomial. Since Y_l is orthogonal on S^{n-1} to all polynomials of lower degree, we conclude that $\mu_{lk} = 0$ if $l > k - n + 1$. Therefore, we introduce *the test function space $S_{sh}(Z)$, which consists of finite linear combinations of the terms $\mu_l(p)Y_l(\alpha)$, where the functions μ_l, for odd n, satisfy the assumptions:*

$$\mu_l \in S(\mathbb{R}), \quad \mu_l(-p) = (-1)^l \mu_l(p), \quad l = 0, 1, 2, \ldots, \tag{10.5.10a}$$

$$\int_{-\infty}^{\infty} \mu_l(p) p^k \, dp = 0, \quad 0 \le k \le l + n - 2. \tag{10.5.10b}$$

Condition (10.5.10a) guarantees that $\mu_l Y_l \in S(Z)$ and $\mu_l Y_l$ is even, and (10.5.10b) ensures that $\mu_l Y_l$ satisfies (10.2.1). Since finite linear combinations $\sum_l \mu_l Y_l$ are dense in $S(Z)$ (cf. Exercise 10.5.1), we conclude that the test function space $S_{sh}(Z)$ is dense in S_t. The subscript 'sh' in the notation $S_{sh}(Z)$ stands for the 'spherical harmonics'.

Now suppose that n is even. Theorem 10.2.1 asserts that $\mu = \mathcal{H}g$, where $g \in S(Z)$ is odd and g satisfies (10.2.1). Since $g = -\mathcal{H}\mu$, we define, similarly to (10.5.9) and (10.5.10), *the test function space $S_{sh}(Z)$, which consists of finite linear combinations of the terms $\mu_l(p)Y_l(\alpha)$, where the functions μ_l, for even n, satisfy the assumptions:*

$$\mu_l \in \mathcal{H}S(\mathbb{R}), \quad \mu_l(-p) = (-1)^l \mu_l(p), \quad l = 0, 1, 2, \ldots, \tag{10.5.11a}$$

$$\int_{-\infty}^{\infty} (\mathcal{H}\mu_l)(p) p^k \, dp = 0, \quad 0 \le k \le l + n - 2. \tag{10.5.11b}$$

We see now that Equation (10.5.5) defines R on distributions $f \in S'$ via the spherical harmonics of f. Indeed, assuming that μ_l satisfy (10.5.10a,b) if n is odd, or - (10.5.11a,b) if n is even, our derivation ensures that the functions ν_l, computed by (10.5.4), belong to the corresponding test spaces $S_l(\mathbb{R}_+), l = 0, 1, \ldots$, the series on the right-hand side of (10.5.5) contains finitely many terms and, therefore, the functional Rf is well defined if $f \in S'$. Since the test function space $S_{sh}(Z)$ is dense in S_t, we conclude that the Radon transform defined in this section is injective.

Exercise 10.5.2. Check directly that ν_l, defined by (10.5.4), where μ_l satisfies either (10.5.10a,b) or (10.5.11a,b), belongs to $S_l(\mathbb{R}_+)$.

Hint. Differentiating (10.5.4) with respect to r, setting $r = 0$ and using the orthogonality of the Gegenbauer polynomials $G_l^{(q)}$ to $p^k, 0 \le$

$k \leq l - 1$, in $L^2([-1,1], (1-p^2)^{q-0.5})$, one establishes that $w_l^{(k)}(0) = 0, 0 \leq k \leq l-1, l \geq 1$. This, combined with (10.5.10a) or (10.5.11a), proves that $w_l(r) = r^l \varphi(r^2)$ for some $\varphi \in C^\infty(\mathbb{R}_+)$. To check that $\varphi \in S(\mathbb{R}_+)$, for odd n, use integral [(7.321), GR]:

$$\int_{-1}^{1} (1-x^2)^{q-0.5} e^{iax} G_l^{(q)}(x) dx = \frac{\pi 2^{1-q} i^l \Gamma(2q+l)}{l! \Gamma(q)} a^{-q} J_{q+l}(a), \quad (10.5.12)$$

and the formula $(F\mu_l)(\lambda) = \lambda^{l+n-1} \psi(\lambda^2), \psi \in S(\mathbb{R}_+)$, which follows from (10.5.10a,b). The desired result follows from the explicit representation of the Bessel functions of the half-integer order [(8.461), GR]. For even n, use integral (10.5.12), the formula $(F\mu_l)(\lambda) = -i\text{sgn}(\lambda)\lambda^{l+n-1} \psi(\lambda^2)$, $\psi \in S(\mathbb{R}_+)$, and the identity $J_m(x) = \frac{(-i)^m}{2\pi} \int_0^{2\pi} \exp(ix\cos\theta + im\theta) d\theta$ [cf. (8.411.1), GR], to reduce the integral in (10.5.4) to the form $\int_0^{2\pi} \int_0^\infty \psi(\lambda^2)[\lambda e^{i\theta}]^m e^{i\lambda r \cos\theta} \lambda d\lambda \, d\theta$, where $m = l + (n/2) - 1$. Representing the last integral as a two-dimensional one, one proves the desired assertion.

As it was already mentioned, the case $f \in \mathcal{D}'$ can be treated similarly. Namely, in condition (10.5.8) we should assume $\varphi \in C_0^\infty(\mathbb{R}_+)$. Also, assumptions $\mu_l \in S(\mathbb{R})$ and $\mu_l \in \mathcal{HS}(\mathbb{R})$ in (10.5.10a) and (10.5.11a) should be replaced by $\mu_l \in C_0^\infty(\mathbb{R})$ and $\mu_l \in \mathcal{HC}_0^\infty(\mathbb{R})$, respectively. The rest of the argument goes without changes.

10.6. When does the Radon transform on distributions coincide with the classical Radon transform?

In this section we prove the following result.

Theorem 10.6.1. *If $f \in L^1(\mathbb{R}^n)$, then $\hat{f}$ in the sense of (2.1.1) and in the sense of definitions given in Section 10.1 are identical.*

Proof. If $f \in L^1(\mathbb{R}^n)$ and $\mu \in S_t$, then $f(x)\mu(\alpha, \alpha \cdot x) \in L^1(\mathbb{R}^n \times S^{n-1})$. Thus

$$< f, R^*\mu > = \int_{\mathbb{R}^n} dx f(x) \int_{S^{n-1}} d\alpha \mu(\alpha, \alpha \cdot x)$$

$$= \int_{-\infty}^{\infty} dp \int_{S^{n-1}} d\alpha \mu(\alpha, p) \int_{\alpha \cdot x = p} ds f(x) = (\hat{f}, \mu). \quad (10.6.1)$$

This equation and (10.1.1) imply $(\hat{f}, \mu) = (Rf, \mu) \, \forall \mu \in S_t$. Thus $\hat{f} = Rf$ in the distributional sense. $\square$

REMARK 10.6.1. If $f \in L^1(\mathbb{R}^n)$, the Fubini theorem implies that $\hat{f}(\alpha, p) \in L^1(\mathbb{R})$ in p for any fixed $\alpha \in S^{n-1}$.

Exercise 10.6.1. Prove the conclusion of Theorem 10.6.1 under a weaker assumption $(1 + |x|)^{-1} f \in L^1(\mathbb{R}^n)$.

Hint. First, prove that if $\mu \in S_t$, then $R^*|\mu| = O(|x|^{-1})$ as $|x| \to \infty$. This and the assumption $(1 + |x|)^{-1} f \in L^1(\mathbb{R}^n)$ imply that the function $\int_{S^{n-1}} |f(x)\mu(\alpha, \alpha \cdot x)| d\alpha$ is absolutely integrable. By the Fubini theorem, $f(x)\mu(\alpha, \alpha \cdot x) \in L^1(\mathbb{R}^n \times S^{n-1})$. Therefore, the calculation identical to (10.6.1) proves that $\hat{f} = Rf$ in the distributional sense. $\square$

REMARK 10.6.2. We do not know if the conclusion of Theorem 10.6.1 holds in the case when f is a tempered distribution which is a locally continuous function absolutely integrable along any hyperplane $l_{\alpha p}$. If the conclusion of Theorem 10.6.1 would hold in this case, then Theorem 10.1.1 would assert that for such f the condition $Rf = 0$ implies $f = 0$. Presently, it is not known if such an implication holds, and the reader should consult Section 2.7.3 in order to appreciate the difficulty of the problem.

10.7. The dual Radon transform on distributions

10.7.1. Definition of R^* on certain classes of distributions

In this section we define the dual Radon transform R^* on distributions and investigate some of the properties of the solution μ to the equation $R^*\mu = h$, where h is a distribution. As it was noted in Section 2.5, this problem is interesting because we want to compute functionals from the unknown function f given its Radon transform Rf using the duality relation

$$< f, R^*\mu >= (Rf, \mu). \qquad (10.7.1)$$

Substituting $\mu = KRh$ into (10.7.1) and using the identity $R^*KR = I$, we get

$$< f, h >= (Rf, KRh). \qquad (10.7.2)$$

Now, the left-hand side of (10.7.2) is well defined for any $f \in S$ and $h \in S'$. Therefore, (10.7.2) defines the functional KRh, which acts on the space RS of Radon transforms of functions from S. Denoting $\mu = KRh$, we assume that $R^*\mu := h$. Thus, we may introduce definitions of $(R^*)^{-1}$ and R^* on distributions:

Definition 10.7.1. Let $h \in S'$. Then $(R^*)^{-1}h := KRh$ is a linear functional acting on the space RS of test functions. The value of KRh on a test function Rf, $f \in S$, is defined to be $(KRh, Rf) :=< h, f >$.

Definition 10.7.2. Let $\mu \in KRS'$, that is $\mu = KRh$ for some $h \in S'$. Then $R^*\mu$ is a linear functional acting on the space S of test functions. The value of $R^*\mu$ on $f \in S$ is defined to be

$$< R^*\mu, f >:= (\mu, Rf) :=< h, f > .$$

It is clear that the domain of definition of R^* is not $S'(Z)$, but the space $(RS)'$. The domain of definition of the operator $(R^*)^{-1} = KR$ is a more familiar space S'. This, however, is quite natural. In view of relation (10.7.1), distribution μ should be a functional on RS, i.e. $\mu \in (RS)'$, and, in view of (10.7.2), distribution h should be a functional on S, i.e. $h \in S'$.

Clearly, similar consideration holds if we use the space $\mathcal{D}$ instead of S.

Thus, we have generalized inversion formula (2.5.7)

$$\mu(\alpha, p) = \frac{1}{2(2\pi)^n} \int_{-\infty}^{\infty} |\lambda|^{n-1} \tilde{h}(\lambda\alpha) \exp(-i\lambda p) d\lambda = KRh \qquad (10.7.3)$$

from $h \in S$ to $h \in S'$.

Example 10.7.1. Compute $\mu = (R^*)^{-1} h$, where $h(x) = \delta(x - x_0)$. Clearly, $\tilde{h}(\lambda\alpha) = e^{i\lambda\alpha \cdot x_0}$, and (10.7.3) implies

$$\mu(\alpha, p) = \frac{1}{2(2\pi)^n} \int_{-\infty}^{\infty} |\lambda|^{n-1} e^{-i\lambda(p - \alpha \cdot x_0)} d\lambda$$

$$= \begin{cases} (-1)^{\frac{n-1}{2}} \pi(2\pi)^{-n} \delta^{(n-1)}(p - \alpha \cdot x_0), & n \text{ odd}, \\ (-1)^{\frac{n}{2}} (n-1)!(2\pi)^{-n} |p - \alpha \cdot x_0|^{-n}, & n \text{ even.} \end{cases} \qquad (10.7.4)$$

Equations (10.7.4) and (10.7.1) are equivalent to inversion formulas (2.2.4) and (2.2.5), because

$$(Rf, \mu) = < f, h > = < f, \delta(x - x_0) > = f(x_0).$$

Example 10.7.2. Let $n = 3$ and D be a ball of radius a centered at the origin. Let us compute $\mu = (R^*)^{-1} h$, where h is the characteristic function of the ball D, $h = \chi_D$. Then, we get

$$\tilde{h}(\lambda\alpha) = \int_0^a r^2 \int_{S^2} e^{i\lambda r \alpha \cdot \beta} d\beta \, dr = \int_0^a r^2 \frac{4\pi \sin(\lambda r)}{\lambda r} dr$$

$$= \frac{4\pi}{\lambda^3} [\sin(\lambda a) - (\lambda a) \cos(\lambda a)].$$

Furthermore, Equation (10.7.3) yields

$$\mu(\alpha, p) = \frac{1}{(2\pi)^2} \int_{-\infty}^{\infty} \frac{[\sin(\lambda a) - \lambda a \cos(\lambda a)]}{\lambda} \exp(-i\lambda p) d\lambda$$

$$= (4\pi)^{-1} [-a\delta(a - |p|) + \phi(p)], \quad \phi(p) := \begin{cases} 1, & |p| < a, \\ \frac{1}{2}, & p = a, \\ 0, & |p| > a. \end{cases} \qquad (10.7.5)$$

Here we have used formulas 8 and 9 from [GS, p.359]. It is clear from (10.7.5) that $\mu(\alpha, p)$ is a distribution, μ is not locally integrable, and $\mu = 0$ for $|p| > a$. Alternatively, Equation (10.7.5) can be derived directly from (2.5.8) with $n = 3$ by noting that the Radon transform of the characteristic function of the ball of radius a centered at the origin is $\hat{h} = \pi(a^2 - p^2), p \le a$, and $\hat{h} = 0, p > a$.

Consider a similar example in the case $n = 2$.

Example 10.7.3. Let $h(x)$ be a characteristic function of a disk in $\mathbb{R}^2$ centered at y with radius a. The solution to the equation $R^*\mu = h$ is given by formula (2.5.7). If $n = 2$, then the solution is

$$\mu(\alpha, p) = \frac{1}{8\pi^2} \int\limits_{-\infty}^{\infty} |t|\tilde{h}(t\alpha) \exp(-itp)dt.$$

Let us calculate

$$\tilde{h}(t\alpha) = \int\limits_{|x-y|<a} \exp(it\alpha \cdot x)dx = \frac{2\pi a}{t} J_1(at) \exp(it\alpha \cdot y),$$

where J_1 is the Bessel function. Thus, using formula 11.4.37 in [AS, p.487], we get

$$\mu(\alpha, p) = \frac{a}{2\pi} \int\limits_{0}^{\infty} J_1(at) \cos[t(\alpha \cdot y - p)]dt$$

$$= \begin{cases} 1/(2\pi), & |\alpha \cdot y - p| < a, \\ -\dfrac{a^2/(2\pi)}{(|\alpha\cdot y - p|^2 - a^2)^{1/2}\left(|\alpha\cdot y - p| + \sqrt{|\alpha\cdot y - p|^2 - a^2}\right)}, & |\alpha \cdot y - p| > a. \end{cases}$$

10.7.2. Singularities and singular support of the solution to the equation $R^*\mu = h$

As Example 10.7.2 shows, if h is not smooth, then $\mu = (R^*)^{-1}h$ is also not smooth. In this section we investigate the relationship between singular supports of μ and h and describe the singularities of μ.

Theorem 10.7.1. *Singular support S_μ of the even solution μ to the equation $R^*\mu = h$ can be calculated from the singular support S_h of h as follows: assume S_h to be a hypersurface and let $x_n = g(x')$ be its equation in the local coordinates, $x' \in \mathbb{R}^{n-1}$. Let $q(\beta) = Lg(x')$, where $\beta \in \mathbb{R}^{n-1}$ and L is the Legendre transform. Then $q = q(\beta)$ is the equation of S_μ in the local coordinates.*

Proof. We have the inversion formula $\mu = KRh = K\hat{h}$. Using (2.2.16), we see that the operator $K = \gamma F^{-1}|t|^{n-1}F$ is elliptic. In particular, K preserves singular support. Thus, singular supports of μ and

$\hat{h}$ coincide. This and Theorem 4.1.2 in Section 4.1 prove the Theorem.
□

In particular, Theorem 10.7.1 asserts that $(\alpha, p) \in S_\mu$ if and only if the hyperplane $l_{\alpha p}$ is tangent to S_h at some point. Therefore, α is normal to S_h at the point of tangency.

Now let us consider a specific important example when h is a characteristic function of a bounded domain D with a smooth boundary. In this case, we describe the leading singular term of μ in a neighborhood of singsupp μ.

Theorem 10.7.2. *Let D be a bounded domain with a smooth convex boundary S whose Gaussian curvature is strictly positive, $h(x) = \chi_D(x)$, and $S_\mu = $ singsupp μ. Fix any $(\alpha, p_0) \in S_\mu$. Let x_0 be the point where the plane $l_{\alpha p_0}$ is tangent to S. Then the leading term of the expansion in smoothness of $\mu(\alpha, p)$ as $p \to p_0$, α being fixed, is:*

$$\mu(\alpha, p) \sim \frac{(-1)^m}{2(2\pi)^m} \mathcal{K}^{-\frac{1}{2}} \delta^{(m-1)} (\pm(p - p_0)), \qquad n = 2m + 1, \qquad (10.7.6)$$

$$\mu(\alpha, p) \sim \begin{cases} \frac{(-1)^m (2m-3)!!}{(2\pi)^m 2^{m-0.5}} \mathcal{K}^{-1/2} (\pm(p_0 - p))_+^{0.5-m}, & m > 1, \\ -\frac{\mathcal{K}^{-1/2}}{2\sqrt{2\pi}} (\pm(p_0 - p))_+^{-0.5}, & m = 1, \end{cases} \qquad n = 2m,$$

$$(10.7.7)$$

where $x_+^\lambda := x^\lambda, x > 0$, and $x_+^\lambda := 0, x < 0$, $\mathcal{K} := (R_1 \cdot \ldots \cdot R_{n-1})^{-1}$ is the Gaussian curvature, and R_j are the principal radii of curvature of S at x_0. In formulas (10.7.6), (10.7.7), one should take '+' or '−' if α is the interior or exterior normal to S at x_0, respectively.

REMARK 10.7.1. One can check that the results obtained in Examples 10.7.2 and 10.7.3 are in agreement with Theorem 10.7.2. Taking $m = 1$ and $\mathcal{K} = a^2$ in (10.7.6), we get

$$\mu(\alpha, p) \sim -\frac{a}{4\pi} \delta(\pm(p - p_0)).$$

If α is the interior normal to the sphere $S = \{x \in \mathbb{R}^3 : |x| = a\}$ at x_0, then $p_0 = -a$ and, since $p < 0$ in a neighborhood of $p_0 = -a$, we get

$$\mu(\alpha, p) \sim -\frac{a}{4\pi} \delta(p - p_0) = -\frac{a}{4\pi} \delta(-|p| + a).$$

If α is the exterior normal to the sphere S at x_0, then $p_0 = a$ and, since $p > 0$ in a neighborhood of $p_0 = a$, we get

$$\mu(\alpha, p) \sim -\frac{a}{4\pi} \delta(-p + p_0) = -\frac{a}{4\pi} \delta(-p + a) = -\frac{a}{4\pi} \delta(-|p| + a).$$

The above two formulas are in agreement, as $p \to p_0$, with (10.7.5)

Analogously, one can easily check that the formula obtained in Example 10.7.3 agrees, as $|\alpha \cdot y - p| \to a$, with (10.7.7) for $m = 1$.

Proof Theorem 10.7.2. Let $n = 2m + 1$. Consider, for example, the case when α is the interior normal. The case when α is the exterior normal can be treated similarly. We have, according to (2.5.8):

$$\mu = \frac{(-1)^m}{2(2\pi)^{2m}} \left(\frac{\partial}{\partial p}\right)^{2m} \hat{h}, \qquad (10.7.8)$$

and, according to the results from Exercise 4.7.1, we get

$$\hat{h}(\alpha, p) \sim \frac{(2\pi)^m}{m!} \mathcal{K}^{-\frac{1}{2}} (p - p_0)_+^m.$$

Substituting this formula into (10.7.8), we get

$$\mu(\alpha, p) \sim \frac{(-1)^m}{2(2\pi)^m} \mathcal{K}^{-\frac{1}{2}} \frac{1}{m!} \left(\frac{\partial}{\partial p}\right)^{2m} (p - p_0)_+^m$$

$$= \frac{(-1)^m}{2(2\pi)^m} \mathcal{K}^{-\frac{1}{2}} \left(\frac{\partial}{\partial p}\right)^m (p - p_0)_+^0 = \frac{(-1)^m}{2(2\pi)^m} \mathcal{K}^{-\frac{1}{2}} \delta^{(m-1)} (p - p_0),$$

which proves (10.7.6) with the '+'-sign.

Now let $n = 2m$. Again, we consider only the case when α is the interior normal. Using (2.5.8) and the results from Exercise 4.7.1, we get

$$\mu = \frac{(-1)^m}{2(2\pi)^{2m-1}} \mathcal{H} \left(\frac{\partial}{\partial p}\right)^{2m-1} \hat{h},$$

$$\hat{h}(\alpha, p) \sim \frac{(2\pi)^{m-0.5}}{\Gamma(m + 0.5)} \mathcal{K}^{-\frac{1}{2}} (p - p_0)_+^{m-0.5}.$$

Therefore,

$$\mu(\alpha, p) \sim \frac{(-1)^m}{2(2\pi)^{m-0.5}} \mathcal{K}^{-\frac{1}{2}} \mathcal{H} \frac{1}{\Gamma(m + 0.5)} \left(\frac{\partial}{\partial p}\right)^{2m-1} (p - p_0)_+^{m-0.5}$$

$$= \frac{(-1)^m}{\sqrt{2}(2\pi)^m} \mathcal{K}^{-\frac{1}{2}} \mathcal{H} \left(\frac{\partial}{\partial p}\right)^{m-1} \frac{1}{\sqrt{(p - p_0)_+}}. \qquad (10.7.9)$$

Since derivatives and the Hilbert transform commute and, as elementary calculations show, $\mathcal{H}(p - p_0)_+^{-0.5} \sim (p_0 - p)_+^{-0.5}$ as $p \to p_0$, we obtain

$$\mu(\alpha, p) \sim \frac{(-1)^m}{\sqrt{2}(2\pi)^m} \mathcal{K}^{-\frac{1}{2}} \left(\frac{\partial}{\partial p}\right)^{m-1} \frac{1}{\sqrt{(p_0 - p)_+}}$$

$$= \begin{cases} \frac{(-1)^m (2m-3)!!}{(2\pi)^m 2^{m-0.5}} \mathcal{K}^{-\frac{1}{2}} (p_0 - p)_+^{0.5-m}, & m > 1, \\ -\frac{\mathcal{K}^{-\frac{1}{2}}}{2\sqrt{2\pi}} (p_0 - p)_+^{-0.5}, & m = 1. \end{cases} \qquad (10.7.10)$$

which proves (10.7.7) with the '+'-sign.

Considering the case when α is the exterior normal, using in this case the formulas

$$\hat{h}(\alpha, p) \sim \frac{(2\pi)^m}{m!} \mathcal{K}^{-\frac{1}{2}} (p_0 - p)_+^m, \quad n = 2m + 1,$$

$$\hat{h}(\alpha, p) \sim \frac{(2\pi)^{m-0.5}}{\Gamma(m + 0.5)} \mathcal{K}^{-\frac{1}{2}} (p_0 - p)_+^{m-0.5}, \quad n = 2m,$$

and arguing as above, we prove Equations (10.7.6) and (10.7.7) with the '−'-sign. □

The nature of the singularity of the distribution $\mu = (R^*)^{-1} h$ remains the same for h piecewise smooth with a jump across S. The jump will enter as a factor on the right-hand sides of (10.7.6) and (10.7.7). For the characteristic function this factor is 1.

It is interesting to note that in the case of even n, the leading singular term of $\mu(\alpha, p)$ as $p \to p_0$ is on the exterior side of S, and not on the interior side. By the interior side, we mean values of p such that the hyperplanes $l_{\alpha p}$ intersect the domain D. Let $n = 2$. Since the Hilbert transforms $\mathcal{H}_{p \to t}(p - p_0)_+^{-0.5}$ and $\mathcal{H}_{p \to t}(p_0 - p)_+^{-0.5}$ are bounded as $t \to p_0, t > p_0$ and $t \to p_0, t < p_0$, respectively, Equation (10.7.9) implies that $\mu(\alpha, p)$ is bounded as p approaches p_0 from the inside.

CHAPTER 11

ABEL-TYPE INTEGRAL EQUATION

11.1. The classical Abel equation

Consider the classical Abel equation

$$Af := \frac{1}{\Gamma(1-a)} \int_0^x \frac{f(t)dt}{(x-t)^a} = g(x), \quad x \ge 0, \qquad (11.1.1)$$

where $\Gamma(x)$ is the gamma function, $g(x)$ is a given function, and $f(t)$ is to be found for $t \ge 0$. In the classical theory, one assumes that $0 < a < 1$, which ensures absolute convergence of the integral in (11.1.1).

Define the distribution

$$\Phi_\lambda := \frac{x_+^{\lambda-1}}{\Gamma(\lambda)}.$$

Then (11.1.1) can be written as a convolution:

$$Af = f * \Phi_{1-a} = g. \qquad (11.1.2)$$

It is easy to check that if

$$g = f * \Phi_n = \frac{1}{(n-1)!} \int_0^x f(t)(x-t)^{n-1}dt,$$

then $g^{(n)} = f$. Therefore, $\Phi_\lambda * \Phi_\mu = \Phi_{\lambda+\mu}$, $\lambda, \mu = 0, 1, \ldots$. If we understand the convolution of distributions as in Section 14.2.1.3, then this relation can be generalized to all real λ and μ:

$$\Phi_\lambda * \Phi_\mu = \Phi_{\lambda+\mu}, \quad \lambda, \mu \in \mathbb{R}. \qquad (11.1.3)$$

In particular, substituting $\lambda = -n$ and $\mu = n$ into (11.1.3), we get the obvious identities

$$\Phi_{-n} * \Phi_n * f = f, \quad \Phi_0 = \delta,$$

because $\Phi_{-n} = \frac{d^n}{dx^n}$ (see Section 14.2.1.2). Suppose λ is not an integer. If $\lambda < 0$, the function $\Phi_\lambda * f$ is called the *fractional order derivative of*

325

f. If $\lambda > 0$, the function $\Phi_\lambda * f$ is called the *fractional order integral of* f. Using (11.1.2), (11.1.3), and the identity $\Phi_0 = \delta$, we obtain the solution to (11.1.1):

$$f = \Phi_{a-1} * g = \Phi_a * \Phi_{-1} * g = \Phi_a * g' = \frac{1}{\Gamma(a)} g' * x_+^{a-1}$$

$$= \frac{1}{\Gamma(a)} \int_0^x \frac{g'(t)dt}{(x-t)^{1-a}}. \tag{11.1.4}$$

Note that if $0 < a < 1$, the integral in (11.1.4) is absolutely convergent provided g is smooth enough.

11.2. Abel-type equations

Consider Equations (2.1.63) and (2.2.39). These equations can be written in the form

$$\int_0^\infty b\left(\frac{p}{r}\right) h(r) \frac{dr}{r} = g(p). \tag{11.2.1}$$

In particular, (2.2.39) is obtained by denoting

$$b(t) = \begin{cases} \gamma_{nl} G_l^{\left(\frac{n-2}{2}\right)}(t)(1-t^2)^{\frac{n-3}{2}}, & 0 \le t \le 1, \\ 0, & t \ge 1, \end{cases} \quad h(r) = f_l(r) r^{n-1}. \tag{11.2.2}$$

Equation (11.2.1) can be solved using the Mellin transform, which is defined as follows:

$$Mh(x) := \int_0^\infty t^{x-1} h(t)dt, \quad x > 0. \tag{11.2.3}$$

Multiply (11.2.1) by p^{x-1} and integrate with respect to p over the positive semi-axis to get

$$Mg(x) = \int_0^\infty p^{x-1} \int_0^\infty b\left(\frac{p}{r}\right) h(r) \frac{dr}{r} dp. \tag{11.2.4}$$

Changing the order of integration in (11.2.4), we get

$$Mg(x) = \int_0^\infty \frac{h(r)}{r} \int_0^\infty p^{x-1} b\left(\frac{p}{r}\right) dp\, dr$$

$$= \int_0^\infty h(r) r^{x-1} \int_0^\infty t^{x-1} b(t)dt\, dr = Mh(x) \cdot Mb(x). \tag{11.2.5}$$

Integral of the type (11.2.1) is called the *convolution associated with the Mellin transform*. Equation (11.2.5) implies that the Mellin transform of a convolution equals to the product of the Mellin transforms:

$$M(h * b) = Mh \cdot Mb, \qquad (11.2.6)$$

which is analogous to the similar property of the Fourier transform.

Exercise 11.2.1. Verify the identities:

$$Mh'(x) = (1 - x)Mh(x - 1), \qquad (11.2.7a)$$

$$M(t^s h)(x) = Mh(x + s), \qquad (11.2.7b)$$

$$M(t^j h^{(j)})(x) = (-1)^j \frac{\Gamma(x + j)}{\Gamma(x)} Mh(x). \qquad (11.2.7c)$$

Using the table of the Mellin transforms from [Sne], we get

$$f(t) = \begin{cases} G_l^{(\lambda)}(t)(1 - t^2)^{\lambda - \frac{1}{2}}, & 0 \le t < 1, \\ 0, & t \ge 1, \end{cases}$$

$$Mf(x) = c_1(\lambda) \frac{\Gamma(x)2^{-x}}{\Gamma\left(\lambda + \frac{x+1+l}{2}\right) \Gamma\left(\frac{x+1-l}{2}\right)}, \quad c_1(\lambda) = \sqrt{\pi}\Gamma(\lambda + 0.5), \ s > 0, \tag{11.2.8}$$

$$\psi(t) = \begin{cases} G_l^{(\lambda)}(1/t)(1 - t^2)^{\lambda - \frac{1}{2}}, & 0 < t < 1, \\ 0, & t \ge 1, \end{cases}$$

$$M\psi(x) = c_2(\lambda) \frac{2^{x-1}\Gamma\left(\lambda + \frac{x+l}{2}\right) \Gamma\left(\frac{x-l}{2}\right)}{\Gamma(x + 2\lambda)}, \quad c_2(\lambda) = \frac{\Gamma(2\lambda)}{\Gamma(\lambda)}, \ s > l. \tag{11.2.9}$$

If $\lambda = 0$, we define $c_2(0) = \lim_{\lambda \to 0} c_2(\lambda)$. Using (11.2.2) and (11.2.8) with $\lambda = \frac{n-2}{2}$, we find

$$\frac{1}{Mb(x-1)} = \frac{1}{\gamma_{nl}c_1((n-2)/2)} \frac{\Gamma\left(\frac{n-2}{2} + \frac{x+l}{2}\right) \Gamma\left(\frac{x-l}{2}\right)}{\Gamma(x-1)2^{-x+1}}$$

$$= \frac{c_2((n-2)/2) \ 2^{x-1}\Gamma\left(\frac{n-2}{2} + \frac{x+l}{2}\right) \Gamma\left(\frac{x-l}{2}\right)}{c_3} \frac{\Gamma(x+n-2)}{\Gamma(x+n-2)} \frac{\Gamma(x+n-2)}{\Gamma(x-1)},$$

$$c_3 = \gamma_{nl}c_1((n-2)/2)c_2((n-2)/2).$$

Using (11.2.9), we can rewrite the last equation as follows

$$\frac{1}{Mb(x-1)} = \frac{1}{c_3} M\psi \frac{\Gamma(x+n-2)}{\Gamma(x-1)}. \qquad (11.2.10)$$

Using consequtively Equations (11.2.7b) with $s = -1$, (11.2.5), (11.2.10), (11.2.7a), and (11.2.7c), we get

$$M(h/r)(x) = Mh(x-1) = \frac{1}{Mb(x-1)}Mg(x-1)$$

$$= \frac{1}{c_3}M\psi\frac{\Gamma(x+n-2)}{\Gamma(x-1)}Mg(x-1)$$

$$= \frac{-1}{c_3}M\psi\frac{\Gamma(x+n-2)}{\Gamma(x)}(1-x)Mg(x-1)$$

$$= \frac{-1}{c_3}M\psi\frac{\Gamma(x+n-2)}{\Gamma(x)}Mg'(x)$$

$$= \frac{(-1)^{n-1}}{c_3}M\psi M\left[t^{n-2}g^{(n-1)}\right].$$

Therefore, (11.2.6) yields

$$\frac{h(r)}{r} = \frac{(-1)^{n-1}}{c_3}\int_0^\infty \psi(r/t)t^{n-2}g^{(n-1)}(t)\frac{dt}{t},$$

which is equivalent to (2.2.42).

11.3. Reduction of Equation (2.2.42) to a more stable one

If $n = 2$, one has $C_l^{\left(\frac{n-2}{2}\right)}(x) = C_l^{(0)}(x)$, $T_l(x) = \frac{C_l^{(0)}(x)}{C_l^{(0)}(1)}, l \geq 0$ (see (14.4.33)). Equations (2.2.39) and (2.2.42) are equivalent to

$$2\int_p^\infty \frac{T_{|l|}\left(\frac{p}{r}\right)}{\sqrt{1-\frac{p^2}{r^2}}}f_l(r)dr = g_l(p), \quad l = 0, \pm 1, \pm 2, \ldots,$$

and

$$f_l(r) = -\frac{1}{\pi}\int_r^\infty \frac{T_{|l|}\left(\frac{p}{r}\right)}{(p^2-r^2)^{\frac{1}{2}}}g_l'(p)dp, \quad l = 0, \pm 1, \pm 2, \ldots. \qquad (11.3.1)$$

Here the coefficients of the spherical harmonics expansion $f_l(r)$ and $g_l(p)$ are defined according to (2.2.34) and (2.2.36). Formula (11.3.1) is not convenient for numerical implementation. Indeed, the values of $T_{|l|}(x)$ grow as $x \to +\infty$:

$$T_l(x) \geq \frac{(x+\sqrt{x^2-1})^l}{2} \quad \text{for} \quad l > 0 \quad \text{and} \quad x \to +\infty.$$

This estimate follows from formula (14.4.30). Indeed, using that cosh is even and $\text{arccosh} x = \pm \ln(x + \sqrt{x^2 - 1})$, we get

$$\cosh(\pm l \ \text{arccosh} \ x) \geq \frac{1}{2} \exp\{\ln(x + \sqrt{x^2 - 1})^l\}$$
$$= \frac{1}{2}(x + \sqrt{x^2 - 1})^l \sim 2^{l-1}x^l, \quad x \to +\infty.$$

Let us transform (11.3.1) using the moment conditions for $g_l(p)$:

$$\int\limits_{-\infty}^{\infty} p^m g'(\alpha, p)dp = -m \int\limits_{-\infty}^{\infty} p^{m-1}g(\alpha, p)dp = -m\mathcal{P}_{m-1}(\alpha),$$

where $\mathcal{P}_{m-1}(\alpha)$ is a homogeneous polynomial of degree $m - 1$ with respect to the variables $\alpha_1 = \cos\theta$ and $\alpha_2 = \sin\theta$. Thus

$$\int\limits_{-\infty}^{\infty} p^m g_l'(p)dp = \frac{-m}{\sqrt{2\pi}} \int\limits_{-\pi}^{\pi} \int\limits_{-\infty}^{\infty} p^{m-1}g(\alpha, p)dp \exp(-il\theta)d\theta$$

$$= \frac{-m}{\sqrt{2\pi}} \int\limits_{-\pi}^{\pi} \exp(-il\theta)\mathcal{P}_{m-1}(\cos\theta, \sin\theta)d\theta = 0 \quad \text{for} \quad |l| \geq m. \quad (11.3.2)$$

Since g is even: $g(\alpha, p) = g(-\alpha, -p)$, we get

$$g_l'(-p) = (-1)^{l+1}g_l'(p).$$

Thus $g_l'(p)$ is even if l is odd and odd if l is even. Let $Q_{|l|}(p)$ be a polynomial of degree $< |l|$ which is even when l is odd and odd when l is even. Then

$$\int\limits_{0}^{\infty} Q_{|l|}(p)g_l'(p)dp = \frac{1}{2} \int\limits_{-\infty}^{\infty} Q_{|l|}(p)g_l'(p)dp = 0, \quad (11.3.3)$$

where Equation (11.3.2) has been used. Let us rewrite (11.3.1) using Equation (11.3.3):

$$f_l(r) = -\frac{1}{\pi r} \int\limits_{r}^{\infty} \left[\left(\frac{p^2}{r^2} - 1\right)^{-\frac{1}{2}} T_{|l|}\left(\frac{p}{r}\right) - Q_{|l|}\left(\frac{p}{r}\right) \right] g_l'(p)dp$$

$$+ \frac{1}{\pi r} \int\limits_{0}^{r} Q_{|l|}\left(\frac{p}{r}\right) g_l'(p)dp. \quad (11.3.4)$$

Let us choose

$$Q_{|l|}(x) = U_{|l|-1}(x), \tag{11.3.5}$$

where $U_{|l|}(x)$ is the Chebyshev polynomial of the second kind (see Section 14.4.3.3). Then

$$(x^2 - 1)^{-\frac{1}{2}} T_{|l|}(x) - U_{|l|-1}(x) = \frac{(x + \sqrt{x^2 - 1})^{-|l|}}{\sqrt{x^2 - 1}}.$$

Using (11.3.4), (11.3.5), and the last equation, we get:

$$f_l(r) = -\frac{1}{\pi r} \int_r^\infty \frac{\left(\frac{p}{r} + \sqrt{\frac{p^2}{r^2} - 1}\right)^{-|l|} g_l'(p)dp}{\sqrt{\frac{p^2}{r^2} - 1}}$$

$$+ \frac{1}{\pi r} \int_0^r U_{|l|-1}\left(\frac{p}{r}\right) g_l'(p)dp. \tag{11.3.6}$$

We see that the coefficient in front of $g_l'(p)$ in (11.3.6) decays as $p^{-|l|-1}$, $p \to \infty$. Therefore, calculations according to (11.3.6) are more stable with respect to small perturbations of $g_l'(p)$ than calculations based on (11.3.1).

11.4. Finding locations and values of jumps of the solution to the Abel equation

In this section we shall consider the problem of finding discontinuities of the solution to the Abel equation (11.1.1), where it is assumed that $0 < a < 1$. Using formula (11.1.4) for the solution to (11.1.1), we fix $\rho > 0$ and introduce functions f_ρ and f_ρ^c:

$$f_\rho(x) = \frac{1}{\Gamma(a)} \int_{x-\rho}^x \frac{g'(\xi)d\xi}{(x-\xi)^{1-a}}, \quad f_\rho^c(x) = \frac{1}{\Gamma(a)} \int_0^{x-\rho} \frac{g'(\xi)d\xi}{(x-\xi)^{1-a}}, \quad x \geq \rho.$$

$$\tag{11.4.1}$$

Lemma 11.4.1. *Let f be piecewise continuous. Then the function $f_\rho^c = f - f_\rho$ is continuous.*

Proof. The second formula in (11.4.1) yields

$$f_\rho^c(x) = \frac{1}{\Gamma(a)} \left(\frac{g(x-\rho)}{\rho^{1-a}} - (1-a) \int_0^{x-\rho} \frac{g(\xi)d\xi}{(x-\xi)^{2-a}} \right). \tag{11.4.2}$$

Clearly, the integral in (11.4.2) is continuous in x. It is well-known that if g is given by (11.1.1), where f is piecewise-continuous, then g is continuous (see [GV, p. 70]). Thus f_ρ^c is continuous. $\square$

Corollary 11.4.1. *Functions f and f_ρ have the same discontinuities and the same sizes of jumps across these discontinuities.*

Lemma 11.4.2. *Suppose one has $f(x) = f_0(x) + D\theta(x - x_0)$ in a neighborhood of a point $x_0 > 0$, where f_0 is smooth and θ is a step function: $\theta(x) := 1, x > 0$, and $\theta(x) := 0, x < 0$. Then, in a neighborhood of x_0, f_ρ is given by*

$$f_\rho(x) = \varphi(x) + D\psi\left(\frac{x - x_0}{\rho}\right),\qquad (11.4.3)$$

where φ is smooth, and

$$\psi(t) = \begin{cases} 0, & t \le 0, \\ 1, & 0 < t \le 1, \\ \frac{\sin(a\pi)}{\pi} \int_0^1 \frac{d\xi}{\xi^{1-a}(t-\xi)^a}, & t \ge 1. \end{cases} \qquad (11.4.4)$$

In particular, ψ is continuous at $t = 1$ and $\psi(t) = O(t^{-a})$, $t \to \infty$.

Proof. If one substitutes $f = f_0 + D\theta(x - x_0)$ into (11.1.1), then the smooth term f_0 is mapped into a smooth term and the jump is mapped into a function which is easily calculated analytically. Thus, in a neighborhood of $x = x_0$, we have

$$g'(x) = \frac{D}{\Gamma(1-a)} \frac{1}{(x - x_0)_+^a} + \varphi_1(x),$$

where φ_1 is smooth. Substitution into the first equation in (11.4.1) yields

$$f_\rho(x) = \frac{D}{\Gamma(1-a)\Gamma(a)} \int_{\max(x-\rho,x_0)}^{x} \frac{d\xi}{(x-\xi)^{1-a}(\xi - x_0)^a} + \varphi(x), \quad (11.4.5)$$

where φ is smooth. Using the identity

$$\int_{x_0}^{x} \frac{d\xi}{(x-\xi)^{1-a}(\xi - x_0)^a} = \Gamma(1-a)\Gamma(a)$$

for the case $x - \rho \le x_0 < x$ and a change of variables in integral (11.4.5) in the case $x - \rho \ge x_0$, we obtain the assertions of the lemma. $\square$

The graph of the function $\psi(t)$ with $a = 0.5$ is presented in Figure 11.4.1.

Corollary 11.4.2. *In a neighborhood of x_0, $f_\rho'(x)$ is given by*

$$f_\rho'(x) = \varphi'(x) + D[\delta(x - x_0) + \rho^{-1}\psi_1((x - x_0)/\rho)], \qquad (11.4.6)$$

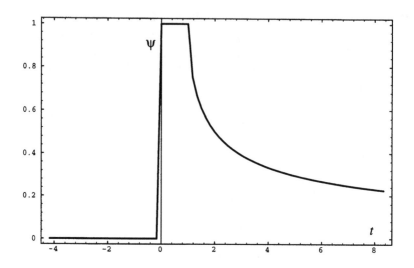

FIGURE 11.4.1. The graph of the function $\psi(t)$ with $a = 0.5$.

where φ' is smooth and $\psi_1(t)$ has the properties

$$\psi_1(t) = 0, \ t < 1; \quad \psi_1(t) \sim -\frac{\sin a\pi}{\pi}(t-1)^{-a}, \ t \to 1^+;$$
$$\psi_1(t) = O(t^{-(a+1)}), \ t \to +\infty. \tag{11.4.7}$$

Using Equations (11.4.6) and (11.4.7), we can construct an algorithm for locating and estimating the jumps of f using f_ρ. Indeed, if we compute the mollified function $f_{\rho\epsilon} := f_\rho * w_\epsilon$, where w_ϵ is a mollifier:

$$w_\epsilon(x) = w_1(x/\epsilon), \quad w_1(x) = 0 \text{ for } |x| \geq 1,$$
$$w_1(x) > 0 \text{ for } |x| < 1, \quad w_1(0) = 1,$$

then in a neighborhood of the jump point x_0 of f, the function $f'_{\rho\epsilon}$ will have two peaks: the larger of them (corresponding to the delta-function in (11.4.6)) will coincide with the point $x = x_0$, and the smaller one (corresponding to the singularity of $\psi_1((x-x_0)/\rho)$) will be situated at $x = x_0 + \rho$. Hence, the location of the larger peak of $f'_{\rho\epsilon}$ determines the discontinuity of f, and the value of $f'_{\rho\epsilon}$ at this peak equals to the value of the jump of f at this point. Note that we do not impose the standard condition $\int_{-\infty}^{\infty} w_\epsilon(x)dx = 1$. Since f'_ρ is a distribution containing delta-function, the normalization condition becomes $w_\epsilon(0) = 1$.

From (11.1.1) and (11.4.1) it follows that

$$f'_{\rho\epsilon}(x) = \frac{1}{\Gamma(a)} \int_{x-\rho}^{x} \frac{1}{(x-\xi)^{1-a}} \frac{\partial(g * w'_\epsilon)}{\partial\xi} d\xi$$

$$= \frac{1}{\Gamma(a)} \int_{x-\rho-\epsilon}^{x+\epsilon} \tilde{w}_\epsilon(x-\xi)g(\xi)d\xi, \qquad (11.4.8)$$

$$\tilde{w}_\epsilon(t) = \int_{\max(0,t-\epsilon)}^{\min(t+\epsilon,\rho)} \frac{w''_\epsilon(t-\xi)}{\xi^{1-a}} d\xi, \quad -\epsilon \le t \le \rho + \epsilon,$$

$$\tilde{w}_\epsilon(t) = 0, \quad t \le -\epsilon \text{ or } t \ge \rho + \epsilon. \qquad (11.4.9)$$

Equation (11.4.3) implies that in a neighborhood of $x = x_0$ we have

$$f'_{\rho\epsilon}(x) = \varphi'(x) * w_\epsilon(x) + D\left[\frac{\partial}{\partial x}\psi\left(\frac{x-x_0}{\rho}\right)\right] * w_\epsilon(x)$$

$$= \varphi(x) * w'_\epsilon(x) + D\psi\left(\frac{x-x_0}{\rho}\right) * w'_\epsilon(x).$$

The graph of a function $\psi(t) * w'_1(t)$ with $w_1(t) = (1-t^2)^4_+$ is presented in Figure 11.4.2.

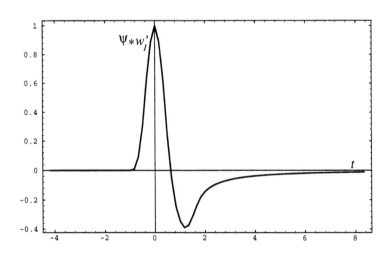

FIGURE 11.4.2. The graph of a function $\psi(t) * w'_1(t)$ with $w_1(t) = (1-t^2)^4_+$.

In Figure 11.4.3 we presented some results on the recovery of discontinuities of f from $f'_{\rho\epsilon}$. The dot-dashed curve is the Abel transform g of the discontinuous function

$$f(x) = \begin{cases} 4, & x < 0.5, \\ 2, & x > 0.5. \end{cases}$$

The Abel transform g was computed at 201 points equispaced on the interval $[0, 1]$. The jump of f at $x_0 = 0.5$ equals $D = f(x_0 + 0) - f(x_0 - 0) = -2$. The big round dot represents the location and the value of the jump of f. The solid curve is the graph of $f'_{\rho\epsilon}$ computed according to (11.4.8) and (11.4.9) with $\rho = \epsilon = 6\Delta x$, where Δx is the discretization step in x. We see a very good agreement between the true and estimated jumps of the unknown function f.

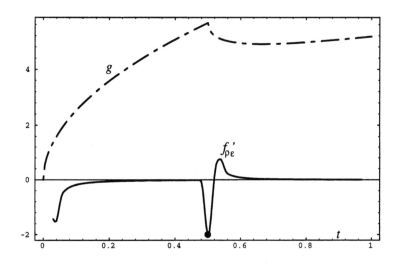

FIGURE 11.4.3. Results on the recovery of discontinuities of f from $f'_{\rho\epsilon}$.

CHAPTER 12

MULTIDIMENSIONAL ALGORITHM
FOR FINDING DISCONTINUITIES OF
SIGNALS FROM NOISY DISCRETE DATA

12.1. Introduction

In this chapter we propose new algorithms for finding discontinuity surface S of a signal $s(x)$ specified with random errors at a regular grid in $\mathbb{R}^p$, $p \geq 1$. Let the unknown standard deviation of the noise be σ. The following assumptions are used:

(1) one is given the radius R with the property $|s(x) - s(x_0)| \ll \sigma$ for any x, $|x - x_0| \leq R$, such that the line segment $[x, x_0]$ does not intersect a discontinuity surface S of the signal $s(x)$;

(2) the errors are independent, identically distributed random variables with common (unknown) distribution function $F(x)$ satisfying the condition: $\exists F^{-1}(\gamma)$ for any γ, $0 < \gamma < 1$;

(3) the desired probability ϵ of the 'false alarm' is given;

(4) the standard deviation of random errors is finite.

Two variants of the general problem of finding discontinuities of a signal from its noisy measurements are considered.

Variant I. Suppose additionally that the discontinuity surface S can be accurately approximated by a plane in the neigborhood $|x - x_0| \leq R$ of any point $x_0 \in S$. The algorithm operates as follows. In a scanning fashion, one moves the center of the ball with radius R (a sliding window) over the grid, and for each position of the ball the decision is made: whether $\tilde{x}_j \in S$ or $\tilde{x}_j \notin S$, where $\tilde{x}_j$ is the current center of the ball. To choose between the two hypotheses, one arranges the observed values from the window in ascending order of magnitude and computes the centroid x_j^+ of the second half of the points, i.e. using the half which contains the points with larger values of the signal. Then the distance $|\tilde{x}_j - x_j^+|$ is compared with the threshold A_0 (which is to be defined below). If $|\tilde{x}_j - x_j^+| \leq A_0$, one decides that $\tilde{x}_j \notin S$. If $|\tilde{x}_j - x_j^+| > A_0$, one decides that $\tilde{x}_j \in S$. The grounds for such a decision rule are quite clear. If $\tilde{x}_j \notin S$, the points with small and large values of the signal are almost uniformly mixed inside the ball, hence the centroid x_j^+ is close

335

to the center $\tilde{x}_j$. If $\tilde{x}_j \in S$, the points with smaller values of the signal (on one side of S) and the points with larger values of the signal (on the other side of S) are spatially separated (although the noise can blur this separation to a certain extent). Hence, the distance $|\tilde{x}_j - x_j^+|$ is large. The threshold value A_0 is computed so that the probability of the 'false alarm' (one decides that $\tilde{x}_j \in S$ when $\tilde{x}_j \notin S$) equals ϵ. If the decision '$\tilde{x}_j \in S$' has been made, then the orientation of S at the point x_j is determined by the normal vector to S, namely the vector joining the points $\tilde{x}_j$ and x_j^+.

Variant II. In this case, the assumption is that the discontinuity surface S is the union of two smooth surfaces S_1 and S_2 situated close to each other at a constant distance $2d < 2R$. Suppose, for example, that the values of the signal between S_1 and S_2 are larger than in the neighborhoods adjoint to S_1 and S_2. Let U be the domain between S_1 and S_2, S_0 be the surface situated at the middle between S_1 and S_2, and M_0 be the number of grid points inside the set $B_R(\tilde{x}_j) \cap U$, $\tilde{x}_j \in S_0$. The algorithm operates as follows. One slides a ball $B_R(\tilde{x}_j)$ over the grid and for each $\tilde{x}_j$ decides between the two hypotheses '$\tilde{x}_j \in S_0$' and '$\tilde{x}_j \notin S_0$'. To this end one arranges the observed values from the window in ascending order. If $\tilde{x}_j \in S_0$, it is likely that the majority of the last M_0 points $\{x_{i_k}\}_{k=2M-M_0+1}^{2M}$ (i.e. the points with the largest values of the signal) belong to $B_R(\tilde{x}_j) \cap U$, hence, they are close to the diameter plane $B_R(\tilde{x}_j) \cap S_0$ and the minimal eigenvalue λ_{min} of the matrix

$$B_{M_0} := (b_{mn}), \quad b_{mn} = h \sum_{k=2M-M_0+1}^{2M} \left(x_{i_k}^{(m)} - \tilde{x}_j^{(m)}\right) \cdot \left(x_{i_k}^{(n)} - \tilde{x}_j^{(n)}\right),$$
$$1 \le m, n \le p.$$

is small. Here $x_{i_k}^{(m)}$ is the m-th coordinate of the point x_{i_k}, $m = 1, \ldots, p$; $h := h_1 \cdot \ldots \cdot h_p$, where h_i is the grid step size along the i-th direction, $i = 1, \ldots, p$. If $B_R(\tilde{x}_j) \cap U = \varnothing$, the points $\{x_{i_k}\}_{k=2M-M_0+1}^{2M}$ are likely to be almost uniformly distributed over $B_R(\tilde{x}_j)$ and λ_{min} is large. Thus, to make a decision, one compares λ_{min} with the threshold Λ_0. If $\lambda_{min} < \Lambda_0$, the hypothesis '$\tilde{x}_j \in S_0$' is accepted. If $\lambda_{min} > \Lambda_0$, the hypothesis '$\tilde{x}_j \notin S_0$' is accepted. The threshold value Λ_0 is computed so that the probability of 'false alarm' equals ϵ. If the decision '$\tilde{x}_j \in S_0$' has been made, the orientation of S_0 at the point $\tilde{x}_j$ is determined by the normal vector to S_0, namely, the eigenvector corresponding to λ_{min}. This algorithm can be easily modified for the case when the values of the signal inside U are smaller than in its neighborhood or for the case when they are larger than values from one neighborhood (say, adjoint to S_1) and smaller than values from the other neighborhood (adjoint to S_2).

The problem of finding discontinuities of a function is very important in many applications. In image analysis, Variant I is known as 'edge detection', and Variant II is known as 'thin line detection'. Since the terminology of image analysis is very convenient, we will use it in what follows.

Chapter 12 is organized as follows. In Section 12.2 the edge detector is described. In Section 12.3 the thin line detector is described. In Section 12.4 the above two algorithms are generalized, so that the resulting scheme can be applied for the solution of other pattern recognition problems. The spot detection problem is considered as an example. In Section 12.5 we prove the consistency of the proposed edge detector: the probability of missing the points from the discontinuity hypersurface S and the error of the recovery of its orientation go to zero as the grid step size goes to zero. In Section 12.6 similar results are proved for the proposed thin line detector. In Section 12.7 we prove the consistency of the general scheme. In Section 12.8 the results of testing the proposed algorithms are presented.

12.2. Edge detection algorithm

Let us consider the following model. Let the signal $s(x)$, $x \in \mathbb{R}^p$, $p \geq 1$ have a discontinuity hypersurface S (which might consist of several disjoint components) and be a smooth function at other points $(x \notin S)$. The known data are the values

$$u_i = s_i + n_i, \quad s_i := s(x_i), \tag{12.2.1}$$

where $x_i := (x_i^{(1)}, \ldots, x_i^{(p)})$. The observation grid is supposed to be rectangular with the step size h_k along the k-th direction, $k = 1, \ldots, p$. The random errors n_i are assumed to be independent, identically distributed. Let us assume that we are given the radius R such that the signal $s(x)$ is approximately constant inside any ball $B_R(x_0) := \{x : |x - x_0| \leq R\}$ not intersecting S, that is $|\nabla s(x)|R \ll \sigma$. Let us also assume that the discontinuity surface S is sufficiently smooth so that inside any ball $B_R(x_0)$, $x_0 \in S$, it can be approximated by a hyperplane. In this case we say that S is locally flat.

Consider the point $\tilde{x}_j := x_j + (h_1/2, \ldots, h_p/2) \in S$. Let x_{i_k}, $k = 1, \ldots, 2M$, be the set of grid points which are situated inside $B_R(\tilde{x}_j)$ and s_{i_k}, $k = 1, \ldots, 2M$, be the corresponding values of the signal. Assume that these values have been arranged in ascending order $s_{i_1} \leq s_{i_2} \leq \cdots \leq s_{i_{2M}}$. Since S is locally flat in the neighborhood of $\tilde{x}_j \in S$, one sees that S divides $B_R(\tilde{x}_j)$ into two equal halfballs $B_R^-(\tilde{x}_j)$ and $B_R^+(\tilde{x}_j)$. Furthermore, one has $x_{i_1}, \ldots, x_{i_M} \in B_R^-(\tilde{x}_j)$, $x_{i_{M+1}}, \ldots, x_{i_{2M}} \in B_R^+(\tilde{x}_j)$. Thus, if one calculates the centroid of the points $x_{i_{M+1}}, \ldots, x_{i_{2M}}$, namely, $x_j^+ = (x_{i_{M+1}} + \cdots + x_{i_{2M}})/M$, then the straight line joining $\tilde{x}_j$ and

x_j^+ is perpendicular to S at the point $\tilde{x}_j$ (this line contains also the centroid of the points $x_{i_1}, \ldots, x_{i_M}$). Since the two sets $\{x_{i_k}\}_{k=1}^M$ and $\{x_{i_k}\}_{k=M+1}^{2M}$ are not spatially mixed ($\{x_{i_k}\}_{k=1}^M \in B_R^-(\tilde{x}_j)$, $\{x_{i_k}\}_{k=M+1}^{2M} \in B_R^+(\tilde{x}_j)$, $B_R^-(\tilde{x}_j) \cap B_R^+(\tilde{x}_j) = \varnothing$), the distance $|\tilde{x}_j - x_j^+|$ is the largest possible (compared with the case when the sets $\{x_{i_k}\}_{k=1}^M$ and $\{x_{i_k}\}_{k=M+1}^{2M}$ are mixed). Hence, the following algorithm is natural:

Step1. Fix a point $\tilde{x}_j$. Arrange the observed values from $B_R(\tilde{x}_j)$ in ascending order $u_{i_1} \le u_{i_2} \le \cdots \le u_{i_{2M}}$ with $u_{i_k} := u(x_{i_k})$ and calculate the centroid x_j^+ of the set $\{x_{i_k}\}_{k=M+1}^{2M}$.

Now, consider any point $\tilde{x}_j$ such that $B_R(\tilde{x}_j) \cap S = \varnothing$ and apply Step 1 to this $B_R(\tilde{x}_j)$. Since the noise and signal are independent and the signal is approximately constant inside $B_R(\tilde{x}_j)$, one sees that the sets $\{x_{i_k}\}_{k=1}^M$ and $\{x_{i_k}\}_{k=M+1}^{2M}$ are almost uniformly mixed and the distance $|\tilde{x}_j - x_j^+|$ is close to zero. Therefore, one can formulate the following decision rule:

Step 2. Select the threshold $A_0 > 0$, as described in Step 0 below. Compare the distance $|\tilde{x}_j - x_j^+|$ with the threshold A_0. If $|\tilde{x}_j - x_j^+| \le A_0$, the hypothesis '$\tilde{x}_j \notin S$' is accepted. If $|\tilde{x}_j - x_j^+| > A_0$, the hypothesis '$\tilde{x}_j \in S$' is accepted and the orientation of S inside $B_R(\tilde{x}_j)$ is determined by the vector $\tilde{x}_j - x_j^+$.

The threshold A_0 is determined from:

$$P\{|\tilde{x}_j - x_j^+| > A_0 \big| B_R(\tilde{x}_j) \cap S = \varnothing\} = \epsilon, \qquad (12.2.2)$$

where $P\{U|V\}$ is the conditional probability of U, given V, and ϵ, $0 < \epsilon < 1$, is an a priori fixed number. A convenient way to compute A_0 from (12.2.2) is by using the Monte Carlo method (see Section 14.10.5) as described in Step 0:

Step 0. Fix an arbitrary ball $B_R(\tilde{x}_j)$. Model a sufficiently large number of the samples of u_i with $s_i \equiv 0$ and an arbitrary fixed common probability density function (PDF) for n_i. Using Step 1, compute the distance $|\tilde{x}_j - x_j^+|$ for each sample. Select A_0 so that $|\tilde{x}_j - x_j^+| > A_0$ in $100\epsilon\%$ cases.

Since in Step 1 we use only the ranks of observations and not the actual values of observations, it is clear that A_0, indeed, does not depend on a chosen PDF.

12.3. Thin line detection algorithm

In this section, we consider the particular case when the discontinuity surface S is the union of two surfaces S_1 and S_2 situated close to each other at a constant distance $\mathrm{dist}(x_1, S_2) = \mathrm{dist}(x_2, S_1) = 2d < 2R$ for any $x_i \in S_i$, $i = 1, 2$. Assume that d is known. The distance here is defined in the usual way $\mathrm{dist}(x, S) := \inf_{y \in S} |x - y|$. We also assume that the surfaces S_i, $i = 1, 2$, are sufficiently smooth so that they can be accurately approximated by a plane inside any ball $B_R(x_0)$, $x_0 \in S_i$. Assume also that the values of the signal at the points between S_1 and S_2 are greater than in the neighborhoods adjoint to S_1 and S_2. The problem is to find S_1 and S_2.

Let S_0 be the surface situated at the middle between S_1 and S_2, that is the midpoint of every interval $[x_1, x_2]$, $x_1 \in S_1$, $x_2 \in S_2$, perpendicular to S_1 and S_2 belongs to S_0. Since the distance between S_1 and S_2 is known, it is sufficient to find the surface S_0. To solve this problem, consider an arbitrary ball $B_R(\tilde{x}_j)$, $\tilde{x}_j \in S_0$. Let M_0 be the number of points from $B_R(\tilde{x}_j)$ situated between S_1 and S_2. Clearly, M_0 is the same for any ball $B_R(\tilde{x}_j)$, $\tilde{x}_j \in S_0$. Thus if one arranges the observed values of the signal from $B_R(\tilde{x}_j)$ in ascending order $u_{i_1} \leq u_{i_2} \leq \cdots \leq u_{i_{2M}}$, it is likely that the last M_0 points $\{x_{i_k}\}_{k=2M-M_0+1}^{2M}$ lie near the diameter plane $S_j^{(0)} = B_R(\tilde{x}_j) \cap S_0$. This means that the quantity

$$\Delta\big(\{x_{i_k}\}_{k=2M-M_0+1}^{2M}, S_j^{(0)}\big) := h \sum_{k=2M-M_0+1}^{2M} \mathrm{dist}^2(x_{i_k}, S_j^{(0)}),$$

$$h := h_1 \cdot \ldots \cdot h_p, \tag{12.3.1}$$

is small compared to the case when the points $\{x_{i_k}\}_{k=2M-M_0+1}^{2M}$ are uniformly distributed over $B_R(\tilde{x}_j)$. Since the actual orientation of $S_j^{(0)}$ is not known, the natural measure of the closeness of the set of points $\{x_{i_k}\}_{k=2M-M_0+1}^{2M}$ to a plane is given by

$$\Delta\big(\{x_{i_k}\}_{k=2M-M_0+1}^{2M}\big) := \min_{S; \tilde{x}_j \in S} \Delta\big(\{x_{i_k}\}_{k=2M-M_0+1}^{2M}, S\big). \tag{12.3.2}$$

The minimum on the right-hand side of (12.3.2) is computed over all planes containing the point $\tilde{x}_j$. Let θ be a unit vector perpendicular to a plane S. Then one obtains from (12.3.1) and (12.3.2)

$$\Delta\big(\{x_{i_k}\}_{k=2M-M_0+1}^{2M}\big) := \min_{\theta \in S^{p-1}} h \sum_{k=2M-M_0+1}^{2M} \langle x_{i_k} - \tilde{x}_j, \theta \rangle^2, \tag{12.3.3}$$

where S^{p-1} is the unit sphere in $\mathbb{R}^p$, $\langle x, y \rangle = \sum_{i=1}^{p} x_i y_i$ is the scalar product in $\mathbb{R}^p$. Define the $p \times p$ matrix B_{M_0} by

$$B_{M_0} := (b_{mn}), \quad b_{mn} = h \sum_{k=2M-M_0+1}^{2M} \big(x_{i_k}^{(m)} - \tilde{x}_j^{(m)}\big) \cdot \big(x_{i_k}^{(n)} - \tilde{x}_j^{(n)}\big),$$

$$1 \leq m, n \leq p. \tag{12.3.4}$$

Note that B_{M_0} is a positive semi-definite symmetric matrix with real-valued entries. Hence (12.3.3) can be rewritten as

$$\Delta\left(\{x_{i_k}\}_{k=2M-M_0+1}^{2M}\right) = \min_{\theta \in S^{p-1}} (B_{M_0}\theta, \theta).$$

The solution to (12.3.3) is $\Delta\left(\{x_{i_k}\}_{k=2M-M_0+1}^{2M}\right) = \lambda_{min}(B_{M_0})$, where $\lambda_{min}(B_{M_0})$ is the minimal eigenvalue of the matrix B_{M_0}. The corresponding eigenvector θ_0 can be used as the estimate of the orientation of S_0 at $\tilde{x}_j$. From (12.3.1) – (12.3.4) it follows that if $\lambda_{min}(B_{M_0})$ is small, the points $\{x_{i_k}\}_{k=2M-M_0+1}^{2M}$ lie close to some plane, and, similarly, if $\lambda_{min}(B_{M_0})$ is large, the points $\{x_{i_k}\}_{k=2M-M_0+1}^{2M}$ are distributed uniformly over $B_R(\tilde{x}_j)$. As a result we came to the following algorithm.

Step 1'. Fix a point $\tilde{x}_j$. Arrange the observed values from $B_R(\tilde{x}_j)$ in ascending order $u_{i_1} \leq u_{i_2} \leq \cdots \leq u_{i_{2M}}$ with $u_{i_k} := u(x_{i_k})$ and calculate the minimal eigenvalue $\lambda_{min}(B_{M_0})$, where the matrix B_{M_0} is given by (12.3.4).

Step 2'. Select the threshold $\Lambda_0 > 0$, as described in Step 0' below. Compare $\lambda_{min}(B_{M_0})$ with the threshold Λ_0. If $\lambda_{min}(B_{M_0}) \geq \Lambda_0$, the hypothesis '$\tilde{x}_j \notin S_0$' is accepted. If $\lambda_{min}(B_{M_0}) < \Lambda_0$, the hypothesis '$\tilde{x}_j \in S_0$' is accepted and the orientation of S_0 inside $B_R(\tilde{x}_j)$ is given by the eigenvector corresponding to $\lambda_{min}(B_{M_0})$.

Similar to Section 12.2.2, the threshold Λ_0 is determined from the relation:

$$P\{\lambda_{min}(B_{M_0}) < \Lambda_0 | B_R(\tilde{x}_j) \cap S_0 = \varnothing\} = \epsilon$$

by the Monte Carlo method according to Step 0':

Step 0'. Fix an arbitrary ball $B_R(\tilde{x}_j)$. Model a sufficiently large number of the samples of u_i with $s_i \equiv 0$ and an arbitrary fixed common PDF for n_i. Using Step 1' compute $\lambda_{min}(B_{M_0})$ for each sample. Select Λ_0 so that $\lambda_{min}(B_{M_0}) < \Lambda_0$ in $100\epsilon\%$ cases.

It is easy to see that the above algorithm (Steps 0', 1', 2') is also applicable in two other cases:

(a) the values of the signal between S_1 and S_2 are smaller than the values from the neighborhoods adjoint to S_1 and S_2. In this case one calculates matrix B_{M_0} using the first M_0 points $\{x_{i_k}\}_{k=1}^{M_0}$ from the window (with smallest values of the signal);

(b) the values of the signal between S_1 and S_2 are smaller than the values from one neighborhood (say, adjoint to S_1) and larger than values from another neighborhood (adjoint to S_2). In this case, one uses M_0 middle points $\{x_{i_k}\}_{k=N-M_0/2+1}^{N+M_0/2}$ in (12.3.4).

12.4. Generalization of the algorithms

The algorithms described in Sections 12.2.2 and 12.2.3 can be generalized in the following way. Suppose one wants to recover a domain U which satisfies the following properties:

(1) in a neighborhood of U one has $s(x_1) > s(x_2)$ if $x_1 \in U$, $x_2 \notin U$, i.e. the boundary of U is the discontinuity surface (or a part of it);

(2) there exists a set of points $S_0 \subset U$ such that for any $\tilde{x}_{j_1}, \tilde{x}_{j_2} \in S_0$ the sets $B_R(\tilde{x}_{j_1}) \cap U$ and $B_R(\tilde{x}_{j_2}) \cap U$ can be obtained from each other by translating and rotating;

(3) $B_R(\tilde{x}_j) \setminus U \neq \varnothing$ for any $\tilde{x}_j \in S_0$;

(4) the set U can be uniquely determined from S_0.

Let us illustrate the geometrical meaning of assumptions (1) – (4) by examples. In the case of edge detection let U be the domain such that the discontinuity surface S is its boundary and let $S_0 = S$. Clearly, Properties 1 and 3 are satisfied. Property 2 also holds if S is locally flat. Furthermore the set U is uniquely determined by its boundary $S(= S_0)$. In the case of thin line detection let U be the domain between S_1 and S_2, and S_0 be the 'middle surface' (see Section 12.2.3). As above, Property 1 holds. Property 3 also holds since $d < R$. Property 2 is satisfied because we assumed that S_1 and S_2 are locally flat. Property 4 is satisfied because d and S_0 determine S_1, S_2 and, hence, U uniquely.

Thus, for each $\tilde{x}_j$ one wishes to have a decision rule for choosing between the two hypotheses: '$\tilde{x}_j \in S_0$' and '$\tilde{x}_j \notin S_0$'. The decision can be made using the following argument. Fix an arbitrary $\tilde{x}_j$ and arrange the observed values of the signal from $B_R(\tilde{x}_j)$ in ascending order $u_{i_1} \leq u_{i_2} \leq \cdots \leq u_{i_{2M}}$. If $\tilde{x}_j \in S_0$, then, by Properties 1, 2 and 3, the set of points $\{x_{i_k}\}_{k=2M-M_0+1}^{2M}$ (M_0 is the number of the grid points inside $B_R(\tilde{x}_j) \cap U$) is likely to exhibit some nonuniformity. Namely, most of these points belong to $B_R(\tilde{x}_j) \cap U$, and only a small part of them lie outside $B_R(\tilde{x}_j) \cap U$. If $B_R(\tilde{x}_j) \cap U = \varnothing$, the points $\{x_{i_k}\}_{k=2M-M_0+1}^{2M}$ are uniformly distributed inside $B_R(\tilde{x}_j)$. Let N be the number of points from $\{x_{i_k}\}_{k=2M-M_0+1}^{2M}$ lying outside $B_R(\tilde{x}_j) \cap U$ (or outside any fixed set obtained from $B_R(\tilde{x}_j) \cap U$ by rotation of this set around $\tilde{x}_j$). Suppose there exists a continuous functional $\lambda(\{x_{i_k}\}_{k=2M-M_0+1}^{2M}, \tilde{x}_j)$ with the properties

$$\lambda(\{x_{i_k}\}_{k=2M-M_0+1}^{2M}, \tilde{x}_j)\big|_{N=N_1} > \lambda(\{x_{i_k}\}_{k=2M-M_0+1}^{2M}, \tilde{x}_j)\big|_{N=N_2},$$
$$N_1 < N_2, \tag{12.4.1}$$

$$\exists \lim_{M \to \infty} \lambda(\{x_{i_k}\}_{k=2M-M_0+1}^{2M}, \tilde{x}_j) := \hat{\lambda}(\gamma, \tilde{x}_j) \quad \text{if} \quad \frac{N}{M_0} \to \gamma, \ 0 \leq \gamma \leq 1, \tag{12.4.2}$$

$$\hat{\lambda}(\gamma_1, \tilde{x}_j) > \hat{\lambda}(\gamma_2, \tilde{x}_j) \quad \text{if} \quad 0 \le \gamma_1 < \gamma_2 \le 1. \tag{12.4.3}$$

Then, on the basis of (12.4.1), one constructs the following algorithm.

Step 1''. Fix a point $\tilde{x}_j$. Arrange the observed values from $B_R(\tilde{x}_j)$ in ascending order $u_{i_1} \le \cdots \le u_{i_{2M}}$ with $u_{i_k} := u(x_{i_k})$ and calculate the functional $\lambda_j = \lambda(\{x_{i_k}\}_{k=2M-M_0+1}^{2M}, \tilde{x}_j)$.

Step 2''. Select the threshold $\Lambda > 0$, as described in Step 0'' below. Compare λ_j with the threshold Λ. If $\lambda_j \le \Lambda$, the hypothesis '$\tilde{x}_j \notin S_0$' is accepted. If $\lambda_j > \Lambda$, the hypothesis '$\tilde{x}_j \in S_0$' is accepted and, if necessary, the orientation of S_0 at $\tilde{x}_j$ is calculated according to the nature of the functional $\lambda(\{x_{i_k}\}_{k=2M-M_0+1}^{2M}, \tilde{x}_j)$.

Similarly, the threshold Λ is determined from:

$$P\{\lambda(\{x_{i_k}\}_{k=2M-M_0+1}^{2M}, \tilde{x}_j) > \Lambda | B_R(\tilde{x}_j) \cap S_0 = \varnothing\} = \epsilon$$

by the Monte Carlo method:

Step 0''. Fix an arbitrary ball $B_R(\tilde{x}_j)$. Model a sufficiently large number of the samples of u_i with $s_i \equiv 0$ and an arbitrary fixed common PDF for n_i. Using Step 1'' compute $\lambda(\{x_{i_k}\}_{k=2M-M_0+1}^{2M}, \tilde{x}_j)$ for each sample. Select Λ so that $\lambda(\{x_{i_k}\}_{k=2M-M_0+1}^{2M}, \tilde{x}_j) > \Lambda$ in $100\epsilon\%$ cases.

It is clear that the algorithms from Sections 12.2.2 and 12.2.3 fit well into the above scheme. In this section, we present one more example which shows that this scheme is sufficiently general and can be used for the solution of other pattern recognition problems. Let us consider the spot detection problem. Suppose one is looking for a 'spot' which can be represented by a ball $B_d(\tilde{x}_{j_0})$ with known radius $d < R$, such that the values of the signal inside $B_d(\tilde{x}_{j_0})$ are larger than in its neighborhood. Then one sees that the Properties 1 – 4 are satisfied. Here $U = B_d(\tilde{x}_{j_0})$, M_0 is the number of grid points inside $B_d(\tilde{x}_{j_0})$ and $S_0 = \tilde{x}_{j_0}$. The functional $\lambda(\{x_{i_k}\}_{k=2M-M_0+1}^{2M}, \tilde{x}_j)$ can be easily constructed:

$$\lambda(\{x_{i_k}\}_{k=2M-M_0+1}^{2M}, \tilde{x}_j) := -h \sum_{k=2M-M_0+1}^{2M} |x_{i_k} - \tilde{x}_j|^2. \tag{12.4.4}$$

It is easy to see that the functional (12.4.4) satisfies (12.4.1) – (12.4.3). We put the minus sign in front of the summation in (12.4.4) so that the value of the functional is the largest when $\tilde{x}_j = \tilde{x}_{j_0}$. Hence, using Steps 0'', 1'', 2'' with the functional defined by (12.4.4), one finds the position of the spot $\tilde{x}_{j_0}$.

Clearly, the algorithm given by Steps 0'', 1'', 2'' can be modified for the case when in a neighborhood of U one has $s(x_1) < s(x_2)$ if $x_1 \in U$, $x_2 \notin U$. The only difference is in Step 1'', where one arranges points in descending order.

12.5. Justification of the edge detection algorithm

In this section, we prove the asymptotic consistency of the edge detection algorithm as $\max(h_1, \ldots, h_p) \to 0$ ($M \to \infty$).

Theorem 12.5.1. *Fix any $\epsilon, \delta, D > 0$ and consider $\tilde{x}_j \in S$. Then:*

(a) *the probability of the decision '$\tilde{x}_j \notin S$' goes to zero as $M \to \infty$;*

(b) *the probability of an angular error larger than δ in determining the orientation of S at the point $\tilde{x}_j$ (by the algorithm described in Steps 1 and 2) goes to zero as $M \to \infty$.*

For the proof of the theorem we need some auxiliary results which are formulated in Lemmas 12.5.1 and 12.5.2 below. Then the proof of Theorem 12.5.1 is given.

Consider an arbitrary point $\tilde{x}_j \in S$. In this case, $B_R(\tilde{x}_j)$ is split by S into two equal parts $B_R^-(\tilde{x}_j)$ and $B_R^+(\tilde{x}_j)$. At the same time, since the signal is corrupted by noise, some of the points from $B_R^-(\tilde{x}_j)$ will be erroneously classified as belonging to the set $\{x_{i_k}\}_{k=M+1}^{2M}$, and equal number of points from $B_R^+(\tilde{x}_j)$ will be erroneously classified as belonging to the set $\{x_{i_k}\}_{k=1}^{M}$. Consequently, the orientation of $S \cap B_R(\tilde{x}_j)$ will be determined with some error. The points which belong to $B_R^-(\tilde{x}_j)$ $(B_R^+(\tilde{x}_j))$ and which are erroneously classified as being from $\{x_{i_k}\}_{k=M+1}^{2M}$ $(\{x_{i_k}\}_{k=1}^{M})$ will be called the misclassified points. To find the probability $P(N, M)$ of misclassifying N or more points, let us order the points from each of the halfballs *separately*:

$$v_{i_1} \leq v_{i_2} \leq \cdots \leq v_{i_M}, \qquad v_{i_{M+1}} \leq v_{i_{M+2}} \leq \cdots \leq v_{i_{2M}},$$

$$x_{i_1}, \ldots, x_{i_M} \in B_R^-(\tilde{x}_j), \qquad x_{i_{M+1}}, \ldots, x_{i_{2M}} \in B_R^+(\tilde{x}_j).$$

$$\tag{12.5.1}$$

Clearly, we do not know how to do this ordering because we do not know which points are in $B_R^+(\tilde{x}_j)$ $(B_R^-(\tilde{x}_j))$. Nevertheless this ordering exists. Here we have used the notation v_{i_k} for the data instead of u_{i_k} to stress the fact that the ranking (12.5.1) is different from the ranking used in Step 1. Note that now we *do not* have the inequality $v_{i_M} \leq v_{i_{M+1}}$. It is easy to see that there will be N or more misclassified points in each halfball if and only if

$$v_{i_{M+N}} < v_{i_{M-N+1}}. \tag{12.5.2}$$

Let $s^- \approx s(x)$, $x \in B_R^-(\tilde{x}_j)$, $s^+ \approx s(x)$, $x \in B_R^+(\tilde{x}_j)$, $D := s^+ - s^-$, that is D is the magnitude of the jump of the signal at the point $\tilde{x}_j$. Let $f(\tau)$ be the common PDF of random errors n_i. Then, as it follows from (12.2.1) and (12.5.1), v_{i_k}, $1 \leq k \leq M$, is the k-th order statistic of the random sample of the size M from a distribution with the PDF $f(\tau - s^-)$. Similarly, $v_{i_{M+k}}$, $1 \leq k \leq M$, is the k-th order statistic of the random sample of the size M from a distribution with the PDF $f(\tau - s^+)$. Using the known formula for the distribution of order satistics [Gal], one obtains from (12.5.2):

$$P(N, M) := P\{v_{iM+N} < v_{iM-N+1}\} = \int F_N(t - D)dF_{M-N+1}(t),$$

$$(12.5.3)$$

where $\int := \int_{-\infty}^{+\infty}$, and

$$F_k(t) = \int_{-\infty}^{t} f_k(\tau)d\tau,$$

$$f_k(\tau) = \frac{M!}{(k-1)!(M-k)!}[F(\tau)]^{k-1}[1 - F(\tau)]^{M-k}f(\tau),$$

$$F(\tau) = \int_{-\infty}^{\tau} f(\xi)d\xi.$$

$$(12.5.4)$$

In (12.5.3) we used the independence of the random variables v_{iM+N} and v_{iM-N+1}. To find the asymptotics of $P(N, M)$ as $M \to \infty$, fix γ, $0 < \gamma < 1$, and let $N = \gamma M$. Using the known asymptotics of the order statistics [Eng, Gal]:

$$\left| F_N(t - D) - \Phi_1(t) \right| \leq \frac{3}{\sqrt{M\gamma(1 - \gamma)}} := C_M,$$

$$\Phi_1(t) := \Phi\left(\sqrt{M}\frac{F(t - D) - \gamma}{\sqrt{\gamma(1 - \gamma)}}\right),$$

$$\left| F_{M-N+1}(t) - \Phi_2(t) \right| \leq C_M, \quad \Phi_2(t) := \Phi\left(\sqrt{M}\frac{F(t) - (1 - \gamma)}{\sqrt{\gamma(1 - \gamma)}}\right),$$

$$(12.5.5)$$

where $\Phi(x) = 1/\sqrt{2\pi}\int_{-\infty}^{x}\exp(-t^2/2)dt$ is the standard normal distribution, one obtains from (12.5.3) and (12.5.5)

$$P(N, M) = \int F_N(t - D)dF_{M-N+1}(t) \leq \int (\Phi_1(t) + C_M)dF_{M-N+1}(t)$$

$$= \int \Phi_1(t)dF_{M-N+1}(t) + C_M$$

$$= \Phi_1(t)F_{M-N+1}(t)\big|_{t=-\infty}^{+\infty} - \int F_{M-N+1}(t)d\Phi_1(t) + C_M$$

$$\leq \Phi\left(\sqrt{M\frac{1 - \gamma}{\gamma}}\right) - \int (\Phi_2(t) - C_M)d\Phi_1(t) + C_M$$

$$\leq \Phi\left(\sqrt{M\frac{1 - \gamma}{\gamma}}\right) - \int \Phi_2(t)d\Phi_1(t) + 2C_M$$

$$\leq \int \Phi_1(t)d\Phi_2(t) + 2C_M + o(\exp(-cM)), \qquad (12.5.6)$$

where $c = \mathrm{const} > 0$ (see (12.9.1)). It follows from (12.5.5) that $C_M \to 0$ as $M \to \infty$. To find the asymptotics of the integral on the right-hand side of (12.5.6), introduce the new variable $y = \sqrt{M}(F(t) - (1 - \gamma))/\sqrt{\gamma(1-\gamma)}$. This yields

$$\int \Phi_1(t)d\Phi_2(t) = \frac{1}{\sqrt{2\pi}} \int_{-\sqrt{M\frac{1-\gamma}{\gamma}}}^{\sqrt{M\frac{\gamma}{1-\gamma}}} \Phi\left(\sqrt{\frac{M}{\gamma(1-\gamma)}}\left\{F\left[F^{-1}\left(1 - \gamma + y\sqrt{\frac{\gamma(1-\gamma)}{M}}\right) - D\right] - \gamma\right\}\right)\exp(-y^2/2)dy$$

$$\to \frac{1}{\sqrt{2\pi}} \int \Phi\left(\sqrt{\frac{M}{\gamma(1-\gamma)}}\{F(\xi_0) - \gamma\}\right)\exp(-y^2/2)dy$$

$$\xrightarrow[M\to\infty]{} \begin{cases} 0, & F(\xi_0) < \gamma \\ 1, & F(\xi_0) > \gamma, \end{cases} \qquad (12.5.7)$$

where

$$\xi_0 := F^{-1}(1 - \gamma) - D. \qquad (12.5.8)$$

The rate of convergence in (12.5.7) is estimated in Section 12.9. Substituting (12.5.8) into (12.5.7), one gets after some calculations

$$P(\gamma M, M) \to \theta\left\{\left[F^{-1}(1 - \gamma) - F^{-1}(\gamma)\right] - D\right\} \quad \text{as} \quad M \to \infty, \qquad (12.5.9)$$

where $\theta(t) = \begin{cases} 0, & t < 0, \\ 1, & t > 0. \end{cases}$ Define γ_0 by

$$\gamma_0 := \inf_{\substack{0 < \gamma < 1 \\ F^{-1}(1-\gamma) - F^{-1}(\gamma) < D}} \gamma. \qquad (12.5.10)$$

Since the function $F^{-1}(1 - \gamma) - F^{-1}(\gamma)$ is strictly decreasing for $0 < \gamma < 1$ and $D > 0$, one obtains from (12.5.9) and (12.5.10)

$$P(\gamma M, M) \to \theta(\gamma_0 - \gamma) \quad \text{as} \quad M \to \infty \quad \text{with} \quad 0 \leq \gamma_0 < 1/2. \quad (12.5.11)$$

Now take any point $\tilde{x}_j$ such that $B_R(\tilde{x}_j) \cap S = \varnothing$ and divide $B_R(\tilde{x}_j)$ by an arbitrary hyperplane (passing through $\tilde{x}_j$) into two equal parts $B_R^-(\tilde{x}_j)$ and $B_R^+(\tilde{x}_j)$. In this case, the asymptotics of the probability of misclassifying γM points can be obtained by putting $D = 0$ in (12.5.9):

$$P(\gamma M, M) \to \theta(F^{-1}(1 - \gamma) - F^{-1}(\gamma)) \quad \text{as} \quad M \to \infty. \qquad (12.5.12)$$

From (12.5.10) it follows that in the case $D = 0$ one has $\gamma_0 = 1/2$ and (12.5.12) reduces to

$$P(\gamma M, M) \to \theta(1/2 - \gamma) \quad \text{as} \quad M \to \infty. \tag{12.5.13}$$

Recall that N is the number of the misclassified points. It is not hard to see that (12.5.11) and (12.5.13) imply

Lemma 12.5.1. *Fix an arbitrary $\delta > 0$. If $\tilde{x}_j \in S$, there exists γ_0, $0 \leq \gamma_0 < 1/2$, such that*

$$P\left\{\left|\frac{N}{M} - \gamma_0\right| > \delta\right\} \to 0 \quad \text{as} \quad \max(h_1, \ldots, h_p) \to 0 \ (M \to \infty).$$

If $B_R(\tilde{x}_j) \cap S = \varnothing$, divide $B_R(\tilde{x}_j)$ arbitrarily into two parts $B_R^-(\tilde{x}_j)$ and $B_R^+(\tilde{x}_j)$ having equal number of points and let N be the number of points from the set $\{x_{i_k}\}_{k=M+1}^{2M}$ lying in $B_R^-(\tilde{x}_j)$. Then one has

$$P\left\{\left|\frac{N}{M} - 1/2\right| > \delta\right\} \to 0 \quad \text{as} \quad \max(h_1, \ldots, h_p) \to 0 \ (M \to \infty).$$

Now let us consider the halfball $B_R^-(\tilde{x}_j)$ and assume that $0 < \gamma_0 \leq 1/2$. Fix an arbitrary nonempty open set $U \subset B_R^-(\tilde{x}_j)$ and let M_U be the number of grid points in U. It is clear that each grid point can be either misclassified or correctly classified, in other words, there cannot be two misclassified points at any given grid node, which is exactly the assumption of the Fermi-Dirac statistics (see Section 14.10.1.2). Hence the probability of the event 'there are N_U misclassified points out of the total number N inside the set U' is given by

$$P(N_U, N; M_U, M) = \frac{C_{N_U}^{M_U} C_{N-N_U}^{M-M_U}}{C_N^M}. \tag{12.5.14}$$

As a direct consequence of the Stirling formula $n! \sim \sqrt{2\pi n} n^n e^{-n}$, one has

$$C_{sn}^n \sim \frac{1}{\sqrt{2\pi n(1-s)s}} \frac{1}{\left((1-s)^{1-s} s^s\right)^n} \quad \text{as} \quad n \to \infty, \tag{12.5.15}$$

where s, $0 < s < 1$, is an arbitrary fixed number. To find the asymptotics of the probabilty (12.5.14) as $M \to \infty$, assume that

$$N_U = \alpha_U N, \quad M_U = \beta_U M, \quad N = \gamma_0 M. \tag{12.5.16}$$

Substituting (12.5.16) into (12.5.14) and using (12.5.15), one obtains

$$P(\alpha_U \gamma_0 M, \gamma_0 M; \beta_U M, M) \to \frac{c}{\sqrt{M}} r^M \quad \text{as} \quad M \to \infty, \tag{12.5.17}$$

where

$$r = \frac{(1-\gamma_0)^{1-\gamma_0}\gamma_0^{\gamma_0}}{\left[(1-s_1)^{1-s_1}s_1^{s_1}\right]^{\beta_U}\left[(1-s_2)^{1-s_2}s_2^{s_2}\right]^{1-\beta_U}},$$

$$s_1 = \frac{\alpha_U\gamma_0}{\beta_U}, \quad s_2 = \frac{(1-\alpha_U)\gamma_0}{1-\beta_U}. \qquad (12.5.18)$$

Here $c > 0$ is some constant which does not depend on M. From (12.5.18) one has

$$\ln r = \beta_U H(s_1) + (1-\beta_U)H(s_2) - H(\beta_U s_1 + (1-\beta_U)s_2), \quad (12.5.19)$$

where $H(s) = -s\ln s - (1-s)\ln(1-s)$ is the entropy [Mor, pp. 49–53]. In (12.5.19), we used the identity $\gamma_0 = \beta_U s_1 + (1-\beta_U)s_2$. Since $-H(s)$ is a strictly convex function, one has $r = 1$ if and only if $s_1 = s_2$ ($\beta_U \neq 0$, $\beta_U \neq 1$, because U was chosen to be a proper nonempty subset of $B_R^-(\tilde{x}_j)$). In all other cases $r < 1$. From (12.5.18) one sees that the condition $s_1 = s_2$ is equivalent to the condition $\alpha_U = \beta_U$. This, together with (12.5.17), almost immediately yields

Lemma 12.5.2. *Fix an arbitrary $\delta > 0$ and consider an arbitrary nonempty open set $U \subset B_R^-(\tilde{x}_j)$. Then one has*

$$P\left\{\left|\frac{N_U}{N} - \frac{Vol(U)}{Vol(B_R^-(\tilde{x}_j))}\right| > \delta\right\} \to 0 \quad as \quad M \to \infty, \qquad (12.5.20)$$

where $Vol(A)$ is the volume of A.

Note that Lemma 12.5.2 guarantees the asymptotically uniform distribution of the misclassified points over the corresponding halfball $B_R^-(\tilde{x}_j)$ or $B_R^+(\tilde{x}_j)$. Hence the distribution of correctly classified points is also asymptotically uniform.

Proof of Theorem 12.5.1. Let M be sufficiently large and consider the misclassified points from the set $\{x_{i_k}\}_{k=M+1}^{2M}$, i.e. the points which lie in $B_R^-(\tilde{x}_j)$. Let N be the number of such points and $K = \{k_1,\ldots,k_N\}$ be the set of the subindices corresponding to them. By Lemma 12.5.1, $N/M \to \gamma_0 < 1/2$, and, by Lemma 12.5.2, these points are uniformly distributed over $B_R^-(\tilde{x}_j)$. Similarly, the number of correctly classified points $M - N$ satisfies $(M - N)/M \to 1 - \gamma_0 > 1/2$, and they are uniformly distributed over $B_R^+(\tilde{x}_j)$. Therefore, the vector $\tilde{x}_j - x_j^+$ must have only one nonzero component, which is perpendicular to S at $\tilde{x}_j$. Hence the distance between $\tilde{x}_j$ and the centroid x_j^+ (with $M \to \infty$) can be calculated as follows

$$
|\tilde{x}_j - x_j^+| = \frac{\left| \sum_{k \in K} \left(x_{i_k}^{(1)} - \tilde{x}_j^{(1)} \right) + \sum_{\substack{k=1 \\ k \notin K}}^{M} \left(x_{i_k}^{(1)} - \tilde{x}_j^{(1)} \right) \right|}{M} \rightarrow
$$

$$
\frac{\left| \gamma_0 \sum_{k \in K} \left(x_{i_k}^{(1)} - \tilde{x}_j^{(1)} \right) \frac{hM}{N} + (1 - \gamma_0) \sum_{\substack{k=1 \\ k \notin K}}^{M} \left(x_{i_k}^{(1)} - \tilde{x}_j^{(1)} \right) \frac{hM}{M-N} \right|}{hM} \rightarrow
$$

$$
\frac{\left| \gamma_0 \int_{\substack{x \in B_R(\tilde{x}_j) \\ x^{(1)} \leq \tilde{x}_j^{(1)}}} \left(x^{(1)} - \tilde{x}_j^{(1)} \right) d\hat{x} + (1 - \gamma_0) \int_{\substack{x \in B_R(\tilde{x}_j) \\ x^{(1)} \geq \tilde{x}_j^{(1)}}} \left(x^{(1)} - \tilde{x}_j^{(1)} \right) d\hat{x} \right|}{Vol(B_R(\tilde{x}_j))} =
$$

$$
\frac{\gamma_0 \int_{-R}^{0} t\omega_{p-1} (R^2 - t^2)^{\frac{p-1}{2}} dt + (1 - \gamma_0) \int_{0}^{R} t\omega_{p-1} (R^2 - t^2)^{\frac{p-1}{2}} dt}{\int_{0}^{R} \omega_{p-1} (R^2 - t^2)^{\frac{p-1}{2}} dt} =
$$

$$
(1 - 2\gamma_0) \frac{\int_0^1 t(1-t^2)^{\frac{p-1}{2}} dt}{\int_0^1 (1-t^2)^{\frac{p-1}{2}} dt} R = (1 - 2\gamma_0) \frac{2}{(p+1)B\left(\frac{p+1}{2}, \frac{1}{2}\right)} R > 0, \tag{12.5.21}
$$

since $\gamma_0 < 1/2$, where $h = h_1 \cdot \ldots \cdot h_p$, $d\hat{x} := dx^{(1)} \ldots dx^{(p)}$, $B(a,b)$ is the beta-function and $\omega_{p-1} = \pi^{(p-1)/2}/S((p+1)/2)$ is the volume of the unit ball in $\mathbb{R}^{p-1}$. The third line in (12.5.21) follows from the second one by observing that the elementary volume per point in $B_R^-(\tilde{x}_j)$ is hM/N, while in $B_R^+(\tilde{x}_j)$ it is $hM/(M-N)$. In (12.5.21) we assumed that the axis $x^{(1)}$ is perpendicular to S. At the same time, since $\gamma_0 \to 1/2$ for arbitrary $\tilde{x}_k$, such that $B_R(\tilde{x}_k) \cap S = \varnothing$, one has $|x_k^+ - \tilde{x}_k| \to 0$ as $M \to \infty$. Thus A_0 (the threshold value determined in Step 0) goes to zero as $M \to \infty$. This and (12.5.21) yield the first assertion of Theorem 12.5.1.

The second assertion of Theorem 12.5.1 follows immediately from Lemma 12.5.2 and (12.5.21), because if the distributions of the misclassified and correctly classified points are uniform, the line through $\tilde{x}_j$ and the centroid x_j^+ is perpendicular to S. $\square$

12.6. Justification of the algorithm for thin line detection

Let us call the points from $\{x_{i_k}\}_{k=2M-M_0+1}^{2M}$ $(\{x_{i_k}\}_{k=1}^{2M-M_0})$ which belong to $B_R(\tilde{x}_j) \setminus U$ $(B_R(\tilde{x}_j) \cap U)$, $\tilde{x}_j \in S_0$ 'the misclassified points'. Here U is the domain between S_1 and S_2. Denote $V_1 := Vol(B_R(\tilde{x}_j) \cap U)$, $V_2 := Vol(B_R(\tilde{x}_j) \setminus U) = Vol(B_R(\tilde{x}_j)) - V_1$, $\eta_0 := V_2/Vol(B_R(\tilde{x}_j))$ and let N be the number of the misclassified points. Let us assume for simplicity that for each pair of neighboring points (x_1, x_2), $x_1 \in S_1$, $x_2 \in S_2$, one has $D(x_1) = D(x_2)$, where $D(x_i)$ is the jump magnitude at the point x_i. Two points $x_1 \in S_1$, $x_2 \in S_2$ are called neighboring if $x_2 - x_1 \perp S_1$, $x_2 - x_1 \perp S_2$. Following the argument used in the proof of Lemma 12.5.1, one can prove

Lemma 12.6.1. *Fix an arbitrary* $\delta > 0$. *If* $\tilde{x}_j \in S_0$, *there exists* γ_0, $0 \leq \gamma_0 < \eta_0$, *such that*

$$P\left\{\left|\frac{N}{M_0} - \gamma_0\right| > \delta\right\} \to 0 \quad as \quad \max(h_1, \ldots, h_p) \to 0 \ (M \to \infty).$$

$$(12.6.1)$$

If $B_R(\tilde{x}_j) \cap U = \varnothing$, *divide* $B_R(\tilde{x}_j)$ *arbitrarily into two parts* U' *and* $B_R(\tilde{x}_j) \setminus U'$ *having* M_0 *and* $2M - M_0$ *grid points correspondingly* $(V_1 = Vol(U'))$. *Let* N *be the number of points from* $\{x_{i_k}\}_{k=2M-M_0+1}^{2M}$ *lying in* $B_R(\tilde{x}_j) \setminus U'$. *Then one has*

$$P\left\{\left|\frac{N}{M_0} - \eta_0\right| > \delta\right\} \to 0 \quad as \quad \max(h_1, \ldots, h_p) \to 0 \ (M \to \infty).$$

$$(12.6.2)$$

Recall that ϵ is the probability of 'false alarm', which is used in Step $0'$.

Theorem 12.6.1. *Fix any pair of neighboring points* (x_1, x_2), $x_1 \in S_1$, $x_2 \in S_2$, $\epsilon, \delta > 0$ *and consider* $\tilde{x}_j = (x_1, x_2) \cap S_0$. *Then, for any* $D(x_1) = D(x_2) = D > 0$, *one has:*

(a) *the probability of the decision '*$\tilde{x}_j \notin S_0'$ *goes to zero as* $M \to \infty$;
(b) *the probability of an angular error larger than* δ *in determining the orientation of* S_0 *at the point* $\tilde{x}_j$ *(by the algorithm described in Steps 1' and 2') goes to zero as* $M \to \infty$.

Proof. In what follows, we assume that M is sufficiently large and $\max(h_1, \ldots, h_p) \ll R$. Without loss of generality one can make the following assumptions: $S_1 = \{x : x^{(1)} = -d\}$, $S_2 = \{x : x^{(1)} = +d\}$, $\tilde{x}_j = 0$. Hence $S_0 = \{x : x^{(1)} = 0\}$. Let $\theta = (\theta^{(1)}, \ldots, \theta^{(p)})$ and recall that $d\hat{x} = dx^{(1)} \ldots dx^{(p)}$. Using Lemmas 12.5.2 and 12.6.1 and replacing the sum in (12.3.3) by the integral, one obtains

$$h \sum_{k=2M-M_0+1}^{2M} \langle x_{i_k} - \tilde{x}_j, \theta \rangle^2 =$$

$$(1-\gamma_0) \int_{B_R(\tilde{x}_j) \cap U} \langle x - \tilde{x}_j, \theta \rangle^2 d\hat{x} + \gamma_0 \frac{V_1}{V_2} \int_{B_R(\tilde{x}_j) \setminus U} \langle x - \tilde{x}_j, \theta \rangle^2 d\hat{x} =$$

$$(1-\gamma_0) \int_{\substack{|x| \le R \\ |x^{(1)}| \le d}} \left(\sum_{i=1}^{p} x^{(i)} \theta^{(i)} \right)^2 d\hat{x} + \gamma_0 \frac{V_1}{V_2} \int_{\substack{|x| \le R \\ |x^{(1)}| > d}} \left(\sum_{i=1}^{p} x^{(i)} \theta^{(i)} \right)^2 d\hat{x} =$$

$$\sum_{i=1}^{p} (\theta^{(i)})^2 \left\{ (1-\gamma_0) \int_{\substack{|x| \le R \\ |x^{(1)}| \le d}} (x^{(i)})^2 d\hat{x} + \gamma_0 \frac{V_1}{V_2} \int_{\substack{|x| \le R \\ |x^{(1)}| > d}} (x^{(i)})^2 d\hat{x} \right\} =$$

$$\sum_{i=1}^{p} (\theta^{(i)})^2 \left\{ (1-\gamma_0 - \gamma_0 \frac{V_1}{V_2}) \int_{\substack{|x| \le R \\ |x^{(1)}| \le d}} (x^{(i)})^2 d\hat{x} \right\} + \gamma_0 \frac{V_1}{V_2} \int_{|x| \le R} \langle x, \theta \rangle^2 d\hat{x}, \quad (12.6.3)$$

where we have used several times that the integrals of the terms proportional to $x^{(i)} x^{(j)}$, $i \ne j$, vanish because of the symmetry. Clearly, the second integral in (12.6.3) does not depend on θ. Since $|\theta| = 1$ and $\gamma_0 < \eta_0 = V_2/(V_1 + V_2)$ (as it is stated in Lemma 12.6.1), it can be easily seen that the right-hand side of (12.6.3) attains minimum when $\theta = \theta_0 := (1, 0, \ldots, 0)$, that is, when $\theta_0 \perp S_0$. This proves the second assertion of Theorem 12.6.1. Furthermore, one sees that to prove the first assertion of this theorem it is sufficient to show that $\Delta(\{x_{i_k}\}_{k=2M-M_0+1}^{2M})|_{\gamma=\gamma_0} < \Delta(\{x_{i_k}\}_{k=2M-M_0+1}^{2M})|_{\gamma=\eta_0}$, i.e.

$$(1-\gamma_0) \int_{\substack{|x| \le R \\ |x^{(1)}| \le d}} (x^{(1)})^2 d\hat{x} + \gamma_0 \frac{V_1}{V_2} \int_{\substack{|x| \le R \\ |x^{(1)}| > d}} (x^{(1)})^2 d\hat{x}$$

$$< \frac{V_1}{V_1 + V_2} \int_{|x| \le R} (x^{(1)})^2 d\hat{x}. \quad (12.6.4)$$

Denoting

$$U_1 := \int_{\substack{|x| \le R \\ |x^{(1)}| \le d}} (x^{(1)})^2 d\hat{x}, \quad U_2 := \int_{\substack{|x| \le R \\ |x^{(1)}| > d}} (x^{(1)})^2 d\hat{x}, \quad (12.6.5)$$

one gets that (12.6.4) is equivalent to

$$(1 - \gamma_0)\frac{U_1}{V_1} + \gamma_0\frac{U_2}{V_2} < \frac{U_1 + U_2}{V_1 + V_2}.$$

Rewrite the last inequality as

$$\left(V_2(1 - \gamma_0) - V_1\gamma_0\right)\frac{U_2V_1 - U_1V_2}{V_1V_2(V_1 + V_2)} > 0.$$

Since $\gamma_0 < V_2/(V_1 + V_2)$, it is sufficient to prove that $U_2V_1 - U_1V_2 > 0$ for $0 < d < R$. Similar to (12.5.21), one has

$$V_1 = 2\int_0^d \omega_{p-1}(R^2 - t^2)^{\frac{p-1}{2}}\,dt, \quad V_2 = 2\int_d^R \omega_{p-1}(R^2 - t^2)^{\frac{p-1}{2}}\,dt. \quad (12.6.6)$$

Using (12.6.5) and (12.6.6), introduce the function

$$\varphi(d) := U_2V_1 - U_1V_2$$

$$= 2\int_d^R t^2\omega_{p-1}(R^2 - t^2)^{\frac{p-1}{2}}\,dt \cdot 2\int_0^d \omega_{p-1}(R^2 - t^2)^{\frac{p-1}{2}}\,dt$$

$$- 2\int_0^d t^2\omega_{p-1}(R^2 - t^2)^{\frac{p-1}{2}}\,dt \cdot 2\int_d^R \omega_{p-1}(R^2 - t^2)^{\frac{p-1}{2}}\,dt, \quad (12.6.7)$$

where we simplified (12.6.5) by taking the integral over the variables $x^{(2)}, \ldots, x^{(p)}$. One has

$$\varphi'(d) = 4\omega_{p-1}^2(R^2 - d^2)^{\frac{p-1}{2}}\left\{\int_0^R t^2(R^2 - t^2)^{\frac{p-1}{2}}\,dt - d^2\int_0^R (R^2 - t^2)^{\frac{p-1}{2}}\,dt\right\}.$$

$$(12.6.8)$$

From (12.6.8) it follows that there exists d^*, $0 < d^* < R$, such that

$$\varphi'(d) > 0 \text{ for } 0 \le d < d^*, \quad \varphi'(d) \le 0 \text{ for } d^* \le d \le R.$$

This, together with $\varphi(0) = \varphi(R) = 0$, shows that $\varphi(d) > 0$ for $0 < d < R$. Theorem 12.6.1 is proved. □

12.7. Justification of the general scheme

The general scheme described in Section 12.2.4 can be justified using arguments similar to that of Sections 12.2.5 and 12.2.6. Let us suppose that for each $\tilde{x}_j \in S_0$ the function $s(x)$ is approximately constant inside $B_R(\tilde{x}_j) \cap U$ and $B_R(\tilde{x}_j) \setminus U$:

$$s_i \approx s(x), \ x \in B_R(\tilde{x}_j) \cap U \quad \text{and} \quad s_o \approx s(x), \ x \in B_R(\tilde{x}_j) \setminus U,$$

with $D := s_i - s_o$ being the jump magnitude in the neighborhood of $\tilde{x}_j$.

Theorem 12.7.1. *Fix any $\epsilon > 0$ and consider $\tilde{x}_j \in S_0$. If properties 1 – 3 and (12.4.1) – (12.4.3) are satisfied, the probability of the decision $\tilde{x}_j \notin S_0$ goes to zero as $M \to \infty$ for any $D > 0$.*

Proof. Let us note that Lemma 12.6.1 holds for arbitrary shape of the set U, provided that properties 1 – 3 from Section 12.2.4 are satisfied. Consider the case $B_R(\tilde{x}_j) \cap U = \emptyset$. Using (12.4.2) and (12.6.2), one gets

$$P\{|\lambda(\{x_{i_k}\}_{k=2M-M_0+1}^{2M}, \tilde{x}_j) - \hat{\lambda}(\eta_0, \tilde{x}_j)| > \delta\} \to 0 \quad \text{as} \quad M \to \infty$$
$$(12.7.1)$$

for any fixed $\delta > 0$. According to the choice of the threshold Λ (see Step 0''), one obtains from (12.7.1)

$$P\{|\Lambda - \hat{\lambda}(\eta_0, \tilde{x}_j)| > \delta\} \to 0 \quad \text{as} \quad M \to \infty. \qquad (12.7.2)$$

Now consider the case $\tilde{x}_j \in S_0$. Using (12.4.2) and (12.6.1), one obtains

$$P\{|\lambda(\{x_{i_k}\}_{k=2M-M_0+1}^{2M}, \tilde{x}_j) - \hat{\lambda}(\gamma_0, \tilde{x}_j)| > \delta\} \to 0 \quad \text{as} \quad M \to \infty.$$
$$(12.7.3)$$

Since $\gamma_0 < \eta_0$, one proves Theorem 12.7.1 using Equations (12.4.3), (12.7.2), (12.7.3), and the triangle inequality. $\square$

12.8. Numerical experiments

Figure 12.8.1a represents a synthetic image of square and circle edges with the jump magnitude $D = 1.5$ specified at a square 101×101 grid with the step size h_0 along each direction. The image is corrupted by noise with uniform distribution on the interval $[-\sqrt{3}/2, \sqrt{3}/2]$ (thus the variance is $\sigma^2 = 0.25$). The radius of the window has been chosen $R = 3.5h_0$, the probability of false alarm has been $\epsilon = 0.002$. Figure 12.8.1b represents the image of detected edges of Figure 12.8.1a.

To test the algorithm on a natural image, we used the black and white photo of a woman's head, which has been scanned with the grey level range $0 - 255$ (a value of 0 means black, a value of 255 means white). Figure 12.8.2a shows the 248×230 image corrupted by discrete noise which takes values $0, \pm 1, \dots, \pm 10$ with equal probability. The image of detected edges of Figure 12.8.2a is presented in Figure 12.8.2b. White pixels correspond to 'no edge' area, grey pixels correspond to places where the edges have been detected. The intensity of grey level at each point is proportional to the estimated jump value at this point. The radius of the window has been also $R = 3.5h_0$, the probability of false alarm has been $\epsilon = 0.001$.

Figure 12.8.3a represents a Synthetic Aperture Radar (SAR) image of an airport collected by Sandia National Laboratories as a part of the project for the 'Open Skies' treaty. The task is to build a sensor

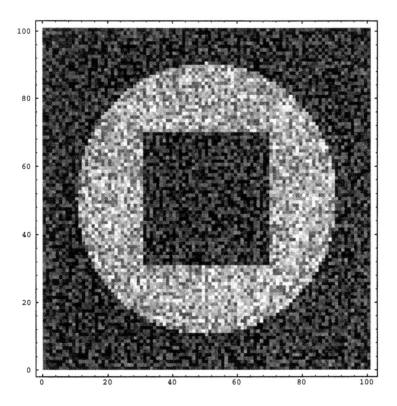

FIGURE 12.8.1A.   A synthetic image of square and circle edges.

that can fly over countries, which are members of the treaty, and collect images that can be processed later by the respective countries. One of the things to do is 'find' airports. There is an enormous amount of data taken in one flight alone, far too much for photointerpreters to look through, so the problem is to detect long, linear 'edges' (runways) in the images. Only these will be queued to the photointerpreter, thus thinning the amount of work the human will have to do. Once an airport is found, photointerpreters may want to perform some other task like count the number of airplanes or check for new construction. Figure 12.8.3b represents the image of detected edges of Figure 12.8.3a. The radius of the window has been $R = 4.0h_0$, the probability of false alarm has been $\epsilon = 0.01$. Note that most standard edge detection algorithms do not perform well on SAR images because there is a substantial noise component.

Figure 12.8.4a represents a synthetic image of thin line edges with the jump magnitude $D = 1.5$ specified at a square 101×101 grid. The

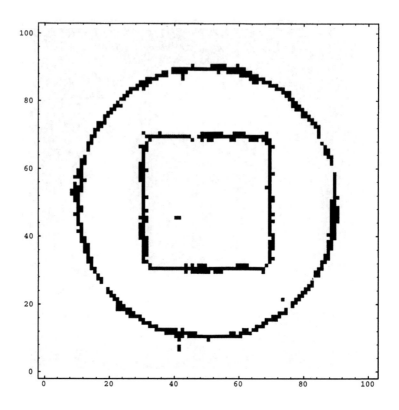

FIGURE 12.8.1B.   The image of detected edges of Figure 12.8.1a.

line edges are two nodes thick. The image is corrupted by noise with
the same parameters as in the first model. The parameters R, ϵ also
are the same as in the first model. Figure 12.8.4b represents the image
of detected edges of Figure 4a.

Note that to supress spurious edge elements in the above examples
we used the following general approach [Rob]: if a detected edge element
does not have a single edge element in its neighborhood, it is likely to
be a false one. In our case, this means that if the decision '$\tilde{x}_j \in S$' (or
'$\tilde{x}_j \in S_0$') has been made, but there are no other edge points in the ball
$B_R(\tilde{x}_j)$, we assume that $\tilde{x}_j \notin S$ ($\tilde{x}_j \notin S_0$).

In conclusion, let us discuss the recommendations for the choice of R
in numerical experiments. From the theory developed in Sections 12.2.5–
12.2.7 it follows that R should not be too small, so that there would be
enough grid points for the asymptotic statistical properties (Lemmas
12.5.1, 12.5.2, 12.6.1, and Theorems 12.5.1–12.7.1) to be satisfied with
sufficient accuracy. Thus, one can write $R \geq R_{min}$, where the particular

FIGURE 12.8.2A.   A natural image corrupted by noise.

value R_{min} depends on the allowable level of deviation from the limiting values of the statistics. The behavior of those statistics for small R (and, hence, for small M) is difficult to study theoretically and such study requires extensive numerical experiments. The limitation on R from above $R \leq R_{max}$ comes from the following assumptions:

(1) the signal does not change much within balls with radius R not intersecting S, more precisely: $\max_{B_R(\tilde{x}_j)} s(x) - \min_{B_R(\tilde{x}_j)} s(x) \ll \sigma$, $B_R(\tilde{x}_j) \cap S = \varnothing$, where σ is the standard deviation of noise; and

(2) the discontinuity surface S can be accurately approximated by a tangent plane within the balls $B_R(\tilde{x}_j)$, $\tilde{x}_j \in S$.

Thus, the upper bound R_{max} depends on the particular problem under consideration and should be specified by the researcher.

12.9. Proof of auxiliary results

Let us denote

$$\sqrt{M\gamma/(1-\gamma)} := t_M, \ \sqrt{M(1-\gamma)/\gamma} := q_M, \ \sqrt{M/\big(\gamma(1-\gamma)\big)} := \tau_M.$$

FIGURE 12.8.2B.   The image of detected edges of Figure 12.8.2a.

Then, using the inequalities $0 \leq \Phi(t) \leq 1$ and

$$\int_a^\infty \exp(-x^2)dx \leq \int_a^\infty \exp(-ax)dx = a^{-1}\exp(-a^2), \quad a > 0, \quad (12.9.1)$$

one gets

$$\left| \int_{t_M}^\infty \Phi\left(\tau_M \left\{ F\left[F^{-1}(1 - \gamma + y\tau_M^{-1}) - D \right] - \gamma \right\} \right) \exp(-y^2/2)dy \right|$$

$$\leq \frac{2}{t_M}\exp(-\frac{t_M^2}{2}),$$

$$\left| \int_{-\infty}^{-q_M} \Phi\left(\tau_M \left\{ F\left[F^{-1}(1 - \gamma + y\tau_M^{-1}) - D \right] - \gamma \right\} \right) \exp(-y^2/2)dy \right|$$

$$\leq \frac{2}{q_M}\exp(-\frac{q_M^2}{2}). \qquad\qquad (12.9.2)$$

FIGURE 12.8.3A.   A Synthetic Aperture Radar (SAR) image of an airport. SAR image courtesy of Sandia National Laboratories' AMPS (Airborne Multisensor Pod System) Synthetic Aperture Radar platform; developed for the United States Department of Energy Office of Intelligence and National Security.

Furthermore, using (12.9.1) again, one obtains with $0 < t < \min(t_M, q_M)$:

$$\int_{-q_M}^{t_M} \Phi\left(\tau_M\left\{F\left[F^{-1}(1 - \gamma + y\tau_M^{-1}) - D\right] - \gamma\right\}\right) \exp(-y^2/2)dy$$

$$= \int_{-t}^{t} \Phi\left(\tau_M\left\{F\left[F^{-1}(1 - \gamma + y\tau_M^{-1}) - D\right] - \gamma\right\}\right) \exp(-y^2/2)dy$$

$$+ O\left(t^{-1}\exp\left(-\frac{t^2}{2}\right)\right). \tag{12.9.3}$$

On the interval $(-t, t)$ one has, with $1 - \gamma < \xi_1 < 1 - \gamma + y\tau_M^{-1}$, $\xi_0 < \xi_2 < \xi_0 + y\tau_M^{-1}/f(\xi_1)$ and $\xi_0 = F^{-1}(1 - \gamma) - D$,

$$\Phi\left(\tau_M\left\{F\left[F^{-1}(1 - \gamma + y\tau_M^{-1}) - D\right] - \gamma\right\}\right)$$

$$= \Phi\left(\tau_M\left\{F\left[\xi_0 + y\tau_M^{-1}/f(\xi_1)\right] - \gamma\right\}\right)$$

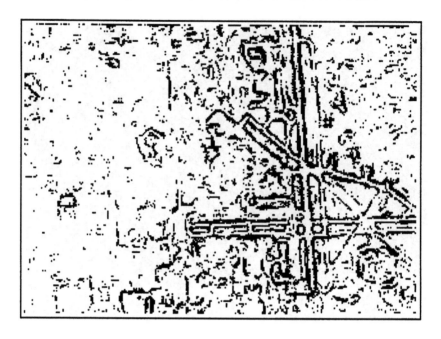

FIGURE 12.8.3B. The image of detected edges of Figure 12.8.3a.

$$= \Phi\left(\tau_M\left\{F(\xi_0) - \gamma + \left(y\tau_M^{-1}/f(\xi_1)\right)f(\xi_2)\right\}\right).$$

$$(12.9.4)$$

Thus, with $b := F(\xi_0) - \gamma$ and $\delta_M := \left(y\tau_M^{-1}/f(\xi_1)\right)f(\xi_2)$, $-t \le y \le t$, inequality (12.9.1) implies

$$\left|\Phi\left(\tau_M\{b + \delta_M\}\right) - \theta(b)\right| \le 2\tau_M^{-1}\exp(-\tau_M^2/2).$$

$$(12.9.5)$$

Estimate (12.9.5) holds if $t = t(M)$, where $t(M)\tau_M^{-1} \to 0$ as $M \to \infty$, since in this case $\xi_0 + y\tau_M^{-1}/f(\xi_1) \to \xi_0$ and $1 - \gamma + y\tau_M^{-1} \to 1 - \gamma$. Thus

$$\frac{1}{\sqrt{2\pi}}\int_{-t(M)}^{t(M)} \Phi\left(\tau_M\left\{F\left[F^{-1}\left(1 - \gamma + y\tau_M^{-1}\right) - D\right] - \gamma\right\}\right)\exp(-y^2/2)dy$$

$$= \frac{1}{\sqrt{2\pi}}\int_{-t(M)}^{t(M)} \left\{\theta(F(\xi_0) - \gamma) + O\left(\tau_M^{-1}\exp(-\tau_M^2/2)\right)\right\}\exp(-y^2/2)dy$$

$$= \theta(F(\xi_0) - \gamma) + O\left\{\tau_M^{-1}\exp(-\tau_M^2/2)\right\} + O\left\{\exp(-t^2(M)/2)/t(M)\right\}.$$

$$(12.9.6)$$

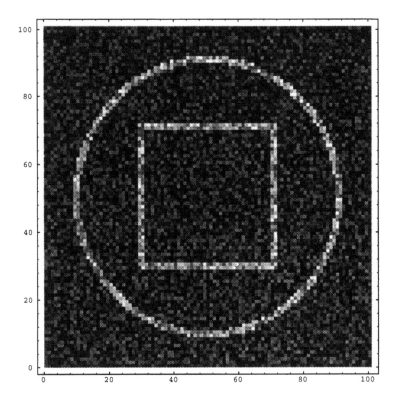

FIGURE 12.8.4A.   A synthetic image of square and circle edges.

Select arbitrarily α, $0 < \alpha < 1/2$, and let $t(M) = M^\alpha$. Then

$$O\{\tau_M^{-1} \exp(-\tau_M^2/2)\} + O\{\exp(-t^2(M)/2)/t(M)\}$$
$$= O\{M^{-\alpha} \exp(-M^{2\alpha}/2)\}. \tag{12.9.7}$$

Also

$$\left(\int_{-q_M}^{-t(M)} + \int_{t(M)}^{t_M} \right) \Phi\left(\tau_M \left\{ F\left[F^{-1}\left(1 - \gamma + y\tau_M^{-1}\right) - D \right] - \gamma \right\} \right)$$
$$\times \exp(-y^2/2)dy$$
$$\leq 2 \int_{t(M)}^{\max(t_M, q_M)} \exp(-y^2/2)dy = O\{\exp(-t^2(M)/2)/t(M)\}$$
$$= O\{M^{-\alpha} \exp(-M^{2\alpha}/2)\}. \tag{12.9.8}$$

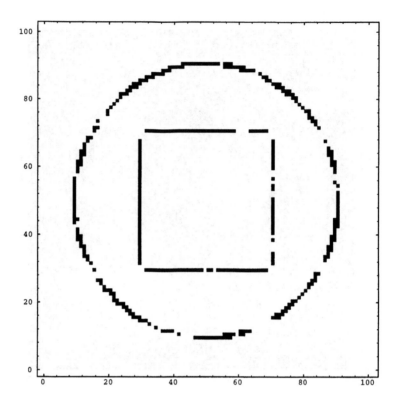

FIGURE 12.8.4B.   The image of detected edges of Figure 12.8.4a.

Since α, $0 < \alpha < 1/2$, was arbitrary, one obtains from (12.9.6) – (12.9.8)

$$\left| \int_{-q_M}^{t_M} \Phi\left(\tau_M\left\{F[F^{-1}(1 - \gamma + y\tau_M^{-1}) - D] - \gamma\right\}\right) \exp(-y^2/2)dy \right.$$

$$\left. - \theta(F(\xi_0) - \gamma) \right| = o\{M^{-\alpha}\exp(-M^{2\alpha}/2)\} \quad \forall \quad \alpha, \ 0 < \alpha < 1/2.$$

CHAPTER 13

TEST OF RANDOMNESS AND ITS APPLICATIONS

13.1. Introduction

In this chapter we consider modifications of a rank test for the one-dimensional case, multidimensional fixed (regular) design case, one-dimensional and multidimensional random design cases. Our main result is the proof of consistency of the test in each of the above cases against the following two general alternatives.

The first alternative: there exists a pairwise disjoint partition $\cup_{i=1}^m D_i = D$, where $D \subset \mathbb{R}^d$, $d \geq 1$, is a bounded domain inside which one makes observations, such that

(1) if an observation point falls inside D_i, then the corresponding observed value is the realization of a random variable ξ_i, $i = 1, \ldots, m$,

(2) there exists an ordering $\{\xi_{i_k}\}_{k=1}^m$, where ξ_{i_k} is stochastically smaller than $\xi_{i_{k+1}}$, $k = 1, \ldots, m - 1$,

(3) the partition is independent of the number N of observation points.

Note that m, *this ordering, and the sets D_i are not known a priori*: one tests only for the existence of such a partition. Thus the alternative we consider is different from the standard m-sample alternatives [Man], where boundaries between samples are supposed to be known. In the one-dimensional case our alternative reduces to the existence of change points, and we do not assume that the initial sequence is stochastically monotone: there can be both jumps up and jumps down. Note also that in the case $m = 2$ there can be any finite number of change points, because the sets D_1 and D_2 can be multiconnected.

The second alternative: there exists an arbitrary 'asymptotically continuous' trend in location. In the fixed design model the assumptions on the trend are the following. In the one-dimensional case the trend $\{\theta_k\}_{k=1}^N$ is supposed to satisfy the condition: there exists the limit $\lim_{N\to\infty, k/N\to t} \theta_k := \phi(t)$, such that $\phi(t) \in C[0,1]$, $\phi \not\equiv$ const. In the multidimensional case let the given data be $x_{k_1,\ldots,k_d}$, $1 \leq k_i \leq \beta_i N$, $0 < \beta_i < \infty$, $i = 1, \ldots, d$. Denote $B_N := \{k = (k_1, \ldots, k_d) \in \mathbb{N}^d : 1 \leq k_i \leq \beta_i N, i = 1, \ldots, d\}$, $D := \{t = (t_1, \ldots, t_d) \in \mathbb{R}^d : 0 \leq$

$t_i \leq \beta_i$, $i = 1, \ldots, d\}$. The condition on the trend $\{\theta_k, k \in B_N\}$ is: $\lim_{N \to \infty, k/N \to t} \theta_k := \phi(t) \in C[D]$, $\phi \not\equiv$ const. In the random design model, we assume that observation points $\{\rho_k\}_{k=1}^{N}$ are randomly chosen inside an open bounded domain $D \subset \mathbb{R}^d$, $d \geq 1$ (that is, $\{\rho_k\}_{k=1}^{N}$ are independent random points distributed in D with constant probability density), and the trend at each point is given by $\theta_k = \phi(\rho_k)$, where $\phi(t) \in C(\bar{D})$, $\phi \not\equiv$ const, and $\bar{D}$ is the closure of D.

In the one-dimensional case, equispaced design model, the statistic we use is

$$\nu_N := \frac{1}{N-1} \sum_{k=1}^{N-1} \left(\frac{R_{k+1} - R_k}{N} \right)^2, \qquad (13.1.1)$$

where R_k is the rank of the k-th element of the sequence to be tested, $k = 1, \ldots, N$. One sees that ν_N is closely related to the rank statistic R introduced by Wald and Wolfowitz [WW]: $R = \sum_{i=1}^{N-1} R_i R_{i+1} + R_1 R_N$. One has

$$N^2(N-1)\nu_N = N(N+1)(2N+1)/3 - 2R - (R_1 - R_N)^2, \quad (13.1.2)$$

thus ν_N and R are asymptotically equivalent. Many results are known concerning the statistic R: the asymptotic normality, consistency against monotone trend, cyclical movement, serial correlation and some other alternatives [WW, No, Aiy, AGA]. Also note that ν_N has the form of the Durbin-Watson statistic [DW] with observations replaced by their ranks. However, we could not find any proofs of consistency against the general alternatives we consider. The most frequently considered alternatives are one change point, monotone trend and serial correlation [AGA, Bh, CsH, KO, KM, KS]. Different rank tests and different results for the case of multiple change points can be found in [Lom].

In the multidimensional case, regular design model, the statistic we use is based on a modification of the Geary statistic [Gea, CO] with observations replaced by their ranks:

$$\nu_N := \frac{1}{M_N} \sum_{k \in B_N} \sum_{l \in L(k)} \left(\frac{R_k - R_l}{\hat{N}} \right)^2, \qquad (13.1.3)$$

where $L(k)$ is the set of lattice points neighboring to a point k, M_N is the number of elements in the double sum (13.1.3), and $\hat{N}$ is the number of lattice points.

In the case of the random design model, the analogue of (13.1.1) and (13.1.3) is

$$\nu_N := \frac{1}{N} \sum_{k=1}^{N} \left(\frac{R_{n(k)} - R_k}{N} \right)^2, \qquad (13.1.4)$$

where $\{\rho_k\}_{k=1}^N$ is a set of random observation points within a bounded domain, and $n(k)$ is the index of the point closest to ρ_k (a nearest neighbor).

The chapter is organized as follows. In Section 13.2 we consider the change points (change surfaces) alternative. In Sections 13.2.1 and 13.2.2 the proof of consistency in the one-dimensional cases: $m = 2$ and $m > 2$ is given. It is based on the following approach. First, we prove that $\nu_N \xrightarrow[N\to\infty]{ms} 1/6$ under the hypothesis of randomness, and that $\nu_N \xrightarrow[N\to\infty]{ms} E_m$ under the alternative. Here E_m is some constant. Notation '$\xrightarrow[N\to\infty]{ms}$' denotes convergence in mean-square. Since the asymptotic behavior of the Wald-Wolfowitz statistic R, and thus that of ν_N, under the hypothesis of randomness is well known [WW], we present the values of the first two moments of ν_N without detailed calculations. Under the alternative, the behavior of ν_N has been studied less extensively, although the convergence $\nu_N \xrightarrow[N\to\infty]{ms} E_m$ can be easily established by a standard technique. Thus, we only sketch the proof of this fact in the case $m = 2$. In the case $m > 2$, it can be proved similarly, so we presented only the final formulas. The main new point is the proof of the inequality $E_m < 1/6$, so it is given in detail. In Section 13.2.3 the case of the data specified at the nodes of a regular d-dimensional grid is considered. Since this case is completely analogous to the one-dimensional case, we only describe the model, statistic, and state the main results without proofs. In Section 13.2.4 the case of data points randomly distributed within a certain bounded domain is considered. A numerical example illustrating the use of the obtained results for image analysis (edge detection) is presented in Section 13.2.5. In Section 13.3 we prove consistency of the test against trend in location.

13.2. Consistency of rank test against change points (change surfaces) alternative

13.2.1. One-dimensional case, $m = 2$

Let $\{x_k\}_{k=1}^N$ be a random sequence of size N. In this case the observation points are $k/N \in D := (0, 1]$, $k = 1, \ldots, N$. The problem is to test the null hypothesis

$$H_0: \quad F_1(x) = F_2(x) = \cdots = F_N(x), \tag{13.2.1}$$

where $F_k(x)$ is a continuous distribution function of the random variable x_k, $k = 1, \ldots, N$, against the alternative

$$H_2: \quad F_k(x) = G_1(x), \ k/N \in D_1, \qquad F_k(x) = G_2(x), \ k/N \in D_2, \tag{13.2.2a}$$

$$G_1(x) \geq G_2(x), \quad x \in \mathbb{R}, \qquad G_1(x_0) > G_2(x_0) \text{ for some } x_0,$$
$$(13.2.2b)$$

where D_1 and D_2 are (unknown) nonintersecting measurable sets such that $D_1 \cup D_2 = (0, 1]$. We assume that $G_1(x)$ and $G_2(x)$ are continuously differentiable, and the ties do not occur with probability 1. Let $p^{(i)}$ be the number of interior points defined by:

$$p^{(i)} := p_1^{(i)} + p_2^{(i)}, \quad p_j^{(i)} := \#\{k : k, k+1 \in K_j\},$$
$$K_j := \{k : k/N \in D_j\}, \quad j = 1, 2. \qquad (13.2.3)$$

All other points are called the boundary (change) points, there are $p^{(b)} := N - p^{(i)}$ of them. Let us also assume that

$$p^{(b)} < \text{const}, \quad p_1^{(i)}/N \to \alpha, \quad p_2^{(i)}/N \to 1 - \alpha \quad \text{as} \quad N \to \infty, \quad 0 < \alpha < 1,$$
$$(13.2.2c)$$

where $\alpha = \lambda(D_1)$ is the Lebesgue measure of D_1. Let R_i be the rank of the element x_i, $i = 1, \ldots, N$. The test criterion is the statistic

$$\nu_N := \frac{1}{N-1} \sum_{k=1}^{N-1} \left(\frac{R_{k+1} - R_k}{N} \right)^2. \qquad (13.2.4)$$

Let us calculate the moments of ν_N when the null hypothesis is true. One has

$$E(\nu_N) = \frac{1}{N-1} \sum_{k=1}^{N-1} E\left(\frac{R_{k+1} - R_k}{N} \right)^2 = \frac{1}{N(N-1)} \sum_{i=1}^{N} \sum_{\substack{j=1 \\ j \neq i}}^{N} \left(\frac{i-j}{N} \right)^2$$

$$= \frac{1}{6}\left(1 + \frac{1}{N}\right), \qquad (13.2.5)$$

where one used the fact that the distribution of the random variable $(R_{k+1} - R_k)^2$ does not depend on k, and that all combinations of ranks in the pair $R_k, R_{k+1} : R_k = i, R_{k+1} = j, i \neq j$, are equiprobable. The second moment of ν_N is given by:

$$E(\nu_N^2) = \frac{(N+1)(5N^4 - 17N^2 + 18)}{180(N-1)^2 N^3} = \frac{1}{36}(1 + O(1/N)). \qquad (13.2.6)$$

From (13.2.5) and (13.2.6) it follows that

$$\text{var}(\nu_N) = 1/(36N) + O(1/N^2), \quad N \to \infty. \qquad (13.2.7)$$

Therefore, ν_N converges to $1/6$ in mean-square as $N \to \infty$ if the null hypothesis holds. To investigate the behavior of ν_N under the alternative hypothesis, let $N \to \infty$. Since the portion of border points among all the points is of order $O(1/N)$ (see (13.2.2c)), we may write

$$E(\nu_N)$$

$$= \frac{1}{N-1} \sum_{k \in K_1} E\left(\frac{R_{k+1} - R_k}{N}\right)^2 + \frac{1}{N-1} \sum_{k \in K_2} E\left(\frac{R_{k+1} - R_k}{N}\right)^2$$

$$= \frac{|K_1|}{N-1} E\left(\frac{R_{l_1+1} - R_{l_1}}{N}\right)^2 + \frac{|K_2|}{N-1} E\left(\frac{R_{l_2+1} - R_{l_2}}{N}\right)^2 + O(\frac{1}{N})$$

$$\xrightarrow[N \to \infty]{} \alpha \int_0^1 \int_0^1 f_1(x)(x-y)^2 f_1(y) dx dy$$

$$+ (1 - \alpha) \int_0^1 \int_0^1 f_2(x)(x-y)^2 f_2(y) dx dy$$

$$= \frac{2}{3} - \frac{2\alpha}{1-\alpha} \left\{ \left(\int_0^1 f_1(x)x\,dx\right)^2 - \int_0^1 f_1(x)x\,dx + \frac{1}{4\alpha} \right\} := E_2.$$
$$(13.2.8)$$

Here l_1 and l_2 are arbitrary fixed indices such that $l_1, l_1 + 1 \in K_1$, $l_2, l_2 + 1 \in K_2$, and $\int_a^b f_j(r) dr$, $0 < a < b < 1$, is the probability as $N \to \infty$ that the random variable R_i/N, $i \in K_j$, lies between a and b. Existence of such functions $f_j(r)$ follows from the Glivenko theorem (see Section 14.10.2). In (13.2.8), one used the fact that R_k and R_j, $k \neq j$, are asymptotically independent, and $\alpha f_1 + (1-\alpha) f_2 = 1$. Similarly, let us show that $\text{var}(\nu_N) \to 0$ if H_2 holds. One has

$$E(\nu_N^2)$$

$$= \frac{1}{(N-1)^2} E\left\{ \left[\sum_{k \in K_1} \left(\frac{R_{k+1} - R_k}{N}\right)^2 + \sum_{k \in K_2} \left(\frac{R_{k+1} - R_k}{N}\right)^2 \right]^2 \right\}$$

$$= \sum_{p,q=1}^2 \frac{1}{(N-1)^2} \sum_{k \in K_p} \sum_{j \in K_q} E_{kj},$$

$$E_{kj} := E\left[\left(\frac{R_{k+1} - R_k}{N}\right)^2 \left(\frac{R_{j+1} - R_j}{N}\right)^2 \right]. \qquad (13.2.9)$$

Since

a) The number of terms for which $k = j$ or $|k-j| = 1$ is proportional to N;

b) The portion of boundary points is of order $O(1/N)$; and

c) $R_{k+1} - R_k$ and $R_{j+1} - R_j$, $|k - j| \geq 2$, are asymptotically independent;

one has

$$\frac{1}{(N-1)^2} \sum_{k \in K_p} \sum_{j \in K_q} E_{kj} = \frac{1}{(N-1)^2} \sum_{k \in K_p} \sum_{\substack{j \in K_q \\ |j-k| \geq 2}} E_{kj} + O\left(\frac{1}{N}\right) =$$

$$\left[\frac{|K_p|}{N-1} E\left(\frac{R_{m+1}-R_m}{N}\right)^2\right]\left[\frac{|K_q|}{N-1} E\left(\frac{R_{n+1}-R_n}{N}\right)^2\right] + O\left(\frac{1}{N}\right), \tag{13.2.10}$$

where m and n are arbitrary fixed indices such that $m, m+1 \in K_p$, $n, n+1 \in K_q, |m-n| \geq 2$. Formulas (13.2.8) – (13.2.10) yield $E(\nu_N^2) = (E(\nu_N))^2 + O(1/N)$, thus $\text{var}(\nu_N) = O(1/N)$ as $N \to \infty$. Collecting (13.2.5), (13.2.7), and (13.2.8), one proves

Theorem 13.2.1. *One has:* $\nu_N \xrightarrow[N \to \infty]{ms} 1/6$ *if* H_0 *is true, and* $\nu_N \xrightarrow[N \to \infty]{ms} E_2$ *if* H_2 *is true.*

REMARK 13.2.1. It is possible to relax the assumption $p^{(b)} < \text{const}$ as $N \to \infty$ and to require only $p^{(b)}/N = o(1)$ as $N \to \infty$. In this case, one has $\nu_N \to E_2$ with $\text{var}(\nu_N) = o(1)$ as $N \to \infty$. For the proof of this fact note that the only difference in the argument is the replacement of the term $O(1/N)$ in (13.2.8) and (13.2.10) by $o(1)$.

Theorem 13.2.2 below shows that the statistic ν_N can be used for testing H_0 against H_2.

Theorem 13.2.2. *Under assumption (13.2.2b) one has* $E_2 < 1/6$.

Proof. One sees from (13.2.8) that it is enough to prove the inequality

$$z^2 - z + \frac{1}{4\alpha} > \frac{1-\alpha}{4\alpha}, \quad z := \int_0^1 f_1(r)r\,dr, \tag{13.2.11}$$

which is equivalent to $(z - 1/2)^2 > 0$. To prove that $z \neq 1/2$, let us calculate the function $f_1(r)$. Denote $g_1(x) := G_1'(x)$, $g_2(x) := G_2'(x)$. Pick an arbitrary index i, $1 \leq i \leq N$, let x be the value of the random variable at this point and let $r = R_i/N$ be the normalized rank of this value. By the Glivenko theorem (see Section 14.10.2), one has for large N

$$r = \alpha G_1(x) + (1-\alpha)G_2(x) := G(x), \tag{13.2.12}$$

or $x = G^{-1}(r)$. Here and below $G^{-1}(r)$ stands for the inverse of $G(x)$ where the inverse function is defined, for example, where $g(x) := G'(x) > 0$. Using (13.2.12) and taking into account that $1/g(G^{-1}(r))$ is the

Jacobian of the transformation $x \to r$, $x = G^{-1}(r)$, one gets, for r such that $r = G(x)$ and $g(x) > 0$,

$$f_1(r) = \frac{g_1(G^{-1}(r))}{g(G^{-1}(r))}. \qquad (13.2.13)$$

Let E' be the set $\{r : 0 \le r \le 1, \; G(x) = r, \; g(x) = 0\}$. By Sard's theorem [Sard], $\lambda(E') = 0$, where $\lambda(E)$ is the usual Lebesgue measure in $\mathbb{R}^1$. Let E be the complement of E' in $[0,1]$, $\lambda(E) = 1$. Fix any $\epsilon > 0$ and denote by E'_ϵ any open covering of E', such that $\lambda(E'_\epsilon) = \epsilon$. Let $E_\epsilon := [0,1] \setminus E'_\epsilon$ and $\mathbb{R}_\epsilon := G^{-1}(E_\epsilon)$. Consider the integral

$$\int_{E_\epsilon} f_1(r)dr = \int_{E_\epsilon} \frac{g_1(G^{-1}(r))}{g(G^{-1}(r))}rdr = \int_{\mathbb{R}_\epsilon} \frac{g_1(x)}{g(x)}G(x)g(x)dx$$

$$= \int_{\mathbb{R}_\epsilon} G(x)g_1(x)dx.$$

Let $\epsilon \to 0$ in the above formula. Then the limit of the left side is denoted by $\int_0^1 f(r)dr$ (since $\lambda(E) = 1$). This limit does exist since the limit of the right side exists and is equal to $\int_{G^{-1}(E)} G(x)g_1(x)dx = \int G(x)g_1(x)dx$, where $\int := \int_{\mathbb{R}}$. The last equation holds because $g_1(x) = 0$ on $G^{-1}(E')$ and $\mathbb{R} = G^{-1}(E) \cup G^{-1}(E')$. Therefore, one concludes that

$$z := \int_0^1 f_1(r)dr = \int G(x)g_1(x)dx = \frac{\alpha}{2} + (1-\alpha)\int G_2 g_1 dx. \quad (13.2.14)$$

Using the inequalities $G_1(x) \ge G_2(x)$, $G_1(x_0) > G_2(x_0)$ and the continuity of $g_1(x)$, one has $\int G_2 g_1 dx < \int G_1 g_1 dx = 1/2$. This, together with (13.2.14), yields $z < 1/2$. Theorem 13.2.2 is proved. $\square$

REMARK 13.2.2. From the proof of Theorem 13.2.2 it follows that $E_2 = 1/6$ if and only if $G_1(x) \equiv G_2(x)$, i.e. when $H_0 = H_2$. Here E_2 is defined in (13.2.8).

REMARK 13.2.3. From the argument under (13.2.14) one sees that to prove the inequality $z < 1/2$ it is sufficient to have only $\int G_2 g_1 dx < 1/2$. Thus Theorem 13.2.2 holds under condition weaker than (13.2.2b), which can be replaced by $P\{\xi_1 \le \xi_2\} > 1/2$, where ξ_k is the random variable with the distribution function G_k, $k = 1, 2$.

Using Theorems 13.2.1 and 13.2.2, let us construct the following test of randomness. Fix the probability ϵ, $0 < \epsilon < 1$, of the type I error and let the rejection region be $\{x : x < A_0\}$, where the threshold A_0

is determined from the equation $P\{\nu_N < A_0|H_0\} = \epsilon$. The consistency of the proposed test easily follows. Indeed, since $E_2 < 1/6$ (Theorem 13.2.2) and $A_0 \to 1/6$ as $N \to \infty$ (according to the choice of A_0 and Theorem 13.2.1), one has using the Chebyshev inequality (see Section 14.10.1) and assuming that H_2 holds

$$P\{\nu_N \geq A_0\} \leq P\{|\nu_N - E_2| \geq A_0 - E_2\}$$
$$\leq \frac{\mathrm{var}(\nu_N)}{(A_0 - E_2)^2} = O(\frac{1}{N}) \text{ as } N \to \infty.$$

If N is sufficiently large, the threshold A_0 is found using the asymptotic normality of ν_N with the mean value $1/6$ and variance $1/(36N)$. If N is small, a convenient way to compute A_0 is by using the Monte-Carlo method (see Section 14.10.5).

13.2.2. One-dimensional case, $m > 2$

In this section we prove that the test based on ν_N is consistent against an alternative more general than H_2. Let us fix $m \geq 2$ and define the alternative hypothesis H_m.

$$H_m: \quad F_k(x) = G_l(x), \ k/N \in D_l, \ l = 1,\ldots,m, \qquad (13.2.15a)$$

where we assume that

$$G_l(x) \geq G_{l+1}(x), \ x \in \mathbb{R}; \quad \exists x_{0l}: \ G_l(x_{0l}) > G_{l+1}(x_{0l}), \ l = 1,\ldots,m-1, \qquad (13.2.15b)$$

$$\bigcup_{l=1}^{m} D_l = D = (0,1], \ D_i \cap D_j = \varnothing, \ i \neq j, \qquad (13.2.15c)$$

$G_l(x)$ are continuously differentiable and D_l are measurable, $l = 1,\ldots, m$. Similarly to (13.2.3), let us introduce the interior and boundary points

$$p_l^{(i)} := \#\{k: k, k+1 \in K_l\}, \ K_l := \{k: \ k/N \in D_l\}, \ l = 1,\ldots,m;$$

$$p^{(b)} := N - \sum_{l=1}^{m} p_l^{(i)}, \qquad (13.2.16)$$

and assume that

$$p_l^{(i)}/N \to \alpha_l = \lambda(D_l) \text{ as } N \to \infty, \ 0 < \alpha_l < 1, \ l = 1,\ldots,m;$$

$$\sum_{l=1}^{m} \alpha_l = 1; \quad p^{(b)} < \text{const}. \qquad (13.2.17)$$

Similarly to (13.2.8), (13.2.12), and (13.2.13), one has

$$E(\nu_N) \xrightarrow[N \to \infty]{} \sum_{l=1}^{m} \alpha_l \int_0^1 \int_0^1 f_l(x)(x-y)^2 f_l(y) dx dy$$

$$:= E_m(\alpha_1, \ldots, \alpha_m; G_1, \ldots, G_m),$$
(13.2.18)

$$f_l(r) := \frac{g_l(G^{-1}(r))}{g(G^{-1}(r))} \quad \text{for almost all } r \in [0,1], \tag{13.2.19a}$$

$$G(x) := \alpha_1 G_1(x) + \cdots + \alpha_m G_m(x), \quad g(x) := G'(x). \tag{13.2.19b}$$

Theorem 13.2.3. *Under assumptions (13.2.15b, c) and (13.2.17) one has*

$$E_m(\alpha_1, \ldots, \alpha_m; G_1, \ldots, G_m)$$
$$< E_{m-1}(\alpha_1, \ldots, \alpha_{m-2}, \alpha_{m-1} + \alpha_m; G_1, \ldots, G_{m-1}).$$
(13.2.20)

Proof. Substitution of (13.2.19) into (13.2.18) and the change of variables $x = G^{-1}(r)$ yields

$$E_m(\alpha_1, \ldots, \alpha_m; G_1, \ldots, G_m) = 2\left\{ \frac{1}{3} - \sum_{l=1}^{m} \alpha_l \left(\int G(x)g_l(x)dx \right)^2 \right\}.$$

Defining

$$\tilde{G}(x) := \alpha_1 G_1(x) + \cdots + \alpha_{m-2} G_{m-2}(x) + (\alpha_{m-1} + \alpha_m)G_{m-1}(x),$$
$$\tilde{g}(x) := \tilde{G}'(x), \tag{13.2.21}$$

one obtains that (13.2.20) is equivalent to

$$\sum_{l=1}^{m} \alpha_l \left(\int G(x)g_l(x)dx \right)^2 > \sum_{l=1}^{m-2} \alpha_l \left(\int \tilde{G}(x)g_l(x)dx \right)^2 +$$

$$(\alpha_{m-1} + \alpha_m)\left(\int \tilde{G}(x)g_{m-1}(x)dx \right)^2. \tag{13.2.22}$$

Define $H(x) := G_m(x) - G_{m-1}(x)$, $h(x) := H'(x)$. Since $G(x) = \tilde{G}(x) + \alpha_m H(x)$ and $g_m(x) = g_{m-1}(x) + h(x)$, one has from (13.2.21) and (13.2.22)

$$2\sum_{l=1}^{m-1} \alpha_l \int \tilde{G}g_l dx \int \alpha_m H g_l dx + \sum_{l=1}^{m-1} \alpha_l \left(\int \alpha_m H g_l dx \right)^2 +$$

$$\alpha_m \left(\int (\tilde{G} + \alpha_m H)(g_{m-1} + h)dx \right)^2 > \alpha_m \left(\int \tilde{G}g_{m-1} dx \right)^2. \tag{13.2.23}$$

One has $H(x) = G_m(x) - G_{m-1}(x) \to 1 - 1 = 0$ as $x \to \infty$ and $H(x) \to 0 - 0 = 0$ as $x \to -\infty$. Therefore, $\int Hh\,dx = \int H\,dH = (H^2(+\infty) - H^2(-\infty))/2 = 0$. Hence

$$\left(\int (\tilde{G} + \alpha_m H)(g_{m-1} + h)\,dx \right)^2$$

$$= \left(\int \tilde{G}g_{m-1}\,dx + \int \alpha_m H g_{m-1}\,dx + \int \tilde{G}h\,dx \right)^2$$

$$= \left(\int \tilde{G}g_{m-1}\,dx \right)^2 + 2 \int \tilde{G}g_{m-1}\,dx \left(\int \alpha_m H g_{m-1}\,dx + \int \tilde{G}h\,dx \right)$$

$$+ \left(\int \alpha_m H g_{m-1}\,dx + \int \tilde{G}h\,dx \right)^2 .$$

Substitution into (13.2.23) gives

$$2 \sum_{l=1}^{m-1} \alpha_l \int \tilde{G}g_l\,dx \int \alpha_m H g_l\,dx$$

$$+ 2\alpha_m \int \tilde{G}g_{m-1}\,dx \left(\int \alpha_m H g_{m-1}\,dx + \int \tilde{G}h\,dx \right) + A > 0, \tag{13.2.24}$$

where

$$A := \sum_{l=1}^{m-1} \alpha_l \left(\int \alpha_m H g_l\,dx \right)^2 + \alpha_m \left(\int \alpha_m H g_{m-1}\,dx + \int \tilde{G}h\,dx \right)^2 \geq 0. \tag{13.2.25}$$

One has

$$\int \alpha_m H g_{m-1}\,dx + \int \tilde{G}h\,dx = \int \alpha_m H g_{m-1}\,dx - \int H\tilde{g}\,dx =$$

$$\int H(\alpha_m g_{m-1} - \tilde{g})\,dx = - \int H(\alpha_1 g_1 + \cdots + \alpha_{m-1} g_{m-1})\,dx.$$

The last equation and (13.2.24) imply that it is sufficient to prove

$$\sum_{l=1}^{m-1} \alpha_l \int \tilde{G}g_l\,dx \int H g_l\,dx$$

$$\geq \int \tilde{G}g_{m-1}\,dx \int H(\alpha_1 g_1 + \cdots + \alpha_{m-1} g_{m-1})\,dx,$$

where one cancelled $2\alpha_m > 0$. This inequality is equivalent to the following one

$$\sum_{l=1}^{m-1} \alpha_l \left\{ \int \tilde{G}(g_l - g_{m-1})\,dx \right\} \int H g_l\,dx \geq 0. \tag{13.2.26}$$

Integrating by parts the expression in braces, one gets

$$\int \tilde{G}(g_l - g_{m-1})dx = \tilde{G}(G_l - G_{m-1})\Big|_{-\infty}^{\infty} - \int (G_l - G_{m-1})\tilde{g}dx =$$

$$\int (G_{m-1} - G_l)\tilde{g}dx \leq 0, \ 1 \leq l \leq m - 1.$$

The last inequality holds because $\tilde{g} = \alpha_1 g_1 + \cdots + (\alpha_{m-1}+\alpha_m)g_{m-1} \geq 0$ and $G_{m-1} \leq G_l$, $1 \leq l \leq m - 1$. Together with the inequality $H = G_m - G_{m-1} \leq 0$, this proves (13.2.26). From (13.2.25) and (13.2.26) one sees that we proved (13.2.24) with '$\geq$' in place of '$>$'. To prove the strict inequality, it is sufficient to prove that $\int H g_{m-1}dx < 0$, which implies $A > 0$. One has, using (13.2.15b)

$$\int H g_{m-1}dx = \int (G_m - G_{m-1})dG_{m-1} < 0$$

Theorem 13.2.3 is proved. □

Applying inequality (13.2.20) repeatedly and using Theorem 13.2.2, one obtains

$$E(\nu_N) \xrightarrow[N\to\infty]{} E_m(\alpha_1,\ldots,\alpha_m;G_1,\ldots,G_m) <$$
$$E_2(\alpha_1,\alpha_2 + \cdots + \alpha_m;G_1,G_2) = E_2(\alpha_1,1-\alpha_1;G_1,G_2) = E_2 < 1/6.$$

As in the previous section, it is easy to prove that $\text{var}(\nu_N) = O(1/N)$ if H_m holds. Thus the test of randomness based on ν_N is also consistent against H_m for any fixed m, $m \geq 2$, with probability of type II error being of order $O(1/N)$, $N \to \infty$.

13.2.3. Multidimensional case, fixed design model

Let the given data be $x_{k_1,\ldots,k_d}$, $1 \leq k_i \leq \beta_i N$, $i = 1,\ldots,d$, where β_i are fixed integers. In this case the observation points are

$$\frac{1}{N}(k_1,\ldots,k_d) \in D := (0,\beta_1] \times \cdots \times (0,\beta_d] \subset \mathbb{R}^d.$$

Let us denote $k := (k_1,\ldots,k_d)$ and $B_N := \{k \in \mathbb{N}^d : 1 \leq k_i \leq \beta_i N, \ i = 1,\ldots,d\}$. The problem is to test the null hypothesis

$$H_0 : \quad F_k(x) = F_j(x) \quad \text{for every } k,j \in B_N, \tag{13.2.27}$$

where $F_k(x)$ is a continuous distribution function of the random variable x_k, $k \in B_N$, against the alternative

$$H_m : \quad F_k(x) = G_j(x), \ k/N \in D_j, \ j = 1,\ldots,m, \tag{13.2.28a}$$

$$\bigcup_{j=1}^{m} D_j = D, \ D_i \cap D_j = \emptyset, \ i \neq j. \qquad (13.2.28b)$$

Here the functions $G_j(x)$ satisfy the same conditions as in Section 13.2.2 (see (13.2.15b) and below), and D_j are measurable, $j = 1, \ldots, m$. For an arbitrary multiindex $k \in B_N$, we define the set of multiindices neighboring to k by the formula $L(k) := \{l \in B_N : l \neq k, \max_{1 \leq i \leq d} |l_i - k_i| = 1\}$. One easily sees that if k is strictly inside B_N, the number of elements in $L(k)$ is independent of k and is equal to $3^d - 1$. Similarly to (13.2.3) and (13.2.16), we introduce the interior and boundary points

$$p_j^{(i)} := \#\{k \in \mathbb{N}^d : \ k, L(k) \in K_j\}, \ K_j := \{k \in \mathbb{N}^d : \ k/N \in D_j\},$$

$$j = 1, \ldots, m; \ p^{(b)} := N^d - \sum_{j=1}^{m} p_j^{(i)},$$

and assume that

$$p_j^{(i)}/\hat{N} \to \alpha_j := \lambda(D_j) \text{ as } N \to \infty, \ 0 < \alpha_j < 1, \ j = 1, \ldots, m;$$

$$\sum_{j=1}^{m} \alpha_j = 1; \ \ p^{(b)} < \text{const} N^{d-1}.$$

Here $\lambda(\cdot)$ is the Lebesgue measure in $\mathbb{R}^d$ and $\hat{N} := \beta_1 \cdot \ldots \cdot \beta_d N^d$ is the number of lattice points. Let R_i be the rank of the element x_i, $i \in B_N$. The test criterion is the statistic

$$\nu_N := \frac{1}{M_N} \sum_{k \in B_N} \sum_{l \in L(k)} \left(\frac{R_k - R_l}{\hat{N}} \right)^2, \qquad (13.2.29)$$

where M_N is the number of elements in double sum (13.2.29). One sees that $M_N = (3^d - 1)\hat{N}(1 + O(1/N))$ as $N \to \infty$. Note also that each pair $(R_k, R_l), k \neq l$, regardless of order, appears in the double sum in (13.2.29) twice. Similarly to (13.2.5) and (13.2.6), the first two moments of ν_N when the null hypothesis is true are given by

$$E(\nu_N) = 1/6 + 1/(6\hat{N}), \qquad E(\nu_N^2) = 1/36 + O(N^{-d}). \qquad (13.2.30)$$

This yields

$$\text{var}(\nu_N) = E(\nu_N^2) - (E(\nu_N))^2 = O(N^{-d}), \ N \to \infty. \qquad (13.2.31)$$

Now let us consider the behavior of ν_N under the alternative hypothesis H_m. Since the portion of border points among all the points is of order

$O(1/N)$, we may write similarly to (13.2.8) and (13.2.18)

$$E(\nu_N) = \sum_{j=1}^{m} \frac{1}{M_N} \sum_{k \in K_j} \sum_{l \in L(k)} E\left(\frac{R_k - R_l}{\hat{N}}\right)^2$$

$$= \sum_{j=1}^{m} \frac{|K_j|}{\hat{N}} E\left(\frac{R_{k_j} - R_{l_j}}{\hat{N}}\right)^2 + O(\frac{1}{N})$$

$$\xrightarrow[N\to\infty]{} \sum_{j=1}^{m} \alpha_j \int_0^1 \int_0^1 f_j(x)(x-y)^2 f_j(y) dx dy$$

$$= E_m(\alpha_1, \ldots, \alpha_m; G_1, \ldots, G_m) = E_m.$$

$$(13.2.32)$$

where k_j and l_j are arbitrary different multiindices such that $k_j, l_j \in K_j$, $j = 1, \ldots m$. As in Section 13.2.1, it is easy to show that $\text{var}(\nu_N) = O(1/N)$ as $N \to \infty$. Collecting (13.2.30) – (13.2.32), one proves

Theorem 13.2.4. *One has:* $\nu_N \xrightarrow[N\to\infty]{ms} 1/6$ *if* H_0 *is true, and* $\nu_N \xrightarrow[N\to\infty]{ms} E_m$ *if* H_m *is true.*

Using Theorem 13.2.2, one sees that the statistic ν_N defined by formula (13.2.29) can be used for testing H_0 against H_m in the multidimensional case. The consistency of the test follows easily from Theorems 13.2.2 and 13.2.4. Note that the probability of a type II error is $O(1/N)$ as $N \to \infty$. The threshold A_0 is determined from the equation $P\{\nu_N < A_0|H_0\} = \epsilon$. If N is sufficiently large, A_0 is found using the asymptotic normality of ν_N [CO]. If N is small, a convenient way to compute A_0 is by using the Monte-Carlo method.

13.2.4. Random design model

Let the observation points $\{\rho_k\}_{k=1}^{N}$ be randomly chosen inside an open bounded domain $D \subset \mathbb{R}^d$, $d \geq 1$, and let $\{x_k\}_{k=1}^{N}$ be a set of corresponding observations. The observation points are called random if they are independent and identically distributed inside D with constant probability density. The problem is to test the null hypothesis

$$H_0: \quad F_1(x) = F_2(x) = \cdots = F_N(x), \qquad (13.2.33)$$

where $F_k(x)$ is a continuous distribution function of the random variable observed at the point ρ_k, $k = 1, \ldots, N$, against the alternative

$$H_m: \quad F_k(x) = G_i(x) \text{ for } \rho_k \in D_i, \qquad (13.2.34a)$$

$$\lambda(D_j) > 0, \ j = 1, \ldots, m, \ \bigcup_{j=1}^{m} D_j = D, \ D_i \cap D_j = \varnothing, \ i \neq j, \quad (13.2.34b)$$

where we assume, as usual, that the partition D_j, $j = 1, \ldots, m$, and $m \geq 2$ are fixed. Here $\lambda(\cdot)$ is the Lebesgue measure in $\mathbb{R}^d$, and the functions G_i satisfy the same conditions as in Section 13.2.2. The test criterion is the statistic

$$\nu_N := \frac{1}{N} \sum_{k=1}^{N} \left(\frac{R_{n(k)} - R_k}{N} \right)^2, \qquad (13.2.35)$$

where $n(k)$ is the index of the point closest to ρ_k:

$$|\rho_{n(k)} - \rho_k| = \min_{\substack{1 \leq j \leq N \\ j \neq k}} |\rho_j - \rho_k|.$$

Note that $n(k)$ is unique with probability 1. The first two moments of ν_N can be easily computed under the assumption that H_0 holds:

$$E(\nu_N) = 1/6 + 1/(6N), \qquad E(\nu_N^2) = 1/36 + O(1/N). \qquad (13.2.36)$$

In (13.2.36), we used that

a) For a fixed index k, there exists a number γ, $0 < \gamma < \infty$, which is independent of k, $\{\rho_k\}_{k=1}^{N}$, and N, such that the number of indices l for which $\{k, n(k)\} \cap \{l, n(l)\} \neq \varnothing$ is bounded by γ; and

b) The random variables $R_{n(k)} - R_k$ and $R_{n(l)} - R_l$ for k and l such that $\{k, n(k)\} \cap \{l, n(l)\} = \varnothing$ are asymptotically independent.

Equations (13.2.36) imply

$$\mathrm{var}(\nu_N) = O(1/N), \quad N \to \infty. \qquad (13.2.37)$$

Now let us study the asymptotic behavior of ν_N under the assumption that H_m holds. Define $S := \bigcup_{i=1}^{m} \partial D_i$, where ∂D_i is the boundary of D_i. Fix any $\epsilon > 0$ and define $V_\epsilon := \{s \in D : \mathrm{dist}(s, S) \leq \epsilon\}$. We assume that S is sufficiently smooth, so that $\lambda(V_\epsilon) \to 0$ as $\epsilon \to 0$. One has $P\{\lim_{N\to\infty} (\max_{1 \leq k \leq N} \min_{\substack{1 \leq j \leq N \\ j \neq k}} |\rho_k - \rho_j|) = 0\} = 1$. Using this, fix any $\delta > 0$ and find N_1 such that

$$P\{\max_{1 \leq k \leq N} \min_{\substack{1 \leq j \leq N \\ j \neq k}} |\rho_k - \rho_j| > \epsilon\} < \delta, \quad N \geq N_1. \qquad (13.2.38)$$

Let us introduce some notation

$$\rho := \{\rho_k\}_{k=1}^{N}, \quad V_i := D_i \cap V_\epsilon, \quad U_i := D_i \setminus V_i,$$

$$d_i := \lambda(D_i)/\lambda(D), \quad v_\epsilon := \lambda(V_\epsilon)/\lambda(D),$$

$$v_i := \lambda(V_i)/\lambda(D), \quad u_i := \lambda(U_i)/\lambda(D),$$

$$\tilde{U}_i := \tilde{U}_i(\rho) := \#\{k : \rho_k \in U_i\}, \quad \tilde{V}_i := \tilde{V}_i(\rho) := \#\{k : \rho_k \in V_i\},$$

$$\tilde{D}_i := \tilde{D}_i(\rho) := \#\{k : \rho_k \in D_i\}, \Delta_N := \Delta_N(\rho) := \max_{1 \le k \le N} \min_{\substack{1 \le j \le N \\ j \ne k}} |\rho_k - \rho_j|,$$

$$P(\delta, N) := \{\rho : |\frac{\tilde{V}_i}{N} - v_i| < \delta, |\frac{\tilde{U}_i}{N} - u_i| < \delta, 1 \le i \le m\}. \qquad (13.2.39)$$

Recall that we assume $N \gg m$. Since the distribution of points within D is assumed to be random, one can easily get $P\{\rho \notin P(\delta, N)\} \to 0$ as $N \to \infty$. Thus,

$$\exists N_2 \quad \text{such that} \quad P\{\rho \notin P(\delta, N)\} < \delta, \ N \ge N_2. \qquad (13.2.40)$$

Using properties of conditional expectation, one gets

$$\begin{aligned} E(\nu_N) =& E(\nu_N|\rho \in P(\delta, N), \ \Delta_N \le \epsilon)P\{\rho \in P(\delta, N), \ \Delta_N \le \epsilon\} \\ &+ E(\nu_N|\rho \notin P(\delta, N), \Delta_N \le \epsilon)P\{\rho \notin P(\delta, N), \Delta_N \le \epsilon\} \\ &+ E(\nu_N|\Delta_N > \epsilon)P\{\Delta_N > \epsilon\}. \end{aligned}$$

The last equation together with (13.2.38) and (13.2.40) implies

$$|E(\nu_N) - E(\nu_N|W)P\{W\}| < 2\delta, \ N \ge N_0 := \max(N_1, N_2), \quad (13.2.41)$$

where the event $\{W\}$ is defined as $\{W\} := \{\rho \in P(\delta, N), \ \Delta_N \le \epsilon\}$. Pick an arbitrary ρ such that $\rho \in P(\delta, N)$ and $\Delta_N \le \epsilon$. Using (13.2.35), we get

$$E(\nu_N|\rho) = \sum_{i=1}^m \frac{1}{N} \sum_{k:\rho_k \in U_i} E\left(\left(\frac{R_{n(k)} - R_k}{N}\right)^2 \middle| \rho\right) +$$

$$\sum_{i=1}^m \frac{1}{N} \sum_{k:\rho_k \in V_i} E\left(\left(\frac{R_{n(k)} - R_k}{N}\right)^2 \middle| \rho\right). \qquad (13.2.42)$$

Since the number of observations inside V_ϵ is bounded by $(v_\epsilon + m\delta)N$, and the distribution of the random variable $\left(\frac{R_{n(k)} - R_k}{N}\right)^2$ does not depend on k, provided that $\rho_k \in U_i$ (according to the choice of ρ and U_i, if $\rho_k \in U_i \subset D_i$, then $\rho_{n(k)} \in D_i$), one obtains from (13.2.39) and (13.2.42):

$$\left| E(\nu_N|\rho) - \sum_{i=1}^m \frac{\tilde{U}_i}{N} E\left(\left(\frac{R_{n(k_i)} - R_{k_i}}{N}\right)^2 \middle| \rho\right) \right| \le v_\epsilon + m\delta. \qquad (13.2.43)$$

Let ρ_0 be the distribution of observation points such that $\tilde{V}_i(\rho_0)/N = v_i$, $\tilde{U}_i(\rho_0)/N = u_i$, $i = 1, \ldots, m$. We will use two facts.

a) The distribution function of the random variable $\left(\frac{R_{n(k_i)} - R_{k_i}}{N}\right)^2$

   for $\rho_{k_i} \in U_i$, does not depend on the location of observation points ρ_k within the sets U_i and V_i, $i = 1, \ldots, m$, it depends only on parameters $\tilde{U}_i(\rho) + \tilde{V}_i(\rho)$, $i = 1, \ldots, m$.

b) Consider the set of observations $\{x_k\}_{k=1}^N$ corresponding to ρ and change arbitrarily the values of observations at no more that δN points in each of the sets U_i and V_i, $i = 1, \ldots, m$. One obtains a new sequence $\{\hat{x}_k\}_{k=1}^N$ and a new set of corresponding ranks $\{\hat{R}_k\}_{k=1}^N$. If the observation has not been changed at the point ρ_k, then $|(R_k - \hat{R}_k)/N| \le 4m\delta$.

According to the choice of ρ and ρ_0, one easily gets using a), b), and the triangle inequality

$$\left| E\left(\left(\frac{R_{n(k_i)} - R_{k_i}}{N}\right)^2 \middle| \rho\right) - E\left(\left(\frac{R_{n(k_i)} - R_{k_i}}{N}\right)^2 \middle| \rho_0\right) \right| \le 32m\delta.$$
(13.2.44)

Definitions (13.2.39) and inequalities (13.2.43) and (13.2.44) imply

$$\left| E(\nu_N|\rho) - \sum_{i=1}^m d_i E\left(\left(\frac{R_{n(k_i)} - R_{k_i}}{N}\right)^2 \middle| \rho_0\right) \right| \le 2v_\epsilon + 34m\delta. \quad (13.2.45)$$

Since the distribution of observations ρ_0 is fixed, one obtains similarly to (13.2.8) and (13.2.18)

$$\lim_{N \to \infty} \sum_{i=1}^m d_i E\left(\left(\frac{R_{n(k_i)} - R_{k_i}}{N}\right)^2 \middle| \rho_0\right) = E_m(d_1, \ldots, d_m; G_1, \ldots, G_m)$$

$$:= E_m.$$

Hence

$$|E(\nu_N|\rho) - E_m| \le 2v_\epsilon + 34m\delta + o(1), \quad N \to \infty. \quad (13.2.46)$$

Note that $o(1)$, $N \to \infty$, in (13.2.46) is independent of ρ. Recalling the definition of the event W (see under (13.2.41)), we get $|E(\nu_N|W) - E_m| \le 2v_\epsilon + 34m\delta + o(1)$. This together with (13.2.41) and an obvious inequality $P\{W\} \ge 1 - P\{\rho \notin \mathcal{P}(\delta, N)\} - P\{\Delta_N > \epsilon\} \ge 1 - 2\delta$ implies

$$|E(\nu_N) - E_m| \le 4\delta + (2v_\epsilon + 34m\delta + o(1))(1 + 2\delta), \quad N \to \infty.$$

Taking the limit as $N \to \infty$ and using that $\epsilon, \delta > 0$ were arbitrary, we conclude

$$\lim_{N \to \infty} E(\nu_N) = E_m.$$

Similarly, one can show that $\text{var}(\nu_N) \to 0$ as $N \to \infty$. Together with (13.2.36) and (13.2.37), this yields

Theorem 13.2.5. *One has:* $\nu_N \xrightarrow[N\to\infty]{ms} 1/6$ *if* H_0 *is true, and* ν_N $\xrightarrow[N\to\infty]{ms} E_m$ *if* H_m *is true.*

Using Theorem 13.2.2, one sees that the statistic ν_N defined by formula (13.2.35) can be used for testing H_0 against H_m in the case of random observation points. The consistency of the test follows easily from Theorems 13.2.2 and 13.2.5.

13.2.5. Numerical experiments

The results obtained in this chapter can be used in many applications, in particular, in image processing for edge detection. Let us describe the algorithm for edge detection based on these results. In image processing, the data are intensities of grey level specified at each pixel, i.e. at the nodes of two-dimensional square grid (image). An edge (discontinuity of a signal) can be defined as follows: the grey level is relatively consistent in each of the two adjacent extensive regions, and changes abruptly as the border between two regions is crossed [Pr, RoKa]. Consider $N \times N$ window B_N sliding over the image. For each position of the window one wants to make a decision: whether $S \cap B_N = \varnothing$ or not, where S is an edge. If $S \cap B_N \neq \varnothing$, then S divides B_N into two sets K_1 and K_2, such that the values of grey level in one set are stochastically larger than in the other set, hence the hypothesis H_2 (or, more generally, H_m) takes place. If $S \cap B_N = \varnothing$, then the grey level is approximately constant inside B_N and the hypothesis H_0 takes place. Thus, the choice between '$S \cap B_N = \varnothing$'(H_0) and '$S \cap B_N \neq \varnothing$'($H_m$) can be made using the test of randomness which is described in Section 13.2.3. If the hypothesis H_m is accepted, the center of the current window is marked as an edge point. Repeating this process for each position of the window, one finds all edge points.

The numerical results of an application of the above algorithm are illustrated by the following example. Figure 13.2.1 represents a synthetic image of square and circle edges with the jump magnitude $D = 1.5$ specified at a square 101×101 grid. The image is corrupted by noise with the uniform distribution and standard deviation $\sigma = 0.75$. The window size has been chosen $N = 7$, the probability of false alarm has been $\epsilon = 0.01$. Figure 13.2.2 represents the image of detected edges of Figure 13.2.1.

13.3. Consistency of rank
test against trend in location

13.3.1. One-dimensional case, equispaced design model

Let $\{x_k\}_{k=1}^{N}$ be a random sequence of size N. The problem is to test the null hypothesis

$$H_0: \quad F_1(x) = F_2(x) = \cdots = F_N(x), \qquad (13.3.1)$$

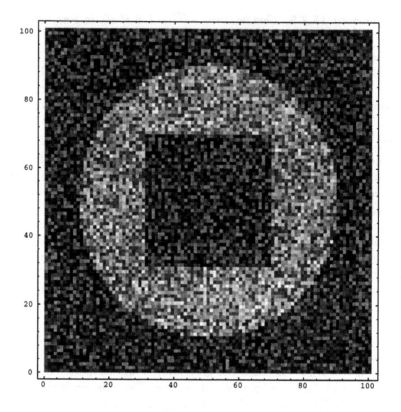

FIGURE 13.2.1.   A 101×101 synthetic image of square and circle step edges corrupted by noise.

where $F_k(x)$ is continuous distribution function of the random variables x_k, $k = 1, \ldots, N$, against the alternative

$$H_1 : \quad F_k(x) = F(x - \theta_k), \ \ k = 1, \ldots, N, \qquad (13.3.2)$$

where F is a continuously differentiable distribution function with

$$f(x) := F'(x), \ \ \sup_{x \in \mathbb{R}^1} f(x) < \infty,$$

and $\theta_k \in \mathbb{R}^1$ are some constants that are not all equal. The test criterion is statistic (13.2.4)

$$\nu_N := \frac{1}{N-1} \sum_{k=1}^{N-1} \left(\frac{R_{k+1} - R_k}{N} \right)^2 . \qquad (13.3.3)$$

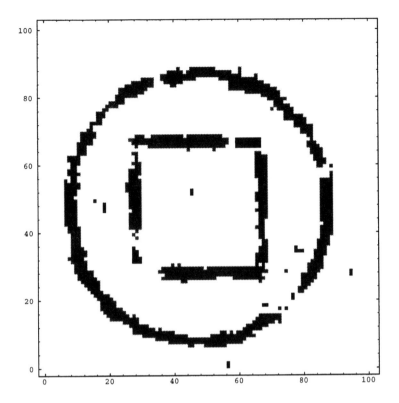

FIGURE 13.2.2. Detected step edges of Figure 13.2.1.

We have $\nu_N \xrightarrow{\text{ms}} 1/6$ as $N \to \infty$ if H_0 is true (see Section 13.2.1). As in Sections 13.2.1 and 13.2.2, to prove consistency of the test against the alternative H_1, it is sufficient to prove that ν_N converges to some constant E_∞ as $N \to \infty$ if H_1 holds, and that $E_\infty < 1/6$. Let us study the asymptotic behavior of ν_N as $N \to \infty$ under H_1. Suppose that the trend θ_k, $k = 1, \ldots, N$, satisfies the condition:

$$\phi(t) := \lim_{N \to \infty, k/N \to t} \theta_k, \quad 0 \le t \le 1; \quad \phi(t) \in C[0,1], \ \phi(t) \not\equiv \text{const.}$$

(13.3.4)

Fix m, $m \ge 2$, and define the intervals Δ_l

$$\Delta_l := [(l-1)/m, l/m), \ l = 1, \ldots, m.$$

(13.3.5)

Let $\nu_N^{(m)}$ be the statistic (13.3.3) calculated in the case when the trend $\tilde{\theta}_k$, $k = 1, \ldots, N$, is constant inside each interval Δ_l:

$$\tilde{\theta}_k = \phi(l/m) \text{ for } l \text{ such that } k/N \in \Delta_l, \ k = 1, \ldots, N.$$

(13.3.6)

Using the results obtained in Section 13.2.1, we have

$$\nu_N^{(m)} \xrightarrow[N\to\infty]{\text{ms}} 2\left\{\frac{1}{3} - \frac{1}{m}\sum_{l=1}^{m}\left(\int_{\mathbb{R}} G^{(m)}(x)g_l^{(m)}(x)dx\right)^2\right\} := E_m < 1/6,$$

(13.3.7a)

$$G^{(m)}(x) = \frac{1}{m}\sum_{l=1}^{m}F(x - \phi(l/m)), \quad g_l^{(m)}(x) = f(x - \phi(l/m)). \quad (13.3.7b)$$

Denote $E_\infty := \lim_{m\to\infty} E_m$. Existence of the limit E_∞ can be established, an analytical expression for this limit is given in formula (13.3.15) below. Let '$\xrightarrow{P}$' denote the convergence in probability.

Theorem 13.3.1. *Under assumptions (13.3.2) and (13.3.4) we have*
$$\nu_N \xrightarrow[N\to\infty]{P} E_\infty.$$

Let $\{y_k\}_{k=1}^N$ be a random sample from the distribution $F(x)$, and let us define two other sequences $\{\hat{y}_k := y_k + \theta_k\}_{k=1}^N$, $\{\hat{y}_k^{(m)} := y_k + \tilde{\theta}_k\}_{k=1}^N$. Let $\hat{R}_k$, $\hat{R}_k^{(m)}$ be the ranks of the elements $\hat{y}_k$, $\hat{y}_k^{(m)}$, respectively. Denote $\hat{r}_k = \hat{R}_k/N$, $\hat{r}_k^{(m)} = \hat{R}_k^{(m)}/N$, and let $\hat{\nu}_N$, $\hat{\nu}_N^{(m)}$ be the values of statistic (13.3.3) calculated for sequences $\{\hat{y}_k\}_{k=1}^N$, $\{\hat{y}_k^{(m)}\}_{k=1}^N$, respectively. First, we prove two auxiliary lemmas.

Lemma 13.3.1. *For every $\epsilon > 0$ there exists M_ϵ such that*

$$\lim_{N\to\infty} P\{|\hat{\nu}_N - \hat{\nu}_N^{(M_\epsilon)}| \geq \epsilon\} = 0.$$

Proof of Lemma 13.3.1. Fix $\epsilon > 0$, denote $h_m := \max_{1\leq k\leq N}|\theta_k - \tilde{\theta}_k|$, where m is the same as in (13.3.5), and find M_ϵ such that

$$\sup_x |F(x + 2h_{M_\epsilon}) - F(x - 2h_{M_\epsilon})| < \epsilon/24. \quad (13.3.8)$$

The existence of such M_ϵ follows the uniform boundedness of $f(x)$ and (13.3.4)-(13.3.6). Denote for brevity in what follows $\hat{\nu}_N^{(\epsilon)} := \hat{\nu}_N^{(M_\epsilon)}$, $\hat{r}_k^{(\epsilon)} := \hat{r}_k^{(M_\epsilon)}$, $h_\epsilon := h_{M_\epsilon}$. From the definitions of $\hat{\nu}_N$, $\hat{\nu}_N^{(\epsilon)}$ we obtain:

$$|\hat{\nu}_N - \hat{\nu}_N^{(\epsilon)}| = \left|\frac{1}{N-1}\sum_{k=1}^{N-1}\left[(\hat{r}_{k+1} - \hat{r}_k)^2 - (\hat{r}_{k+1}^{(\epsilon)} - \hat{r}_k^{(\epsilon)})^2\right]\right| \leq$$

$$\frac{4}{N-1}\sum_{k=1}^{N-1}|(\hat{r}_{k+1} - \hat{r}_{k+1}^{(\epsilon)}) - (\hat{r}_k - \hat{r}_k^{(\epsilon)})| \leq \frac{8}{N-1}\sum_{k=1}^{N}|\hat{r}_k - \hat{r}_k^{(\epsilon)}|.$$

(13.3.9)

Let us estimate the probability $P\{|\hat{r}_k - \hat{r}_k^{(\epsilon)}| > \epsilon\}$. From the definition of h_ϵ we have

$$P\{|\hat{r}_k - \hat{r}_k^{(\epsilon)}| > \epsilon\}$$

$$\leq P\left\{\frac{\#\{l : y_l + \tilde{\theta}_l \in [y_k + \tilde{\theta}_k - 2h_\epsilon, y_k + \tilde{\theta}_k + 2h_\epsilon]\}}{N} > \epsilon\right\}.$$
(13.3.10a)

Similarly,

$$P\left\{\frac{1}{N}\sum_{k=1}^{N}|\hat{r}_k - \hat{r}_k^{(\epsilon)}| > \epsilon\right\}$$

$$\leq P\left\{\sup_{y \in \mathbb{R}} \frac{\#\{k : y_k + \tilde{\theta}_k \in [y - 2h_\epsilon, y + 2h_\epsilon]\}}{N} > \epsilon\right\}.$$
(13.3.10b)

Together with (13.3.9), this yields

$$P\{|\hat{\nu}_N - \hat{\nu}_N^{(\epsilon)}| > \epsilon\}$$

$$\leq P\left\{\sup_{y \in \mathbb{R}} \frac{\#\{k : y_k + \tilde{\theta}_k \in [y - 2h_\epsilon, y + 2h_\epsilon]\}}{N} > \tilde{\epsilon}/8\right\}$$

$$\leq P\left\{\sup_{y \in \mathbb{R}} \frac{\sum_{l=1}^{M_\epsilon} \#\{k : k/N \in \Delta_l, y_k + \phi(l/M_\epsilon) \in [y - 2h_\epsilon, y + 2h_\epsilon]\}}{N}\right.$$

$$\left. > \tilde{\epsilon}/8\right\}$$

$$\leq \sum_{l=1}^{M_\epsilon} P\left\{\sup_{y \in \mathbb{R}} \frac{\#\{k : k/N \in \Delta_l, y_k \in [y - 2h_\epsilon, y + 2h_\epsilon]\}}{N/M_\epsilon} > \tilde{\epsilon}/8\right\}$$

$$= \sum_{l=1}^{M_\epsilon} P\left\{\sup_{y \in \mathbb{R}}(F_l(y + 2h_\epsilon) - F_l(y - 2h_\epsilon)) > \tilde{\epsilon}/8\right\},$$
(13.3.11)

where $\tilde{\epsilon} = \epsilon(N-1)/N$, and $F_l(y)$ is the empirical distribution function of the sample $\{y_k\}_{k \in \Delta_l}$, $1 \leq l \leq M_\epsilon$. In (13.3.11), we used the inequality $P\{\xi_1 + \cdots + \xi_M > M\epsilon\} \leq P\{\xi_1 > \epsilon\} + \cdots + P\{\xi_M > \epsilon\}$, where ξ_i are arbitrary random variables. From (13.3.8) and (13.3.11) we obtain

$$P\{|\hat{\nu}_N - \hat{\nu}_N^{(\epsilon)}| > \epsilon\} \leq \sum_{l=1}^{M_\epsilon} P\left\{\sup_{y \in \mathbb{R}}|F_l(y + 2h_\epsilon) - F(y + 2h_\epsilon)|+\right.$$

$$\sup_{y \in \mathbb{R}}|F(y + 2h_\epsilon) - F(y - 2h_\epsilon)| + \sup_{y \in \mathbb{R}}|F_l(y - 2h_\epsilon) - F(y - 2h_\epsilon)| > \tilde{\epsilon}/8\right\}$$

$$\leq \sum_{l=1}^{M_\epsilon} P\left\{\sup_{y \in \mathbb{R}}|F_l(y) - F(y)| > \epsilon(1 - 3/(2N))/24\right\}.$$

382 THE RADON TRANSFORM AND LOCAL TOMOGRAPHY

Taking the limit as $N \to \infty$ on both sides of the last inequality and using the Glivenko theorem (see Section 14.10.2), we complete the proof of Lemma 13.3.1. $\square$

Lemma 13.3.2. *Fix an arbitrary $\epsilon > 0$ and find M_ϵ such that inequality (13.3.8) holds. Then for any $a \notin [E_{M_\epsilon} - \epsilon, E_{M_\epsilon} + \epsilon]$, where E_{M_ϵ} is the same as in (13.3.7a) with $m = M_\epsilon$, the following equation holds*

$$\lim_{N \to \infty} \left| P\{\nu_N < a\} - P\{\nu_N^{(M_\epsilon)} < a\} \right| = 0.$$

Proof of Lemma 13.3.2. Denote $\nu_N^{(\epsilon)} := \nu_N^{(M_\epsilon)}$ and write the dependence of $\hat{\nu}_N$ and $\hat{\nu}_N^{(\epsilon)}$ on $y_1, \ldots, y_N$ implicitly as $\hat{\nu}_N = \hat{\nu}_N(y)$ and $\hat{\nu}_N^{(\epsilon)} = \hat{\nu}_N^{(\epsilon)}(y)$. From the definitions of ν_N, $\nu_N^{(\epsilon)}$, $\hat{\nu}_N$, and $\hat{\nu}_N^{(\epsilon)}$, we obtain

$$\left| P\{\nu_N < a\} - P\{\nu_N^{(\epsilon)} < a\} \right| = \left| \int \theta(a - \hat{\nu}_N(y)) f_N(y) dy - \right.$$

$$\left. \int \theta(a - \hat{\nu}_N^{(\epsilon)}(y)) f_N(y) dy \right| \le \int \left| \theta(a - \hat{\nu}_N(y)) - \theta(a - \hat{\nu}_N^{(\epsilon)}(y)) \right| f_N(y) dy, \tag{13.3.12}$$

where

$$\int := \int_{\mathbb{R}^N}, \quad dy := dy_1 \ldots dy_N, \quad y := y_1 \ldots y_N, \quad f_N(y) := f(y_1) \ldots f(y_N),$$

and

$$\theta(t) = \begin{cases} 1, & t > 0, \\ 0, & t < 0. \end{cases}$$

The integral on the right-hand side of (13.3.12) can be estimated as follows

$$\int \left| \theta(a - \hat{\nu}_N(y)) - \theta(a - \hat{\nu}_N^{(\epsilon)}(y)) \right| f_N(y) dy$$

$$\le \int_{\{y : |\hat{\nu}_N(y) - \hat{\nu}_N^{(\epsilon)}(y)| < \epsilon\}} \left| \theta(a - \hat{\nu}_N(y)) - \theta(a - \hat{\nu}_N^{(\epsilon)}(y)) \right| f_N(y) dy$$

$$+ P\{|\hat{\nu}_N - \hat{\nu}_N^{(\epsilon)}| > \epsilon\}$$

$$= \int_{\substack{\{y : |\hat{\nu}_N^{(\epsilon)}(y) - a| < \epsilon, \\ |\hat{\nu}_N(y) - \hat{\nu}_N^{(\epsilon)}(y)| < \epsilon\}}} 1 \cdot f_N(y) dy + P\{|\hat{\nu}_N - \hat{\nu}_N^{(\epsilon)}| > \epsilon\}$$

$$\le P\{|\hat{\nu}_N^{(\epsilon)} - a| < \epsilon\} + P\{|\hat{\nu}_N - \hat{\nu}_N^{(\epsilon)}| > \epsilon\}. \tag{13.3.13}$$

From (13.3.7a) we have $\lim_{N\to\infty} P\{|\hat{\nu}_N^{(\epsilon)} - a| < \epsilon\} = 0$ for $a \notin [E_{M_\epsilon} - \epsilon, E_{M_\epsilon} + \epsilon]$, and from Lemma 13.3.1 we have $\lim_{N\to\infty} P\{|\hat{\nu}_N - \hat{\nu}_N^{(\epsilon)}| > \epsilon\} = 0$. This, together with (13.3.12) and (13.3.13) proves Lemma 13.3.2. □

Proof of Theorem 13.3.1. Pick an arbitrary $\epsilon > 0$ and find M_ϵ satisfying (13.3.8) such that $|E_{M_\epsilon} - E_\infty| < \epsilon$. Thus, using Lemma 13.3.2 and (13.3.7a), we get

$$\lim_{N\to\infty} P\{\nu_N < E_\infty - 2\epsilon\} = \lim_{N\to\infty} P\{\nu_N^{(\epsilon)} < E_\infty - 2\epsilon\} = 0. \quad (13.3.14a)$$

Similarly,

$$\lim_{N\to\infty} P\{\nu_N < E_\infty + 2\epsilon\} = \lim_{N\to\infty} P\{\nu_N^{(\epsilon)} < E_\infty + 2\epsilon\} = 1. \quad (13.3.14b)$$

Combining (13.3.14a) and (13.3.14b) proves Theorem 13.3.1. □

Now we prove that the statistic ν_N can be used for testing H_0 against H_1 for arbitrary trend satisfying (13.3.4), where $\phi(t) \not\equiv \text{const}$ is an arbitrary continuous function.

Theorem 13.3.2. *Under assumptions (13.3.2) and (13.3.4) we have* $E_\infty < 1/6$.

From (13.3.4) – (13.3.7a) it follows that

$$G^{(m)}(x) \xrightarrow[m\to\infty]{} \int_0^1 F(x - \phi(t))dt$$

and

$$E_\infty := \lim_{m\to\infty} E_m$$

$$= 2\left\{\frac{1}{3} - \int_0^1 \left(\int_{\mathbb{R}}\left[\int_0^1 F(x - \phi(t))dt\right]f(x - \phi(s))dx\right)^2 ds\right\}. \quad (13.3.15)$$

Thus we need to prove the inequality

$$I(\phi) := \int_0^1\left\{\int_{\mathbb{R}}\left[\int_0^1 F(x - \phi(t))dt\right]f(x - \phi(s))dx\right\}^2 ds > \frac{1}{4}. \quad (13.3.16)$$

Let us prove (13.3.16). Denote $a := \min_{t\in[0,1]}\phi(t)$, $b := \max_{t\in[0,1]}\phi(t)$. Since $\phi(t) \not\equiv \text{const}$, we have $a < b$. Pick an arbitrary y such that $a < y < b$, denote $I_y := \{t : t \in [0, 1], \phi(t) \geq y\}$, and let $\psi(t)$ be any localy integrable nonnegative function, $\psi(t) \not\equiv 0$, $\text{supp}\psi(t) \subseteq I_y$. Denote also

$$\tilde{\phi}_y(t) := \tilde{\phi}(t) := \begin{cases} \phi(t), & t \notin I_y, \\ y, & t \in I_y. \end{cases} \quad (13.3.17)$$

First we prove an auxiliary Lemma, then the proof of Theorem 13.3.2 is given.

Lemma 13.3.3. *One has*

$$\frac{\partial}{\partial\alpha}I(\tilde{\phi}+\alpha\psi)\Big|_{\alpha=0} > 0.$$

Proof of Lemma 13.3.3. Changing variables $x' = x - \phi(s)$ and differentiating $I(\tilde{\phi}+\alpha\psi)$ with respect to α, we get

$$\frac{\partial I}{\partial\alpha}\Big|_{\alpha=0} = 2\int_0^1\Big\{\int_0^1\int_{\mathbb{R}} F(x+\tilde{\phi}(s)-\tilde{\phi}(t))f(x)dxdt\cdot$$

$$\cdot\int_0^1\int_{\mathbb{R}} f(y+\tilde{\phi}(s)-\tilde{\phi}(\tau))f(y)dy(\psi(s)-\psi(\tau))d\tau\Big\}ds. \tag{13.3.18}$$

Note that the above change of variables makes it clear that $I(\phi)$ is Gateaux differentiable in $C[0,1]$. This observation will be used later. Since $\psi(t)$ vanishes outside I_y and $\tilde{\phi}(t) = y$ inside I_y, we may write

$$\frac{\partial I}{\partial\alpha}\Big|_{\alpha=0} = 2\Big\{\int_0^1 A(y-\tilde{\phi}(t))dt\int_0^1 B(y-\tilde{\phi}(\tau))d\tau\int_{I_y}\psi(s)ds-$$

$$\int_0^1\int_0^1 A(\tilde{\phi}(s)-\tilde{\phi}(t))dtB(\tilde{\phi}(s)-y)ds\int_{I_y}\psi(\tau)d\tau\Big\},$$

where

$$A(u) := \int_{\mathbb{R}} F(x+u)f(x)dx, \quad B(u) := \int_{\mathbb{R}} f(x+u)f(x)dx. \tag{13.3.19}$$

Since $\int_{I_y}\psi(t)dt > 0$, we see that the assertion of Lemma 13.3.3 is equivalent to

$$\int_0^1 A(y-\tilde{\phi}(t))dt\int_0^1 B(y-\tilde{\phi}(\tau))d\tau > \int_0^1\int_0^1 A(\tilde{\phi}(s)-\tilde{\phi}(t))dtB(\tilde{\phi}(s)-y)ds.$$

Since $B(u)$ is an even function (see (13.3.19)), we can rewrite the last inequality as

$$\int_0^1\int_0^1 B(\tilde{\phi}(s)-y)\{A(y-\tilde{\phi}(t)) - A(\tilde{\phi}(s)-\tilde{\phi}(t))\}dtds > 0. \tag{13.3.20}$$

Since $B(u) \geq 0$, $A(u)$ is nondecreasing function, and $\tilde{\phi}(s) \leq y$, inequality (13.3.20) is established with '$\geq$' in place of '$>$'. Now let us prove the strict inequality (13.3.20). First, consider the integral over s. Clearly, there exists an s_0 such that $|y - \tilde{\phi}(s_0)| \ll 1$, $y > \tilde{\phi}(s_0)$ and $B(y - \tilde{\phi}(s_0)) > 0$, because $B(0) = \int_{\mathbb{R}} f^2(x)dx > 0$. Fix $s, t = s_0$ in the expression in braces in (13.3.20). From (13.3.19) we obtain

$$A(y - \tilde{\phi}(s_0)) - A(0) = \int_{-\infty}^{\infty} [F(x + (y - \tilde{\phi}(s_0))) - F(x)]f(x)dx =$$

$$\int_{-\infty}^{\infty} F(x + (y - \tilde{\phi}(s_0)))f(x)dx - \frac{1}{2} > 0,$$

since $F(x)$ is nondecreasing, continuously differentiable, has points of growth, and $y - \tilde{\phi}(s_0) > 0$. Thus the integrand in (13.3.20) is strictly positive in a neighborhood of the point (s_0, s_0). This, together with the continuity of $A(u)$ and $B(u)$, proves Lemma 13.3.3. $\square$

Proof of Theorem 13.3.2. Fix $n \geq 2$ and consider a partition of the interval $[a, b]$: $a = y_1 < y_2 < \cdots < y_{n+1} = b$, with $y_{k+1} - y_k = h$, $k = 1, \ldots, n$, $h = (b - a)/n$. Define the functions

$$\psi_{y_l}(t) = \begin{cases} 1, & \phi(t) \geq y_l, \\ 0, & \phi(t) < y_l \end{cases}, \quad \phi_n(t) = a + h \sum_{l=1}^{n} \psi_{y_l}(t),$$

$$\phi_{n,k}(t) = a + h \sum_{l=1}^{k} \psi_{y_l}(t), \quad k = 1, \ldots, n. \tag{13.3.21}$$

Clearly, one has $I(\phi_{n,k+1}) = I(\phi_{n,k}) + \frac{\partial I(\phi_{n,k} + \alpha \psi_{y_k})}{\partial \alpha}\Big|_{\alpha=0} h + o(h)$. Thus, one gets

$$I(\phi_n) = \frac{1}{4} + h \sum_{k=1}^{n} \frac{\partial I(\phi_{n,k} + \alpha \psi_{y_k})}{\partial \alpha}\Big|_{\alpha=0} + o(1), \tag{13.3.22}$$

where we have used the equation $I(\text{const}) = 1/4$, which follows from the definition (13.3.16). Note that

$$\max_{t \in [0,1]} |\phi(t) - \phi_n(t)| \to 0, \quad \max_{t \in [0,1]} |\tilde{\phi}_y(t) - \phi_{n,[yn]}(t)| \to 0$$

as $n \to \infty$, where the function $\tilde{\phi}_y$ was defined in (13.3.17), and $[u]$ is the integer part of u. Using this and the continuity of the functional $I(\phi)$

and its Gateaux derivative in the space $C[0, 1]$, we obtain from (13.3.22), by taking $n \to \infty$ ($h \to 0$), the following formula:

$$I(\phi) = \frac{1}{4} + \int_a^b \left. \frac{\partial I(\tilde{\phi}_y + \alpha \psi_y)}{\partial \alpha} \right|_{\alpha=0} dy. \qquad (13.3.23)$$

Using Lemma 13.3.3, we see that the integrand in (13.3.23) is strictly positive for all y, $a < y < b$. Therefore, $I(\phi) > 1/4$, and Theorem 13.3.2 is proved. $\square$

13.3.2. Multidimensional case, regular design model

Let the given data be $x_{k_1,\ldots,k_d}$, $1 \le k_i \le \beta_i N$, $0 < \beta_i < \infty$, $i = 1, \ldots, d$. Without loss of generality we may assume β_i to be integers. Denote $k := (k_1, \ldots, k_d) \in \mathbb{N}^d$, $t := (t_1, \ldots, t_d) \in \mathbb{R}^d$, $B_N := \{k : 1 \le k_i \le \beta_i N, i = 1, \ldots, d\}$, $D := \{t : 0 \le t_i \le \beta_i, i = 1, \ldots, d\}$. The problem is to test the null hypothesis

$$H_0 : \quad F_k(x) = F_j(x) \quad \forall\, k, j \in B_N \qquad (13.3.24)$$

against the alternative

$$H_1 : \quad F_k(x) = F(x - \theta_k), \ k \in B_N, \qquad (13.3.25)$$

where F is a continuously differentiable distribution function with

$$f(x) := F'(x), \quad \sup_{x \in \mathbb{R}^1} f(x) < \infty,$$

and $\theta_k \in \mathbb{R}^1$ are some constants satisfying the condition:

$$\phi(t) := \lim_{N \to \infty, k/N \to t} \theta_k, \ t \in D; \quad \phi(t) \in C[D], \ \phi(t) \not\equiv \text{const}. \quad (13.3.26)$$

For an arbitrary multiindex $k \in B_N$ we define the set of multiindices neighboring to k by the formula $L(k) := \{l \in B_N : l \neq k, \max_{1 \le i \le d} |l_i - k_i| = 1\}$. Let R_i be the rank of the element x_i, $i \in B_N$. The test criterion is statistic (13.2.29):

$$\nu_N := \frac{1}{M_N} \sum_{k \in B_N} \sum_{l \in L(k)} \left(\frac{R_k - R_l}{\hat{N}} \right)^2, \qquad (13.3.27)$$

where $\hat{N} := \beta_1 \ldots \beta_d N^d$, and M_N is the number of elements in double sum (13.3.27). We have $\nu_N \xrightarrow{\text{ms}} 1/6$ as $N \to \infty$ if H_0 is true (see Section 13.2.3). Similarly to the one-dimensional case, to prove the consistency of the test against the alternative H_1, it is sufficient to prove that ν_N converges to some constant E_∞ as $N \to \infty$ if H_1 holds, and that $E_\infty < 1/6$. Let us study the asymptotic behavior of ν_N as $N \to \infty$ under H_1. Fix m, $m \geq 2$, and consider the following sublattices

$$\Delta_l := [\beta_1 \frac{l_1 - 1}{m}, \beta_1 \frac{l_1}{m}) \times \cdots \times [\beta_d \frac{l_d - 1}{m}, \beta_d \frac{l_d}{m}),$$
$$l := (l_1, \ldots, l_d) \in \mathbb{N}^d, \ 1 \leq l_i \leq m, \ 1 \leq i \leq d.$$

Let $\nu_N^{(m)}$ be the statictic (13.3.27) calculated in the case when the trend $\tilde{\theta}_k$, $k \in B_N$, is constant inside each sublattice Δ_l

$$\tilde{\theta}_k = \phi(l/m) \text{ for } k/N \in \Delta_l, \ 1 \leq l_i \leq m, \ 1 \leq i \leq d.$$

Using the results obtained in Section 13.2.3, we have

$$\nu_N^{(m)} \xrightarrow[N \to \infty]{ms} 2\left\{ \frac{1}{3} - \frac{1}{m^d} \sum_{i=1}^{d} \sum_{l_i=1}^{m} \left(\int_{\mathbb{R}} G^{(m)}(x) g_l^{(m)}(x) dx \right)^2 \right\}$$
$$:= E_m < 1/6,$$

$$G^{(m)}(x) = \frac{1}{m^d} \sum_{i=1}^{d} \sum_{l_i=1}^{m} F(x - \phi(l/m)), \quad g_l^{(m)}(x) = f(x - \phi(l/m)).$$

Denoting $dt := dt_1 \ldots dt_d$, $ds := ds_1 \ldots ds_d$ and taking the limit as $m \to \infty$, we obtain

$$E_\infty := \lim_{m \to \infty} E_m = 2\left\{ \frac{1}{3} - \int_D \left(\int_{\mathbb{R}} \left[\int_D F(x - \phi(t)) dt \right] f(x - \phi(s)) dx \right)^2 ds \right\}.$$

Theorem 13.3.1'. *Under assumptions (13.3.25) and (13.3.26) we have* $\nu_N \xrightarrow[N \to \infty]{p} E_\infty$.

Theorem 13.3.2'. *Under assumptions (13.3.25) and (13.3.26) we have* $E_\infty < 1/6$.

The proofs of Theorems 13.3.1' and 13.3.2' are omitted because they are very similar to those of Theorems 13.3.1 and 13.3.2.

13.3.3. Random design model

Let the observation points $\{\rho_k\}_{k=1}^{N}$ be randomly chosen inside an open bounded domain $D \subset \mathbb{R}^d$, $d \geq 1$, and let $\{x_k\}_{k=1}^{N}$ be a set of corresponding observations. The problem is to test the null hypothesis

$$H_0: \quad F_1(x) = F_2(x) = \cdots = F_N(x), \tag{13.3.28}$$

where $F_k(x)$ is a continuous distribution function of the random variable observed at the point ρ_k, $k = 1, \ldots, N$, against the alternative

$$H_1: \quad F_k(x) = F(x - \phi(\rho_k)), \quad \phi(t) \in C[\bar{D}], \quad \phi(t) \not\equiv \text{const}, \quad (13.3.29)$$

where F is a continuously differentiable distribution function with $f(x) := F'(x)$, $\sup_{x \in \mathbb{R}^1} f(x) < \infty$, and $\bar{D}$ is the closure of D. The test criterion is statistic (13.2.35)

$$\nu_N := \frac{1}{N} \sum_{k=1}^{N} \left(\frac{R_{n(k)} - R_k}{N} \right)^2, \qquad (13.3.30)$$

where $n(k)$ is the index of the point closest to ρ_k. As in previous sections, $\nu_N \xrightarrow{ms} 1/6$ as $N \to \infty$ if H_0 is true (Section 13.2.4). Therefore, to prove the consistency of the test against the alternative H_1, it is sufficient to prove that ν_N converges to some constant E_∞ as $N \to \infty$ if H_1 holds, and that $E_\infty < 1/6$. Let us study the asymptotic behavior of ν_N as $N \to \infty$ under H_1. Fix $m \geq 2$ and consider a finite, disjoint, measurable partition of D: $\bigcup_{j=1}^{m} D_j = D$, $D_i \cap D_j = \varnothing$, $i \neq j$, $\lambda(D_j) = \lambda(D)/m$, $j = 1, \ldots, m$, where $\lambda(D)$ is the Lebesgue measure in $\mathbb{R}^d$. Fix arbitrarily points $t_j \in D_j$ and define a piecewise-constant trend function $\tilde{\phi}(t) := \phi(t_j)$ if $t \in D_j$. Let $\nu_N^{(m)}$ be the statistic (13.3.30) calculated for such a trend. Using the results obtained in Section 13.2.4, we have

$$\nu_N^{(m)} \xrightarrow[N \to \infty]{ms} 2 \left\{ \frac{1}{3} - \frac{1}{m} \sum_{i=1}^{m} \left(\int_{\mathbb{R}} G^{(m)}(x) g_i^{(m)}(x) dx \right)^2 \right\} := E_m < 1/6,$$

$$G^{(m)}(x) = \frac{1}{m} \sum_{i=1}^{m} F(x - \phi(t_i)), \quad g_i^{(m)}(x) = f(x - \phi(t_i)).$$

As in Section 13.3.2, we get

$$E_\infty := \lim_{m \to \infty} E_m = 2 \left\{ \frac{1}{3} - \int_D \left(\int_{\mathbb{R}} \left[\int_D F(x - \phi(t)) dt \right] f(x - \phi(s)) dx \right)^2 ds \right\}.$$

Theorem 13.3.1''. *Under assumptions (13.3.29) we have* $\nu_N \xrightarrow[N \to \infty]{P} E_\infty$.

Proof. Since the proofs of Theorems 13.3.1 and 13.3.1'' are similar, we present here only a brief discussion of differences between them.

Let $\{\rho_k\}_{k=1}^{N}$ be a set of random observation points and let $\{y_k\}_{k=1}^{N}$ be a random sample from the distribution $F(x)$. As in Section 13.2.4, we define two other sequences $\{\hat{y}_k := y_k + \phi(\rho_k)\}_{k=1}^N$, $\{\hat{y}_k^{(m)} := y_k + \tilde{\phi}(\rho_k)\}_{k=1}^N$. Thus, the statistics $\hat{\nu}_N$ and $\hat{\nu}_N^{(m)}$ calculated for sequences $\{\hat{y}_k\}_{k=1}^N$ and $\{\hat{y}_k^{(m)}\}_{k=1}^N$, respectively, depend jointly on two random sets $\rho := \{\rho_k\}_{k=1}^N$ and $\{y_k\}_{k=1}^N$. It is not hard to see that Lemma 13.3.1 still holds for such $\hat{\nu}_N$ and $\hat{\nu}_N^{(m)}$. The differences between the proofs are

a) In the definition of h_m (see above (13.3.8)): here it should be $h_m := \max_{t \in D} |\phi(t) - \tilde{\phi}(t)|$; and

b) In the fact that each set D_i contains now a random number of observation points.

Since the number of observation points inside each D_i goes to infinity with probability 1 as $N \to \infty$, one concludes that the empirical distribution functions inside each D_i converge to $F(x)$ with probability 1 and the conclusion of Lemma follows.

Lemma 13.3.2 also holds in this case. Indeed, the inequality

$$\left|P\{\nu_N < a\} - P\{\nu_N^{(\epsilon)} < a\}\right| \le P\{|\hat{\nu}_N^{(\epsilon)} - a| < \epsilon\} + P\{|\hat{\nu}_N - \hat{\nu}_N^{(\epsilon)}| > \epsilon\}$$

easily follows if we combine (13.3.12) and (13.3.13) and write the resulting inequality using the notation of this Section as

$$\left|P\{\nu_N < a|\rho\} - P\{\nu_N^{(\epsilon)} < a|\rho\}\right|$$
$$\le P\{|\hat{\nu}_N^{(\epsilon)} - a| < \epsilon|\rho\} + P\{|\hat{\nu}_N - \hat{\nu}_N^{(\epsilon)}| > \epsilon|\rho\}.$$

The rest of the argument goes without changes. □

Theorem 13.3.2''. *Under assumptions (13.3.29) we have* $E_\infty <$ 1/6.

The proof of Theorem 13.3.2'' is the same as that of Theorem 13.3.2.

CHAPTER 14

AUXILIARY RESULTS

14.1. Abstract and functional spaces

14.1.1. Abstract spaces

In this section, X denotes a linear vector space.

A mapping $d : X \times X \to \mathbb{R}$, is called *metric on* X if

(1) $d(x, y) \geq 0$ for all $x, y \in X$, and $d(x, y) = 0$ if and only if $x = y$;
(2) $d(x, y) = d(y, x)$ for all $x, y \in X$; and
(3) $d(x, y) \leq d(x, z) + d(z, y)$ for all $x, y, z \in X$.

A space X with the metric is called *a metric space*.

A mapping $|| \cdot || : X \to \mathbb{R}$ is called *a norm on* X if

(1) $||x|| \geq 0$, $||x|| = 0$ if and only if $x = 0$;
(2) $||\lambda x|| = |\lambda| \, ||x||$ for all $\lambda \in \mathbb{C}$ and all $x \in X$;
(3) $||x + y|| \leq ||x|| + ||y||$.

A space X with a norm $|| \cdot ||$ is called *Banach space* if X is complete with respect to the topology defined by the norm.

A map $X \times X \to \mathbb{C}$ is called an inner product and is denoted (x, y), if it has the properties:

(1) $(x, x) \geq 0$, $(x, x) = 0$ if and only if $x = 0$;
(2) $(x, y) = \overline{(y, x)}$, the bar stands for complex conjugate;
(3) $(\lambda x, y) = \lambda(x, y)$ for any $\lambda \in \mathbb{C}$.

A space X with the inner product is called *Hilbert space* if X is complete with respect to the norm $||x|| = (x, x)^{1/2}$.

A real-valued function $p(x)$, defined on X, is called a semi-norm if $p(x + y) \leq p(x) + p(y)$ and $p(\lambda x) = |\lambda| p(x)$ for all $x, y \in X$ and $\lambda \in \mathbb{C}$.

A set M is called *convex* if $x_1, x_2 \in M$ implies $\lambda_1 x_1 + \lambda_2 x_2 \in M$ for all $\lambda_1, \lambda_2 \geq 0$, such that $\lambda_1 + \lambda_2 = 1$.

A set M is called *balanced* if $x \in M$ and $|\lambda| \leq 1$ imply $\lambda x \in M$.

A set M is called *absorbing* if for any $x \in X$ there is $\lambda > 0$, such that $x \in \lambda M$.

A space X is called a *linear locally convex topological space* (or just *locally convex space*) if any of its open sets containing the origin, contains a convex, balanced, and absorbing open set.

A *Frechet space* is a complete locally convex topological vector space whose topology is induced by a translation invariant metric $d(x, y)$ (that is, $d(x, y) = d(x + z, y + z)$ for all $x, y, z \in X$).

A convex, balanced, and absorbing closed set in a locally convex linear topological space X is called a *barrel*. X is a *barreled space* if every barrel is a neighborhood of the origin.

An operator $T : X \to Y$, where Y is a Banach space, is called *nuclear* if it is a uniform limit of a sequence of finite-rank operators T_n, $T_n(x) := \sum_{j=1}^{n} c_j f_j(x) y_j$. Here $y_j \in Y$, $f_j \in X^*$ are bounded linear functionals on X, and $c_j \geq 0$ are numbers such that $\sum_{j=1}^{\infty} c_j < \infty$. A nuclear operator maps a bounded neighborhood of X into a relatively compact set of Y.

A linear locally convex topological space X is called *nuclear* if for any convex balanced neighborhood V of the origin there exists a neighborhood U of the origin such that $U \subset V$, U is convex and balanced, and the mapping $T : X_U \to \hat{X}_V$ is nuclear. Here $\hat{X}_V$ is the closure of the normed space X_V. The space X_V is defined as follows. If V is a convex balanced neighborhood of the origin, define $p_V(x) = \inf_{\frac{x}{\lambda} \in V, \lambda > 0} \lambda$, and let $N_V := \{x : p_V(x) = 0\}$. Then N_V is closed linear subspace of X and $X_V := X / N_V$ is the quotient space with the norm $\|\tilde{x}\| = p_V(x)$, $\tilde{x} \in X_V$ is the residue class containing x.

A locally convex topological space is called *reflexive* if $X = (X^*)^*$, where X^* is the space of bounded linear functionals on X.

14.1.2. Lebesgue and Sobolev spaces

14.1.2.1. Main definitions. Let $U \subset \mathbb{R}^n$ be a domain. By $C^k(U)$ and $C^k(\bar{U})$, $0 \leq k \leq \infty$, we mean the spaces of functions which are k times continuously differentiable in U and $\bar{U}$, respectively. Here $\bar{U}$ is the closure of U. Also, we define $C_0^k(U) := \{f : f \in C^k(U), \operatorname{supp} f \subset U\}$. If $k = \infty$, the corresponding functions are assumed to be infinitely differentiable. If $k = 0$, the corresponding functions are assumed to be continuous.

Consider the space $L^p(U)$ of Lebesgue measurable functions, defined on U, such that

$$\|f\|_{L^p(U)} := \left(\int_U |f(x)|^p dx \right)^{1/p} < \infty, \quad p \geq 1. \qquad (14.1.1)$$

The space of *locally integrable functions*, which is denoted by $L_{loc}^p(U)$, consists of functions f such that $f \in L^p(U_1)$ for any $U_1 \subsetneq U$.

Fix any $f \in L_{loc}^p(U)$ and let α be a multiindex. Then a locally integrable function g is called the α-th *weak derivative* of f if

$$\int_U \varphi(x)g(x)dx = (-1)^{|\alpha|} \int_U f(x)\partial^\alpha\varphi(x)dx \text{ for any } \varphi \in C_0^{|\alpha|}(U).$$

(14.1.2)

Here $\partial^\alpha = \partial_{x_1}^{\alpha_1}\dots\partial_{x_n}^{\alpha_n}$ and $|\alpha| = \alpha_1 + \cdots + \alpha_n$. A function f is k *times weakly differentiable* if all its weak derivatives exist for orders up to and including k. The space of k times weakly differentiable functions is denoted by $W^k(U)$. Evidently, $C^k(U) \subset W^k(U)$. The Sobolev space $W^{k,p}(U)$ is defined as follows:

$$W^{k,p}(U) := \{f \in W^k(U) : \partial^\alpha f \in L^p(U) \text{ for all } |\alpha| \le k\}. \quad (14.1.3)$$

$W^{k,p}(U)$ is a Banach space with the norm

$$\|f\|_{W^{k,p}(U)} = \left(\int_U \sum_{|\alpha|\le k} |\partial^\alpha f|^p dx\right)^{1/p}. \quad (14.1.4)$$

Taking the closure of $C_0^k(U)$ in the norm (14.1.4), we obtain the Sobolev space $W_0^{k,p}(U)$. If U is bounded, $W_0^{k,p}(U) \subsetneq W^{k,p}(U)$. Local Sobolev spaces $W_{loc}^{k,p}(U)$ consist of functions f such that $f \in W^{k,p}(U_1)$ for any $U_1 \subsetneq U$.

An important role is played by the spaces $H^k(U) := W^{k,2}(U)$ and $H_0^k(U) := W_0^{k,2}(U)$, which are the Hilbert spaces with the inner product

$$< f, g >_k := \int_U \sum_{|\alpha|\le k} \partial^\alpha f \, \overline{\partial^\alpha g} dx. \quad (14.1.5)$$

14.1.2.2. Approximation results. Fix any nonnegative function $\varphi \in C_0^\infty(\mathbb{R}^n)$ such that $\int_{\mathbb{R}^n} \varphi(x)dx = 1$ and $\varphi(x) = 0$ for $|x| \ge 1$. For any $f \in L^p(U)$, $f = 0$ in $\mathbb{R}^n \setminus U$, define

$$f_\epsilon(x) := \epsilon^{-n} \int_{\mathbb{R}^n} \varphi\left(\frac{x - y}{\epsilon}\right) f(y)dy. \quad (14.1.6)$$

Then

$$f_\epsilon \in C_0^\infty(\mathbb{R}^n), \|f - f_\epsilon\|_{L^p(U)} \to 0 \text{ as } \epsilon \to 0,$$
$$\|f_\epsilon\|_{L^p(\mathbb{R}^n)} \le \|f\|_{L^p(U)}.$$

The functions f_ϵ are called *mollifications* of f. The identity

$$\partial^\alpha f_\epsilon(x) = (\partial^\alpha f)_\epsilon(x) \quad (14.1.7)$$

shows that (14.1.6) can be generalized to approximation of weak derivatives. Therefore, the subspace $C^\infty(U) \cap W^{k,p}(U)$ is dense in $W^{k,p}(U)$.

Suppose U is starshaped with respect to some point $x \in U$, that is any ray from this point has a unique common point with the boundary of U. Then $C^\infty(\bar{U})$ is dense in $W^{k,p}(U)$.

If the boundary of U is sufficiently smooth: for instance, if it satisfies the cone condition (that is, for each point $x \in \Gamma := \partial U$ there exists a cone K with vertex x such that the set $K \cap B_\epsilon(x)$ belongs to U if $\epsilon > 0$ is sufficiently small), then the set $C^\infty(\bar{U})$ is dense in $W^{k,p}(U)$.

14.1.2.3. Imbedding theorems. A Banach space $\mathcal{B}_1$ is continuously imbedded in a Banach space $\mathcal{B}_2$, if there exists a bounded, linear, one-to-one mapping $\mathcal{B}_1 \to \mathcal{B}_2$.

The following mappings are continuous compact imbeddings:

$$W_0^{k,p}(U) \to L^{\frac{np}{n-p}}(U), \quad \text{if } kp < n, \tag{14.1.8}$$

and

$$W_0^{k,p}(U) \to C^m(\bar{U}), \quad \text{if } 0 \le m < k - (n/p). \tag{14.1.9}$$

In general, $W_0^{k,p}(U)$ cannot be replaced by $W^{k,p}(U)$ in (14.1.8) and (14.1.9). Suppose that the boundary ∂U of the domain U is Lipschitz continuous, that is in a neighborhood of every point $x \in \partial U$, the boundary ∂U can be described in local coordinates by the equation $x_n = f(x')$, $x' = (x_1, \ldots, x_{n-1})$, with $f(x')$ satisfying the Lipschitz condition: $|f(x') - f(y')| \le c|x' - y'|$, $c = \text{const} > 0$. Then one has

$$W^{k,p}(U) \to L^{\frac{np}{n-p}}(U), \quad \text{if } kp < n, \tag{14.1.8'}$$

and

$$W^{k,p}(U) \to C_B^m(U), \quad \text{if } 0 \le m < k - (n/p), \tag{14.1.9'}$$

where $C_B^m(U) := \{f \in C^m(U) : \sup_{x \in U} |\partial^\alpha f(x)| < \infty \text{ for all } |\alpha| \le m\}$.

14.1.2.4. Sobolev spaces via the Fourier transform. Sobolev spaces $H^s(\mathbb{R}^n)$, $s \in \mathbb{R}$, are defined as a completion of the space $C_0^\infty(\mathbb{R}^n)$ in the norm

$$\|f\|_s := \left(\frac{1}{(2\pi)^n} \int_{\mathbb{R}^n} (1 + |\xi|^2)^s |\mathcal{F}f(\xi)|^2 d\xi \right)^{1/2}. \tag{14.1.10}$$

$H^s(\mathbb{R}^n)$ is the Hilbert space with the inner product

$$<f, g>_s := \frac{1}{(2\pi)^n} \int_{\mathbb{R}^n} (1 + |\xi|^2)^s \mathcal{F}f(\xi)\overline{\mathcal{F}g(\xi)} d\xi. \tag{14.1.11}$$

For an open $U \subset \mathbb{R}^n$, $H_0^s(U)$ is a completion of the space $C_0^\infty(U)$ in the norm (14.1.10).

One can see that $H^k(\mathbb{R}^n) = W^{k,2}(\mathbb{R}^n)$, $H_0^k(U) = W_0^{k,2}(U)$, and the norms (14.1.4) (with $p = 2$) and (14.1.10) are equivalent.

14.2. Distribution theory

14.2.1. Spaces of test functions and distributions

14.2.1.1. Main definitions. In this section we use the word 'smooth' for 'infinitely differentiable'. The following three spaces of test functions are frequently used.

(1) The space $\mathcal{D} := C_0^\infty(\mathbb{R}^n)$, which consists of smooth functions with compact support. A sequence of functions $\phi_n \in \mathcal{D}$ converges to zero in $\mathcal{D}$ if there exists a compact set K such that supp $\phi_n \subset K, n \geq 0$, and $\sup_{x \in K, |\alpha| \leq m} |\partial^\alpha \phi_n(x)| \to 0$ as $n \to \infty$ for any integer m. Here $\alpha = (\alpha_1, \ldots, \alpha_n)$ is a multiindex, and ∂^α denotes derivative with respect to x.

(2) The space $\mathcal{S} := \mathcal{S}(\mathbb{R}^n)$, which consists of smooth functions which rapidly decay with all their derivatives. More precisely, $\phi \in \mathcal{S}$ if and only if all the norms

$$\|\phi\|_{m,k} = \sup_{x \in \mathbb{R}^n, |\alpha| \leq k} |(1 + |x|)^m \partial^\alpha \phi(x)|, \quad m, k \in \mathbb{N}, \qquad (14.2.1)$$

are finite. A sequence of functions $\phi_n \in \mathcal{S}$ converges to zero in $\mathcal{S}$ if $\|\phi_n\|_{m,k} \to 0$ as $n \to \infty$ for any integers m and k.

(3) The space $\mathcal{E} := C^\infty(\mathbb{R}^n)$, which consists of smooth functions without any restriction on their growth at infinity. A sequence of functions $\phi_n \in \mathcal{E}$ converges to zero in $\mathcal{E}$ if for any compact $K \subset \mathbb{R}^n$ and any integer k, one has

$$\sup_{x \in K, |\alpha| \leq k} |\partial^\alpha \phi_n(x)| \to 0, \quad n \to \infty.$$

The elements of $\mathcal{D}$, $\mathcal{S}$, and $\mathcal{E}$ are called *test functions*. The corresponding dual spaces (that is, the spaces of bounded linear functionals) are denoted by $\mathcal{D}'$, $\mathcal{S}'$, and $\mathcal{E}'$, respectively. Elements of $\mathcal{D}'$ are called *distributions*, elements of $\mathcal{S}'$ are called *tempered distributions*, and elements of $\mathcal{E}'$ are called *distributions with compact support*.

The imbeddings $\mathcal{D} \to \mathcal{S}$ and $\mathcal{S} \to \mathcal{E}$ are continuous with dense images. Therefore, the dual maps $\mathcal{E}' \to \mathcal{S}'$ and $\mathcal{S}' \to \mathcal{D}'$ are also continuous with dense images.

Let $f \in \mathcal{D}'(\mathbb{R}^n)$. We say that $f = 0$ in $U \subset \mathbb{R}^n$ if $< f, \phi >= 0$ for any $\phi \in C_0^\infty(U)$. The complement of the largest open set U such that $f = 0$ in U is called *the support* of f and is denoted by supp f. It can be proved that if $f \in \mathcal{E}'$, then f is compactly supported. This justifies the name 'distributions with compact support' for the space $\mathcal{E}'$.

We say that $f = g$ in $U \subset \mathbb{R}^n$ if $f - g = 0$ in U, $f, g \in \mathcal{D}'(\mathbb{R}^n)$. *The singular support of f is the complement of the largest open set U such that f coincides in U with a smooth function. The singular support of a distribution f is denoted by singsupp f.*

Each locally integrable function f defines a distribution, i.e., a functional on $C_0^\infty(\mathbb{R}^n)$, such that the value of this functional on a test function $\phi \in C_0^\infty(\mathbb{R}^n)$ is given by

$$< f, \phi >= \int_{\mathbb{R}^n} f(x)\overline{\phi(x)}dx. \qquad (14.2.2)$$

If f is polynomially bounded at infinity (that is $|f(x)| < c(1+|x|^m)$ for some positive constants c and m), then f defines a tempered distribution, because integral (14.2.2) converges for $\phi \in S(\mathbb{R}^n)$.

14.2.1.2. Examples. Define the distribution δ by the formula $< \delta, \phi > = \phi(0)$. Usually, δ is called the *delta function*. One has: $\operatorname{supp}\delta = \{0\}$ and $\operatorname{singsupp}\delta = \{0\}$.

In the book we frequently used the function x_+^λ, i.e., $x_+^\lambda := x^\lambda$ for $x > 0$ and $x_+^\lambda := 0$ for $x < 0$, $x \in \mathbb{R}$. Clearly, $\operatorname{supp}x_+^\lambda = [0, \infty)$ and $\operatorname{singsupp}x_+^\lambda = \{0\}$. This function is locally integrable for $\operatorname{Re}\lambda > -1$ and is polynomially bounded at infinity, so it defines a tempered distribution. This distribution is an analytic function of λ, that is, for each $\phi \in S(\mathbb{R})$ the value of the functional x_+^λ is an analytic function of λ in the half plane $\operatorname{Re}\lambda > -1$. This function can be extended to $\mathbb{C}$ as a meromorphic function with simple poles at the points $\lambda = -n$, $n \in \mathbb{N}$. The value of this functional on a test function ϕ for λ such that $\operatorname{Re}\lambda > -n - 1$, $\lambda \neq -1, -2, \ldots, -n$, is given by the formula

$$< x_+^\lambda, \phi >= \int_0^1 x^\lambda \left(\phi(x) - \phi(0) - x\phi'(0) - \cdots - x^{n-1}\frac{\phi^{(n-1)}(0)}{(n-1)!} \right) dx$$
$$+ \int_1^\infty x^\lambda \phi(x)dx + \sum_{k=1}^n \frac{\phi^{(k-1)}(0)}{(\lambda+k)(k-1)!}. \qquad (14.2.3)$$

The residue of x_+^λ at $\lambda = -n$ equals $\frac{(-1)^{n-1}\delta^{(n-1)}(x)}{(n-1)!}$, $n \in \mathbb{N}$. This leads to the formula

$$\frac{x_+^\lambda}{\Gamma(\lambda+1)}\bigg|_{\lambda=-n} = \delta^{(n-1)}(x), \quad n = 1, 2, \ldots. \qquad (14.2.4)$$

The distribution x_-^λ is defined similarly to x_+^λ. In particular, we have

$$\frac{x_-^\lambda}{\Gamma(\lambda+1)}\bigg|_{\lambda=-n} = (-1)^{(n-1)}\delta^{(n-1)}(x), \quad n = 1, 2, \ldots. \qquad (14.2.5)$$

Using the distributions x_+^λ and x_-^λ, we get $x^{-k} = x_+^{-k} + (-1)^k x_-^{-k}$ and $|x|^\lambda = x_+^\lambda + x_-^\lambda$. The distributions x^{-k} for $k = 1$ and $k = 2$, which are used in this book, are defined by the following formulas:

$$\int_{-\infty}^{\infty} \frac{\phi(x)}{x} dx := \int_0^\infty \frac{\phi(x) - \phi(-x)}{x} dx, \qquad (14.2.6)$$

and

$$\int_{-\infty}^{\infty} \frac{\phi(x)}{x^2} dx := \int_0^\infty \frac{\phi(x) + \phi(-x) - 2\phi(0)}{x^2} dx.$$

Consider the following pair of functions:

$$(x \pm i0)^\lambda = \lim_{y \to \pm 0} (x + iy)^\lambda = \begin{cases} e^{\pm i\lambda\pi} |x|^\lambda, & x < 0, \\ x^\lambda, & x > 0. \end{cases}$$

To these ordinary functions, we can associate the distributions

$$(x \pm i0)^\lambda = x_+^\lambda + e^{\pm i\lambda\pi} x_-^\lambda, \quad \lambda \neq -1, -2, \ldots,$$

$$(x \pm i0)^{-n} = x^{-n} \pm \frac{i\pi(-1)^n}{(n-1)!} \delta^{(n-1)}(x), \quad n = 1, 2, \ldots.$$

Let $f \in \mathcal{D}'$. Suppose that for some $p \in \mathbb{R}$ one has $< f(\lambda x), \phi >= \lambda^p < f, \phi >$ for all $\lambda > 0$, where $< f(\lambda x), \phi >:= \lambda^{-n} < f, \phi(x/\lambda) >$. Then f is called a *distribution homogeneous of degree p*. The distribution $\delta(x), x \in \mathbb{R}^n$, is homogeneous of degree $-n$, and $\partial^\alpha \delta$ is homogeneous of degree $-n - |\alpha|$. The distribution x_+^λ is homogeneous of degree λ.

14.2.1.3. Operations with distributions. Distributions can be multiplied by C^∞ functions. Let $f \in \mathcal{D}'(\mathbb{R}^n)$ and $\varphi \in C^\infty(\mathbb{R}^n)$. Then the distribution $f\varphi \in \mathcal{D}'(\mathbb{R}^n)$ is defined by the formula $< f\varphi, \phi >=< f, \varphi\phi >$.

Let f and ϕ be $C_0^\infty(\mathbb{R}^n)$ functions. Then, integrating by parts, we get $< \frac{\partial f}{\partial x_j}, \phi >= - < f, \frac{\partial \phi}{\partial x_j} >$. Let us use this identity to define the derivative of a distribution. The *partial derivative* $\frac{\partial f(x)}{\partial x_j}$ *of a distribution* f is a linear bounded functional on $\mathcal{S}(\mathbb{R}^n)$ defined by the equation $< \frac{\partial f}{\partial x_j}, \phi >= - < f, \frac{\partial \phi}{\partial x_j} >$. Higher order derivatives of a distribution are defined similarly, and the ordinary rules of differentiation apply also to distributions. In particular, the derivative of a sum is the sum of derivatives. Any distribution has derivatives of all orders.

Every distribution supported at a point a is a finite linear combination of $\delta(x-a)$ and its derivatives. Every distribution with compact support K can be represented as a finite linear combination of derivatives of continuous functions vanishing outside an ϵ-neighborhood of K, where $\epsilon > 0$ is an arbitrary small number. One can prove that any tempered distribution is a derivative of some order of a polynomially bounded locally integrable function.

Let f and g be smooth functions, and one of them be compactly supported. Then the convolution $f * g$ is defined by the formula

$$(f * g)(x) = \int_{\mathbb{R}^n} f(x - y)g(y)dy.$$

If $f \in \mathcal{D}'$ and $\phi \in C_0^\infty(\mathbb{R}^n)$, then

$$(f * \phi)(x) := < f(y), \phi(x - y) >,$$

where the right-hand side denotes the value of the functional f on a test function $\phi(x - y)$ with respect to the variable y. The convolution operation can be generalized to pairs of distributions if one of the distributions is compactly supported. Let $f \in \mathcal{E}'$ and $g \in \mathcal{D}'$. Then $f * g \in \mathcal{D}'$ is a distribution defined by the formula $< f * g, \phi > = < f, < g, \phi(x + y) >>$. One has: $f * g = g * f$, $f * (g * h) = (f * g) * h$ (at least two of the distributions should be compactly supported), $\partial^\alpha(f * g) = (\partial^\alpha f) * g$, $\delta * f = f$, $\operatorname{supp} f * g \subset \operatorname{supp} f + \operatorname{supp} g := \{x \in \mathbb{R}^n : x = y + z, y \in \operatorname{supp} f, z \in \operatorname{supp} g\}$, $\operatorname{singsupp} f * g \subset \operatorname{singsupp} f + \operatorname{singsupp} g$.

14.2.2. Fourier transform of distributions

14.2.2.1. Fourier transform of integrable functions. Let $f \in L^1(\mathbb{R}^n)$. Define the Fourier transform of f by the formula

$$\tilde{f}(\xi) := \mathcal{F}f(\xi) := \int_{\mathbb{R}^n} f(x)e^{ix \cdot \xi}dx. \tag{14.2.7}$$

Lemma 14.2.1. *If $f \in L^1(\mathbb{R}^n)$, then $\tilde{f}$ is bounded, uniformly continuous function and $\tilde{f}(\xi) \to 0$ as $|\xi| \to \infty$.*

If it turns out that if $\tilde{f} \in L^1(\mathbb{R}^n)$, then the inversion formula

$$f(x) = \frac{1}{(2\pi)^n} \int_{\mathbb{R}^n} \tilde{f}(\xi)e^{-ix \cdot \xi}d\xi \tag{14.2.8}$$

holds for almost all $x \in \mathbb{R}^n$. The following theorem extends the Fourier transform to functions from $L^2(\mathbb{R}^n)$.

Theorem 14.2.1 (Parseval-Plancherel). *The restriction of the Fourier transform to $L^1(\mathbb{R}^n) \cap L^2(\mathbb{R}^n)$ can be uniquely extended to a bounded linear operator on $L^2(\mathbb{R}^n)$. Let us denote this extension also by $\mathcal{F}$. Then $(2\pi)^{-n/2}\mathcal{F}$ is an isometry of $L^2(\mathbb{R}^n)$ onto $L^2(\mathbb{R}^n)$. Moreover, for $f, g \in L^2(\mathbb{R}^n)$, one has*

$$< f, g >= (2\pi)^{-n} < \tilde{f}, \tilde{g} >, \qquad (14.2.9)$$

and inversion formula (14.2.8) holds also for $f \in L^2(\mathbb{R}^n)$.

Fix any $f \in L^p(\mathbb{R}^n)$, $1 \le p \le 2$. Then the Hausdorff-Young inequality asserts that $\tilde{f} \in L^q(\mathbb{R}^n)$, where $q^{-1} + p^{-1} = 1$, and

$$\|\tilde{f}\|_{L^q} \le (2\pi)^{n(1-\frac{1}{p})}\|f\|_{L^p}.$$

Let us consider the action of $\mathcal{F}$ on the Schwartz space of functions. It can be shown that $\mathcal{F}$ maps $S(\mathbb{R}^n)$ onto itself isomorphically: if $f \in S(\mathbb{R}^n)$, then $\mathcal{F}f \in S(\mathbb{R}^n)$, and for any $\varphi \in S(\mathbb{R}^n)$ there is a unique $f \in S(\mathbb{R}^n)$ such that $\mathcal{F}f = \varphi$. Moreover, the mapping $\mathcal{F} : S(\mathbb{R}^n) \to S(\mathbb{R}^n)$ is continuous (see Section 14.2.1 about the topology of $S(\mathbb{R}^n)$). Let $f, g \in S(\mathbb{R}^n)$. The following properties can be easily verified.

$$\mathcal{F}(\partial_x^\alpha f) = (-i)^{|\alpha|}\xi^\alpha \tilde{f}(\xi), \quad \mathcal{F}(x^\alpha f) = i^{|\alpha|}\partial_\xi^\alpha \tilde{f}(\xi); \qquad (14.2.10)$$

$$\mathcal{F}(f * g) = \tilde{f}\,\tilde{g}, \quad \mathcal{F}(fg) = (2\pi)^{-n}\tilde{f} * \tilde{g}. \qquad (14.2.11)$$

Clearly, the Parseval equality (14.2.9) holds for $f, g \in S(\mathbb{R}^n)$.

14.2.2.2. Fourier transform of distributions. Let f be a tempered distribution and $\phi \in S(\mathbb{R}^n)$. Having the Parseval equality in mind and using that $\mathcal{F} : S(\mathbb{R}^n) \to S(\mathbb{R}^n)$ is an isomorphism, the Fourier transform of f is defined to be a functional $\mathcal{F}f$ such that the equality

$$< \mathcal{F}f, \phi >= (2\pi)^n < f, \mathcal{F}^{-1}\phi > . \qquad (14.2.12)$$

holds for all $\phi \in S(\mathbb{R}^n)$. It can be checked that identities (14.2.10) hold also for $f \in S'(\mathbb{R}^n)$, where the derivatives of distributions and multiplication by a C^∞ function are understood as in Section 14.2.1. It can be shown that if $f \in \mathcal{E}'(\mathbb{R}^n)$, then $\mathcal{F}f$ is a C^∞ function. The convolution operation may be extended by continutity to pairs of distributions $f_1 \in S'(\mathbb{R}^n)$ and $f_2 \in \mathcal{E}'(\mathbb{R}^n)$. Denote $g = f_1 * f_2$. Then one can prove that $g \in S'(\mathbb{R}^n)$ and $\mathcal{F}g = \mathcal{F}f_1\mathcal{F}f_2$. Since $\mathcal{F}f_2$ is smooth, this product is well defined.

Example 14.2.1. The following formula holds:

$$(Fx_{\pm}^{\lambda})(\sigma) = \Gamma(\lambda+1)e^{\pm \frac{i\pi(\lambda+1)}{2}}(\sigma \pm i0)^{-(\lambda+1)}, \quad \lambda \neq -1, -2, \ldots,$$
$$(14.2.13)$$

where F denotes one-dimensional Fourier transform.

One can prove that if f is a tempered distribution which is homogeneous of degree p, then the Fourier transform of f is a tempered distribution which is homogeneous of degree $-p - n$.

14.2.2.3. The Paley-Wiener-Schwartz theorems. Let $U \subset \mathbb{R}^n$ be a non-empty set. The support function of U is defined by

$$H_U(\xi) := \sup_{x \in U}(x \cdot \xi).$$

Clearly, H_U is homogeneous of degree 1 and $H_U(0) = 0$. In particular, if $U = B_a = \{x \in \mathbb{R}^n : |x| \leq a\}$, then $H_U(\xi) = a|\xi|$, and if $U := \{x\}$, then $H_U(\xi) = (x \cdot \xi)$.

Theorem 14.2.2 (Paley-Wiener-Schwartz). *Take an arbitrary $f \in \mathcal{E}'(\mathbb{R}^n)$, and let H be the support function of $\operatorname{supp} f$. Then $\tilde{f}(\xi), \xi \in \mathbb{C}^n$, is an entire analytic function. Moreover, there are a constant $c > 0$ and an integer N such that*

$$|\tilde{f}(\xi)| \leq c(1 + |\xi|)^N e^{H(\operatorname{Im}\xi)}, \quad \xi \in \mathbb{C}^n.$$

Conversely, suppose $F(\xi), \xi \in \mathbb{C}^n$, is an entire analytic function, H is a support function of a compact convex set $K \subset \mathbb{R}^n$, and

$$|F(\xi)| \leq c(1 + |\xi|)^N e^{H(\operatorname{Im}\xi)}, \quad \xi \in \mathbb{C}^n,$$

for some constants c and N. Then there exists a unique $f \in \mathcal{E}'(\mathbb{R}^n)$ such that $\mathcal{F}f = F$. Moreover, $\operatorname{supp} f \subset K$.

Theorem 14.2.3 (Paley-Wiener-Schwartz). *Let K be a compact convex subset of $\mathbb{R}^n$, $f \in C_0^{\infty}(K)$, and let H be the support function of K. Then $\tilde{f}(\xi), \xi \in \mathbb{C}^n$, is an entire analytic function, and for any $N > 0$ there is a constant $c_N > 0$ such that*

$$|\tilde{f}(\xi)| \leq c_N(1 + |\xi|)^{-N} e^{H(\operatorname{Im}\xi)}, \quad \xi \in \mathbb{C}^n. \quad (14.2.14)$$

Conversely, any entire analytic function, which satisfies an estimate of the type (14.2.14) for any $N > 0$, is the Fourier transform of some function $f \in C_0^{\infty}(K)$.

14.2.3. Wave front of a distribution

Let $f \in \mathcal{D}'(U)$, where $U \subset \mathbb{R}^n$ is a domain. Let $(x_0, \xi_0) \in U \times (\mathbb{R}^n \setminus \{0\})$. We say that (x_0, ξ_0) does not belong to the *wave front* of f, if there exists a function $\phi \in C_0^\infty(U)$ such that $\phi(x_0) \neq 0$ and

$$|\mathcal{F}(\phi f)(\xi)| \leq c_N (1 + |\xi|)^{-N} \quad \text{if} \quad \left| \frac{\xi}{|\xi|} - \frac{\xi_0}{|\xi_0|} \right| < \epsilon$$

for a sufficiently small $\epsilon > 0$ and all $N > 0$, that is $\mathcal{F}(\phi f)(\xi)$ decreases rapidly in a conic neighborhood of ξ_0 as $|\xi| \to \infty$. Wave front of f is denoted by $WF(f)$. The following properties of the wave front can be established:

(1) $WF(f)$ is a closed conic set in $U \times (\mathbb{R}^n \setminus \{0\})$;
(2) Let $\pi : U \times (\mathbb{R}^n \setminus \{0\}) \to U$ be the natural projection and $f \in \mathcal{D}'(U)$. Then

$$\pi WF(f) = \text{singsupp } f;$$

(3) $WF(\partial^\alpha f) = WF(f)$.

Example 14.2.2.

(1) Let $f(x) = \delta(x)$, then $WF(f) = \{(0, \xi) : \xi \neq 0\}$.
(2) Let $f(x) = (x + i0)^\lambda \in \mathcal{D}'(\mathbb{R})$, $\lambda \neq 1, 2, \ldots$, then $WF(f) = \{(0, \xi) : \xi < 0\}$.
(3) Let $f(x) = x_+^\lambda \in \mathcal{D}'(\mathbb{R})$, then $WF(f) = \{(0, \xi) : \xi \neq 0\}$.

14.3. Pseudodifferential and Fourier integral operators

14.3.1. Oscillatory integrals

Let $U \subset \mathbb{R}^n$ be a domain.

Definition 14.3.1. The function $a(x, \xi)$ is called an *amplitude function of class* $S_{\rho, \delta}^m$ if $a \in C^\infty(U \times \mathbb{R}^N)$ and there exist real numbers m, ρ, and δ, $0 \leq \rho, \delta \leq 1$, such that

$$|\partial_\xi^\alpha \partial_x^\beta a(x, \xi)| \leq c(\alpha, \beta, K)(1 + |\xi|)^{m - \rho|\alpha| + \delta|\beta|}, \quad x \in K, \; \xi \in \mathbb{R}^N, \quad (14.3.1)$$

for arbitrary multiindices α, β and an arbitrary compact $K \subset U$.

In what follows, we will always assume that $\rho > 0$ and $\delta < 1$.

Definition 14.3.2. The function $\Phi(x, \xi) \in C^\infty(U \times (\mathbb{R}^N \setminus 0))$ is called a *phase function* if

(1) Φ is positively homogeneous of degree 1, that is $\Phi(x, t\xi) = t\Phi(x, \xi)$ $\forall x \in U, \xi \in \mathbb{R}^N$, and $t > 0$;
(2) Φ, as a function of (x, ξ), does not have critical points in $U \times \mathbb{R}^N$, that is $\nabla_{x, \xi} \Phi(x, \xi) \neq 0$ for $x \in U$ and $\xi \in \mathbb{R}^N \setminus 0$.

Consider the integral

$$I(u) := \int\limits_{\mathbb{R}^N} \int\limits_{\mathbb{R}^n} \exp[i\Phi(x,\xi)]a(x,\xi)u(x)dxd\xi, \qquad (14.3.2)$$

where $u \in C_0^\infty(U)$. If $a(x,\xi)$ is an amplitude function, and $\Phi(x,\xi)$ is a phase function, then (14.3.2) is called an *oscillatory integral*.

In general, this integral does not converge absolutely. Therefore, it has to be defined. Choose a cutoff function $\eta(\xi) \in C_0^\infty(\mathbb{R}^n)$ such that $\eta = 1$ in a neighborhood of the origin, and let

$$I_\epsilon(u) := \int\limits_{\mathbb{R}^N} \int\limits_{\mathbb{R}^n} \eta(\epsilon\xi)\exp[i\Phi(x,\xi)]a(x,\xi)u(x)dxd\xi. \qquad (14.3.3)$$

It can be proved that the limit of $I_\epsilon(u)$ as $\epsilon \to 0$ exists and is independent of the cutoff function. Then, by the definition,

$$I(u) := \lim_{\epsilon \to 0} I_\epsilon(u). \qquad (14.3.4)$$

Let us introduce the following notation

$$C_\Phi := \{(x,\xi) : x \in U, \xi \in \mathbb{R}^N \setminus 0, \nabla_\xi \Phi(x,\xi) = 0\} \qquad (14.3.5)$$

and

$$S_\Phi := \pi C_\Phi, \quad R_\Phi := U \setminus S_\Phi, \qquad (14.3.6)$$

where $\pi : U \times (\mathbb{R}^N \setminus 0) \to U$ is the natural projection. Define a distribution A by the formula

$$(A,u) := I(u), \qquad (14.3.7)$$

where $I(u)$ is given by (14.3.2), (14.3.4). Then $A \in C^\infty(R_\Phi)$. If $a \in S_{\rho\delta}^m$ and $a = 0$ in a conical neighborhood of C_Φ, then $A \in C^\infty(U)$.

Assume that $\Phi(x,\xi)$ is nondegenerate, that is, the rank of the $N \times (N+n)$ matrix $\left(\frac{\partial^2\Phi}{\partial\xi_j\partial\tilde{x}_m}\right)$, $1 \le j \le N$, $1 \le m \le N+n$, equals N. Here $\tilde{x} := (\xi_1, \ldots \xi_N, x_1, \ldots, x_n)$. The implicit function theorem and the nondegeneracy of Φ imply that C_Φ is a submanifold of dimension n in $U \times (\mathbb{R}^N \setminus 0)$.

14.3.2. Fourier integral operators

Let $U_j, j = 1, 2$, be some domains in $\mathbb{R}^{n_j}$. Consider an oscillatory integral of the form:

$$(Au_1, u_2) = \int\limits_{\mathbb{R}^N} \int\limits_{U_2} \int\limits_{U_1} \exp[i\Phi(x,y,\xi)]a(x,y,\xi)u_1(y)u_2(x)dydxd\xi,$$

$$\qquad (14.3.8)$$

where $u_j \in C_0^\infty(U_j), j = 1, 2$, Φ is a phase function on $U \times \mathbb{R}^N$, $U = U_1 \times U_2$, $a(x, y, \xi) \in S_{\rho,\delta}^m(U \times \mathbb{R}^N)$, $\rho > 0$, $\delta < 1$. Clearly, (14.3.8) defines a linear operator

$$A : C_0^\infty(U_1) \to \mathcal{D}'(U_2), \qquad (14.3.9)$$

which is called a *Fourier integral operator* (FIO) of order m. The distribution on $C_0^\infty(U)$, defined by (14.3.8), is called the *Schwartz kernel* of A:

$$K_A := \exp\left[i\Phi(x, y, \xi)\right] a(x, y, \xi). \qquad (14.3.10)$$

14.3.3. Pseudodifferential operators

14.3.3.1. Definitions and some properties. Assume that $U_1 = U_2 = U \subset \mathbb{R}^n$, and

$$\Phi(x, y, \xi) = (y - x) \cdot \xi. \qquad (14.3.11)$$

Assume also that $a(x, y, \xi) \in S_{\rho\delta}^m$. Then the corresponding FIO is called a *pseudodifferential operator* (PDO) of the class $L_{\rho\delta}^m := L_{\rho\delta}^m(U)$.

Example 14.3.1. If $A = \sum_{|j| \le m} a_j(x) i^{|j|} \partial_x^j$, where $a_j \in C^\infty(U)$, then

$$Au(x) = \frac{1}{(2\pi)^n} \iint \exp[i(y - x) \cdot \xi] \sigma_A(x, \xi) u(y) dy d\xi, \qquad \int := \int_{\mathbb{R}^n}.$$
$$(14.3.12)$$

Here the function σ_A is called the *symbol* of A:

$$\sigma_A(x, \xi) = \sum_{|j| \le m} a_j(x) \xi^j. \qquad (14.3.13)$$

Let a PDO be given by the formula:

$$Au(x) = \frac{1}{(2\pi)^n} \iint \exp[i\xi \cdot (y - x)] a(x, y, \xi) u(y) dy d\xi, \qquad (14.3.14)$$

and $A \in L_{\rho,\delta}^m$. Then

$$\text{singsupp}\, Au \subset \text{singsupp}\, u. \qquad (14.3.15)$$

This property is called *pseudolocality*. If

$$\text{supp}\, Au \subset \text{supp}\, u, \quad u \in C_0^\infty(U), \qquad (14.3.16)$$

then A is called *local*.

Exercise 14.3.2. If $A : C_0^\infty(U) \to C_0^\infty(U)$ is a local linear bounded operator, then A is a linear differential operator on $C_0^\infty(U_1)$ for any U_1, such that $\overline{U_1}$ is compact in U.

A PDO A is called *properly supported* if the natural projections $\pi_j :$ supp $K_A \to U_j, j = 1, 2$, are both proper maps. Here K_A is the Schwartz kernel of A, and $\pi_j(U_1 \times U_2) = U_j, j = 1, 2$. A continuous map $\psi : M \to N$ between topological spaces M and N is called proper if for any compact $K \subset N$ the inverse image $\psi^{-1}(N)$ is compact in M.

If $A \in L_{\rho,\delta}^m(U)$ is a properly supported PDO, then one can define the symbol of A by the formula

$$\sigma_A(x, \xi) = \exp(i\xi \cdot x) A[\exp(-i\xi \cdot x)]. \qquad (14.3.17)$$

The symbol defines A uniquely, and if $A \in L_{\rho\delta}^m(U)$, then its symbol can be calculated by the formula:

$$\sigma_A(x, \xi) \sim \sum_\alpha \frac{1}{\alpha!} \left[\partial_\xi^\alpha \left(\frac{1}{-i} \partial_y \right)^\alpha a(x, y, \xi) \right] \Bigg|_{y=x}, \qquad (14.3.18)$$

where α is a multiindex and $\alpha! = \alpha_1! \cdot \ldots \cdot \alpha_n!$. The asymptotic sum in (14.3.18) has the following meaning.

Let $a_j(x, \xi) \in S_{\rho,\delta}^{m_j}(U \times \mathbb{R}^n), m_j \to -\infty$ as $j \to \infty$, and let $a(x, \xi) \in C^\infty(U \times \mathbb{R}^n)$. One writes

$$a(x, \xi) \sim \sum_{j=1}^\infty a_j(x, \xi) \qquad (14.3.19)$$

if

$$a(x, \xi) - \sum_{j=1}^{r-1} a_j(x, \xi) \in S_{\rho,\delta}^{\bar{m}_r}, \quad \bar{m}_r := \max_{j \geq r} m_j. \qquad (14.3.20)$$

14.3.3.2. Classes of PDO. A PDO A is called *infinitely smoothing* (or, simply, smoothing) if it maps $\mathcal{E}'(U)$ into $C^\infty(U)$. In particular, if the amplitude $a(x, y, \xi)$ of A has zero of infinite order at $x = y$, then $K_A \in C^\infty(U \times U)$ and A is smoothing.

A function $\sigma(x, \xi)$ is called a *hypoelliptic symbol* if

$$c_1|\xi|^{m_0} \leq |\sigma(x, \xi)| \leq c_2|\xi|^m, \quad |\xi| > R, \quad x \in K \qquad (14.3.21)$$

where $c_j > 0, j = 1, 2, m_0, m \in \mathbb{R}, K \subset U$ is an arbitrary compact set, and the following estimate holds

$$\left| \frac{1}{\sigma(x, \xi)} \partial_\xi^\alpha \partial_x^\beta \sigma(x, \xi) \right| \leq c(\alpha, \beta, K)|\xi|^{-\rho|\alpha|+\delta|\beta|}, \quad |\xi| > R, x \in K,$$
$$(14.3.22)$$

where α and β are arbitrary multiindices, and $K \subset U$ is an arbitrary compact set. The class of such symbols is denoted $HL^{m,m_0}_{\rho,\delta}(U \times \mathbb{R}^n)$. Clearly $HL^{m,m_0}_{\rho,\delta} \subset S^m_{\rho,\delta}$.

A PDO A is called *hypoelliptic* if $A = A_1 + R_1$, where $A_1 \in HL^{m,m_0}_{\rho,\delta}$ and R_1 is infinitely smoothing. If A is hypoelliptic, then $WF(f) = WF(Af)$, where $WF(f)$ is the wave front of f (see Section 14.2.5).

A differential operator $A = \sum_{|j| \leq m} a_j(x) i^{|j|} \partial^j_x$ is called *elliptic* if

$$a_m(x,\xi) := \sum_{|j|=m} a_j(x)\xi^j \neq 0 \quad \text{for} \quad (x,\xi) \in U \times (\mathbb{R}^n \setminus 0). \quad (14.3.23)$$

A PDO is called *elliptic of order* m_0, if its symbol satisfies estimate (14.3.21) with $m = m_0$. For example, the symbol $|\xi|^m g(\xi_0)$ is elliptic of order m if g is a strictly positive C^∞ function on S^{n-1} and $\xi_0 := \xi/|\xi|$.

If A is a hypoelliptic PDO, then

$$\text{singsupp}\, Au = \text{singsupp}\, u, \quad u \in \mathcal{E}'(U). \quad (14.3.24)$$

This means that if $U \subset \mathbb{R}^n$ is a domain, then the restriction of Au on U is $C^\infty(U)$ if and only if $u \in C^\infty(U)$.

If $A \in L^m_{\rho,\delta}(U), \delta < \rho$, then $A : H^s_0(U) \to H^{s-m}_{loc}(U)$, where $H^s(U)$ is the Sobolev space of order s, $H^s_0(U)$ is the space of compactly supported in U elements of $H^s(U)$, $H^s_{loc}(U)$ is the set of functions which have the restrictions on any compact $K \subset U$ belonging to $H^s(K)$.

Consider the operator (14.3.14) with $a(x,y,\xi) = a(x,\xi)$, so that

$$Au(x) = \frac{1}{(2\pi)^n} \int_{\mathbb{R}^n} \exp(-ix \cdot \xi) a(x,\xi) \tilde{u}(\xi) d\xi, \quad \tilde{u} = \mathcal{F}u, \quad (14.3.25)$$

and $a(x,\xi) \in S^m_{\rho,\delta}$. Then $a(x,\xi) = \sigma(x,\xi)$ by formula (14.3.17).

A symbol $a(x,\xi) \in C^\infty(U \times \mathbb{R}^n)$ is called the *classical symbol* if for some, possibly complex, m:

$$a(x,\xi) \sim \sum_{j=0}^{\infty} \psi(\xi) a_{m-j}(x,\xi), \quad (14.3.26)$$

where $\psi \in C^\infty(\mathbb{R}^n)$, $\psi(\xi) = 0$ if $|\xi| \leq 0.5$, and $\psi(\xi) = 1$ if $|\xi| \geq 1$,

$$a_k(x,t\xi) = t^k a_k(x,\xi) \quad \text{if} \quad t > 0, \quad x \in U, \quad \xi \in \mathbb{R}^n \setminus 0. \quad (14.3.27)$$

The class of PDO with such symbols is denoted $CL^m(U)$, and the class of the corresponding symbols is denoted $CS^m(U \times \mathbb{R}^n)$.

If $A_1 \in CL^{m_1}$, $A_2 \in CL^{m_2}$, then $A_1 A_2$ and $A_2 A_1$ belong to $CL^{m_1+m_2}$ provided that $A_j, j = 1, 2$, are properly supported.

Suppose that $K(x, y) \in C^\infty(U \times (\mathbb{R}^n \setminus 0))$, $K(x, ty) = t^{-n} K(x, y)$ for $t > 0$, and

$$\int\limits_{S^{n-1}} K(x, y) dy = 0. \qquad (14.3.28)$$

Then, for any $u \in C_0^\infty(U)$, there exists the limit:

$$Au(x) := \lim_{\epsilon \to 0} \int\limits_{|x-y| \ge \epsilon} K(x, x - y) u(y) dy \qquad (14.3.29)$$

and $A \in CL^0(U)$. The operator A, defined by (14.3.29), is called a *singular integral operator*.

14.4. Special functions

14.4.1. Gamma and beta functions
Definition:

$$\Gamma(z) = \int\limits_0^\infty \exp(-t) t^{z-1} dt, \quad \operatorname{Re} z > 0. \qquad (14.4.1)$$

Recurrence formula and other relations:

$$\Gamma(z+1) = z\Gamma(z), \quad \Gamma(z)\Gamma(1-z) = \frac{\pi}{\sin(\pi z)}, \qquad (14.4.2)$$

$$\Gamma(2z) = \frac{2^{2z-1}}{\sqrt{\pi}} \Gamma(z)\Gamma\left(z + \frac{1}{2}\right). \qquad (14.4.3)$$

Asymptotics:

$$\Gamma(z) = \sqrt{2\pi} e^{-z} z^{z-0.5} (1 + O(|z|^{-1})), \quad |z| \to \infty, |\arg z| < \pi. \quad (14.4.4)$$

Specific values:

$$\Gamma(1) = 1, \quad \Gamma(m+1) = m!,$$

$$\Gamma\left(\frac{1}{2}\right) = \sqrt{\pi}, \quad \Gamma\left(m + \frac{1}{2}\right) = \frac{(2m-1)!!}{2^m} \sqrt{\pi}. \qquad (14.4.5)$$

Definition of the beta function:

$$B(x, y) = \int\limits_0^1 t^{x-1} (1-t)^{y-1} dt, \quad \operatorname{Re} x > 0, \quad \operatorname{Re} y > 0. \qquad (14.4.6)$$

Relation to the gamma function:

$$B(x, y) = \frac{\Gamma(x)\Gamma(y)}{\Gamma(x+y)}. \qquad (14.4.7)$$

14.4.2. Bessel functions

The solutions to the differential equation

$$z^2 u'' + zu' + (z^2 - \nu^2)\, u = 0 \tag{14.4.8}$$

are called Bessel functions of the first kind $J_\nu(z)$, of the second kind $N_\nu(z)$, and of the third kind $H_\nu^{(1)}(z)$ and $H_\nu^{(2)}(z)$ (the latter are also called Hankel's functions). Definition of $J_\nu(z)$:

$$J_\nu(z) = \left(\frac{z}{2}\right)^\nu \sum_{k=0}^\infty \frac{(-z^2/4)^k}{k!\,\Gamma(\nu + k + 1)}. \tag{14.4.9}$$

Generating function:

$$\exp\left[\frac{z}{2}\left(t - \frac{1}{t}\right)\right] = J_0(z) + \sum_{n=1}^\infty J_n(z)\left[t^n + (-1)^n t^{-n}\right]$$

$$= \sum_{n=-\infty}^\infty J_n(z) t^n. \tag{14.4.10}$$

Recurrence and other relations:

$$J_{\nu-1}(z) + J_{\nu+1}(z) = \frac{2\nu}{z} J_\nu(z), \ \ J_{\nu-1}(z) - J_{\nu+1}(z) = 2J_\nu'(z), \tag{14.4.11a}$$

$$\frac{d}{dz}\left(z^\nu J_\nu(z)\right) = z^\nu J_{\nu-1}(z), \ \ \frac{d}{dz}\left(z^{-\nu} J_\nu(z)\right) = -z^{-\nu} J_{\nu+1}(z). \tag{14.4.11b}$$

Bessel functions of the second kind:

$$N_\nu(z) = \frac{J_\nu(z)\cos(\nu\pi) - J_{-\nu}(z)}{\sin(\nu\pi)}, \ \ N_n(z) = \lim_{\nu \to n} N_\nu(z). \tag{14.4.12}$$

Hankel functions:

$$H_\nu^{(1)}(z) = J_\nu(z) + iN_\nu(z), \ \ H_\nu^{(2)} = J_\nu(z) - iN_\nu(z). \tag{14.4.13}$$

Bessel functions of imaginary argument:

$$I_\nu(z) = \left(\frac{z}{2}\right)^\nu \sum_{k=0}^\infty \frac{(z^2/4)^k}{k!\,\Gamma(\nu + k + 1)}, \tag{14.4.14}$$

$$K_\nu(z) = \frac{\pi}{2} \frac{I_{-\nu}(z) - I_\nu(z)}{\sin(\nu\pi)}, \ \ K_n(z) = \lim_{\nu \to n} K_\nu(z). \tag{14.4.15}$$

$$I_\nu(z) = \exp\left(-\frac{i\nu\pi}{2}\right) J_\nu(iz), \quad -\pi < \arg z \le \frac{\pi}{2}, \qquad (14.4.16)$$

$$K_\nu(z) = \frac{i\pi}{2} \exp\left(\frac{i\nu\pi}{2}\right) H_\nu^{(1)}(iz), \quad -\pi < \arg z \le \frac{\pi}{2}. \qquad (14.4.17)$$

Bessel functions of the half-integer order $J_{n+\frac{1}{2}}(z)$:

$$J_{\frac{1}{2}}(z) = \sqrt{\frac{2}{\pi z}} \sin z, \quad J_{-\frac{1}{2}}(z) = \sqrt{\frac{2}{\pi z}} \cos z,$$

$$J_{n+\frac{1}{2}}(z) = (-1)^n \sqrt{\frac{2}{\pi}} z^{n+\frac{1}{2}} \left(\frac{1}{z}\frac{d}{dz}\right)^n \left(\frac{\sin z}{z}\right).$$

$$(14.4.18)$$

Asymptotics for large values of the argument:

$$J_\nu(x) = \sqrt{\frac{2}{\pi x}} \left(\cos\left(x - \frac{\nu\pi}{2} - \frac{\pi}{4}\right) + O(x^{-1})\right), \quad x \to +\infty,$$

$$N_\nu(z) = \sqrt{\frac{2}{\pi x}} \left(\sin\left(x - \frac{\nu\pi}{2} - \frac{\pi}{4}\right) + O(x^{-1})\right), \quad x \to +\infty,$$

$$H_\nu^{(s)}(x) = \sqrt{\frac{2}{\pi x}} \exp\left\{\pm i \left(x - \frac{\nu\pi}{2} - \frac{\pi}{4}\right)\right\} (1 + O(x^{-1})), \quad x \to +\infty,$$

where '+' is taken if $s = 1$, and '−' is taken if $s = 2$. Asymptotics for small values of the argument:

$$J_\nu(z) = \frac{z^\nu}{2^\nu \Gamma(1+\nu)}(1 + O(|z|^2)), \quad z \to 0, \quad \nu \ne -1, -2, \ldots,$$

$$N_\nu(z) \sim -\frac{2^\nu \Gamma(\nu)}{\pi z^\nu}, \quad z \to 0, \quad \mathrm{Re}\,\nu > 0,$$

$$N_0(z) \sim \frac{2}{\pi} \ln z, \quad z \to 0,$$

Asymptotics for large values of the order:

$$J_\nu(z) \sim \frac{1}{\sqrt{2\pi\nu}} \left(\frac{ez}{2\nu}\right)^\nu, \quad \nu \to +\infty,$$

$$N_\nu(z) \sim -\sqrt{\frac{2}{\pi\nu}} \left(\frac{ez}{2\nu}\right)^{-\nu}, \quad \nu \to +\infty.$$

Plane wave expansion:

$$\exp(iz\cos\varphi) = \sum_{k=-\infty}^{\infty} i^k J_k(z) \exp(ik\varphi).$$

14.4.3. Orthogonal polynomials

14.4.3.1. Gegenbauer or ultraspherical polynomials. Definition (the Rodrigues formula):

$$C_m^{(p)}(t)$$

$$= \frac{(-1)^m \Gamma(p+0.5)\Gamma(m+2p)(1-t^2)^{0.5-p}}{2^m m! \Gamma(2p)\Gamma(m+p+0.5)} \frac{d^m}{dt^m}\left[(1-t^2)^{m+p-0.5}\right]. \tag{14.4.19}$$

Differential equation leading to $C_m^{(p)}(t)$:

$$(1-t^2)y'' - (2p+1)ty' + m(2p+m)y = 0.$$

Generating function:

$$(1-2tw+t^2)^{-p} = \sum_{m=0}^{\infty} C_m^{(p)}(w)t^m, \quad |t| < 1, p > 0;$$

$$-\ln|1-2tw+t^2| = \sum_{m=0}^{\infty} C_m^{(0)}(w)t^m, \quad |t| < 1. \tag{14.4.20}$$

Recurrence relations:

$$\frac{d^k}{dt^k}C_m^{(p)}(t) = \frac{2^k \Gamma(p+k)}{\Gamma(p)}C_{m-k}^{(p+k)}(t), \tag{14.4.21a}$$

$$(m+1)C_{m+1}^{(p)}(t) = 2(p+m)tC_m^{(p)}(t) - (m+2p-1)C_{m-1}^{(p)}(t), \tag{14.4.21b}$$

$$(m+p)C_m^{(p)}(t) = p\left[C_m^{(p+1)}(t) - C_{m-2}^{(p+1)}(t)\right]. \tag{14.4.21c}$$

Special cases and particular values:

$$C_0^{(p)}(t) = 1; \quad C_1^{(p)}(t) = 2pt, \ p > 0; \quad C_1^{(0)}(t) = 2t;$$

$$C_m^{(p)}(-t) = (-1)^m C_m^{(p)}(t), \tag{14.4.22}$$

$$C_m^{(p)}(1) = \frac{\Gamma(2p+m)}{m!\Gamma(2p)}, \ p > 0; \quad C_m^{(0)}(1) = \frac{2}{m}, \ m > 0. \tag{14.4.23}$$

Orthogonality:

$$\int_{-1}^{1} C_m^{(p)}(t)C_{m'}^{(p)}(t)(1-t^2)^{p-\frac{1}{2}}dt = \delta_{mm'}\frac{\pi\Gamma(2p+m)}{2^{2p-1}m!(m+p)\Gamma^2(p)}$$

$$:= \delta_{mm'}\Lambda_{m,n}, \quad p > -0.5, p \neq 0. \tag{14.4.24}$$

Fourier transform:

$$\int_{-1}^{1} (1-t^2)^{p-\frac{1}{2}}C_m^{(p)}(t)e^{iat}dt = \frac{\pi 2^{1-p}i^m\Gamma(2p+m)}{m!\Gamma(p)}\frac{J_{m+p}(a)}{a^p}. \tag{14.4.25}$$

14.4.3.2. Legendre polynomials. Definition and differential equation:

$$P_m(t) = \frac{1}{2^m m!} \frac{d^m}{dt^m} (t^2 - 1)^m; \quad (1 - t^2)P_m'' - 2tP_m' + m(m+1)P_m = 0.$$
$$(14.4.26)$$

Generating function

$$(1 - 2tw + t^2)^{-\frac{1}{2}} = \sum_{k=0}^{\infty} t^k P_k(w), \quad |t| < \min|w \pm \sqrt{w^2 - 1}|;$$

$$(1 - 2tw + t^2)^{-\frac{1}{2}} = \sum_{k=0}^{\infty} t^{-(k+1)} P_k(w), \quad |t| > \min|w \pm \sqrt{w^2 - 1}|.$$

Selected properties:

$$P_m(-t) = (-1)^m P_m(t), \quad P_m(1) = 1, \quad |P_m(t)| \le 1 \text{ for } |t| \le 1.$$

Orthogonality:

$$\int_{-1}^{1} P_m(t) P_{m'}(t) dt = \frac{2}{2m+1} \delta_{mm'}. \quad (14.4.27)$$

The Legendre polynomials are a particular case of the Gegenbauer polynomials:

$$P_m(t) = C_m^{(\frac{1}{2})}(t). \quad (14.4.28)$$

Asymptotics:

$$P_m(\cos\theta) \sim \left(\frac{2}{\pi m \sin\theta}\right)^{\frac{1}{2}} \sin\left[\left(m + \frac{1}{2}\right)\theta + \frac{\pi}{4}\right],$$
$$\text{as} \quad m \to \infty, \quad 0 < \delta \le \theta \le \pi - \delta.$$

14.4.3.3. Chebyshev polynomials. Chebyshev polynomials of the first and of the second kinds are denoted by T_m and U_m, respectively. Definition:

$$T_m(t) = \cos(m \arccos t), \quad U_m(t) = \frac{\sin[(m+1)\arccos t]}{\sqrt{1 - t^2}}, \quad |t| \le 1;$$
$$(14.4.29)$$

and

$$T_m(t) = \cosh(m \operatorname{arccosh} t), \quad U_m(t) = \frac{\sinh[(m+1)\operatorname{arccosh} t]}{\sinh(\operatorname{arccosh} t)}, \quad |t| > 1.$$
$$(14.4.30)$$

The functions $T_m(t)$ and $\sqrt{1-t^2}U_{m-1}(t)$ are two linearly independent solutions of the differential equation:

$$(1-t^2)y'' - ty' + m^2y = 0.$$

Recurrence relation:

$$y_{m+1}(t) = 2ty_m(t) - y_{m-1}(t), \qquad (14.4.31)$$

where y_m is either T_m or U_m. Special cases:

$$T_0(t) = 1, \quad T_1(t) = t, \quad T_2(t) = 2t^2 - 1;$$
$$U_0(t) = 1, \quad U_1(t) = 2t, \quad U_2(t) = 4t^2 - 1.$$

Orthogonality property:

$$\int_{-1}^{1} T_m(t)T_{m'}(t)\frac{dt}{\sqrt{1-t^2}} = \begin{cases} 0, & m \neq m' \\ \frac{\pi}{2}, & m = m' \neq 0 \\ \pi, & m = m' = 0, \end{cases}$$

$$\int_{-1}^{1} U_m(t)U_{m'}(t)\sqrt{1-t^2}\,dt = \begin{cases} 0, & m \neq m' \\ \frac{\pi}{2}, & m = m'. \end{cases} \qquad (14.4.32)$$

Relation to the Gegenbauer polynomials:

$$T_m(t) = \frac{C_m^{(0)}(t)}{C_m^{(0)}(1)} = \frac{m}{2}C_m^{(0)}(t), \; U_m(t) = C_m^{(1)}(t). \qquad (14.4.33)$$

Extremal property: for any polynomial $\tilde{\theta}_m(t)$ of degree m with leading coefficient 1, one has

$$2^{1-m} = \max_{-1 \leq t \leq 1} |\tilde{T}_m| \leq \max_{-1 \leq t \leq 1} |\tilde{\theta}_m(t)|, \; 2^{1-m} = \int_{-1}^{1} |\tilde{U}_m|dt \leq \int_{-1}^{1} |\tilde{\theta}_m(t)|dt,$$

where $\tilde{T}_m$ and $\tilde{U}_m$ denote Chebyshev polynomials normalized so that their leading coefficients equal 1.

14.4.3.4. Hermite polynomials. Definition:

$$H_m(t) = (-1)^m \exp(t^2)\frac{d^m \exp(-t^2)}{dt^m}. \qquad (14.4.34)$$

Differential equation:

$$H_m'' - 2tH_m' + 2mH_m = 0. \qquad (14.4.35)$$

Generating function:

$$\exp(2wt - t^2) = \sum_{m=0}^{\infty} \frac{H_m(w)}{m!} t^m. \qquad (14.4.36)$$

Orthogonality relation:

$$\int_{-\infty}^{\infty} \exp(-t^2) H_m(t) H_{m'}(t) dt = \delta_{mm'} 2^m m! \sqrt{\pi}. \qquad (14.4.37)$$

Asymptotics:

$$H_m(t) \sim 2^{\frac{m+1}{2}} m^{\frac{m}{2}} \exp\left(-\frac{m}{2}\right) \exp\left(\frac{t^2}{2}\right) \cos\left(\sqrt{2m+1}\,t - \frac{m\pi}{2}\right),$$
$$m \to \infty.$$

Let us give four formulas which might be useful in view of Exercise 2.1.9:

$$x^{2m} = \sum_{j=0}^{m} c_{2j} H_{2j}(x), \ c_{2j} = \frac{1}{2^{2j}(2j)!\sqrt{\pi}} \int_{-\infty}^{\infty} x^{2m} \exp(-x^2) H_{2j}(x) dx$$

$$\exp(ax) = \sum_{j=0}^{\infty} \frac{\exp\left(\frac{a^2}{4}\right) a^j}{2^j j!} H_j(x)$$

$$\exp(-a^2 x^2) = \sum_{j=0}^{\infty} c_{2j} H_{2j}(x), \ c_{2j} = \frac{(-1)^j a^{2j}}{2^{2j} j! (1+a^2)^{j+\frac{1}{2}}}, \ \mathrm{Re}\,a^2 > -1.$$

$$\mathrm{sgn}\,x = \frac{1}{\sqrt{\pi}} \sum_{j=0}^{\infty} \frac{(-1)^j}{2^{2j}(2j+1)j!} H_{2j+1}(x).$$

14.4.4. Integration over spheres

The spherical coordinates in $\mathbb{R}^n$ are defined by the formulas:

$$x_1 = r \sin\theta_{n-1} \ldots \sin\theta_3 \sin\theta_2 \sin\theta_1$$
$$x_2 = r \sin\theta_{n-1} \ldots \sin\theta_3 \sin\theta_2 \cos\theta_1$$
$$x_3 = r \sin\theta_{n-1} \ldots \sin\theta_3 \cos\theta_2$$
$$\ldots \qquad\qquad (14.4.38)$$
$$x_{n-1} = r \sin\theta_{n-1} \cos\theta_{n-2}$$
$$x_n = r \cos\theta_{n-1}$$

where $0 \le r < \infty, 0 \le \theta_1 < 2\pi$, and $0 \le \theta_j \le \pi$ for $2 \le j \le n - 1$. The numbers $(\theta_1, \ldots, \theta_{n-1})$ can be considered as coordinates of a point on S^{n-1}. The element of the surface area dx on S^{n-1} is given by the formula:

$$dx = (\sin \theta_{n-1})^{n-2} \ldots (\sin \theta_3)^2 \sin \theta_2 d\theta_1 d\theta_2 \ldots d\theta_{n-1},$$

therefore,

$$\int_{S^{n-1}} f(x) dx$$

$$= \int_0^\pi \cdots \int_0^\pi \int_0^{2\pi} f(x)(\sin \theta_{n-1})^{n-2} \ldots (\sin \theta_3)^2 \sin \theta_2 d\theta_1 d\theta_2 \ldots d\theta_{n-1}, \tag{14.4.39}$$

where $x = (x_1, \ldots, x_n)$ depends on $(\theta_1, \ldots, \theta_{n-1})$ according to (14.4.38). In particular,

$$\int_{S^{n-1}} dx := |S^{n-1}| = \frac{2\pi^{\frac{n}{2}}}{\Gamma\left(\frac{n}{2}\right)}, \quad |S^0| = 2, \ |S^1| = 2\pi, \ |S^2| = 4\pi.$$

$$\tag{14.4.40}$$

If $f(x) = f(\omega \cdot x)$, we get

$$\int_{S^{n-1}} f(\omega \cdot x) dx = |S^{n-2}| \int_{-1}^1 f(t)(1 - t^2)^{\frac{n-3}{2}} dt. \tag{14.4.41}$$

14.4.5. Spherical harmonics

A polynomial f is called harmonic if $\Delta f = 0$, where Δ is the Laplacian:

$$\Delta = \sum_{j=1}^n \frac{\partial^2}{\partial x_j^2}.$$

Let $R^{n,l}$ be a linear space of homogeneous polynomials of degree l in n variables. One can check that if $h \in R^{n,l}$ and $\Delta h = 0$, then $x_j h - \frac{1}{n+2l-2} r^2 \frac{\partial h}{\partial x_j}$ is a homogeneous harmonic polynomial of degree $l+1$. Let $H^{n,l}$ be the space of homogeneous harmonic polynomials of degree l in n variables. One can prove that

$$R^{n,l} = H^{n,l} \dotplus r^2 R^{n,l-2}$$

where the sum is direct.

Let us write the Laplacian in spherical coordinates:

$$\Delta = \frac{1}{r^{n-1}}\frac{\partial}{\partial r}\left(r^{n-1}\frac{\partial}{\partial r}\right) + \frac{1}{r^2}\Delta_\alpha, \qquad (14.4.42)$$

where Δ_α is the angular part of the Laplacian. Let P_l be a homogeneous harmonic polynomial of degree l. Then, $P_l(r\alpha) = r^l Y_l(\alpha), \alpha \in S^{n-1}$. Substituting into (14.4.42) and using that $\Delta P_l = 0$, we get that Y_l satisfies the equation

$$(\Delta_\alpha + \lambda_l)Y_l = 0, \quad \lambda_l = l(n-2+l). \qquad (14.4.43)$$

Nontrivial solutions to (14.4.43), normalized so that $\int_{S^{n-1}} Y_l^2(\alpha)d\alpha = 1$, are called the *spherical harmonics of degree l*. Spherical harmonics are restrictions on S^{n-1} of the homogeneous harmonic polynomials of degree l. Any two spherical harmonics of different degrees are orthogonal on S^{n-1}. There are

$$N(n,l) := (2l + n - 2)\frac{(n+l-3)!}{(n-2)!l!}, \quad N(n,0) = 1, \qquad (14.4.44)$$

linearly independent spherical harmonics of degree l in $\mathbb{R}^n$. For example, $N(2,l) = 2\ (l \geq 1)$, $N(3,l) = 2l + 1$.
   The following formula is useful:

$$\sum_m Y_{l,m}(\alpha)\overline{Y_{l,m}(\beta)} = C_l^{(\frac{n-2}{2})}(\alpha\cdot\beta)\frac{C_l^{(\frac{n-2}{2})}(1)}{|S^{n-2}|\Lambda_{l,n}}, \qquad (14.4.45)$$

where there summation is over $N(n,l)$ linearly independent spherical harmonics of degree l, $\Lambda_{l,n}$ is defined in (14.4.24), and $|S^{n-2}|$ is the area of the unit sphere in $\mathbb{R}^{n-1}$. If $n = 3$ one has:

$$\sum_{m=-l}^{l} Y_{l,m}(\alpha)\overline{Y_{l,m}(\beta)} = \frac{2l+1}{4\pi}P_l(\alpha\cdot\beta),$$

where $P_l(x)$ are the Legendre polynomials (14.4.26), and $Y_{l,m}(\alpha)$ are defined for $n = 3$ under formula (14.4.48).
   The Funk-Hecke theorem can be stated as follows:

$$\int_{S^{n-1}} h(\alpha\cdot\theta)Y_l(\alpha)d\alpha = c(n,l)Y_l(\theta), \qquad (14.4.46)$$

where $h(t) \in L^2\left([-1,1],(1-t^2)^{\frac{n-3}{2}}\right)$,

$$c(n,l) := \frac{|S^{n-2}|}{C_l^{(\frac{n-2}{2})}(1)}\int_{-1}^{1} h(t)C_l^{(\frac{n-2}{2})}(t)(1-t^2)^{\frac{n-3}{2}}dt, \qquad (14.4.47)$$

and $C_l^{(p)}(t)$ are the Gegenbauer polynomials. The Funk-Hecke theorem can be proved using the relation $\triangle(r^l Y_l(\alpha)) = 0$, Green's formula for the harmonic functions, and Equation (14.4.20). Alternatively, the Funk-Hecke theorem can be derived from (14.4.45).

Using (14.4.25) and the Funk-Hecke theorem, we obtain another useful identity

$$\int\limits_{S^{n-1}} \exp(i\sigma\alpha\cdot\beta)Y_l(\alpha)d\alpha = (2\pi)^{\frac{n}{2}} i^l \frac{J_{l+\frac{n-2}{2}}(\sigma)}{\sigma^{\frac{n-2}{2}}} Y_l(\beta), \qquad (14.4.48)$$

where $J_\nu(\lambda)$ is the Bessel function.

In $\mathbb{R}^3$ the spherical harmonics are:

$$Y_{l,m}(\alpha) = (-1)^m i^l \left[\frac{2l+1}{4\pi} \frac{(l-m)!}{(l+m)!}\right]^{\frac{1}{2}} \exp(im\varphi) P_{l,m}(\cos\theta), \quad |m| \le l.$$

Here $l = 0, 1, 2, \ldots$, $\alpha := \alpha(\theta, \varphi) := (\sin\theta\cos\varphi, \sin\theta\sin\varphi, \cos\theta)$.

$$P_{l,m}(\cos\theta) := (\sin\theta)^m \left(\frac{d}{d\cos\theta}\right)^m P_l(\cos\theta), \quad 0 \le m \le l,$$

$P_{l,m}$ are associated Legendre functions, and P_l are the Legendre polynomials. One has

$$P_{l,-m}(t) = (-1)^m \frac{(l-m)!}{(l+m)!} P_{l,m}(t), \quad t = \cos\theta,$$

$$P_{l,m}(-t) = (-1)^{l+m} P_{l,m}(t), \quad t = \cos\theta,$$

$$Y_{l,m}(-\alpha) = (-1)^l Y_{l,m}(\alpha), \quad \alpha = (\alpha_1, \alpha_2, \alpha_3),$$

$$\bar{Y}_{l,m}(\alpha) = (-1)^{l+m} Y_{l,-m}(\alpha), \quad \alpha = (\alpha_1, \alpha_2, \alpha_3),$$

where the overbar stands for complex conjugate.

14.4.6. The Hankel transform and the Fourier transform

Let us write inverse and direct Fourier transforms in spherical coordinates:

$$\tilde{f}(\lambda\alpha) = \int\limits_0^\infty dr\, r^{n-1} \int\limits_{S^{n-1}} d\beta f(r\beta) \exp(i\lambda r\alpha\cdot\beta), \quad x = r\beta, \quad \xi := \lambda\alpha,$$

$$f(r\beta) = \frac{1}{(2\pi)^n} \int\limits_0^\infty d\lambda\, \lambda^{n-1} \int\limits_{S^{n-1}} d\alpha \exp(-i\lambda r\alpha\cdot\beta)\tilde{f}(\lambda\alpha). \qquad (14.4.49)$$

From (14.4.48), we get

$$\exp(i\lambda r\alpha \cdot \beta) = (2\pi)^{\frac{n}{2}} \sum_{l=0}^{\infty} i^l \frac{J_{l+\frac{n-2}{2}}(\lambda r)}{(\lambda r)^{\frac{n-2}{2}}} Y_l(\alpha)\overline{Y_l(\beta)}.$$

Define

$$f_l(r) := \int_{S^{n-1}} f(r\beta)\overline{Y_l(\beta)}d\beta, \quad \tilde{f}_l(\lambda) := \int_{S^{n-1}} \tilde{f}(\lambda\alpha)\overline{Y_l(\alpha)}d\alpha.$$

Then

$$\tilde{f}(\lambda\alpha) = (2\pi)^{\frac{n}{2}} \sum_{l=0}^{\infty} i^l \int_0^{\infty} drr^{n-1} \frac{J_{l+\frac{n-2}{2}}(\lambda r)}{(\lambda r)^{\frac{n-2}{2}}} f_l(r)Y_l(\alpha)$$

and

$$\tilde{f}_l(\lambda) = (2\pi)^{\frac{n}{2}} i^l \int_0^{\infty} drr^{n-1} \frac{J_{l+\frac{n-2}{2}}(\lambda r)}{(\lambda r)^{\frac{n-2}{2}}} f_l(r) := H_{ln}f_l. \qquad (14.4.50)$$

Similarly, formula (14.4.49) implies

$$f_l(r) = \frac{(-i)^l}{(2\pi)^{n/2}} \int_0^{\infty} d\lambda\lambda^{n-1} \frac{J_{l+\frac{n-2}{2}}(\lambda r)}{(\lambda r)^{\frac{n-2}{2}}} \tilde{f}_l(\lambda). \qquad (14.4.51)$$

Formula (14.4.50) defines the Hankel transform of $f_l(r)$, and formula (14.4.51) is the inverse Hankel transform.

Formula (14.4.50) is valid for $f_l \in L^2([0,\infty), r^{n-1})$, and H_{ln} maps isomorphically $L^2([0,\infty), r^{n-1})$ onto itself because the Fourier transform maps isomorphically $L^2(\mathbb{R}^n)$ onto itself.

14.5. Asymptotic expansions

14.5.1. Definitions

If $\lim_{x\to a} \frac{f(x)}{g(x)} = 1$, one writes $f(x) \sim g(x)$ as $x \to a$.

If $\lim_{x\to a} \frac{f(x)}{g(x)} = 0$, one writes $f(x) = o(g(x))$ as $x \to a$.

If $|f(x)| \le c|g(x)|$ in a neighborhood of a point a, one writes $f(x) = O(g(x))$ as $x \to a$.

In these definitions x may belong to a certain set, e.g., the above relations may hold only for $x > a$.

Let a sequence of functions $\{\varphi_j(x)\}$ be given such that

$$\varphi_{j+1}(x) = o(\varphi_j(x)), \quad x \to a. \qquad (14.5.1)$$

Suppose

$$f(x) - \sum_{j=1}^{m} a_j\varphi_j(x) = o(\varphi_m(x)), \quad x \to a, \ m = 1, 2, \dots. \qquad (14.5.2)$$

Then we say that the sum $\sum_{j=1}^{\infty} a_j\varphi_j(x)$ as the *asymptotic expansion* of f as $x \to a$ and denote $f(x) \sim \sum_{j=1}^{\infty} a_j\varphi_j(x)$, $x \to a$.

Exercise 14.5.1. Prove uniqueness of asymptotic expansion (14.5.2)

Hint. Assume $f \sim \sum_{j=1}^{\infty} a_j \varphi_j$ and $f \sim \sum_{j=1}^{\infty} b_j \varphi_j$. Then $(a_1 - b_1)\varphi_1 = o(\varphi_1)$, $x \to a$. Thus $a_1 = b_1$. Repeat the argument and get $a_j = b_j$ for all j. Two different functions may have the same asymptotic expansion, e.g., $f_1(x) = 0$ and $f_2(x) = \exp(-x)$ as $x \to +\infty$, where $\varphi_j(x) = x^{-j}$.

Exercise 14.5.2. Let $f \in C([0,1])$ and suppose $f(0) \neq 0$. Define the function

$$F(s,a,b) := \int_0^1 t^{b-1}(t+s)^a f(t)dt, \quad b > 0, \ a \in \mathbb{R}, \ s > 0, \qquad (14.5.3)$$

and suppose $a \neq 1, 2, \dots$. Find the asymptotics of F as $s \to 0$.

Hint. The asymptotics of F as $s \to 0$ is the same as that of

$$f(0) \int_0^1 t^{b-1}(t+s)^a dt := f(0)F_0(s,a,b).$$

If $a + b < 0$, then

$$F_0 = \int_0^{\infty} t^{b-1}(t+s)^a dt - \int_1^{\infty} t^{b-1}(t+s)^a dt := F_1 + F_2.$$

One has

$$F_1 = s^{a+b} \int_0^{\infty} t^{b-1}(1+t)^a dt = s^{a+b}B(b, -a-b),$$

where $B(x,y)$ is the beta function, and

$$F_2 = O(1) \quad \text{as } s \to 0.$$

Thus

$$F \sim f(0)s^{a+b}B(b, -a-b) \text{ as } s \to 0 \text{ if } a + b < 0. \qquad (14.5.4)$$

Assume now that $a+b > 0$, $a+b \neq 1, 2, \dots$. Differentiate F with respect to s several times and get the previous case. Integrate with respect to s and get (14.5.4) in the case $a + b > 0$. If $a + b = N$, where N is a nonnegative integer, one needs a limiting argument.

14.5.2. Laplace's method

In this section we present the asymptotics of the integral

$$F(t) = \int_A^B f(x) \exp(th(x)) dx \qquad (14.5.5)$$

as $t \to +\infty$, where h is a real-valued function. First, consider the integral

$$\Phi_1(t) = \int_0^1 x^{b-1} f(x) \exp(-tx^a) dx, \quad a > 0, \quad b > 0, \qquad (14.5.6)$$

where $f \in C^\infty([0,1])$. Note that

$$\int_0^\infty x^{b-1} \exp(-tx^a) dx = \frac{1}{a} t^{-\frac{b}{a}} \Gamma\left(\frac{b}{a}\right), \quad t > 0. \qquad (14.5.7)$$

Lemma 14.5.1. *One has*

$$\Phi_1(t) \sim \frac{1}{a} \sum_{k=0}^\infty t^{-(k+b)/a} \Gamma\left(\frac{k+b}{a}\right) \frac{f^{(k)}(0)}{k!}, \quad t \to +\infty. \qquad (14.5.8)$$

This asymptotic expansion can be differentiated with respect to t any number of times. The main term in (14.5.8) is:

$$\Phi_1(t) \sim \frac{1}{a} t^{-\frac{b}{a}} \Gamma\left(\frac{b}{a}\right) f(0), \quad t \to +\infty. \qquad (14.5.9)$$

Lemma 14.5.2. *Assume that*

$$h \in C^\infty(\mathbb{R}), \quad h'(x_0) = \cdots = h^{(N-1)}(x_0) = 0, \quad h^{(N)}(x_0) \neq 0.$$

Then there exist open neighborhoods of x_0 and of the point $y = 0$, and a function $x = \varphi(y)$ such that $h(\varphi(y)) = h(x_0) + sy^N$, $s := \operatorname{sgn} h^{(N)}(x_0)$, $\varphi(y) \in C^\infty$ in a neighborhood of the origin, and

$$\varphi(0) = 0, \quad \varphi'(0) = \left(\frac{N!}{|h^{(N)}(x_0)|}\right)^{1/N}.$$

Using Lemmas 14.5.1 and 14.5.2 one can prove

Theorem 14.5.1. *Assume that*

(1) $\max_{A \leq x \leq B} h(x)$ *is attained at a single point* x_0, $A < x_0 < B$,
(2) $f(x)$ *and* $h(x)$ *are* C^∞ *smooth in a neighborhood of* x_0 *and are continuous on* $[A, B]$,
(3) $h''(x_0) \neq 0$.

Then

$$F(t) \sim \frac{\exp[th(x_0)]}{\sqrt{t}} \sum_{k=0}^{\infty} c_k t^{-k}, \qquad (14.5.10)$$

where

$$c_k = \frac{\Gamma\left(k + \frac{1}{2}\right)}{(2k)!} \left(\frac{d}{dx}\right)^{2k} \left[f(x) \left(\frac{-(h(x) - h(x_0))}{(x - x_0)^2}\right)^{-k-\frac{1}{2}} \right]\Bigg|_{x=x_0}. \qquad (14.5.11)$$

The main term in (14.5.10) is

$$F(t) \sim t^{-\frac{1}{2}} \left(\frac{-2\pi}{h''(x_0)}\right)^{\frac{1}{2}} f(x_0) \exp[th(x_0)], \quad t \to \infty. \qquad (14.5.12)$$

Lemma 14.5.3. *If* $f \in C^1$ *in a neighborhood of the origin,* $f \in C([0,1])$, $b > 0$, $\gamma \in \mathbb{R}$. *Then*

$$\int_0^1 x^{b-1} |\ln x|^\gamma f(x) \exp(-tx) dx \sim t^{-b} (\ln t)^\gamma \sum_{k=0}^{\infty} a_k (\ln t)^{-k},$$

$$t \to +\infty, \qquad (14.5.13)$$

where

$$a_0 = \Gamma(b) f(0), \quad a_k = (-1)^k \binom{\gamma}{k} f(0) \left(\frac{d}{db}\right)^k \Gamma(b). \qquad (14.5.14)$$

and $\binom{\gamma}{k} = \frac{\gamma(\gamma-1)\ldots(\gamma-k+1)}{k!}$ *if* $k \geq 1$, $\binom{\gamma}{0} = 1$.

Consider now the case of a single degenerate critical point x_0, $A < x_0 < B$:

$$h'(x_0) = \cdots = h^{(2m-1)}(x_0) = 0, \quad h^{(2m)}(x_0) \neq 0, \quad m \geq 1. \quad (14.5.15)$$

In this case

$$F(t) \sim t^{-\frac{1}{2m}} \exp[th(x_0)] \sum_{k=0}^{\infty} a_k t^{-\frac{k}{m}}, \quad t \to \infty, \qquad (14.5.16)$$

where

$$a_k = \frac{1}{m}\frac{(2m)^{2k}}{(2k)!}\Gamma\left(\frac{2k+1}{2m}\right)$$

$$\times \left\{\left(h(x,x_0)\frac{d}{dx}\right)^{2k}[f(x)h(x,x_0)]\right\}\bigg|_{x=x_0}, \quad (14.5.17)$$

and

$$h(x,x_0) = \frac{(h(x_0)-h(x))^{1-\frac{1}{2m}}}{h'(x)}. \quad (14.5.18)$$

14.5.3. The stationary phase method
Consider the integral

$$G(t) := \int_A^B f(x)\exp[ith(x)]dx, \quad (14.5.19)$$

where $h(x) \in C^1([A,B])$, h is real-valued, $f(x)$ may be complex-valued, and $t \to +\infty$. If h does not have critical points, that is, $h'(x) \neq 0$, $A \leq x \leq B$, then the aymptotics of $G(t)$ can be easily calculated using integration by parts: if $f \in C^{N+1}$, $h \in C^{N+2}$, then

$$G(t) = \sum_{k=0}^{N}(-1)^k(it)^{-k-1}\exp[ith(x)]$$

$$\times \left[\left(\frac{1}{h'(x)}\frac{d}{dx}\right)^k\left(\frac{f(x)}{h'(x)}\right)\right]\bigg|_A^B + o\left(t^{-N-1}\right), \quad t \to \infty. \quad (14.5.20)$$

The integration by parts technique can be easily generalized to integrals more general than (14.5.19). More precisely, suppose that the function f depends also on t and has the property

$$\frac{\partial^k}{\partial t^k}\frac{\partial^j}{\partial x^j}f(t,x) = O(t^{\gamma-k}), \quad k,j = 0,1,\ldots, \quad (14.5.21)$$

for some $\gamma \in \mathbb{R}$. Thus, integrating by parts, we get similarly to (14.5.20):

$$G(t) = \sum_{k=0}^{N}(-1)^k(it)^{-k-1}\exp[ith(x)]$$

$$\times \left[\left(\frac{1}{h'(x)}\frac{\partial}{\partial x}\right)^k\left(\frac{f(t,x)}{h'(x)}\right)\right]\bigg|_A^B + o\left(t^{-N-1}\right), \quad t \to \infty. \quad (14.5.20')$$

Property (14.5.21) ensures that expansion (14.5.20') makes sense and can be differentiated with respect to t any number of times.

REMARK 14.5.1. Derivation of all asymptotic expansions of integrals presented in this section are based on the integration by parts technique. Therefore, the above argument shows that these expansions can be used for integrals, where $f(x)$ is replaced by $f(t, x)$, provided $f(t, x)$ satisfies (14.5.21) (if $x \in \mathbb{R}^n, n > 1$, then j in (14.5.21) is a multiindex). This results in replacing $f(x)$ by $f(t, x)$ in the corresponding expansions.

Let

$$\Phi_2(t) = \int_0^1 x^{b-1} (\ln x) f(x) \exp(itx^a) dx, \quad b > 0, a \geq 1. \qquad (14.5.22)$$

Assume that $f \in C^\infty([0, 1])$ and $f(x)$ vanishes with all derivatives for $x \geq \frac{1}{2}$. Then the following result, known as *the Erdélyi lemma*, holds.

Lemma 14.5.4. *One has*

$$\Phi_2(t) \sim \sum_{k=0}^\infty a_k(t) t^{-\frac{k+b}{a}}, \quad t \to \infty, \qquad (14.5.23)$$

where

$$a_k(t) = \frac{1}{a^2} \Gamma\left(\frac{b+k}{a}\right) \left[-\ln t + \psi\left(\frac{b+k}{a}\right) + \frac{i\pi}{2}\right]$$
$$\times \exp\left[\frac{i\pi(k+b)}{2a}\right] \frac{f^{(k)}(0)}{k!}, \qquad (14.5.24)$$

and $\psi(x) = \frac{\Gamma'(x)}{\Gamma(x)}$. *The main term in (14.5.23) is*

$$\Phi_2(t) = -\frac{1}{a^2} \Gamma\left(\frac{b}{a}\right) \exp\left(\frac{i\pi b}{2a}\right) t^{-\frac{b}{a}} \ln t \left[f(0) + O\left(\frac{1}{\ln t}\right)\right]. \qquad (14.5.25)$$

Let

$$\Phi_3(t) = \int_0^1 x^{b-1} f(x) \exp(itx^a) dx, \quad b > 0, \ a \geq 1. \qquad (14.5.26)$$

Under the same assumptions as above one can prove

Lemma 14.5.5. *One has*

$$\Phi_3(t) \sim \sum_{k=0}^{\infty} a_k t^{-\frac{k+b}{a}}, \quad t \to \infty, \tag{14.5.27}$$

where

$$a_k = \frac{1}{a} \frac{f^{(k)}(0)}{k!} \Gamma\left(\frac{k+b}{a}\right) \exp\left[\frac{i\pi(k+b)}{2a}\right]. \tag{14.5.28}$$

One can differentiate (14.5.27) any number of times. The main term in (14.5.27) is

$$\Phi_3(t) \sim \frac{f(0)}{a} \Gamma\left(\frac{b}{a}\right) \exp\left(\frac{i\pi b}{2a}\right) t^{-\frac{b}{a}}, \quad t \to \infty. \tag{14.5.29}$$

Consider integral (14.5.19) and assume that

$$f(x) \in C_0^{\infty}([A,B]), \quad h(x) \in C^{\infty}([A,B]),$$

$h(x)$ has a single stationary point x_0, $A < x_0 < B$, $h''(x_0) \neq 0$.

Theorem 14.5.2. *Under the above assumptions one has*

$$G(t) \sim \exp\left[ith(x_0) + \frac{i\pi}{4} sgnh''(x_0)\right] t^{-\frac{1}{2}} \sum_{k=0}^{\infty} a_k t^{-k}, \quad t \to \infty, \tag{14.5.30}$$

where

$$a_k = \exp\left[\frac{i\pi k}{4} sgnh''(x_0)\right] \frac{\Gamma\left(k+\frac{1}{2}\right)}{(2k)!}$$

$$\times \left[\left(h(x,x_0)\frac{d}{dx}\right)^{2k}(f(x)h(x,x_0))\right]\Bigg|_{x=x_0+0} \tag{14.5.31}$$

and

$$h(x,x_0) = \frac{2[(h(x)-h(x_0))sgnh''(x_0)]^{\frac{1}{2}}}{h'(x)sgnh''(x_0)}. \tag{14.5.32}$$

The main term in (14.5.30) is:

$$G(t) = \left(\frac{2\pi}{|h''(x_0)|}\right)^{1/2} \exp\left[ith(x_0) + \frac{i\pi}{4} sgnh''(x_0)\right]$$

$$\times t^{-\frac{1}{2}}\left[f(x_0) + O\left(\frac{1}{t}\right)\right], \quad t \to \infty. \tag{14.5.33}$$

Expansion (14.5.30) can be differentiated any number of times.

Consider now the multidimensional case:

$$G(t) = \int_U f(x) \exp[ith(x)]dx, \quad U \subset \mathbb{R}^n, \tag{14.5.34}$$

where $f \in C_0^\infty(U)$, $h \in C^\infty(U)$, $h(x)$ is real-valued and has a single non-degenerate stationary point x_0 in U:

$$\nabla h(x_0) = 0, \quad \det h''(x_0) \neq 0, \quad h'' := (h''_{jm}) = \left(\frac{\partial^2 h}{\partial x_j \partial x_m}\right). \tag{14.5.35}$$

Theorem 14.5.3. *Under the above assumptions one has*

$$G(t) \sim t^{-\frac{n}{2}} \exp[ith(x_0)] \sum_{k=0}^{\infty} a_k t^{-k}, \quad t \to \infty, \tag{14.5.36}$$

where $a_k = $const, and one can differentiate formula (14.5.36) any number of times. The main term in (14.5.36) is:

$$G(t) = t^{-\frac{n}{2}} \exp[ith(x_0)]$$
$$\times \frac{(2\pi)^{\frac{n}{2}} \exp\left(\frac{i\pi}{4} sgn h''(x_0)\right)}{|\det h''(x_0)|^{\frac{1}{2}}} \left[f(x_0) + O\left(\frac{1}{t}\right)\right], \quad t \to \infty. \tag{14.5.37}$$

Here sgn h'' is the difference between the number of positive and negative eigenvalues of the Hessian matrix $h''(x_0)$.

14.5.4. The Morse lemma

Let $U, V \subset \mathbb{R}^n$ be open sets. Consider a map $\psi : U \to \mathbb{R}^n$ and suppose $\psi(U) = V$. The map ψ is called a *diffeomorphism* if

(1) ψ is differentiable,
(2) there exists an inverse map $\psi^{-1} : V \to \mathbb{R}^n$ such that $\psi \circ \psi^{-1}$ and $\psi^{-1} \circ \psi$ are identical maps on V and U, respectively, and
(3) ψ^{-1} is differentiable.

The following lemma is very useful in the derivation of asymptotic expansions of multidimensional integrals. It is also used in Chapter 4 for the analysis of asymptotic behavior of the Radon transform $\hat{f}$ near its singular support singsupp $\hat{f}$.

Lemma 14.5.6 (Morse). *Let x_0 be a non-degenerate critical point (see (14.5.35)) of a real-valued C^∞ function $h(x)$. Then there exist neighborhoods U and V of the points $y = 0$ and x_0, respectively, and a C^∞ diffeomorphism $\psi : U \to V$ such that*

$$h(\psi(y)) = h(x_0) + \frac{1}{2}\sum_{j=1}^n \mu_j y_j^2, \tag{14.5.38}$$

where μ_j are eigenvalues of the Hessian matrix $h''(x_0)$ (see (14.5.35)).
Furthermore, the Jacobian of the transformation satisfies

$$\det\left(\partial\psi_j(y)/\partial y_m)|_{y=0}\right) = 1. \qquad (14.5.39)$$

Occasionally, one uses a diffeomorphism $\tilde{\psi} : \tilde{U} \to \mathbb{R}^n$ such that

$$h(\tilde{\psi}(z)) = h(x_0) + z_1^2 + \cdots + z_r^2 - z_{r+1}^2 - \cdots - z_n^2, \qquad (14.5.40)$$

where r is the number of positive eigenvalues of $h''(x_0)$, and does not require condition (14.5.39).

14.6. Linear equations in Banach spaces

14.6.1. Closedness. Normal solvability

Let $A : X \to Y$ be a linear map from X into Y, where X and Y are Banach spaces. Consider the equation

$$Ax = y. \qquad (14.6.1)$$

Let $D(A)$ and $R(A)$ be the domain and range of A, respectively, $N(A) = \{x : Ax = 0\}$ be the null space of A, and $A^* : Y^* \to X^*$ be the adjoint operator. If X is a Hilbert space and $A = A^*$, then A is called *selfadjoint*. If $AA^* = A^*A$, then A is called *normal*. Equation (14.6.1) is uniquely solvable on $R(A)$ if and only if $N(A) = \{0\}$. In this case A is called *injective*, the inverse operator A^{-1} is well defined on $R(A)$, and $x = A^{-1}y$ is the unique solution to (14.6.1) for $y \in R(A)$. Equation (14.6.1) is called *normally solvable* if and only if $R(A)$ is closed, and is called *correctly solvable* on $R(A)$ if and only if

$$||x|| \le k||Ax||, \qquad (14.6.2)$$

where $k > 0$ is a constant independent of $x \in D(A)$. If (14.6.2) holds, then $N(A) = \{0\}, R(A)$ is closed and $||A^{-1}|| \le k, D(A^{-1}) = R(A)$. Recall that

$$||A|| := \sup_{||x||=1} ||Ax||, \qquad (14.6.3)$$

and A is called *bounded* if and only if $||A|| < \infty$.

Equation (14.6.1) is called *densely solvable* if $\overline{R(A)} = Y$. The overbar denotes closure. If $R(A) = Y$, then Equation (14.6.1) is called *everywhere solvable*, and A is called *surjective*.

A linear operator is called *closed* if $\{x_n \to x$ and $Ax_n \to y\}$ implies $\{x \in D(A)$ and $Ax = y\}$.

Theorem 14.6.1 (Banach). *If a linear closed operator A is defined on all of X, then A is bounded.*

One has

$$N(A^*) = R(A)^\perp, \qquad (14.6.4)$$

where the sign $\perp$ stands for the orthogonal complement. A set $M^\perp, M \subset X$, consists of all elements $x^* \in X^*$ such that $(x^*, x) = 0$ for any $x \in M$. If $N \subset Y$, then $N^\perp$ consists of all elements $y^* \in Y^*$ such that $(y^*, y) = 0$ for all $y \in N$. By (x^*, x) the value of the linear functional x^* at the element x is denoted.

If $\|A\| \leq \infty$, then $N(A) =^\perp R(A^*)$. The symbol $^\perp M^*$, where $M^* \subset X^*$, denotes the set of all $x \in X$ such that $(m^*, x) = 0$ for any $m^* \in M^*$.

If A is not bounded, then $R(A^*) \subset N(A)^\perp$, but, in general, $R(A^*) \neq N(A)^\perp$.

14.6.2. Conditions for surjectivity

Theorem 14.6.2. *Equation (14.6.1) with a closed operator A is everywhere solvable if and only if the equation*

$$A^*g = f \qquad (14.6.5)$$

is correctly solvable on $R(A^*)$.

Theorem 14.6.3. *Equation (14.6.5) is everywhere solvable if and only if Equation (14.6.1) is correctly solvable on $R(A)$.*

Proof of Theorem 14.6.3. Let us prove the necessity. Assume that Equation (14.6.5) is everywhere solvable. Take $x \in D(A)$. Then

$$|(f, x)| = |(A^*g, x)| = |(g, Ax)| \leq \|g\| \cdot \|Ax\|. \qquad (14.6.6)$$

Let

$$h_x(f) := \frac{(f, x)}{\|Ax\|}, \quad |h_x(f)| \leq \|g\|. \qquad (14.6.7)$$

The functional $h_x(f)$ for any fixed $x \in D(A)$ is a linear functional of f, uniformly bounded for all $f \in X^*$. Therefore there exists $k = \text{const} > 0$ such that

$$|h_x(f)| \leq k\|f\|, \quad \forall f \in X^*. \qquad (14.6.8)$$

Thus

$$|(f, x)| \leq k\|f\| \|Ax\|. \qquad (14.6.9)$$

Therefore,

$$\|x\| \leq k\|Ax\|. \qquad (14.6.10)$$

The necessity is proved.

Let us prove the sufficiency. Assume (14.6.10). Fix an arbitrary $f \in X^*$ and define a functional on $R(A)$ by the formula

$$\varphi(y) = f(x), \qquad y = Ax. \tag{14.6.11}$$

Since (14.6.10) implies $N(A) = \{0\}$, the functional $\varphi(y)$ is uniquely defined by formula (14.6.11) and is a linear bounded functional on $R(A)$:

$$|\varphi(y)| = |f(x)| \leq ||f|| \cdot ||x|| \leq k||f|| \cdot ||Ax|| = k||f|| \cdot ||y||. \tag{14.6.12}$$

Therefore, $\varphi(y)$ is uniquely extendable to $\overline{R(A)}$ by continuity, and then from $\overline{R(A)}$ to all of Y (non-uniquely but with the same norm) by the Hahn-Banach Theorem. Let us denote this extension by g. Then

$$g(Ax) = \varphi(Ax) = \varphi(y) = f(x) \quad \forall x \in D(A). \tag{14.6.13}$$

Thus, the sufficiency is proved. Theorem 14.6.3 is proved. □

Proof of Theorem 14.6.2. Necessity. Let $R(A) = Y$. One has

$$|g(y)| = |(g, Ax)| = |(A^* g, x)| \leq ||A^* g|| \cdot ||x||. \tag{14.6.14}$$

Therefore, for any fixed $g \in D(A^*)$, one has $h_g(y) := \frac{|g(y)|}{||A^* g||} \leq ||x||$. By the uniform boundedness principle, there exists a constant $k > 0$, such that $|(g, y)| \leq k||A^* g|| \cdot ||y||$. Thus

$$||g|| \leq k||A^* g||. \tag{14.6.15}$$

Therefore, Equation (14.6.5) is correctly solvable. In this argument the closedness of A was not used. The necessity is proved.

Sufficiency. Assume (14.6.15). Then $N(A^*) = \{0\}$ and, therefore, $\overline{R(A)} = Y$. Since A is closed and (14.6.5) holds, one concludes that $R(A)$ is closed. Thus $R(A) = Y$, and the sufficiency is proved. Theorem 14.6.2 is proved. □

Corollary 14.6.1. *If A is closed, then estimate (14.6.15) is sufficient for the surjectivity of A.*

14.6.3. Compact operators

A linear operator $A : X \to Y$ is called *compact* if it maps bounded sets into precompact sets, that is, the sets whose closures are compact. A set K in a Banach space X is called compact if any sequence $x_n \in K$ contains a subsequence converging to an element of K. A set N_ϵ is called an ϵ-net for a set B if and only if for any $b \in B$ there is an $x \in N_\epsilon$ such that $||x - b|| < \epsilon$.

A set $K \subset X$ is precompact if and only if it has a finite ϵ-net for any $\epsilon > 0$. In a finite-dimensional Banach space X a set is precompact if and only if it is bounded. If $\dim X = \infty$, then the unit ball in X is not precompact.

The set of all linear compact operators $A : X \to Y$ is denoted $\mathcal{K}(X, Y)$.

If A is compact, so is A^*, its adjoint, $A^* : Y^* \to X^*$. If A is injective, X is infinite-dimensional Banach space, and $A \in \mathcal{K}(X, X)$, then A^{-1} is unbounded. The operator AB is compact if one of the operators is compact and the other is bounded. If $X = H$, where H is a Hilbert space, and A is a linear compact operator from H into a Hilbert space $Y = H_1$, then the operator $|A| := (A^*A)^{\frac{1}{2}}$ is compact, selfadjoint, and non-negative definite in the sense

$$(|A|x, x) \geq 0 \qquad \forall x \in H. \qquad (14.6.16)$$

Let λ_j be its eigenvalues:

$$\lambda_1 \geq \lambda_2 \cdots \geq 0. \qquad (14.6.17)$$

The *singular values* (or, simply, *s*-values) of A are defined to be

$$s_j = s_j(A) := \lambda_j(|A|). \qquad (14.6.18)$$

One has

$$|A| = \sum_{j=1}^{\infty} s_j(\cdot, \varphi_j)\varphi_j, \qquad (14.6.19)$$

where φ_j are the normalized eigenvectors of $|A|$:

$$|A|\varphi_j = \lambda_j \varphi_j. \qquad (14.6.20)$$

Any bounded linear operator A admits polar representation

$$A = U|A|, \qquad (14.6.21)$$

where U is an isometry from $R(A^*)$ onto $R(A)$. Applying U to (14.6.19) one gets

$$A = \sum_j s_j(\cdot, \varphi_j)\psi_j, \quad \psi_j := U\varphi_j \qquad (14.6.22)$$

Formula (14.6.22) is called a *singular value decomposition* of A.

14.6.4. Resolution of the identity

Definition 14.6.1. A resolution of the identity is a one-parameter family of projection operators E_t, where t runs through a (possibly infinite) interval $[a, b]$, which satisfy the following conditions:

(1) $E_a = 0$, $E_b = I$, where I is the identity operator;
(2) $\lim_{s \to t, s < t} E_s = E_t$, $a < t < b$; and
(3) $E_u E_v = E_s$, where $s = \min(u, v)$.

If the interval $[a, b]$ is infinite, we define $E_{-\infty} = \lim_{t \to -\infty} E_t$ and $E_{+\infty} = \lim_{t \to +\infty} E_t$.

Let H be a Hilbert space. Using the definition, it is easy to check that for an arbitrary fixed $f \in H$ the function $\sigma(t) := (E_t f, f)$ is non-decreasing, has bounded variation, is continuous from the left, and

$$\sigma(a) = 0, \quad \sigma(b) = (f, f).$$

Let $\varphi(t), t \in [a, b]$, be a piecewise continuous function. The expression $A = \int_a^b \varphi(t) dE_t$ means that for any

$$f \in D(\varphi(A)) := \{f : \int_a^b |\varphi^2(t)| d(E_t f, f) < \infty\}$$

and $g \in H$, we have

$$(\varphi(A)f, g) = \int_a^b \varphi(t) d(E_t f, g). \tag{14.6.23}$$

Similarly, given $f \in D(\varphi(A))$, the expression $\varphi(A)f = \int_a^b \varphi(t) dE_t f$ is understood in the sense that (14.6.23) holds for any $g \in H$.

Theorem 14.6.4. *To each resolution of the identity $E_t, -\infty \leq t \leq \infty$, there corresponds a uniquely defined self-adjoint operator A such that $A = \int_{-\infty}^{\infty} t dE_t$. The domain $D(A)$ of A consists of all $f \in H$ such that*

$$\|Af\|_H^2 = \int_{-\infty}^{\infty} t^2 d(E_t f, f) < \infty. \tag{14.6.24}$$

Conversely, for any self-adjoint operator A, there exists a resolution of the identity $E_t, -\infty \leq t \leq \infty$, belonging to A in the sense that $D(A)$ consists of all $f \in H$ for which (14.6.24) holds and, for any $f \in D(A)$, one has

$$Af = \int_{-\infty}^{\infty} t dE_t f.$$

Given the resolution of the identity of a self-adjoint operator A, one can represent the resolvent operator $R_z = (A - zI)^{-1}, z \in \mathbb{C}$, as follows:

$$R_z = \int_{-\infty}^{\infty} \frac{1}{t - z} dE_t. \tag{14.6.25}$$

14.7. Ill-posed problems

14.7.1. Basic definitions

Let $A : X \to Y$ be an operator from a Banach space X into a Banach space Y. Consider the equation

$$Af = g. \tag{14.7.1}$$

Suppose that A is a bijection, that is, A is surjective: the range of A coincides with Y, and A is injective: if $Af_1 = g$ and $Af_2 = g$, then $f_1 = f_2$. Assume also that A^{-1} is continuous. If all of these assumptions hold, then problem (14.7.1) is called *well-posed*. If some of these assumptions are not satisfied, then problem (14.7.1) is called *ill-posed*. The above notion of well-posedness is due to J. Hadamard.

The problem can be ill-posed, for example, if A is not injective. In this case one can consider (for linear operators A) the operator $\tilde{A}$ which acts from the factor space $X/N(A)$ into Y.

The operator $\tilde{A}$ is injective. If the null-space $N(A)$ of A is known, then the solution to the equation $\tilde{A}f = g$ gives a complete description of the set of solutions to Equation (14.7.1).

The other reason for Equation (14.7.1) to be ill-posed is the lack of surjectivity of A: $\overline{R(A)} \neq Y$. Suppose that A is linear normally solvable injective operator, that is $R(A) = \overline{R(A)}$ and $R(A) \neq Y$. Then $Y_1 := R(A)$ is a Banach space and the operator $A : X \to Y_1$ is surjective. Therefore, Equation (14.7.1) can be easily studied: if $g \notin R(A)$, this equation is not solvable, and if $g \in R(A)$ – it is solvable and the solution is unique. The inverse operator A^{-1} is linear, is defined on all of the space Y_1 and, by the Banach theorem, is bounded. This means that the solution to (14.7.1) under the above assumptions is a continuous function of $g \in Y_1$. The third possible reason for problem (14.7.1) to be ill-posed is the lack of normal solvability: $R(A) \neq \overline{R(A)}$. If A is closed and injective and A^{-1} is unbounded, then $R(A)$ cannot be closed because in this case A^{-1} has to be bounded by the Banach theorem. Therefore, Equation (14.7.1) is more difficult to study when A^{-1} is an unbounded operator. This case however is very common and important in applications.

14.7.2. Examples of ill-posed problems

14.7.2.1. Differentiation. Let $g \in C^1(\mathbb{R})$ be given. The problem is to find g'. This problem is equivalent to solving the equation

$$Af = \int\limits_a^x f(t)dt = g(x), \ A : C(\mathbb{R}) \to C^1(\mathbb{R}). \qquad (14.7.2)$$

The operator $A^{-1}g = \frac{dg}{dx} : C^1(\mathbb{R}) \to C(\mathbb{R})$ is defined on all $C^1(\mathbb{R})$ and is bounded. However, in practice, one is given not the function g, but its approximation g_δ such that

$$||g_\delta - g||_{L^\infty} < \delta, \qquad (14.7.3)$$

where $\delta > 0$ is some small number. Therefore, we have to consider A as a mapping $A : L^\infty(\mathbb{R}) \to L^\infty(\mathbb{R})$ and, in this case, finding $A^{-1}g$, given the data g_δ, is an ill-posed problem. First, the function g_δ may not be in the domain of A^{-1}, that is, g_δ may not be differentiable. Even if g_δ is differentiable, its derivative g_δ' may differ as much as one wishes from g' in the $L^\infty(\mathbb{R})$ norm. As an example we may consider the function

$$g_\delta(x) = g(x) + \delta \sin(Nx).$$

Although g_δ satisfies (14.7.3) for any real N, we have

$$||g_\delta' - g'||_{L^\infty} = O(N) \to \infty \ \text{ as } \ N \to \infty.$$

14.7.2.2. Inversion of the Radon transform. Consider the equation

$$Rf = g, \qquad (14.7.4)$$

which has to be solved for f. Results from Section 3.2 imply that the operator $R : L^2(B_a) \to H^{1/2}(Z_a)$ is correctly solvable. In practice, one solves Equation (14.7.4) with a perturbed right-hand side g_δ, such that

$$||g - g_\delta||_{L^2(B_a)} \leq \delta. \qquad (14.7.5)$$

Therefore, we have to consider R as a mapping $R : L^2(B_a) \to L^2(Z_a)$, which makes the problem of finding $f \in L^2(B_a)$, given $g_\delta \in L^2(Z_a)$, ill-posed. First, the operator $R^{-1} : L^2(Z_a) \to L^2(B_a)$ is unbounded and, second, the function g_δ may fail to satisfy the moment conditions.

The number of examples of ill-posed problems can be easily increased.

14.7.3. Methods for stable solution of ill-posed problems

1. Let us assume that $A : X \to Y$ is a linear injective operator, $R(A) \neq \overline{R(A)}$, $g \in R(A)$ is not known, and the data are the elements g_δ such that

$$\|g - g_\delta\|_Y < \delta. \tag{14.7.6}$$

The problem is to find f_δ such that

$$\|f_\delta - f\|_X \leq \nu(\delta) \to 0 \quad \text{as} \quad \delta \to 0. \tag{14.7.7}$$

Such a sequence f_δ is called a *stable solution* to Equation (14.7.1) with the perturbed data.

If $R_{\gamma,\delta} : Y \to X$ is a mapping, which depends on two parameters γ and δ, and if for some choice of γ, $\gamma = \gamma(\delta)$, the operator $R_\delta := R_{\gamma(\delta),\delta}$ has the property

$$\|R_\delta g_\delta - A^{-1}g\|_X \to 0 \quad \text{as} \quad \delta \to 0, \tag{14.7.8}$$

where $A^{-1}g = f$, then $R_{\gamma,\delta}$ is called a *regularizing family* for the problem (14.7.1). If such a family is known, then the function

$$f_\delta := R_\delta g_\delta \tag{14.7.9}$$

satisfies (14.7.7) in view of (14.7.8).

2. There are various methods for constructing a regularizing family. One of them is called the *regularization method*. This method consists of solving the variational problem

$$F(u) := \|Au - g_\delta\|_Y^2 + \gamma h(u) \to \min. \tag{14.7.10}$$

Here $\gamma > 0$ is a scalar parameter, and $h(u)$ is a strictly convex weakly lower semicontinuous functional, such that the set

$$K := \{u : h(u) \leq c\} \tag{14.7.11}$$

is precompact in X, and $f \in D(h)$. The functional h is said to be strictly convex if

$$h\left(\frac{u_1 + u_2}{2}\right) < \frac{h(u_1) + h(u_2)}{2} \quad \text{for any} \quad u_1, u_2 \in D(h)$$

provided $\quad u_1 \neq \lambda u_2, \quad \lambda = \text{const.} \tag{14.7.12}$

The functional h is called weakly lower semicontinuous if

$$u_n \rightharpoonup u \Longrightarrow \liminf_{n \to \infty} F(u_n) \geq F(u). \tag{14.7.13}$$

Here $\rightharpoonup$ denotes weak convergence in X.

The following theorem gives a method for constructing a regularizing family for problem (14.7.1).

Theorem 14.7.1. *Assume that A is a linear bounded injective opera-tor defined on a reflexive Banach space X, and $h(u)$ is a strictly convex weakly lower semicontinuous functional such that the set (14.7.11) is precompact in X. Then problem (14.7.10) has a unique solution $u_{\delta,\gamma}$ for any $\gamma > 0$. Moreover, if $\gamma = \gamma(\delta)$ is chosen such that*

$$\gamma(\delta) \to 0, \quad \frac{\delta^2}{\gamma(\delta)} \le m < \infty \quad as \quad \delta \to 0, \quad m = const > 0, \quad (14.7.14)$$

then $f_\delta := u_{\gamma(\delta),\delta}$ satisfies (14.7.7).

3. Another method for constructing a regularization family for prob-lem (14.7.1) is a variational method with constraints, a *method of qua-sisolutions*. This method consists of finding the solution to the problem:

$$||Au - g_\delta||_Y \to \min, \quad u \in K, \quad (14.7.15)$$

where $K \subset X$ is a compact set which contains the exact solution f to (14.7.1).

In fact, the method of quasisolutions does not differ essentially from the regularization method. Indeed, the compact K in most cases can be described by formula (14.7.11), and the Lagrange multipliers method reduces optimization with constraints problem (14.7.15) to the uncondi-tional optimization problem (14.7.10), the parameter in (14.7.10) being the Lagrange multiplier.

4. There are iterative methods for constructing a regularization fam-ily. The role of the regularization parameter γ is played in these methods by the stopping rule.

5. Let $A : H \to H$ be a closed injective bounded operator which has a spectral representation, for example, a selfadjoint or normal operator. Assume that $g \in R(A)$, A^{-1} is unbounded, and the solution to (14.7.1) is given by the formula

$$f = \int_\sigma t^{-1} dE_t g, \quad (14.7.16)$$

where σ is the spectrum of A, and E_t is the resolution of the identity for A. For instance, if A is selfadjoint, then σ is a subset of the segment $[-||A||, ||A||]$, $0 \in \sigma$ since A^{-1} is unbounded. Consider the function

$$f_{\gamma\delta} := \int_\sigma t^{-1} \exp\left(-\frac{\gamma}{|t|}\right) dE_t g_\delta. \quad (14.7.17)$$

Let us assume for simplicity that A is selfadjoint and, without loss of generality, that $||A|| = 1$. Then $g \in R(A)$ implies:

$$\int_{-1}^{1} t^{-2} d(E_t g, g) < \infty. \quad (14.7.18)$$

Fix any function $\gamma(\delta)$ such that

$$\gamma = \gamma(\delta) \to 0, \quad \frac{\delta^2}{\gamma^2(\delta)} \to 0 \quad \text{as} \quad \delta \to 0. \tag{14.7.19}$$

Define

$$f_\delta = \int_{-1}^{1} t^{-1} \exp\left(-\frac{\gamma(\delta)}{|t|}\right) dE_t g_\delta \tag{14.7.20}$$

and assume that

$$\int_{-1}^{1} |t|^{-2-b} d(E_t g, g) \leq M \quad \text{for some} \ \ b > 0. \tag{14.7.21}$$

Assumption (14.7.21) means that $g \in R\left(A^{1+\frac{b}{2}}\right)$.

Theorem 14.7.2. *If (14.7.18)-(14.7.20) hold, then*

$$\|f_\delta - f\| \to 0 \quad \text{as} \quad \delta \to 0. \tag{14.7.22}$$

If (14.7.21) holds and $\gamma(\delta) = c_0 \delta^{\frac{2}{s+2}}$, *where* $s := \min(b, 2)$, *then*

$$\|f_\delta - f\| \leq c_1 \delta^{\frac{s}{s+2}}. \tag{14.7.23}$$

Here

$$c_0 = \left(\frac{0.56}{sH}\right)^{\frac{1}{s+2}}, \tag{14.7.24}$$

$H \geq 1$ *is some constant, and*

$$c_1 := (0.28 c_0^{-2} + H c_0^s)^{\frac{1}{2}} \tag{14.7.25}$$

Proof. One has

$$||f_{\gamma,\delta} - f||^2 = || \int_{-1}^{1} t^{-1} \exp\left(-\frac{\gamma}{|t|}\right) dE_t g_\delta - \int_{-1}^{1} t^{-1} dE_t g ||^2$$

$$\leq 2|| \int_{-1}^{1} t^{-1} \exp\left(-\frac{\gamma}{|t|}\right) dE_t (g_\delta - g) ||^2$$

$$+ 2|| \int_{-1}^{1} t^{-1} \left[1 - \exp\left(-\frac{\gamma}{|t|}\right)\right] dE_t g ||^2$$

$$= 2 \int_{-1}^{1} t^{-2} \exp\left(-\frac{2\gamma}{|t|}\right) dE_t ((g_\delta - g), g_\delta - g)$$

$$+ 2 \int_{-1}^{1} t^{-2} \left[1 - \exp\left(-\frac{\gamma}{|t|}\right)\right]^2 d(E_t g, g)$$

$$:= \mathcal{J}_1 + \mathcal{J}_2 := \mathcal{J}. \tag{14.7.26}$$

Note that

$$\sup_{0<|t|\leq 1} 2t^{-2} \exp\left(-\frac{2\gamma}{|t|}\right) \leq 2e^{-2}\gamma^{-2} \leq 0.28\gamma^{-2}. \tag{14.7.27}$$

Therefore,

$$\mathcal{J}_1 \leq 0.28\delta^2\gamma^{-2}. \tag{14.7.28}$$

Equation (14.7.22) follows from (14.7.28), (14.7.18), (14.7.19) and (14.7.26).

If (14.7.21) holds, then

$$\mathcal{J}_2 \leq 2 \int_{-1}^{1} |t|^b \left[1 - \exp\left(-\frac{\gamma}{|t|}\right)\right]^2 \frac{d(E_t g, g)}{|t|^{2+b}}$$

$$\leq 2M \sup_{|t|\leq 1} \left\{ |t|^b \left[1 - \exp\left(-\frac{\gamma}{|t|}\right)\right]^2 \right\}. \tag{14.7.29}$$

Note that

$$\sup_{0<t\leq 1} t^b \left[1 - \exp\left(-\frac{\gamma}{t}\right)\right]^2 = \gamma^b \sup_{u\geq\gamma} \frac{[1 - \exp(-u)]^2}{u^b}$$

$$= \gamma^b \left(\sup_{u\geq\gamma} \frac{1 - \exp(-u)}{u^{\frac{b}{2}}} \right)^2 := \mathcal{J}_3. \tag{14.7.30}$$

If $b \geq 2$, then the function $h(u) := \frac{1-\exp(-u)}{u^{\frac{b}{2}}}$ is decreasing on $[\gamma, \infty)$. Therefore,

$$\sup_{u \geq \gamma} \frac{1 - \exp(-u)}{u^{\frac{b}{2}}} \leq \frac{1 - \exp(-\gamma)}{\gamma^{\frac{b}{2}}}, \quad b \geq 2, \qquad (14.7.31)$$

and

$$\mathcal{J}_3 \leq [1 - \exp(-\gamma)]^2 \leq \gamma^2, \quad b \geq 2. \qquad (14.7.32)$$

If $0 < b < 2$, the function $h(u)$ attains its maximum at a point u_m. Denote this maximum h_m. Then

$$\mathcal{J}_3 = h_m \gamma^b, \quad b < 2. \qquad (14.7.33)$$

From (14.7.32), (14.7.33), (14.7.26), and (14.7.28) one obtains

$$\mathcal{J} \leq 0.28 \delta^2 \gamma^{-2} + H \gamma^s, \quad s = \min(2, b), \qquad (14.7.34)$$

$$H = \max(1, h_m). \qquad (14.7.35)$$

Minimizing the right-hand side of (14.7.34) with respect to γ, with $\delta > 0$ being fixed, one finds that the minimum is obtained at

$$\gamma = \left(\frac{0.56}{sH}\right)^{\frac{1}{s+2}} \delta^{\frac{2}{2+s}} := c_0 \delta^{\frac{2}{2+s}}, \qquad (14.7.36)$$

and the value of this minimum is

$$0.28 c_0^{-2} \delta^{\frac{2s}{2+s}} + H c_0^s \delta^{\frac{2s}{2+s}} := c_1^2 \delta^{\frac{2s}{s+2}}. \qquad (14.7.37)$$

Theorem 14.7.2 is proved. $\square$

REMARK 14.7.1. There can be other methods for constructing a regularizing family for Equation (14.7.1), based on the idea presented in this section. The advantage of the result formulated in Theorem 14.7.1 consists of the fully explicit expressions for the error estimate (14.7.23) and for the constants in this estimate.

14.7.4. Asymptotics of singular values of the Radon transform

Let us consider the Radon transform operator R acting on $X = L^2(B_a)$. Formula (2.1.27) shows that $R^*R := |R|^2$ is a selfadjoint operator in $L^2(B_a)$ with the kernel $|S^{n-2}| |x - y|^{-1}$. What is the asymptotics of the eigenvalues of this operator? The answer to this question can be obtained from the following result.

Let $D \subset \mathbb{R}^n$ be a bounded domain, and $A(x) = |x|^{-m}$, $0 < m < n$. Define

$$\gamma_\pm(A) := (2\pi)^{-n} \operatorname{meas} \{\xi : \pm\tilde{A}(\xi) > 1\}, \quad \xi \in \mathbb{R}^n, \qquad (14.7.38)$$

where

$$\tilde{A}(\xi) := \lim_{r \to \infty} \int_{|x| \le r} \left(1 - \frac{|x|^2}{r^2}\right)^s A(x) \exp(-ix \cdot \xi) dx, \qquad (14.7.39)$$

and the exponent $s > 0$ is sufficiently large. Note that for $A(x) = |x|^{-m}$ one can calculate $\gamma_\pm(A)$ analytically using the formula

$$\tilde{A}(\xi) = 2^{n-m} \pi^{\frac{n}{2}} \frac{\Gamma(\frac{n-m}{2})}{\Gamma(m/2)} |\xi|^{m-n},$$

which follows from (2.1.33). Define

$$q := \frac{n}{n-m}, \quad 0 \le m < n, \qquad (14.7.40)$$

and let

$$Au := \int_D A(x - y)u(y)dy. \qquad (14.7.41)$$

Let λ_j^+ and $-\lambda_j^-$ denote positive and negative eigenvalues of A arranged so that $\lambda_1^\pm \ge \lambda_2^\pm \ge \ldots$, taking multiplicity into account.

Lemma 14.7.1. *Under the above assumptions one has*

$$\lim_{\lambda \to 0} \lambda^q N_\pm(\lambda, A) = \gamma_\pm |D|, \qquad (14.7.42)$$

where $|D| = \operatorname{meas} D$, and

$$N_\pm(\lambda, A) := \sum_{\lambda_j^\pm(A) > \lambda} 1, \quad \lambda > 0. \qquad (14.7.43)$$

If $0 < m < n$, then $\tilde{A}(\xi)$, defined by formula (14.7.39), coincides with the usual Fourier transform. In particular, if $m = 1$, then

$$|S^{n-2}| \mathcal{F}\left(\frac{1}{|x|}\right) = \frac{2\pi^{\frac{n-1}{2}}}{\Gamma\left(\frac{n-1}{2}\right)} 2^{n-1} \pi^{\frac{n}{2}} \frac{\Gamma\left(\frac{n-1}{2}\right)}{\Gamma\left(\frac{1}{2}\right)} |\xi|^{1-n} = 2^n \pi^{n-1} |\xi|^{1-n}. \qquad (14.7.44)$$

Thus $\gamma_- = 0$ and

$$\gamma_n := \gamma_+ = (2\pi)^{-n} \text{ meas } \{\xi : 2^n \pi^{n-1} |\xi|^{1-n} > 1\} = 2^{\frac{n}{n-1}} |B_1|, \quad (14.7.45)$$

and Equation (14.7.42) with $m = 1$ and $q = \frac{n}{n-1}$ yields:

$$N_+(\lambda, R^* R) \sim \frac{\gamma_n |D|}{\lambda^{\frac{n}{n-1}}}, \quad \lambda \to 0. \quad (14.7.46)$$

Note that (14.7.42) is equivalent to

$$\lim_{j \to \infty} j^{\frac{1}{q}} \lambda_j^{\pm}(A) = (\gamma_\pm |D|)^{\frac{1}{q}}, \quad (14.7.47)$$

that is

$$\lambda_j^{\pm}(A) \sim \frac{(\gamma_\pm |D|)^{\frac{1}{q}}}{j^{\frac{1}{q}}}, \quad j \to \infty. \quad (14.7.48)$$

Therefore, formula (14.7.46) is equivalent to

$$\lambda_j(R^* R) \sim \frac{(\gamma_n |D|)^{\frac{n-1}{n}}}{j^{\frac{n-1}{n}}} = \frac{2|B_1|^{\frac{2(n-1)}{n}} a^{n-1}}{j^{\frac{n-1}{n}}}, \quad (14.7.49)$$

when D is a ball of radius a. In particular if $n = 2$, then $|B_1| = \pi$ and

$$\lambda_j(R^* R) \sim \frac{2\pi a}{j^{\frac{1}{2}}}. \quad (14.7.50)$$

If $n = 3$, then $|B_1| = 4\pi/3$ and

$$\lambda_j(R^* R) \sim \frac{2(4\pi/3)^{\frac{4}{3}} a^2}{j^{\frac{2}{3}}}. \quad (14.7.51)$$

Note that the rate of decay in formulas (14.7.50) and (2.3.17) is the same. The leading coefficients, however, are different, because in Section 2.3 the Radon transform was considered as a mapping $R : L^2(B_1) \to L^2(Z_1, \omega^{1-n})$. Therefore, the adjoint operator to R in Section 2.3 was different from that in this section.

One can measure the degree of ill-posedness of the problem of solving the equation

$$Af = g \quad (14.7.52)$$

by the rate of decay of the singular values of the compact operator A. Estimates (14.7.50) and (14.7.51) show that inversion of the Radon transform is not a very ill-posed problem.

14.8. Examples of regularization of ill-posed problems

14.8.1. Stable differentiation

14.8.1.1. Statement of the problem and the result. Suppose

$$f \in C^2(\mathbb{R}), \quad \sup_{x \in \mathbb{R}} |\partial^2 f(x)| < M,$$

$$f_\delta(x) \in L^\infty(\mathbb{R}), \quad \sup_{x \in \mathbb{R}} |f_\delta(x) - f(x)| < \delta, \qquad (14.8.1)$$

where δ is a given small number. If $g \in L^\infty(\mathbb{R})$, then $\sup_{x \in \mathbb{R}} |g|$ is understood as the *essential supremum*, that is

$$\sup_{x \in \mathbb{R}} |g| := \inf\{a > 0 : |g(x)| < a \text{ almost everywhere}\}.$$

The problem is: given $\{\delta, M, f_\delta(x)\}$, find $F_\delta(x)$ such that

$$\sup_{x \in \mathbb{R}} |f'(x) - F_\delta(x)| \le \eta(\delta) \to 0 \text{ as } \delta \to 0. \qquad (14.8.2)$$

In other words, find a stable approximation to $f'(x)$.

One can prove the following result, which contains a solution to the above problem.

Theorem 14.8.1. *Let*

$$F_\delta(x) := \frac{f_\delta(x+h) - f_\delta(x-h)}{2h}, \quad h = \sqrt{\frac{2\delta}{M}}. \qquad (14.8.3)$$

Then (14.8.2) holds with $\eta(\delta) = \sqrt{2M\delta}$, that is

$$\sup_{x \in \mathbb{R}} |f'(x) - F_\delta(x)| \le \sqrt{2M\delta}. \qquad (14.8.4)$$

Estimate (14.8.3) is the best possible estimate of $f'(x)$ for the class of functions $f(x)$ and the data $\{\delta, f_\delta, M\}$ defined by (14.8.1). This means that if $G_\delta(x; f_\delta)$ is any other estimate, then there exists an $f(x) \in C^2(\mathbb{R})$, $\sup_{x \in \mathbb{R}} |f''| \le M$, and a point $x_0 \in \mathbb{R}$, such that

$$|f'(x_0) - G_\delta(x_0, f_\delta)| \ge \sqrt{2M\delta}. \qquad (14.8.5)$$

One can improve estimates (14.8.3) and (14.8.4) if more a priori information is known about $f(x)$. For instance, if $f \in C^m(\mathbb{R})$ and

$$\sup_{x \in \mathbb{R}} |f^{(m)}(x)| \le M_m,$$

then $f'(x)$ can be estimated with an error of higher order of smallness than in (14.8.4). See [R3, p.85] for the proof of Theorem 14.8.1 and related details.

14.8.1.2. Random perturbation of the data. Assume that the data are the values

$$u_i = s_i + n_i, \tag{14.8.6}$$

where

$$s_i = s(ih), \quad i \in \mathbb{Z},$$

n_i are errors which are assumed to be independent and identically distributed with zero mean value and variance σ^2. Assume that $|s''(t)| \le M$, $-\infty < t < \infty$. The problem is to estimate $s'(t)$. Without loss of generality take $t = 0$. Define

$$\hat{s}' := \frac{1}{h} \sum_{j=-p}^{p} c_j u_j, \tag{14.8.7}$$

where $p > 0$ is a fixed integer, $h > 0$ will be chosen later, and the hat over s' in this section denotes an estimate of s'. We wish to find c_j from the conditions:

$$\overline{\hat{s}'} = s'(0), \quad \overline{|\hat{s}' - s'(0)|^2} = \min. \tag{14.8.8}$$

One has, using Taylor's formula:

$$\hat{s}' = \frac{1}{h} \sum_{j=-p}^{p} c_j n_j + \frac{1}{h} \sum_{j=-p}^{p} c_j s(0)$$

$$+ \frac{h}{h} \sum_{j=-p}^{p} c_j j s'(0) + \frac{h^2}{h} \sum_{j=-p}^{p} c_j \frac{j^2}{2} s''(\xi_j). \tag{14.8.9}$$

From (14.8.8) and (14.8.9) we get

$$\sum_{j=-p}^{p} c_j = 0, \quad \sum_{j=-p}^{p} j c_j = 1, \tag{14.8.10}$$

$$\frac{\sigma^2}{h^2} \sum_{j=-p}^{p} c_j^2 + h^2 \left(\sum_{j=-p}^{p} |c_j| j^2 \frac{M}{2} \right)^2 = \min. \tag{14.8.11}$$

Let us find c_j from conditions (14.8.10) and the condition

$$\sum_{j=-p}^{p} c_j^2 = \min, \tag{14.8.12}$$

and then minimize (14.8.11) with respect to h. Using Lagrange multipliers λ and ν, we get:

$$2c_j - \lambda - \nu j = 0, \quad c_j = \frac{\lambda + \nu j}{2}, \quad -p \le j \le p. \tag{14.8.13}$$

Conditions (14.8.10) imply $\lambda = 0$, and

$$\nu = \frac{6}{p(p+1)(2p+1)}.$$ (14.8.14)

So

$$c_j = \frac{3j}{p(p+1)(2p+1)},$$

$$\sum_{j=-p}^{p} c_j^2 = 2 \cdot 9 \sum_{j=1}^{p} \frac{j^2}{[p(p+1)(2p+1)]^2} = \frac{3}{p(p+1)(2p+1)} := a_p.$$ (14.8.15)

Thus (14.8.11) becomes

$$\frac{\sigma^2 a_p}{h^2} + h^2 M^2 b_p = \min,$$

$$b_p := \left(\sum_{j=1}^{p} \frac{3j^3}{p(p+1)(2p+1)} \right)^2 = \frac{9p^2(p+1)^2}{16(2p+1)^2}.$$ (14.8.16)

Minimizing in h, we find:

$$-2\sigma^2 a_p h^{-3} + 2hM^2 b_p = 0, \quad h_{\mathrm{opt}} = \sqrt{\frac{\sigma}{M}\left(\frac{a_p}{b_p}\right)^{\frac{1}{2}}}.$$ (14.8.17)

Let the minimum at $h = h_{\mathrm{opt}}$ be denoted ϵ^2. Then

$$\epsilon^2 = \sigma^2 a_p \frac{M}{\sigma}\left(\frac{b_p}{a_p}\right)^{\frac{1}{2}} + M^2 b_p \frac{\sigma}{M}\left(\frac{a_p}{b_p}\right)^{\frac{1}{2}} = 2\sigma M \sqrt{a_p b_p}.$$ (14.8.18)

Thus the standard deviation of the optimal estimate is

$$\sqrt{|\hat{s}' - s'(t)|^2} \leq \sqrt{2\sigma M}(a_p b_p)^{\frac{1}{4}}.$$ (14.8.19)

Note that

$$(a_p b_p)^{\frac{1}{4}} \sim \left(\frac{27}{16 \cdot 8p}\right)^{\frac{1}{4}} \approx \frac{0.67}{p^{\frac{1}{4}}}, \quad p \to \infty.$$ (14.8.20)

Let $p = 16$. Then $\epsilon \leq \sqrt{2\sigma M} \cdot 0.34$. Therefore, there is a gain in accuracy compared with the estimate (14.8.4), which holds in the deterministic setting. Estimates (14.8.19) and (14.8.20) show that, in principle, it is possible to attain an arbitrary accuracy of the approximation by taking p sufficiently large.

14.8.2. Stable summation of the Fourier series

In Section 3.5.4 one needs to solve the following problem. Assume that $f(x)$ is a smooth 2π-periodic function

$$f(x) = (2\pi)^{-\frac{1}{2}} \sum_{l=-\infty}^{\infty} f_l \exp(ilx), \qquad (14.8.21)$$

where

$$f_l = (2\pi)^{-\frac{1}{2}} \int_{-\pi}^{\pi} f(x)\exp(-ilx)dx, \qquad (14.8.22)$$

and

$$(1+l^2)^{\frac{s}{2}}|f_l| \le M_s, \quad s > \frac{3}{2}. \qquad (14.8.23)$$

Assume that $f_{\delta l}$ are given such that

$$|f_{\delta l} - f_l| < \delta, \quad l = 0, \pm 1, \pm 2, \ldots, \qquad (14.8.24)$$

and f_l are not known.

The problem is: given $f_{\delta l}$, calculate stably $f'(x)$. In other words, calculate F_δ such that

$$\|F_\delta - f'\| < \eta(\delta) \to 0 \quad \text{as} \quad \delta \to 0. \qquad (14.8.25)$$

The norm here can depend on the particular problem. Let us assume that this is the $L^2[(-\pi, \pi)]$ norm. Let us look for F_δ of the form:

$$F_\delta(x) = (2\pi)^{-1/2} \sum_{l=-N}^{N} il f_{\delta l} \exp(ilx). \qquad (14.8.26)$$

One has, using Parseval's equality,

$$\|F_\delta - f'\|^2 = \sum_{l=-N}^{N} l^2|f_{\delta l} - f_l|^2 + \sum_{|l|>N} l^2|f_l|^2$$

$$\le \delta^2 \frac{N(N+1)(2N+1)}{3} + 2M_s^2 \sum_{l=N+1}^{\infty} \frac{l^2}{(1+l^2)^s}$$

$$\le \frac{2}{3}(N+1)^3\delta^2 + \frac{2M_s^2}{2s-3}(N+1)^{-2s+3}$$

$$:= c_0\nu^3\delta^2 + c_1\nu^{-2s+3}, \quad \nu := N+1,$$

$$(14.8.27)$$

where the constants c_0 and c_1 are defined by the last equation. Minimizing the right-hand side of (14.8.27) with respect to $\nu > 1$, with $\delta > 0$ being fixed, we find the optimal $\nu := \nu(\delta)$:

$$\nu(\delta) = \left\{ \frac{(2s-3)c_1}{3c_0} \right\}^{\frac{1}{2s}} \delta^{-\frac{1}{s}} := c_2 \delta^{-\frac{1}{s}}, \tag{14.8.28}$$

and the error estimate:

$$\|F_\delta - f'\| \le c_3 \delta^{\frac{2s-3}{s}} := \eta(\delta). \tag{14.8.29}$$

If $s = 2$, then $\eta(\delta) = c_3 \delta^{\frac{1}{2}}$. Let us summarize the result.

Theorem 14.8.2. *Assume that $f_{\delta l}$ are given such that (14.8.23) and (14.8.24) hold. Define F_δ by formula (14.8.26) with $N = N(\delta) = \nu(\delta) - 1$, where $\nu(\delta)$ is defined in (14.8.28). Then (14.8.25) holds with $\eta(\delta)$ defined by (14.8.29).*

14.9. Radon transform and PDE

14.9.1. Fundamental solutions of elliptic equations

In this section we use the ideas related to the Radon transform for finding fundamental solutions to partial differential equations. Let $L(\partial)$, $\partial = (\partial_1, \ldots, \partial_n)$, $\partial_j = \frac{\partial}{\partial x_j}$, be a linear formal differential operator of order $2m$ with constant coefficients. Let $L_0(\partial)$ be the principal part of $L(\partial)$, which consists of the terms containing derivatives of order $2m$. Ellipticity of $L(\partial)$ implies that

$$L_0(\xi) \ne 0 \quad \text{if} \quad \xi \ne 0, \quad \xi \in \mathbb{R}^n.$$

Let

$$L(\partial)u = \delta(x), \quad x \in \mathbb{R}^n. \tag{14.9.1}$$

Consider the distribution

$$\frac{2r^\lambda}{|S^{n-1}|\Gamma\left(\frac{\lambda+n}{2}\right)} = \frac{1}{|S^{n-1}|\pi^{\frac{n-1}{2}}\Gamma\left(\frac{\lambda+1}{2}\right)} \int_{S^{n-1}} |\alpha \cdot x|^\lambda d\alpha. \tag{14.9.2}$$

Formula (14.9.2) is proved in Section 14.9.3. Recall that

$$\left. \frac{2r^\lambda}{|S^{n-1}|\Gamma\left(\frac{\lambda+n}{2}\right)} \right|_{\lambda=-n} = \delta(x), \quad x \in \mathbb{R}^n. \tag{14.9.3}$$

Therefore, one wants to solve the equation:

$$L(\partial)v = \frac{|\alpha \cdot x|^\lambda}{|S^{n-1}|\pi^{\frac{n-1}{2}}\Gamma\left(\frac{\lambda+1}{2}\right)}, \tag{14.9.4}$$

and find

$$\nu = \nu(\alpha, \alpha \cdot x). \qquad (14.9.5)$$

If such a ν is found, then the solution to (14.9.1) is

$$u(x) = \int_{S^{n-1}} \nu(\alpha, \alpha \cdot x) d\alpha. \qquad (14.9.6)$$

Let $p := \alpha \cdot x$. Then

$$\frac{\partial}{\partial x_j} = \alpha_j \frac{d}{dp}. \qquad (14.9.7)$$

Thus

$$L(\partial)\nu = L\left(\alpha_1 \frac{d}{dp}, \ldots, \alpha_n \frac{d}{dp}\right) \nu(\alpha, p) = \frac{|p|^\lambda}{|S^{n-1}|\pi^{\frac{n-1}{2}} \Gamma\left(\frac{\lambda+1}{2}\right)}. \qquad (14.9.8)$$

Equation (14.9.8) is an ordinary differential equation for ν. Suppose that $g(\alpha, p)$ is the solution to the equation

$$L\left(\alpha_1 \frac{d}{dp}, \ldots, \alpha_n \frac{d}{dp}\right) g(\alpha, p) = \delta(p), \qquad (14.9.9)$$

which vanishes at infinity. Then the solution to (14.9.8) is

$$\nu(\alpha, p) = \frac{1}{|S^{n-1}|\pi^{\frac{n-1}{2}} \Gamma\left(\frac{\lambda+1}{2}\right)} \int_{-\infty}^{\infty} g(\alpha, p - t)|t|^\lambda dt. \qquad (14.9.10)$$

The solution to Equation (14.9.1) is given by (14.9.6) and (14.9.10) with $\lambda = -n$. If $n = 2m + 1$, then

$$\left.\frac{|t|^\lambda}{\Gamma\left(\frac{\lambda+1}{2}\right)}\right|_{\lambda=-2m-1} = \frac{(-1)^m \delta^{(2m)}(t)}{(2m)(2m-1)\ldots(m+1)}. \qquad (14.9.11)$$

Using the formula $|S^{n-1}| = \frac{2\pi^{\frac{n}{2}}}{\Gamma\left(\frac{n}{2}\right)}$, Equation (14.9.10) with $\lambda = -2m - 1$, and (14.9.11), we get:

$$\begin{aligned}
\nu &= \frac{\Gamma\left(\frac{2m+1}{2}\right)(-1)^m m!}{2\pi^{\frac{2m+1}{2}}\pi^m (2m)!} \frac{\partial^{2m} g(\alpha, p)}{\partial p^{2m}} \\
&= \frac{(-1)^m (2m-1)!!\pi^{\frac{1}{2}} m!}{2\pi^{2m+\frac{1}{2}} 2^m (2m)!} \frac{\partial^{2m} g(\alpha, p)}{\partial p^{2m}}. \\
&= \frac{(-1)^m}{2(2\pi)^{2m}} \frac{\partial^{2m} g(\alpha, p)}{\partial p^{2m}}
\end{aligned} \qquad (14.9.12)$$

In particular, if $n = 3$, $m = 1$, and $L(\partial) = \nabla^2$, then

$$\nu = \frac{-1}{8\pi^2} \frac{\partial^2 g}{\partial p^2},$$ (14.9.13)

and Equation (14.9.9) yields:

$$g'' = \delta(p).$$ (14.9.14)

Therefore, (14.9.13) yields

$$\nu = -\frac{\delta(p)}{8\pi^2}.$$ (14.9.15)

Equations (14.9.6) and (14.9.15) imply

$$u(x) = -\frac{1}{8\pi^2} \int_{S^2} \delta(\alpha \cdot x) d\alpha = -\frac{1}{8\pi^2} \frac{2\pi}{|x|} = -\frac{1}{4\pi|x|}.$$ (14.9.16)

This is the well known fundamental solution of the Laplace equation:

$$\Delta \frac{1}{4\pi|x|} = -\delta(x).$$ (14.9.17)

The general formula which follows from Equations (14.9.6), (14.9.10), and (14.9.12) is

$$u(x) = \frac{(-1)^m}{2(2\pi)^{2m}} \int_{S^{n-1}} \left. \frac{\partial^{2m} g(\alpha, p)}{\partial p^{2m}} \right|_{p=\alpha \cdot x} d\alpha, \quad n = 2m + 1.$$ (14.9.18)

Exercise 14.9.1. Derive Equation (14.9.18) directly from (14.9.1) using (2.1.15), (2.2.4), and (10.3.1).

Hint. Take the Radon transform of (14.9.1). Applying the first equation in (2.1.15) to each term of the differential operator L and using (10.3.1), we get (14.9.9) with $g = Ru$. Hence, using (2.2.4), we prove the desired assertion.

According to Exercise 14.9.1, the basic result of this section can be derived in a much shorter way. We decided to give the longer derivation as well in order to illustrate the usage of the plane wave decomposition (14.9.2). This decomposition allows one to reduce some multidimensional problems for PDE to one- or two-dimensional problems, as we have just seen (see also Section 14.9.2).

14.9.2. Fundamental solution of the Cauchy problem
Let

$$P(\partial_0, \partial)u = 0, \tag{14.9.19}$$

where P is a polynomial with constant coefficients of order m with respect to $\partial_0 := \frac{\partial}{\partial t}$, and

$$u(0, x) = \frac{\partial u(0, x)}{\partial t} = \cdots = \frac{\partial^{m-2} u(0, x)}{\partial t^{m-2}} = 0, \quad \frac{\partial^{m-1} u(0, x)}{\partial t^{m-1}} = \delta(x). \tag{14.9.20}$$

Let us look for the solution to the problem:

$$P\left(\partial_0, \alpha_1 \frac{\partial}{\partial p}, \ldots, \alpha_n \frac{\partial}{\partial p}\right) w(t, \alpha, p) = 0, \tag{14.9.21}$$

with the initial data

$$w(0, \alpha, p) = \cdots = \frac{\partial^{m-2} w(0, \alpha, p)}{\partial t^{m-2}} = 0,$$

$$\frac{\partial^{m-1} w(0, \alpha, p)}{\partial t^{m-1}} = \frac{1}{|S^{n-1}| \pi^{\frac{n-1}{2}}} \frac{|p|^\lambda}{\Gamma\left(\frac{\lambda+1}{2}\right)}. \tag{14.9.22}$$

If the solution $w(t, \alpha, p)$ to problem (14.9.21), (14.9.22) is found, then the solution to problem (14.9.19), (14.9.20) is given by the formula

$$u(t, x) = \frac{1}{|S^{n-1}| \pi^{\frac{n-1}{2}}} \int_{S^{n-1}} d\alpha \int_{-\infty}^{\infty} w(t, \alpha, \alpha \cdot x - z) \frac{|z|^\lambda}{\Gamma\left(\frac{\lambda+1}{2}\right)} dz \bigg|_{\lambda=-n}. \tag{14.9.23}$$

If $n = 2m + 1$, then (14.9.23) becomes

$$u(t, x) = \frac{(-1)^m}{2(2\pi)^{2m}} \int_{S^{n-1}} \frac{\partial^{n-1} w(t, \alpha, p)}{\partial p^{n-1}} \bigg|_{p=\alpha \cdot x} d\alpha. \tag{14.9.24}$$

14.9.3. Proof of identity (14.9.2)
Consider the distribution

$$\mathcal{J} := \int_{S^{n-1}} |\alpha \cdot x|^\lambda d\alpha. \tag{14.9.25}$$

Clearly $\mathcal{J}$ is a homogeneous of order λ function of $r = |x|$. Thus

$$\mathcal{J} = j(n, \lambda) r^\lambda. \tag{14.9.26}$$

To find $j(n, \lambda)$, take α along the x_n-axis and use the spherical coordinates:

$$J = |S^{n-2}| \int\limits_0^\pi d\theta \cos^\lambda \theta \sin^{n-2} \theta d\theta = \frac{2\pi^{\frac{n-1}{2}}}{\Gamma\left(\frac{n-1}{2}\right)} \cdot 2 \int\limits_0^{\frac{\pi}{2}} \cos^\lambda \theta \sin^{n-2} \theta d\theta$$

$$= \frac{2\pi^{\frac{n-1}{2}}}{\Gamma\left(\frac{n-1}{2}\right)} 2 \cdot \frac{1}{2} \frac{\Gamma\left(\frac{\lambda+1}{2}\right) \Gamma\left(\frac{n-1}{2}\right)}{\Gamma\left(\frac{n+\lambda}{2}\right)} = 2\pi^{\frac{n-1}{2}} \frac{\Gamma\left(\frac{\lambda+1}{2}\right)}{\Gamma\left(\frac{\lambda+n}{2}\right)},$$

$$(14.9.27)$$

where we have used formula [(3.621.5) GR] to compute the value of the trigonometric integral. Equations (14.9.25)-(14.9.27) imply (14.9.2).

The above argument is valid for $\mathrm{Re}\lambda > -1$, but formula (14.9.2) is valid for all λ by analytic continuation.

14.10. Statistics

14.10.1. Random variables and some of their basic properties

14.10.1.1. Some probability concepts. In probability theory, a random experiment is represented by a triple $(\Omega, \mathcal{A}, P)$, where Ω is the space of all possible outcomes of the experiment, $\mathcal{A}$ is a family of subsets of Ω, which form a σ-field, and P is a probability measure.

Definition 14.10.1. A family $\mathcal{A}$ is called σ-field if

(1) $\Omega \in \mathcal{A}$;
(2) $A_1, A_2 \in \mathcal{A} \Rightarrow A_1 \cap A_2 \in \mathcal{A}, A_1 \cup A_2 \in \mathcal{A}$;
(3) if $A_1, A_2, \ldots$ are pairwise disjoint sets in $\mathcal{A}$, then $\cup_{i=1}^\infty A_i \in \mathcal{A}$; and
(4) $A \in \mathcal{A} \Rightarrow A^c := \Omega \setminus A \in \mathcal{A}$.

Any element $A \in \mathcal{A}$ is called an *event*.

Definition 14.10.2. A nonnegative function $P\{A\}$ defined on elements in $\mathcal{A}$ is called a probability measure (or, simply, probability) if

(1) $0 \leq P\{A\} \leq 1 \; \forall A \in \mathcal{A}$;
(2) $P\{\Omega\} = 1$;
(3) for pairwise disjoint events $A_1, A_2, \ldots$, one has $P\{\cup_{i=1}^\infty A_i\} = \sum_{i=1}^\infty P\{A_i\}$.

Events $A_1, \ldots, A_n$ are said to be mutually independent if for any subset $\{A_{i_1}, \ldots, A_{i_k}\} \subset \{A_1, \ldots, A_n\}$, $P\{A_{i_1} \cap \cdots \cap A_{i_k}\} = P\{A_{i_1}\} \cdot \ldots \cdot P\{A_{i_k}\}$.

The *Borel field* $\mathcal{B}$ in $\mathbb{R}$ is the smallest σ-field containing all open intervals $\{x \in \mathbb{R} : a < x < b\}$. The Borel field $\mathcal{B}^n$ in $\mathbb{R}^n$ is the smallest σ-field containing all open n-dimensional rectangles $\{(x_1, \ldots, x_n) \in \mathbb{R}^n :$

$a_i < x_i < b_i, 1 \leq i \leq n$. A *random variable* ξ is a function from Ω to $\mathbb{R}$ such that the set $\{\omega \in \Omega : \xi(\omega) \in B\}$ is in $\mathcal{A}$ for every $B \in \mathcal{B}$. A vector $\xi^{(n)} = (\xi_1, \ldots, \xi_n)$ is called a *random vector* if $\xi_1, \ldots, \xi_n$ are random variables. Probability of the event $\xi < a$ is denoted by $F(a)$: $F(a) :=$ $P\{\xi < a\}$, and is called the *cumulative distribution function* (or just the distribution). Similarly, $F(a_1, \ldots, a_n) := P\{\xi_1 < a_1, \ldots, \xi_n < a_n\}$ is called the joint distribution of $\xi_1, \ldots, \xi_n$.

Let ξ_1 and ξ_2 be two random variables with distributions $F_1(x)$ and $F_2(x)$, respectively. Then ξ_1 is called *stochastically less* than ξ_2 if $F_1(x) \geq F_2(x)$ for all $x \in \mathbb{R}$.

A random variable ξ is said to be *normally distributed* if its distribution $F(x)$ is given by

$$F(x) = \frac{1}{\sigma} \Phi\left(\frac{x - \mu}{\sigma}\right), \quad \sigma > 0; \quad \Phi(x) := \frac{1}{\sqrt{2\pi}} \int\limits_{-\infty}^{x} e^{-y^2/2} dy,$$

where μ and σ are some constants, and $\Phi(x)$ is *the standard normal distribution*. A random variable ξ is said to be *uniformly distributed* on an interval (a, b) if $P\{\xi \in (c, d)\} = (d - c)/(b - a)$ for any $c, d, a \leq c \leq d \leq b$. In this case,

$$F(x) = \begin{cases} 0, & x \leq a, \\ \frac{x-a}{b-a}, & a \leq x \leq b, \\ 1, & x \geq b. \end{cases}$$

Let $\xi^{(n)}$ be a random vector. Its *expectation* (or *mean*) is defined as the n-dimensional Lebesgue-Stieltjes integral

$$E(\xi^{(n)}) := \int\limits_{\mathbb{R}^n} x dF(x).$$

Occasionaly, expectation is denoted by $\overline{\xi^{(n)}} := E(\xi^{(n)})$. Let ξ be a random variable. The *r-th moment* α_r of ξ, if it exists, is defined by

$$\alpha_r := E(\xi^r) = \int\limits_{\mathbb{R}} x^r dF(x).$$

Let $\mu := E(\xi)$, then the central moments are defined by the formula:

$$\mu_r = E\big((\xi - \mu)^r\big).$$

Clearly, $\mu_1 = 0$. The second central moment μ_2 is called *variance* and is denoted σ^2, where $\sigma = \sqrt{\mu_2}$ is called *standard deviation*.

Let μ and σ be mean and standard deviation of a random variable ξ. Then the *Chebyshev inequality* holds:

$$P\{|\xi - \mu| \leq h\sigma\} \geq 1 - \frac{1}{h^2}, \quad h > 0.$$

14.10.1.2. The occupancy problem. Suppose one seeks to determine the equilibrium state of a physical system composed of a very large number m of particles of the same nature: electrons, protons, etc. Let us assume that there are M microscopic states in which each of the particles can be (for example, there can be M energy levels). Before one finds the equilibrium state of the system, one has to compute the number of ways N in which m particles may be distributed into M energy levels. To compute N, an assumption must be made as to whether or not the particles obey the Pauli exclusion principle (which states that there cannot be more than one particle in any of the microscopic states) and whether or not the particles themselves are indistinguishable. If the indistinguishable particles obey the exclusion principle, then they are said to possess *Fermi-Dirac statistics* and in this case $N = \binom{M}{m}$. If the indistinguishable particles do not obey the exclusion principle, they are said to possess *Bose-Einstein statistics* and $N = \binom{M+m-1}{m}$. If the particles are distinguishable and do not obey the exclusion principle, they are said to possess *Maxwell-Boltzmann statistics* and $N = M^m$.

14.10.2. Modes of convergence and limit theorems

Definition 14.10.3. A sequence of random variables $\{\xi_i\}$ is said to converge in probability to a random variable ξ if for every $\epsilon > 0$, $\lim_{i \to \infty} P\{|\xi_i - \xi| > \epsilon\} = 0$.

Definition 14.10.4. A sequence of random variables $\{\xi_i\}$ is said to converge in the r-th mean, $r > 0$, to a random variable ξ if for every $\epsilon > 0$, $\lim_{i \to \infty} P\{|\xi_i - \xi|^r > \epsilon\} = 0$. In particular, for $r = 2$, it is said to converge in mean-square.

Definition 14.10.5. A sequence of random variables $\{\xi_i\}$ with corresponding distributions $\{F_i\}$ is said to converge in distribution to a random variable ξ with distribution F if $\lim_{i \to \infty} F_i(x) = F(x)$ at each continuity point x of F.

Definition 14.10.6. A sequence of random variables $\{\xi_i\}$ is said to converge with probability one (also, sometimes, said to converge almost surely or strongly) to a random variable ξ if $P\{\lim_{i \to \infty} \xi_i \to \xi\} = 1$.

Theorem 14.10.1. *Convergence with probability one implies convergence in probability, and convergence in probability implies convergence in the distribution. Convergence in the r-th mean implies convergence in probability.*

Theorem 14.10.2 (The weak law of large numbers). *Let $\{\xi_i\}$ be a sequence of independent random variables with finite means $\mu_1, \mu_2, \ldots$, and variances $\sigma_1^2, \sigma_2^2, \ldots$, respectively. If*

$$\lim_{N \to \infty} \frac{1}{N^2} \sum_{i=1}^{N} \sigma_i^2 = 0,$$

then

$$\frac{1}{N} \sum_{i=1}^{N} \xi_i \to \frac{1}{N} \sum_{i=1}^{N} \mu_i \quad \text{in probability as } N \to \infty.$$

Theorem 14.10.3 (The strong law of large numbers). *Let $\{\xi_i\}$ be a sequence of independent random variables with finite means $\mu_1, \mu_2, \ldots$ and variances $\sigma_1^2, \sigma_2^2, \ldots$, respectively. If*

$$\lim_{N \to \infty} \sum_{i=1}^{N} \frac{\sigma_i^2}{i^2} < \infty,$$

then

$$\frac{1}{N} \sum_{i=1}^{N} \xi_i \to \frac{1}{N} \sum_{i=1}^{N} \mu_i \quad \text{with probability 1 as } N \to \infty.$$

Theorem 14.10.4 (The central limit theorem). *Let $\{\xi_i\}$ be a sequence of independent and identically distributed random variables, each with mean μ and variance $\sigma^2 > 0$. Suppose $E(|\xi_1 - \mu|^3) < \infty$. Let $G_N(x)$ denote the distribution of $\sqrt{N}\left(N^{-1} \sum_{i=1}^{N} \xi_i - \mu\right)/\sigma$, and let Φ denote the standard normal distribution. Then*

$$\sup_{x \in \mathbb{R}} |G_N(x) - \Phi(x)| \le \frac{A}{\sqrt{N}} \frac{E(|\xi_1 - \mu|^3)}{\sigma^3}, \quad N \ge 1,$$

where $A < 0.8$ is a fundamental constant independent of N.

The estimate on the right-hand side of the last equation is known as the Berry-Esseen bound.

Let $(x_1, \ldots, x_N)$ be N realizations of random variable ξ. The function $\hat{F}_N(x) := \#\{x_i : x_i \le x\}$, where $\#\{\cdot\}$ denotes a number of elements in a set $\{\cdot\}$, is called the *empirical distribution*.

Theorem 14.10.5 (Glivenko). *Let $F(x)$ be a distribution of random variable ξ, and let $\hat{F}_N(x)$ be its empirical distribution. Then*

$$\sup_{x \in \mathbb{R}} |\hat{F}_N(x) - F(x)| \to 0 \quad \text{with probability 1 as } N \to \infty.$$

Theorem 14.10.6 (Kolmogorov-Smirnov). *Let $F(x)$ be a continuous distribution of random variable ξ, and let $\hat{F}_N(x)$ be its empirical distribution. Denote*

$$D_N := \sup_{x \in \mathbb{R}} |\hat{F}_N(x) - F(x)|.$$

Then

$$P\{\sqrt{N}D_N \le t\} \to 1 - 2\sum_{k=1}^{\infty}(-1)^{k-1}e^{-2k^2t^2}, \ t > 0, \ N \to \infty.$$

14.10.3. Hypothesis testing

Let ξ be a random variable (vector) and F be its distribution. Suppose that F is not known a priori. Any hypothesis about the probability distribution F of ξ is called a *statistical hypothesis*. If the distribution F is taken to be of a certain form (e.g., normal) and the hypothesis is about value of one or several parameters F depends on (e.g., about only mean or about both mean and variance), then such a hypothesis is called *parametric*. If hypothesis does not involve specific parameters of F (e.g., is F normal with unspecified mean and variance? or is distribution F of random variable ξ the same as distribution G of random variable η?), it is called a *nonparametric hypothesis*.

To test a hypothesis on the basis of a random sample of observations $x = (x_1, \ldots, x_N)$ (which is a realization of random variable ξ), we must divide the space X of all possible sets of observations (the sample space) into two regions: X_a and X_c, $X_a \cup X_c = X$. If the observed sample point x falls into X_a: $x \in X_a$, we accept the hypothesis. If x falls into the complimentary region X_c: $x \in X_c$, we reject the hypothesis. The regions X_a and X_c are called the acceptance region and the critical (rejection) region, respectively. Frequently, the hypothesis being tested is called *the null hypothesis* and is denoted by H_0. Usually, we test the null hypothesis H_0 against some alternative H_1, which we call *the alternative hypothesis*. When performing a test, we may arrive at the correct decision, or we may commit one of the two errors:

(1) rejecting H_0 when it is true (type I error or the false alarm error);
(2) accepting H_0 when it is false (type II error).

Fix a small number ϵ, $0 < \epsilon < 1$. Then we can find the critical region so that

$$P\{x \in X_c | H_0\} = \epsilon,$$

where $P\{A|H\}$ denotes a conditional probability of event A if hypothesis H holds. The number ϵ is called *the level of significance*. Clearly, ϵ is the probability of a Type I error. The probability of a Type II error is a function of the alternative hypothesis H_1 and is denoted by δ. Thus

$$P\{x \in X_a | H_1\} = \delta, \quad P\{x \in X_c | H_1\} = 1 - \delta.$$

The complimentary probability $1 - \delta$ is called *the power of the test* of H_0 against H_1. The specification of H_1 is essential, because power is a function of H_1. It seems natural to make ϵ as small as possible. However, if H_0 and H_1 are fixed, decreasing Type I error practically always increases Type II error. Let N be a size of the sample $x = (x_1, \ldots, x_N)$ which is used in the test of H_0 against H_1. It is reasonable to expect that

$$\delta = P\{x \in X_a | H_1\} \to 0 \quad \text{as } N \to \infty$$

for any fixed ϵ, $0 < \epsilon < 1$. If this property holds, we say that the given test of H_0 is *consistent* against H_1.

Quite frequently, a statistical test operates as follows. Instead of checking whether $x \in X_a$ or not, one uses an appropriately chosen functional $\varphi = \varphi(x_1, \ldots, x_N) : X \to \mathbb{R}$, and checks whether $\varphi(x) \in \varphi(X_a)$ or not. Similarly, if value of the functional φ on the observed sample point x, $\varphi(x)$, falls into $\varphi(X_a)$, the null hypothesis H_0 is accepted. If $\varphi(x) \in \mathbb{R} \setminus \varphi(X_a)$, H_0 is rejected. Therefore, $\varphi(X_a)$ and $\mathbb{R} \setminus \varphi(X_a)$ are also called the acceptance region and the critical region, respectively. Any functional φ defined on a sample space X is called *statistic*.

14.10.4. Randomness and deviations from it, ranks, rank tests, order statistics

A nonparametric statistical procedure may be defined as any statistical procedure which has certain properties holding true under very few assumptions made about the underlying population from which the data are obtained. By a *distribution-free test* it is usually meant that the distribution of the statistic, on which the test is based, under the null hypothesis is the same for all distributions in some well-defined class of distributions. One of the main classes of nonparametric problems which can be solved by distribution-free tests is testing for randomness in the data.

Definition 14.10.7. Let $x_1, \ldots, x_N$ be a series collected sequentially over time or, more generally, according to the level of presence of some other factor. The series is *random* if $x_1, \ldots, x_N$ are *independent and identically distributed (iid)* random variables.

In other words, the series is random if each observation comes independently from the same distribution and, hence, the time order or the intensity of the concomitant factor has no effect on the response. The above definition of randomness can be generalized to the multidimensional case.

Definition 14.10.8. Let $D \subset \mathbb{R}^n$ be a bounded domain. Let $x_1, \ldots,$ x_N be a data set collected at points $\rho_1, \ldots, \rho_N \in D$, respectively. The data set is *random* if $x_1, \ldots, x_N$ are iid random variables.

In a majority of tests of randomness, one assumes that randomness is the null hypothesis, and the alternative hypothesis is that the given data set is not random in some sense. There can be many deviations from randomness, and we will briefly describe only two types of nonrandom behavior.

Let $x_1, \ldots, x_N$ be a series of observations, and let $F_i(x)$ be the distribution function of the i-th observation $i = 1, \ldots, N$. The *trend in location* arises when $F_i(x) = F(x - \theta_i)$, where $F(x)$ is certain unknown distribution function and θ_i, $i = 1, \ldots, N$ is a nonstochastic unknown sequence of numbers which is called *trend*. In particular, if $\theta_1 \leq \theta_2 \leq \cdots \leq \theta_N$ or $\theta_1 \geq \theta_2 \geq \cdots \geq \theta_N$, the trend is monotone. In multidimensional case, the trend in location arises when the distribution function of an observation at a point $\rho \in D$ is $F_\rho(x) = F(x - \theta(\rho))$.

Another type of nonrandomness occurs when one suspects that an unforeseen one-time disturbance may have occurred at an unknown time point m causing the distribution of random variable being observed to remain the same at $i = 1, \ldots, m - 1$, shift up (or down) and remain stable afterwards. An alternative hypothesis of this type is known as hypothesis of *jump at an unknown time point* or just hypothesis of *change point*. Frequently, there can be many change points in an observed sequence. In multidimensional case, the *change surfaces* alternative occurs when there exists a pairwise disjoint partition $\cup_{k=1}^{K} D_k = D$ of the domain D such that if an observation point ρ falls inside D_k, then the corresponding observed value is the realization of a random variable ξ_k, $k = 1, \ldots, K$. The boundaries between subdomains D_k are called change surfaces. Usually, for the test of randomness to be consistent against such an alternative, one has to make additional assumptions about random variables ξ_k. For example, one can suppose that there exists an unknown ordering $\{\xi_{k_l}\}_{l=1}^{K}$, where ξ_{k_l} is stochastically smaller than $\xi_{k_{l+1}}$, $l = 1, \ldots, K - 1$.

One can choose observation points ρ_i inside D in several fashions. If ρ_i are chosen in D according to some nonstochastic rule, we have the *nonrandom design model*. If ρ_i are located at the nodes of some regular grid covering D, we have the *regular design model*. If ρ_i are chosen in D completely randomly, that is probability of any observation point ρ to fall inside any subdomain $V \subset D$ equals $\text{Vol}(V)/\text{Vol}(D)$, then we have the *random design model*.

Suppose that distribution function (or, functions) underlying an observed sample $x_1, \ldots, x_N$ (either in $\mathbb{R}$ or $\mathbb{R}^n, n \geq 2$) are continuous. Then, clearly, the probability that two observations will be equal $x_i = x_j$

for some $i \neq j$ is zero. The *rank* of an observation x_i is defined as one plus the number of observations in the sample which are less than x_i: $R_i := 1 + \#\{x_j : x_j < x_i\}$, or equivalently, $R_i := \#\{x_j : x_j \leq x_i\}$. The statistical test which is based on ranks of observations and not on their actual values is called the *rank test*. Rank tests constitute a large portion of distribution-free tests and are very popular tools in nonparametric statistical procedures.

Let $R_1, \ldots, R_N$ be a sequence of ranks corresponding to some sample. Clearly, the sequence of ranks is the permutation of the set $\{1, \ldots, N\}$. If the sample is random, occurrence of any sequence of ranks and, hence, of any permutation is equiprobable and does not depend on a distribution underlying the sample (provided it is continuous). Using this observation, one establishes some elementary properties of a sequence of ranks corresponding to a random sample:

$$P\{R_i = k\} = \frac{1}{N}, \; \forall i, k, \, 1 \leq i, k \leq N;$$

$$E(R_i) = \frac{1}{N} \sum_{k=1}^{N} k = \frac{N+1}{2}, \; 1 \leq i \leq N;$$

$$E(R_i R_j) = \frac{1}{N(N-1)} \sum_{k=1}^{N} \sum_{\substack{l=1 \\ l \neq k}}^{N} kl = \frac{1}{N(N-1)} \sum_{k=1}^{N} k \left(\sum_{l=1}^{N} l - k \right)$$

$$= \frac{1}{N(N-1)} \left(\left(\sum_{k=1}^{N} k \right)^2 - \sum_{k=1}^{N} k^2 \right)$$

$$= \frac{1}{N(N-1)} \left(\frac{N^2(N+1)^2}{4} - \frac{N(N+1)(2N+1)}{6} \right)$$

$$= \frac{(N+1)(3N+2)}{12}, \quad i \neq j.$$

Let us consider again a random sample $x_1, \ldots, x_N$ and rearrange it in an increasing order $x_{1,N} \leq x_{2,N} \leq \cdots \leq x_{N,N}$. The r-th member $x_{r,N}$ of the new sequence is called the *r-th order statistic*. The two terms $x_{1,N} = \min(x_1, \ldots, x_N)$ and $x_{N,N} = \max(x_1, \ldots, x_N)$ are called extremes. If the original observations are iid with distribution function $F(x)$, one can compute the distribution function of the r-th order statistic:

$$F_{r,N}(x) = P\{x_{r,N} < x\} = \sum_{k=r}^{n} \binom{n}{r} F^k(x) [1 - F(x)]^{n-k}$$

$$= r \binom{n}{r} \int_{0}^{F(x)} t^{r-1} (1-t)^{n-r} dt.$$

In particular,

$$F_{1,N}(x) = 1 - \left[1 - F(x)\right]^N, \quad F_{N,N}(x) = F^N(x).$$

The following inequality is useful for obtaining asymptotic results for $F_{r,N}$:

$$\left| F_{r,N}(x) - \Phi\left(\frac{NF(x) - r}{[r(1 - r/N)]^{1/2}}\right) \right| \le \frac{3}{[r(1 - r/N)]^{1/2}}.$$

Setting $r = \gamma N$, where $\gamma, 0 < \gamma < 1$, is fixed and letting $N \to \infty$, we obtain

$$\left| F_{\gamma N,N}(x) - \Phi\left(\sqrt{N}\frac{F(x) - \gamma}{[\gamma(1 - \gamma)]^{1/2}}\right) \right| \le \frac{3}{[N\gamma(1 - \gamma)]^{1/2}} = O(N^{-1/2}).$$

14.10.5. The Monte Carlo method

The Monte Carlo method (or method of statistical trials) consists of solving various problems of computational mathematics by means of the construction of some random process for each problem, with the parameters of the process equal to the required quantities of the problem. These quantities are then determined approximately by means of observations of the random process and the computation of its statistical characteristics, which are approximately equal to the required parameters [Shr].

Example 14.10.1. Let the problem be to compute the expectation $E(\xi)$ of a certain random variable ξ. The Monte Carlo method for solving this problem consists of modeling ξ N times: $\xi_1, \ldots, \xi_N$, for a sufficiently large N, and then computing their mean value

$$x_N := \frac{\xi_1 + \cdots + \xi_N}{N}.$$

According to the law of large numbers, $E(\xi) \approx x_N$. The accuracy of such an approximation can be estimated using the Chebyshev inequality. Let σ be the standard deviation of ξ. Then, for any fixed $\epsilon > 0$, the following estimate

$$|E(\xi) - x_N| \le \frac{\sigma}{\sqrt{\epsilon N}}$$

holds with a probability not less than $1 - \epsilon$.

REMARK 14.10.1. In the above example, the accuracy of the Monte Carlo method was of the order $O(1/\sqrt{N})$. This, in fact, is a fundamental property of the Monte Carlo method and it remains valid in all cases where the method is used.

In Chapters 12 and 13, the Monte Carlo method is used for solving a problem of the following type. One observes a random variable ξ and the observed value of ξ is used to choose between the null and alternative hypotheses. If the null hypothesis holds, the random variable ξ can be simulated on a computer. Suppose that under the null hypothesis ξ is very unlikely to take small values and, therefore, the critical region is $\{\xi \leq a\}$ for some a. In other words, if x is the observed value of ξ and $P\{\xi \leq x\}$ is less than a threshold, the null hypothesis is rejected. Thus the problem is to estimate the distribution $F(x)$ of ξ. This can be done by simulating ξ a sufficient number of times and using the empirical distribution $F_N(x)$ in place of the exact one $F(x)$. The validity of such an approach follows from the Glivenko theorem. An accuracy of the procedure can be estimated using the Kolmogorov-Smirnov theorem. Since

$$P\{\sqrt{N}D_N > t\} \sim 2e^{-2t^2}, \ D_N = \sup_{x \in \mathbb{R}}|\hat{F}_N(x) - F(x)|,$$

we have

$$P\{\sup_{x \in \mathbb{R}}|\hat{F}_N(x) - F(x)| \geq \delta\} \sim 2e^{-2\delta^2 N}, \ N \to \infty.$$

The above equation implies that for any fixed $\delta > 0$, the probability of deviation of $\hat{F}_N(x)$ from $F(x)$ larger than δ becomes exponentially small as $N \to \infty$. In a similar fashion we obtain that the inequality

$$\sup_{x \in \mathbb{R}}|\hat{F}_N(x) - F(x)| \leq \sqrt{\frac{0.5\ln(2/\epsilon)}{N}}$$

holds with probability which is approximately equal to $1 - \epsilon$, where $\epsilon, 0 < \epsilon < 1$, is fixed.

14.10.6. Image processing

Mathematically, an image is a function $f(x_1, x_2)$ of two variables corresponding to spatial coordinates (coordinates in the image plane). Values of a function may be either brightnesses or k-tuples of brightnesses in several spectral bands (in the latter case, f is a vector-valued function). An image function f is always compactly supported and takes nonnegative, bounded values. Usually a support of f is rectangular. When an image is *digitized*, one samples values of f on a rectangular grid $(n_1 d, n_2 d)$ covering the image. Here n_1, n_2 are integers $0 \leq n_i \leq N_i, i = 1, 2$, and d is a step size. Next, the image sample is *quantized* to a set of discrete gray level values. In other words, the range of f (*gray scale*) is divided into a certain number of intervals of equal length, say $I_0, \ldots, I_M$, and if the value of f at a grid point (n_1, n_2) falls into the interval I_j: $f(n_1 d, n_2 d) \in I_j$, this value is changed to the number of the corresponding interval: $f_{n_1 n_2} = j$. Now it is standard to divide gray scale into 256 intervals (gray levels). The result of sampling and quantizing is a *digital image*, which is a matrix with integer entries. An element of a digital image is called a *pixel*.

Image processing is concerned with the manipulation and analysis of images by computer. Its major subareas are [RoKa]:

(1) *Digitization and compression*: converting images to digital form; coding of images so as to save storage space or channel capacity;

(2) *Enhancement, restoration, and reconstruction*: improving degraded images; reconstructing images from sets of projections (using e.g., computed tomography);

(3) *Matching, description, and recognition*: comparing images to one another; segmenting images into parts, measuring properties of and relationships among the parts, and comparing the resulting descriptions to models that define classes of images.

An important element of image description and recognition is image segmentation. *Segmentation* is the process that subdivides an image into its constituent parts and objects [GoWi]. Segmentation algorithms generally are based on one of two basic properties of gray-level values: discontinuity and similarity. In the first category, known as *local feature detection*, a partitioning is based on abrupt changes in gray level. The principal areas of interest within this category are the detection of edges, thin lines, and isolated points (spots) in an image. These features may be defined as follows [RoKa]:

(1) an *edge* (discontinuity of a signal): the grey level is relatively consistent in each of the two adjacent extensive regions, and changes abruptly as the border between two regions is crossed;

(2) a *thin line*: the gray level is relatively constant except along a thin strip;

(3) a *spot*: the gray level is relatively constant except at one location.

1. Injectivity of the classical Radon transform

If $\hat{f}(\alpha, p) = 0$ for almost all $\alpha \in S^{n-1}$ and $p \in \mathbb{R}$, and if $f \in L^1(\mathbb{R}^n)$, then $f(x) = 0$. The example in Section 2.7.3 shows that $\hat{f}(\alpha, p) = 0$ does not imply $f(x) = 0$ if no restrictions are made on the growth rate of $f(x)$ at infinity.

Problem 1. What are the natural restrictions on the growth of $f(x)$ at infinity which imply the injectivity of R understood in the classical sense?

For example:

(1) is it true that the following three assumptions a),b), and c):
   a)$\hat{f}(\alpha, p) = 0, \quad \forall (\alpha, p) \in Z$;
   b) $\int_{l_{\alpha p}} |f(x)|ds < \infty, \quad \forall (\alpha, p) \in Z$;
   c)$|f(x)| \leq c(1 + |x|)^m, \quad \forall x \in \mathbb{R}^n$,
   where $m > 0$ is a number and $c > 0$ is a constant, imply $f(x) = 0$?
(2) is it true that R, understood in the classical sense, is injective on $S' \cap DR$, where DR is the domain of R which consists of smooth functions absolutely integrable along any hyperplane?

This problem can be also interpreted as a problem of finding the maximal functional space on which the classical Radon transform defined by formula (2.1.1) coincides with the Radon transform defined in Chapter 10 on various spaces of distributions.

2. Stable differentiation of piecewise-smooth functions

Problem 2. It is of practical and theoretical interest to develop a theory for stable differentiation of piecewise-smooth functions $f(x)$, including those whose limits are not necessarily finite when x approaches the discontinuity surface S of f, when the noisy values of these functions are given.

The theory given in Section 14.8.1 deals with the stable differentiation of smooth functions. The derivatives of the Radon transform of piecewise-smooth functions in $\mathbb{R}^2$ do not have finite limits when x approaches S, the discontinuity surface of f.

3. Asymptotics of Bf near singular support of f

Problem 3. It is of practical and theoretical interest to generalize the results of Chapters 5 and 6 to the case of piecewise-smooth discontinuity surfaces or curves S, in particular, fractal surfaces and curves.

BIBLIOGRAPHICAL NOTES

The definition of the Radon transform goes back at least to Radon's paper [Ra] in which the Radon transform is introduced on the plane. In [Fu] it was proved that the integrals of a smooth function on S^2 over the large circles define the function uniquely (a large circle is the intersection of S^2 and a plane through the origin). This result is close to Radon's original result, because there exists a natural mapping which sends S^2 onto $R^2 \cup \{\infty\}$ and which maps large circles onto straight lines. This idea is studied in [GGG]. The Radon transform on $\mathbb{R}^n$ was studied in [J1-3]. The results in Sections 2.1.3 – 2.1.10 are standard. They are nearly immediate consequences of the definition of the Radon transform and can be found in [GGV] and other books on the Radon transform [De2], [Hel2], [Herm], [KaSl], [Nat3]. Moment conditions (2.1.60) appeared in [GGV], [Hel2], [Lu]. The hole theorem (called also 'the support theorem' in [Hel2, p.13]) appeared in the literature several times (see [GGV], [Hel2], and [Lu] for discussions).

Inversion formulas from Section 2.2.1 appeared in [J3], [GGV], [Hel4], and [Lu]. Our derivation is shorter and is taken from [R16]. Inversion formulas from Section 2.2.2 were obtained in [SSW]. In Section 2.2.4 we presented Radon's inversion formula as a corollary to the results from Sections 2.2.1 - 2.2.1. This formula can, of course, be derived independently, as it was done by Radon [Ra] (see also Sections 6.2 - 6.4 for related results). Relation (2.2.39) can be found in [Lu], and inversion formula (2.2.42) belongs to Cormack [Co] for $n = 2$ and to Deans for $n > 2$ [De1].

The singular value decomposition (SVD) of the Radon transform, described in Section 2.3, was derived in [Da1] (where the case of more general weights was considered). Estimates (2.4.3) can be found in [Nat3], where references are made to A. Louis's thesis and [Her1]. Remarks 2.4.1 and 2.4.2 are due to AR [R18]. Exposition in Section 2.5 is taken from [R17], but there are some intersections with [Sol2], where a general inversion formula for R^* was first derived. Less general inversion formula was obtained in [Hel4].

X-ray transform was studied in [Sol1,2], [SSW] and other papers. The inversion formula from Theorem 2.6.1 is standard (see [Nat3, Thrm 1.1, Sec. II.1]), and that in Theorem 2.6.2 was proved in [SSW]. The results from Section 2.7 can be found in [Cor1], [Hel4], [HSSW], [LSS], and, in a systematic way, in [Nat3]. Our proof of Theorem 2.7.4 is slightly different from the one in [Nat3] and is taken from [R14]. Example of the lack of injectivity from Section 2.7.3 belongs to Zalcman [Zal]. It is a consequence of the Arakelyan theorem [Ar] on the approximation by entire functions on closed sets. This example is presented also in [R4, p. 282], and we follow this presentation. In [AG] a similar example in $\mathbb{R}^n$ is given.

The results in Section 2.8 are known (e.g. see [Nat3]). Inversion formula from Theorem 2.8.1 was obtained by Tretyak and Metz [TM]. More general inversion formula for the exponential Radon transform, which takes into account finite resolution of collimators, was obtained in [Kuch]. The injectivity of the exponential Radon transform in the case of linear attenuation was established by Hertle [Her3]. In general, however, invertibility is not known if attenuation is not constant. In [MQ], local invertibility for the generalized and attenuated Radon transforms was proved under assumptions that attenuation is strictly positive and C^2. The problem of finding constant attenuation coefficient μ, given the attenuated Radon transform data (f is unknown also), was solved in [Her4] and [Sol3]. The dependence of the operator R^*R on the defining measures was studied in [Qu1]. In the case of integration over hypersurfaces, the operator R^*R was studied in [Be]. In Section 2.9 the results of the authors are formulated [KR8].

The range Theorem 3.1.1 is given in [GGV], [Hel2], and [Lu]. The derivation here is from [R18] and is based on an idea of Helgason [Hel2]. The range of the Radon transform acting on smooth functions which do not decay fast at infinity was found in [K4]. The range of the exponential Radon transform was described in [KL]. Lemma 3.1.1 belongs to AK and is published here for the first time. Theorem 3.1.2 was first proved in [Sol2]. Its proof, presented in Section 3.1.2, belongs to AK. This proof is shorter than the one in [Sol2]. Related results and some generalizations are given in [SM]. Lemma 3.1.2 was first proved under more stringent assumptions ($1 \leq s \leq n/(n-1)$) in [Sol2]. The proof presented in Section 3.1.2 is much simpler than in [Sol2]. The results in Section 3.2 are based on the well-known Sobolev estimates for the Radon transform. The exposition follows the lines of [R18], but is improved and belongs to both authors. Theorem 3.2.2 is proved by AR and is published here for the first time. It has common points with the discussion in [Nat3, pp. 166-167]. Other range theorems in terms of the spherical harmonics were obtained in [Lu] and [Hel2]. Theorem 3.3.1, in its present form, appeared here for the first time. Related results can be found in [GGV], [Lu], and [Hel2].

In [Lu] and [Hel2,3] the ranges of R and R^* were studied on various spaces of distributions and test functions. In particular, it was found that $R : \mathcal{D}(\mathbb{R}^n) \to \mathcal{D}(Z)$ has no continuous inverse, that $R : \mathcal{E}'(\mathbb{R}^n) \to \mathcal{E}'(Z)$ has continuous inverse and is an isomorphism, $R^* : \mathcal{E}(Z) \to \mathcal{E}(\mathbb{R}^n)$ is surjective. Here $\mathcal{D}, \mathcal{E}$ are the Schwartz spaces of test functions, and $\mathcal{D}'$, $\mathcal{E}'$ are the corresponding spaces of distributions.

Range theorem for X-ray transform in case $f \in L^2(\mathbb{R}^n)$ was proved by Solmon [Sol1]. The version of Theorem 3.4.1 that we presented ($f \in \mathcal{S}(\mathbb{R}^n)$) was proved by Helgason [Hel2]. The discussion in Section 3.5 belongs to AR. The discussion in Sections 3.6 and 3.7 uses the standard material (see e.g. [Nat3]). The idea to use approximation (3.6.14) instead of exact Equation (3.6.13) to improve the efficiency of the filtered back-projection algorithm in case of the fan-beam data was proposed in [Lak].

The results in Chapter 4 are from [RZ1-5], [RSZ], and [R4, Russian edition, Section 6.5], except Section 4.6, which contains new material due to AR. Section 4.10 is based on the paper [RSZ]. Additional results on the material of Section 4.7 can be found in [Z]. Some applications to Fourier analysis are given in [LRZ]. A different approach to microlocal analysis of singularities of X-ray transform is developed in [Qu 4].

First local tomography formula was proposed in [VKK]. Later works in this area include [SK], [VKF], [VF], [FRS]. The family of local tomography functions from Section 5.2 was introduced in [R16]. Section 5.3 contains the material from [R15]. Algorithm for finding values of density jumps, described in Section 5.4, was proposed in [KR11,12]. The results from Section 5.4.3 were obtained by AK. Numerical experiments presented in Section 5.5 are taken from [KR11,12]. Results from Sections 5.6 - 5.8 were obtained by AK in [K1–3].

Local tomography for the generalized Radon transform was proposed also in [KLM]. Our new results compared to [KLM] are as follows: the analysis of the behavior of local tomography function near its singular support, the regularization scheme (see Equations (5.7.14)–(5.7.16)), more general weight function (in [KLM] it is assumed that $\Phi = \Phi(x, \alpha)$, while in Chapter 5 we have $\Phi = \Phi(x, \alpha, p)$), and the weight function is not necessarily even (although it was not stated in [KLM], their derivations are valid only if Φ is even: $\Phi(x, -\alpha) = \Phi(x, \alpha)$). Exposition in Section 5.9 is based on ideas from [KR7] and [KR9], but the presentation here is improved and belongs to AK. AR recently gave a necessary and sufficient condition for a pseudodifferential operator to be a local tomography operator [R22].

Sections 6.2 – 6.4 are based on the paper [KR8], where pseudolocal tomography was introduced for the first time. Note that removing the interval $[x \cdot \theta - d, x \cdot \theta + d]$, where the Cauchy kernel is singular, can be considered as a possible method of regularizing the inversion formula. Therefore, some of the results obtained in Sections 6.2 and 6.3 can be considered as investigation of convergence of a regularized convolution and backprojection (CB) algorithm to the original density function. General results on convergence of CB algorithms which take into account both regularization (which is different from ours) and discretization are obtained in [Po]. Sections 6.5 and 6.6 are based on [KR11,12], where the relation between the local and pseudolocal tomography functions was found. Section 6.7 is based on the paper [K1].

The results in Chapter 7 are from [KR5].

The results in Chapter 8 are from [R1,8,11-13,20,21], [RK], and [KR1]. Theorem 8.1.1 originally was proved in [R8] and since then was used many times in various applications to optics, electromagnetic theory and geophysics (see [R1,2,4]). Formula (8.3.2) was used in [Pal], but similar ideas, based on interpolation theory for entire functions, were used earlier, e.g., in [R20,21]. Equation (8.2.66) was discussed in [BG2], and the properties of the eigenvalues of the operator (8.2.67) were studied in [Sl] and [SlS]. One can find in [Aiz] the Carleman approach to the recovery of an analytic function from its values on a subset of points from its analyticity region. The Davison-Grunbaum algorithm was proposed in [DG]. Not all the approaches to inversion of limited-angle data have been mentioned. In particular, we didn't mention the projection on convex sets methods (e.g., see [PS], [Sez]).

The presentation in Section 9.1 uses the paper by Natterer [Nat4], who proposed a unified approach to deriving different cone-beam data inversion formulas. Condition (9.1.3) is called the Tuy condition. It was formulated by Tuy [Tuy], see also [Smi1]. Identity (9.1.4) is from [HSSW]. Equation (9.1.5) was obtained in [Gr] and later, independently, in [RZ5]. Formulas (9.1.7) and (9.1.10) were first obtained in [Smi1] and [GGo], respectively. In [Smi1], the cone-beam data inversion formula based on (9.1.7) was proposed. This formula, however, contains an error (which was reproduced in [Smi2]). The error was corrected in [DCL]. Singular integral Equation (9.1.9) may have different solutions, and the choice of the solution to (9.1.9) was not discussed in [GGo]. Our choice is based on the study of the singularities of the Radon transform from Chapter 4, which shows that the solution $\hat{f}_p$ is bounded on both ends of the interval $[a(\alpha), b(\alpha)]$. The version of the algorithm, based on formula (9.1.6), for inversion of the cone-beam data, which is computationally more efficient than the one in [Gr], is proposed in [DC]. Exposition in Section 9.2 follows the paper [RZ5]. The algorithm from Section 9.3 was proposed in [Nat4]. An algorithm for the approximate recovery of f given the 3-D X-ray data corresponding to the circular orbit (which does not satisfy the Tuy condition) is presented in [FDK]. A discussion of a computational procedure for approximate reconstruction of locations of density jumps from 3-D cone beam data is given in [LM]. Theoretical analysis of γ-ray tomography, presented in Section 9.4, is taken from [FKR].

Definitions 10.1.1–10.1.3 of the Radon transform on distributions are from [Lu], [Hel2], and [GGV], respectively. Definition 10.1.4 is given by AR. Lemma 10.1.1 and description of the test-function space in Theorem 10.2.1 are from [R19]. The first characterization of the test-function space in case of odd n is given in [GGV]. Distributions, which annihilate all test functions of the space S_t, defined in Sec.10.2, were described in [GGV] in the case of odd n (they are called 'unessential distributions' in [GGV]). For odd n, examples from Section 10.3 can also be found in [GGV]. However, Equation (10.3.12b) in Example 10.3.4 differs from formula 5 in [GGV, p.71] (this formula is given in [GGV] without proof).

Range theorem for the Radon transform on $\mathcal{E}'$ was proved by Helgason [Hel2] and Hertle [Her1]. The latter paper contains also the range theorem for R on the space of rapidly decaying distributions. Definition of R on distributions via spherical harmonics expansion, presented in Section 10.5, appears to be new and belongs to both authors (although there are some intersections with the paper [Lu]). The material in Sections 10.6 and 10.7 belongs to AR and is taken from [R19]. The presentation in Section 10.7 is improved and belongs to both authors.

The material in Sections 11.1–11.3 is standard. A recent survey monograph on the Abel equation is [GV]. The material from Section 11.1 can be found in [GS]. Solution formula for Equation (11.2.1), (11.2.2) is classical [Tit]. Reduction of Equation (2.2.42) to a more stable is standard and can be found in [Nat3]. The result in Section 11.4 was obtained by AK. An approach to solving the Abel equation, which is somewhat close to the idea in Section 11.4, is described in [GR].

The results in Chapters 12 and 13 are from [KR8] and [KR9,10], respectively. The literature on image processing and, in particular, on edge detection is enormous (see e.g. [GoWi], [Pr], [RoKa], and references therein). Many different approaches to edge detection exist. In Chapter 12 we gave a new approach, which appears to be suitable for tomographic and many other applications. Bibliographical comments to Chapter 13 can be found in Section 13.1.

The material in Chapter 14 is, for the most part, taken from the literature. However, some of the results are original: the results in Section 14.8 and Theorems 14.7.1, 14.7.2 are from [R3], [R5-7,9]. In [R3-4] many applications of these results are given; Theorems 14.6.2-3 are from [Kr]; the approach to singular value decomposition of R based on Lemma 14.7.1 belongs to AR, the above lemma is from [BS].

The results from Section 14.1 are from [KA], in Section 14.2 — from [GS], in Section 14.3 — from [Shu] and [Hor], in Section 14.4 — from [Ak], [Leb], and [Vil], in Section 14.5 — from [Fed1,2] (some of the formulas taken from [Fed1,2] are corrected), in Section 14.6 — from [Kr] and [AkGl], in Sections 14.7 and 14.8 — from [R3]. The stable differentiation method presented in Section 14.8.1.1 first appeared in

[R5] and then was applied in [R6,9,10]. It is now widely used. The material in Section 14.8.1.2 belongs to AR and was not published earlier. The material in Section 14.8.2 is from [R7], in section in Section 14.9 — from [GS] and [J3]. The material in Section 14.10 is standard, and was taken from [HS], [Man], [Par], [KW], [Shr].

Due to the limitations of the volume we were not able to present many questions of interest: the sampling and estimates of errors in numerical implementation [Po], [Nat3], [Far], [Des], wavelet localization of the Radon transform [OS], the usage of the Poisson formula [VG], recovery of homogeneous objects from few projections [GaM], [Fal], nonuniqueness and uniqueness results of integral geometry related to tomography [Bo], [LRS], the Radon transform on non-euclidean spaces [GuSt], [Hel4], etc. Some numerical reconstruction methods, such as ART and the Kaczmarz method, are described very briefly.

REFERENCES

[AS]   M. Abramowitz and I. Stegun, *Handbook of Mathematical functions*, Dover, New York, 1970.

[Aiy]   R.J. Aiyar, Asymptotic efficiency of rank tests of randomness against autocorrelation, *Annals of the Institute of Statistical Mathematics* **33** (1981), 255 – 262.

[AGA]   R.J. Aiyar, C.L. Gouillier, and W. Albers, Asymptotic relative efficiencies of rank tests for trend alternatives, *Journal of the American Statistical Association* **74** (1979), 226 – 231.

[Aiz]   L. Aizenberg, *Carleman's formulas in complex analysis*, Nauka, Novosibirsk, 1990. (Russian)

[Ak]   N.I. Akhiezer, *Lectures on Integral Transforms*, Amer. Math. Soc., Providence RI, 1988.

[AkGl]   N.I. Akhiezer and I.M. Glazman, *Theory of Linear Operators in Hilbert Space*, Dover, New York, 1993.

[Ar]   N. Arakelyan, On uniform approximation by entire functions on closed sets, *Izvest. Acad. of Sci. USSR Math.* **28** (1964), no. 5, 1187-1205. (Russian)

[AG]   D. Armitage and M. Goldstein, Nonuniqueness for the Radon transform, *Proc. Amer. Math. Soc.* **117** (1993), no. 1, 175-178.

[BG]   A. Bakushinsky and A. Goncharsky, *Iterative Methods for Solving Ill-posed Problems*, Nauka, Moscow, 1989. (Russian)

[Be]   G. Beylkin, The inversion problem and applications of the generalized Radon transform, *Comm. on Pure and Appl. Math.* **37** (1984), 579-599.

[Bh]   G. K. Bhattacharyya, Tests of randomness against trend or serial correlation, *Handbook of Statistics, vol.4. Nonparametric Methods* (P. R. Krishnaiah and P. K. Sen, eds.), North-Holland, Amsterdam, 1984, pp. 89 – 111.

[BS]   M. Birman and M. Solomyak, Quantitative analysis in the Sobolev imbedding theorems and applications to spectral theory, *Amer. Math. Soc. Transl.* **114** (1980), 1-132.

[BH]   N. Bleistein and R. Handelsman, *Asymptotic Expansions of Integrals*, Dover, Mineola, New York, 1986.

[Bo]   J. Boman, Example of nonuniqueness for a generalized Radon transform, *J. Anal. Math.* **61** (1993), 395-401.

[CS]   A. Calderon and A. Sigmund, Singular integral operators and differential equations, *Amer. J. Math.* **79** (1957), no. 4, 901-921.

[ChH]   S. Chow and J. Hale, *Methods of Bifurcation Theory*, Springer - Verlag, New York, 1982.

[CO]   A.D. Cliff and J.K. Ord, *Spatial Processes: Models and Applications*, Pion, London, 1981.

[Cor1]   A. Cormack, Representation of a function by its line integrals with some radiological applications, *J. Appl. Phys.* **34** (1963), 2722-2727.

[Cor2]   _____, Representation of a function by its line integrals with some radiological applications II, *J. Appl. Phys.* **35** (1964), 2908-2912.

[CsH]   M. Csörgő and L. Horváth, Nonparametric methods for change-point problems, *Handbook of Statistics, vol. 7. Quality Control and Reliability* (P. R. Krishnaiah and P. K. Sen, eds.), North-Holland, Amsterdam, 1988, pp. 403 – 426.

[Da1]   M. Davison, A singular value decomposition for the Radon transform in n-dimensional Euclidean space, *Numer. Func. Anal. and Opimiz.* **3** (1981), 321-340.

[Da2]   _____, The ill-conditioned nature of the limited angle tomography problem, *SIAM J. Appl. Math.* **43** (1981), 428–448.

[DG]   M. Davison and F. Grünbaum, Tomographic reconstruction with arbitrary directions, *Comm. Pure Appl. Math.* **34** (1983), 77-120.

[De1]   S. Deans, Gegenbauer transforms via the Radon transform, *SIAM J. Math. Anal.* **10** (1979), 577-585.

[De2]   S. Deans, *The Radon Transform and Some of Its Applications*, Wiley, New York, 1983.

[DC]   M. Defrise and R. Clack, A cone-beam reconstruction algorithm using shift-variant filtering and cone-beam backprojection, *IEEE Trans. on Medical Imaging* **13** (1994), 186–195.

[DCL]   M. Defrise, R. Clack, and R. Leahy, A note on Smith's reconstruction algorithm for cone beam tomography, *IEEE Transactions on Medical Imaging* **12** (1993), no. 3, 622–623.

[Des]   L. Desbat, Efficient sampling on coarse grids in tomography, *Inverse Problems* **9** (1993), 251–269.

[DW]   J. Durbin and G.S. Watson, Testing for serial correlation in least squares regression I, *Biometrika* **37** (1950), 409 – 428.

[Eng]   G. Englund, Remainder term estimates for the asymptotic normality of order statistics, *Scand. J. Stat.* **7** (1980), 197–202.

[FKR]   V. Faber, A.I. Katsevich, and A.G. Ramm, Inversion of cone-bean data and helical tomography, *Journal of Inverse and Ill-Posed Problems* (to appear).

[Fal]   K. Falconer, X-ray problems for point souces, *Proc. Lond. Math. Soc.* **46** (1983), 241-262.

[Far]   A. Faridani, An application of multidimensional sampling theorem to computed tomography, *Contemporary Mathematics, vol. 113*, Amer. Math. Soc., Providence, RI, 1990, pp. 65-80.

[FKN]   A. Faridani, F. Keinert, F. Natterer, E.L. Ritman, and K.T. Smith, Local and global tomography, *Signal Processing, IMA Vol. Math. Appl., Vol. 23*, Springer - Verlag, New York, 1990, pp. 241 – 255.

[FRS]   A. Faridani, E.L. Ritman, and K.T. Smith, Local tomography, *SIAM J. Appl. Math.* **52** (1992), no. 2, 459 – 484, 1193 – 1198.

[Fed1] M.V. Fedoriuk, *Metod Perevala*, Nauka, Moscow, 1977. (Russian)

[Fed2] M.V. Fedoriuk, Asymptotic methods in analysis, *Encyclopaedia of Mathematical Sciences, Vol. 13. Analysis I. Integral Representations and Asymptotic Methods* (R.V. Gamkrelidze, ed.), Springer-Verlag, New York, 1989.

[FDK]   L.A. Feldkamp, L.C. Davis, and J.W. Kress, Practical cone-beam algorithm, *J. Opt. Soc. Am.* **1** (1984), 612–619.

[Fi]   D. Finch, Cone beam reconstruction with sources on a curve, *SIAM J. Appl. Math.* **45** (1985), 665-673.

[Fu]   P. Funk, Ueber eine geometrische Anwendung der Abelschen Integralgleichungen, *Math. Ann.* **77** (1916), 129-135.

[Gal]   J. Galambos, Order statistics, *Handbook of Statistics, Vol. 4. Nonparametric Methods* (P.R. Krishnaiah and P.K. Sen, eds.), North-Holland, Amsterdam, 1984, pp. 359–382.

[GaM]   R. Gardner and P. McMullen, On Hammer's X-ray problem, *J. Lond. Math. Soc.* **21** (1980), 171-175.

[Gea]   R.C. Geary, The contiguity ratio and statistical mapping, *The Incorporated Statistitian* **5** (1954), 115 – 145.

[GGG]   I.M. Gelfand, S.G. Gindikin, and M.I. Graev, Integral geometry in affine and projective spaces, *Contemporary Problems of Mathematics, Vol. 16*, Nauka, Moscow, pp. 53–226, English translation: J. Soviet Math. **18** (1982), no. 1, 39–167.

[GGo]   I.M. Gelfand and A. Goncharov, Recovery of a compactly supported function starting from its integrals over lines intersecting a given set of points in space, *Sov. Math. Doklady* **34** (1987), 373-376.

[GGr]   I.M. Gelfand and M. Graev, Integrals of test functions and generalized functions over hyperplanes, *Sov. Math. Doklady* **135** (1960), 1307–1309.

[GGV]   I. Gelfand, M. Graev, and N. Vilenkin, *Integral Geometry and Representation Theory*, Academic Press, New York, 1965.

[GS]   I.M. Gelfand and G.E. Shilov, *Generalized Functions, Volume I*, Academic Press, New York, 1964.

[GiMi] S. Gindikin and P. Michor (eds.), *75 Years of the Radon Transform*, Intern. Press, Boston, 1994.

[GoWi] R.C. Gonzalez and P. Wintz, *Digital Image Processing*, 2nd ed., Addison-Wesley, Reading, Massachusetts, 1987.

[GoRu] R. Gorenflo, B. Rubin, Locally controllable regularization of fractional derivatives, *Inverse Problems* **10** (1994), 881-893.

[GV]   R. Gorenflo, S. Vessella, *Abel Integral Equations, Lect. Notes in Math. Vol. 1461*, Springer-Verlag, Berlin, 1991.

[GR]   I.S. Gradshteyn and I.M. Ryzhik, *Table of Integrals, Series, and Products*, 5th ed., Academic Press, Boston, 1994.

[Gr]   P. Grangeat, Mathematical framework of cone beam 3D reconstruction via the first derivative of the Radon transform, *Mathematical Methods in Tomography. Lecture Notes in Math., Vol. 1497* (Herman et al., eds.), Springer, 1991.

[GQ]   E. Grinberg and E. Quinto (eds.), *Integral Geometry and Tomography*, Contemp. Math., Vol. 113, Amer. Math. Soc., Providence, RI, 1990.

[GuSt] V. Guillemin and S. Sternberg, *Geometric Asymptotics*, Amer. Math. Soc., Providence, RI, 1977.

[HSSW] C. Hamaker, K. Smith, D. Solmon, and S. Wagner, The divergent beam X-ray transform, *Rocky Mount. J. Math.* **10** (1980), 253-283.

[Hel1] S. Helgason, *Differential Geometry, Lie Groups and Symmetric Spaces*, Acad. Press, New York, 1978.

[Hel2] ———, *The Radon Transform*, Birkhauser, Boston, 1980.

[Hel3] ———, Ranges of Radon transform, *Proceedings of Symposia in Applied Mathematics, Vol. 27* (L. Shepp, ed.), AMS, Providence, RI, 1982, pp. 63-70.

[Hel4] ———, The Radon transform on euclidean spaces, compact two-point homogeneous spaces and Grassmann manifolds, *Acta Mathematica* **113** (1965), 153–179.

[Herm] G.T. Herman, *Image Reconstruction From Projections: Implementation and Applications*, Acad Press, New York, 1980.

[Her1] A. Hertle, Continuity of the Radon transform and its inverse on Euclidean spaces, *Math.Z.* **184** (1983), 165-192.

[Her2] A. Hertle, On the range of Radon transform and its dual, *Math. Ann.* **267** (1984), 91-99.

[Her3] A. Hertle, On the injectivity of the attenuated Radon transform, *Proceedings of the Amer. Math. Soc.* **92** (1984), 201–205.

[Her4]   A. Hertle, The identification problem for the constantly atten-
          uated Radon transform, *Math. Z.* **197** (1988), 13–19.

[Hor]   L. Hörmander, *The Analysis of Linear Partial Differential Op-
          erators, Vols I-IV*, Springer Verlag, New York, 1983.

[J1]   F. John, Bestimmung einer Function aus ihren Integralen ueber
          gewisse Mannigfaltigkeiten, *Math. Ann.* **100** (1934), 488-520.

[J2]   _____, Abhaengigkeit zwischen den Flaechenintegralen einer
          stetigen Funktion, *Math. Ann.* **3** (1935), 541-559.

[J3]   _____, *Plane Waves and Spherical Means*, Wiley Interscience,
          New York, 1955.

[KaSl]   A.C. Kak and M. Slaney, *Principles of computerized tomo-
          graphic imaging*, IEEE Press, New York, 1988.

[KW]   M. H. Kalos and P. A. Whitlock, *Monte Carlo Methods. Vol-
          ume I: Basics*, Wiley, New York, 1986.

[KA]   L. Kantorovich and G. Akilov, *Functional Analysis*, Pergamon
          Press, Oxford, 1982.

[Kato]   T. Kato, *Perturbation Theory for Linear Operators*, 3rd ed.,
          Springer-Verlag, New York, 1995.

[K1]   A.I. Katsevich, Local reconstructions in exponential tomogra-
          phy, Submitted.

[K2]   _____, Local tomography for the generalized Radon trans-
          form, Submitted.

[K3]   _____, Local tomography for the limited-angle problem, Sub-
          mitted.

[K4]   _____, Range of the Radon transform on functions which do
          not decay fast at infinity, Submitted.

[KR1]   A.I. Katsevich and A.G. Ramm, FBP method for inversion of
          incomplete tomographic data, *Appl. Math. Lett.* **5** (1992),
          no. 3, 77-80.

[KR2]   _____, Multidimensional algorithm for finding discontinuities
          of functions from noisy data, *Math. Comp. Modelling* **18**
          (1993), no. 1, 89-108.

[KR3]   _____, Consistency of rank tests against some general alter-
          natives, *Math. Comp. Modelling* **18** (1993), no. 12, 49-56.

[KR4]   _____, Nonparametric estimation of the singularities of a sig-
          nal from noisy measurements, *Proceedings Amer. Math. Soc.*
          120 (1994), no. 8, 1121-1134.

[KR5]   _____, A method for finding discontinuities of functions from
          the tomographic data, *Tomography, Impedance Imaging, and
          Integral Geometry* (M. Cheney, P. Kuchment, and E.T. Quinto,
          eds.), Lectures in Applied Mathematics, Vol. 30, Amer. Math.
          Soc., New York, 1994, pp. 115–123.

[KR6]   _____, Mathematical results in signal and image processing,

Doklady, Russian Academy of Sciences **339** (1995), no. 1, 11–14. (Russian)

[KR7]   _____, Finding singular support of a function from its tomographic data, *Proceedings of the Japan Academy, Series A: Mathematical Sciences* **71** (1995), no. 3, 62–67.

[KR8]   _____, Pseudolocal tomography, *SIAM J. Appl. Math.* **56** (1996), no. 1, 167–191.

[KR9]   _____, Asymptotics of PDO on discontinuous functions near singular support, *Applicable Analysis* **58** (1995), no. 3-4, 383–390.

[KR10]   _____, Consistency of rank tests against two general alternatives, *Acta Applicandae Mathematicae* **42** (1996), no. 2, 105–137.

[KR11]   _____, Finding jumps of a function using local tomography, Submitted.

[KR12]   _____, New methods for finding values of the jumps of a function from its local tomographic data, *Inverse Problems* **11** (1995), no. 5, 1005–1023.

[KR13]   _____, Inverse geophysical and potential scattering on a small body, *Proc. Experimental and Numerical Methods for Solving Ill-Posed Inverse Problems: Medical and Nonmedical Applications*, vol. SPIE-2570, 1995, pp. 151–162.

[KR14]   _____, Approximate inverse geophysical scattering on a small body, *SIAM J. Appl. Math.* **56** (1996), no. 1, 192–218.

[Kein]   F. Keinert, Inversion of k-plane transforms and applications in computer tomography, *SIAM Review* **31** (1989), 273–298.

[KO]   M. Kendall and J.K. Ord, *Time Series*, 3rd ed., Edward Arnold, UK, 1990.

[KS]   M. Kendall and A. Stuart, *The Advanced Theory of Statistics, Vol. 2*, 4th ed., Charles Griffin, London, 1979.

[Kir]   A. Kirillov, On a problem of I. Gelfand, *Sov. Math. Doklady* **137** (1961), 276-277.

[KK]   G.A. Korn and T.M. Korn, *Mathematical Handbook*, 2nd ed., McGraw-Hill, New York, 1968.

[Kr]   S. Krein, *Linear Equations in Banach Spaces*, Birkhauser, Boston, 1982.

[KM]   P. R. Krishnaiah and B. Q. Miao, Review about estimation of change points, *Handbook of Statistics, vol. 7. Quality Control and Reliability* (P. R. Krishnaiah and P. K. Sen, eds.), North-Holland, Amsterdam, 1988, pp. 375 – 402.

[Kuch]   P. Kuchment, On inversion and range characterization of one transform arising in emission tomography, *Proceedings of the Conference '75 years of the Radon Transform'*, International Press, 1994, pp. 240–248.

[KLM]   P. Kuchment, K. Lancaster, and L. Mogilevskaya, On local tomography, *Inverse Problems* **11** (1995), 571–589.

[KL]   P. Kuchment and Ya. L'vin, Paley-Wiener theorem for the exponential Radon transform, *Acta Appl. Math.* **18** (1990), 251–260.

[Lak]   A.V. Lakshminarayanan, *Reconstruction from Divergent X-ray Data*, Suny Tech. Report 32, State University of New York, Buffalo, New York, 1975.

[LRS]   M. Lavrentiev, V. Romanov, and S. Shishatsky, *Ill-posed Problems of Mathematical Physics and Analysis*, Amer. Math. Soc., Providence, R.I., 1986.

[LP]   P. Lax and R. Phillips, The Paley-Wiener theorem for the Radon transform, *Comm. Pure Appl. Math.* **23** (1970), 409-424.

[LSS]   J.V. Leahy, K.T. Smith, and D.C. Solmon, Uniqueness, non-uniqueness and inversion in the X-ray and Radon transform problems, *Proc. International Symposium on Ill-posed Problems, Newark, DE, October*, 1970.

[Leb]   N.N. Lebedev, *Special Functions and Their Applications*, Prentice - Hall, Englewood Cliffs, N.J., 1965.

[LRZ]   E. Liflyand, A.G. Ramm, and A. Zaslavsky, Estimates from below for lebesgue constants, *Jour. Fourier Anal. Appl.* **2** (1996) (to appear).

[Log]   B. Logan, The uncertainty principle in reconstructiong functions from projections, *Duke Math. Jour.* **42** (1975), 661–706.

[LS]   B. Logan and L. Shepp, Optimal reconstruction of a function from projections, *Duke Math. J.* **42** (1975), 645-659.

[Lom]   F. Lombard, Rank tests for changepoint problems, *Biometrika* **74** (1987), 615 – 624.

[LM]   A.K. Louis and P. Maass, Contour reconstruction in 3-D X-ray CT, *IEEE Trans. on Medical Imaging* **12** (1993), 764–769.

[Lud]   D. Ludwig, The Radon transform on Euclidean spaces, *Comm. Pure Appl. Math.* **23** (1960), 49-81.

[Man]   E.B. Manoukian, *Mathematical Nonparametric Statistics*, Gordon and Breach, New York, 1986.

[MQ]   A. Markoe and E.T. Quinto, An elementary proof of local invertibility for generalized and attenuated Radon transforms, *SIAM J. Math. Anal.* **16** (1985), no. 5, 1114 – 1119.

[Mor]   P.A.P. Moran, *An Introduction to Probability Theory*, Claredon Press, Oxford, 1968.

[Mus]   N. Muskhelishvili, *Singular Integral Equations*, Dover, New York, 1992.

[Nat1] F. Natterer, On the inversion of the attenuated Radon transform, *Numer. Math.* **32** (1979), 431–438.

[Nat2]   _____, Computed tomography with unknown sources, *SIAM J. Appl. Math.* **43** (1983), 1201–1210.

[Nat3]   _____, *The Mathematics of Computerized Tomography*, J. Wiley & and Sons, New York, 1986.

[Nat4]   _____, Recent developments in X-ray tomography, *Lectures in Appl. Math.* **30** (1994), 177-198.

[Ne]   D. Newman, An entire function bounded in every direction, *Amer. Math. Soc. Monthly* **83** (1976), no. 3, 192-193.

[No]   G.E. Noether, Asymptotic properties of the Wald-Wolfowitz test of randomness, *Annals of Math. Statistics* **21** (1950), 231 – 246.

[OS]   T. Olson and J. de Stefano, Wavelet localization of the Radon transform, *IEEE Trans. Sig. Proc.* **42** (1992), 2055–2067.

[Pal]   V.P. Palamodov, Some singular problems in tomography, *Mathematical Problems of Tomography* (I.M. Gelfand and S.G. Gindikin, eds.), Amer. Math. Soc., Providence, Rhode Island, 1990, pp. 123 – 140.

[Par]   E. Parzen, *Modern Probability Theory and Its Applications*, Wiley, New York, 1960.

[PS]   H. Peng and H. Stark, Image recovery in computer-tomography from partial fan-beam projections, *IEEE Trans. Medical Imaging* **11** (1992), no. 4, 470–478.

[Po]   D.A. Popov, On convergence of a class of algorithms for the inversion of the numerical Radon transform, *Mathematical Problems of Tomography* (I.M. Gelfand and S.G. Gindikin, eds.), Amer. Math. Soc., Providence, Rhode Island, 1990, pp. 7 – 65.

[Pr]   W.K. Pratt, *Digital Image Processing*, 2nd ed, Wiley, New York, 1991.

[PBM]   A.P. Prudnikov, Yu.A. Brychkov, and O.I. Marichev, *Integrals and Series, Vol. 2. Special Functions*, Gordon and Breach, New York, 1986.

[Qu1]   E.T. Quinto, The dependence of the generalized Radon transform on defining measures, *Trans. Amer. Math. Soc.* **257** (1980), 331–346.

[Qu2]   _____, The invertibility of rotation invariant Radon transforms, *J. Math. Anal. Appl.* **91** (1983), 510–522.

[Qu3]   _____, Singular value decompositions and inversion methods for the exterior Radon transform and a spherical transform, *J. Math. Anal. Appl.* **95** (1985), 437–448.

[Qu4]   _____, Singularities of the X-ray transform and limited data tomography in $\mathbb{R}^2$ and $\mathbb{R}^3$, *SIAM J. Math. Anal.* **24** (1993), no. 5, 1215 – 1225.

[Ra]   J. Radon, Ueber die Bestimmung von Funktionen durch ihre Integralwaerte laengs gewisser Maennigfaltigkeiten, *Ber. Saechs. Akad. Wiss* **69** (1917), 262–277.

[R1]   A.G. Ramm, *Theory and Applications of Some New Classes of Integral Equations*, Springer-Verlag, New York, 1980.

[R2]   _____, *Scattering by Obstacles*, Reidel, Dordrecht, 1986.

[R3]   _____, *Random Fields Estimation Theory*, Longman/Wiley, New York, 1990.

[R4]   _____, *Multidimensional Inverse Scattering Problems*, Longman/Wiley, New York, 1992; Expanded Russian edition, MIR, Moscow, 1994, pp. 1–496.

[R5]   _____, On numerical differentiation, *Izvestija Vuzov, Ser. Mathematics* **11** (1968), 131-135. (Russian)

[R6]   _____, Simplified optimal differentiators, *Radiotech. i Electron.* **17** (1972), 1325-1328, 1034–1037. (Russian)

[R7]   _____, Optimal harmonic synthesis of generalized Fourier series and integrals with randomly perturbed coefficients, *Radiotechnika* **28** (1973), 44–49. (Russian)

[R8]   _____, An approximation by entire functions, *Izvestija Vusov, Ser. Mathematics* **10** (1978), 72–76. (Russian)

[R9]   _____, Stable solutions of some ill-posed problems., *Math. Meth. in the Appl. Sci.* **3** (1981), 336-363.

[R10]   _____, Estimates of the derivatives of random functions, *J. Math. Anal. Appl.* **102** (1984), 244-250.

[R11]   _____, Inversion of limited-angle tomographic data, *Comp. and Math. with Applic.* **22** (1991), no. 4/5, 101-112.

[R12]   _____, Inversion of the Radon transform with incomplete data, *Math. Methods in the Appl. Sci.* **15** (1992), no. 3, 159-166.

[R13]   _____, Inversion of limited-angle tomographic data II, *Appl. Math. Lett.* **5** (1992), no. 2, 47-49.

[R14]   _____, Uniqueness and inversion of cone-beam data, *Appl. Math. Lett.* **6** (1993), no. 1, 35-38.

[R15]   _____, Optimal local tomography formulas, *PanAmer. Math. Journ.* **4** (1994), no. 4, 125-127.

[R16]   _____, Finding discontinuities from tomographic data, *Proc. Amer. Math. Soc.* **123** (1995), no. 8, 2499-2505.

[R17]   _____, Inversion formulas for the backprojection operator in tomography, *Proc. Amer. Math. Soc.* **124** (1996), no. 2, 567–577.

[R18]   _____, The Radon transform is an isomorhism between $L^2(B)$ and $H_e(Z_a)$, *Appl. Math. Lett.* **8** (1995), no. 1, 25-29.

[R19]   _____, Radon transform on distributions, Submitted.

[R20]   _____, On the Kotelnikow's theorem, *Electrocommunication*
          10 (1962), 71–72. (Russian)

[R21]   _____, Reconstruction of a signal from its values in discrete
          sequence of time moments, *Radiotech. i Electron.* **11** (1965),
          1957–1959. (Russian)

[R22]   _____, Necessary and sufficient conditions for a PDO to be
          a local tomography operator, *Compt. Rend. Acad. Sci, Paris*
          (to appear).

[RK]   A.G. Ramm and A.I. Katsevich, Inversion of incomplete Radon
          transform, *Appl. Math. Lett.* **5** (1992), no. 2, 41-46.

[RM]   A.G. Ramm and T. Miller, Estimates of the derivatives of ran-
          dom functions II, *J. Math. Anal. Appl.* **110** (1985), 429-435.

[RZ1]   A.G. Ramm and A.I. Zaslavsky, Singularities of the Radon
          transform, *Bull. Amer. Math. Soc.* **25** (1993), no. 1, 109 –
          115.

[RZ2]   _____, Reconstructing singularities of a function given its
          Radon transform, *Math. and Comput. Modelling* **18** (1993),
          no. 1, 109 – 138.

[RZ3]   _____, Asymptotics of the Fourier transform of piecewise-
          smooth functions, *Compt. Rend. Acad. Sci, Paris* **316** (1993),
          no. 1, 541 – 545.

[RZ4]   _____, Inversion of incomplete cone-beam data, *Appl. Math.
          Lett.* **5** (1992), no. 4, 91-94.

[RZ5]   _____, X-ray transform, the Legendre transform and envelo-
          pes, *J. Math. Anal. Appl.* **183** (1994), no. 3, 528-546.

[RSZ]   A.G. Ramm, A. Steinberg, and A.I. Zaslavsky, Stable calcula-
          tion of the Legendre transform of noisy data,, *J. Math. Anal.
          Appl.* **178** (1993), no. 2, 592-602.

[Ren]   A. Renyi, *Probability Theory*, North-Holland, Amsterdam,
          1970.

[Rob]   G.S. Robinson, Detection and coding of edges using directional
          masks, *Proc. SPIE Conf. on Advances in Image Transmission
          Techniques*, August 1976, San Diego.

[RoKa]   A. Rosenfeld and A.C. Kak, *Digital Picture Processing*, Vol. 2,
          2nd ed. Academic Press, New York, 1982.

[Sard]   A. Sard, The measure of critical values of differentiable maps,
          Bull. Amer. Math. Soc. **48** (1942), 883 – 890.

[Sez]   M.I. Sezan, An overview of convex projections theory and its
          applications to image recovery problems, *Ultramicroscopy* **40**
          (1992), no. 1, 55-67.

[Shr]   Yu.A. Shreider (ed.), *The Monte Carlo Method*, Pergamon,
          New York, 1966.

[Shu]   M.A. Shubin, *Pseudodifferential Operators and Spectral The-
          ory*, Springer-Verlag, Berlin, 1980.

[Sl] D. Slepian, Prolate spheroidal wave functions, *Bell Syst. Tech. J.* **43** (1964), 3009.

[SlS] D. Slepian and E. Sonneblick, Eigenvalues associated with prolate spheroidal wave functions, *Bell Syst. Tech. J.* **44** (1965), 1745.

[Smi1] B. Smith, Image reconstruction from cone-beam projections: necessary and sufficient conditions and reconstruction methods, *IEEE Trans. Med. Imag.* **4** (1985), 14 – 25.

[Smi2] B. Smith, Cone-beam tomography: recent advances and a tutorial review, *Optical Engineering* **29** (1990), 524 – 534.

[SK] K.T. Smith and F. Keinert, Mathematical foundations of computed tomography, *Appl. Optics* (1985), 3950 – 3957.

[Sne] I.H. Sneddon, *The Use of Integral Transforms*, McGraw-Hill, New York, 1972.

[Sol1] D. Solmon, The X-ray transform, *J. Math. Anal. Appl.* **56** (1976), 61-83.

[Sol2] ———, Asymptotic formulas for the dual Radon transform and applications, *Math. Z.* **195** (1987), 321-343.

[Sol3] ———, The identification problem for the exponential Radon transform, *Mathematical Methods in the Applied Sciences* **18** (1995), 687-695.

[SM] D. Solmon and W. Madych, A range theorem for the Radon transform, *Proceedings of the Amer. Math. Soc.* **104** (1988), 79-85.

[SSW] K. Smith, D. Solmon, and S. Wagner, Practical and mathematical aspects of the problem of reconstructing objects from radiographs, *Bull of Amer. Math. Soc.* **83** (1977), 1227-1270.

[Tit] E. Titchmarsh, *Theory of Fourier Integrals*, Oxford University Press, Oxford, 1948.

[TM] O.J. Tretiak and C.L. Metz, The exponential Radon transform, *SIAM J. Appl. Math.* **39** (1980), 341 – 354.

[Tr] F. Treves, *Introduction to Pseudodifferential and Fourier Integral Operators, Vol. I: Pseudodifferential Operators*, Plenum Press, New York, 1980.

[Tuy1] H.K. Tuy, Reconstruction of a three-dimensional object from a limited range of views, *J. Math. Anal. Appl.* **80** (1981), 598-616.

[Tuy2] ———, An inversion formula for cone-beam reconstruction, *SIAM J. Appl. Math.* **43** (1983), 546 – 552.

[VKK] E.I. Vainberg, I.A. Kazak, and V.P. Kurczaev, Reconstruction of the internal three-dimensional structure of objects based on real-time integral projections, *Soviet J. Nondest. Test.* **17** (1981), 415-423.

[VKF]   E.I. Vainberg, I.A. Kazak, and M.L. Faingoiz, X-ray comput-
        erized back-projection tomography with filtration by double
        differentiation, *Soviet J. Nondest. Test.* **21** (1985), 106–113.
[VF]   E.I. Vainberg and M.L. Faingoiz, Increasing the spatial reso-
        lution of computerized tomography, *Problems in Tomographic
        Reconstruction* (A. Alekseev, M.M. Lavrentiev, and G.N. Pre-
        obrazhensky, eds.), Siberian Branch of Acad. Sci. USSR,
        Novosibirsk, 1985, pp. 28–35.
[Vil]   N.J. Vilenkin, *Special Functions and the Theory of Group Rep-
        resentations*, Amer. Math. Soc., Providence, RI, 1968.
[VG]   N.D. Vvedenskaya and S.G. Gindikin, Discrete Radon trans-
        form and image reconstruction, *Mathematical Problems of To-
        mography* (I.M. Gelfand and S.G. Gindikin, eds.), Amer.
        Math. Soc., Providence, Rhode Island, 1990, pp. 141– 188.
[WW]   A. Wald and J. Wolfowitz, An exact test for randomness in
        the nonparametric case based on serial correlation, *Annals of
        Mathematical Statistics* **14** (1943), 378 – 388.
[Yin]   Y.Q. Yin, Detection of the number, locations, and magnitudes
        of the jumps, *Comm. Statis. Stoch. Models* **4** (1988), no. 3,
        445-455.
[Zal]   L. Zalcman, Uniqueness and nonuniqueness for the Radon
        transform, *Bull. Lond. Math. Soc.* **14** (1982), 241–245.
[Z]   A. Zaslavsky, Multidimensional analog of the Erdelyi lemma
        and the Radon transform, *Tomography, Impedance Imaging,
        and Integral Geometry*, Lectures in Applied Mathematics, Vol.
        30, Amer. Math. Soc., New York, 1994, pp. 259–278.

LIST OF NOTATIONS

$\|A\| := (A^*A)^{1/2}$	modulus of an operator A, 426
$a(x, y, \xi)$	amplitude function of a PDO, 400
$\mathcal{B}$	PDO, 177
$\mathcal{B}_K$	truncated operator, 191
B_1^n, B_1	unit ball in $\mathbb{R}^n$
$\|B_1^n\|$	volume of the unit ball, 17
$B(x, y)$	beta function, 405
$\mathbb{C}$	complex plane
$\mathbb{C}^n$	n-dimensional space of complex vectors
$C^k(U)$	the space of k times continuously differentiable functions, 391
$C_0^k(U)$	the space of compactly supported $C^k(U)$ functions, 391
$C_m^{(p)}$	Gegenbauer polynomials, 408
D	domain in $\mathbb{R}^n$
$\|D\|$	measure of D
$\mathcal{D}$	$:= C_0^\infty(\mathbb{R}^n)$, the space of compactly supported smooth functions, 394
$\mathcal{D}'$	the space of distributions, 394
$D(A)$	domain of operator A, 423
$\det(A)$	determinant of a matrix A
Df	cone-beam transform, 13
$d(x, y)$	metric, the distance between x and y, 390
$\mathcal{E}$	$:= C^\infty(\mathbb{R}^n)$, the space of all smooth functions, 394
$\mathcal{E}'$	the space of compactly supported distributions, 394
E_t	resolution of the identity of a selfadjoint operator, 427
$\mathcal{F}$	Fourier transform in $\mathbb{R}^n$, 15, 397
F	Fourier transform in $\mathbb{R}^1$ (with respect to the second argument), 15
FBP	Filtered Back Projection, 91
FIO	Fourier Integral Operator, 402
f_ρ	PLT function, 206
f_ρ^c	$:= f - f_\rho$, 207
$\mathcal{H}$	Hilbert transform, 29
H_m	Hermite polynomials, 410
$H^{n,l}$	the space of homogeneous harmonic polynomials of degree l in n variables, 412
$H^s(\mathbb{R}^n)$	Sobolev space, 393
$H_0^s(U)$	space of $H^s(\mathbb{R}^n)$ functions with compact support in U, 393

I^a	Riesz potential in $\mathbb{R}^1$ (with respect to the second argument), 28
$\mathcal{I}^a$	Riesz potential in $\mathbb{R}^n$, 28
j	$:= (j_1, \ldots, j_n)$, multiindex
$|j|$	$:= j_1 + \cdots + j_n$, modulus of a multiindex
$j!$	$:= j_1! \cdot \cdots \cdot j_n!$, factorial of a multiindex
$J_m(x)$	Bessel function, 406
K	operator acting on the second argument, 30
$\mathcal{K}(X, Y)$	the set of linear compact operators from X to Y, 426
$l_{\alpha p}$	plane defined by the equation $\alpha \cdot x = p$, 11
L	Legendre map (transform), 100
$L^p(U)$	Lebesgue space, 391
$L_0^p(U)$	Space of $L^p(U)$ functions with compact support in U
LT	local tomography
$N(A)$	null-space of operator A, 423
$N_\pm(\lambda; A)$	$:= \sum_{\lambda_j^\pm(A)>\lambda} 1$, number of positive (negative) eigenvalues of operator A, modulus of which is larger than λ, 435
$N(n, l)$	number of linearly independent spherical harmonics of degree l in $\mathbb{R}^n$, 413
$P\{\cdot\}$	probability of an event, 445
PDO	pseudodifferential operator, 402
PLT	pseudolocal tomography
$\mathcal{P}_j$	homogeneous polynomial of degree j
$P_m(x)$	Legendre polynomials, 409
$P_{l,m}(x)$	associated Legendre functions, 414
$p_V(x)$	Minkowski functional, 391
$\mathrm{R}(A)$	range of operator A, 423
$Rf = \hat{f}(\alpha, p)$	Radon transform, 11
$R_1 f = \check{f}(\alpha, p)$	Radon transform, 12
R^*	operator adjoint to R (the back-projection operator), 16
R_μ	attenuated Radon transform, 57
R_Φ	generalized Radon transform, 63
R_Φ^*	operator adjoint to R_Φ, 63
$\mathbb{R}^n$	n-dimensional Euclidean space
$R^{n,l}$	the space of homogeneous polynomials of degree l in n variables, 412
$S = \partial D$	boundary of D, 98
$\hat{S}$	dual variety, 98

$\mathcal{S}$	the space of rapidly decaying smooth functions, 394
$\mathcal{S}'$	the space of tempered distributions, 394
$s_j(A)$	singular values of operator A, 426
$S^m_{\rho,\delta}$	class of symbols, 400
S^{n-1}	unit sphere in $\mathbb{R}^n$
$\|S^{n-1}\|$	area of the unit sphere, 412
singsupp f	singular support of f, 394
supp f	support of f, 394
T_l	Chebyshev polynomial of the first kind, 409
U_l	Chebyshev polynomial of the second kind, 409
W_ϵ	point-spread function (mollifier), 91
$WF(f)$	wave front of f, 400
$Xf = g(x,\alpha)$	X-ray transform, 13
$X_m f$	X_m transform, 13
X^*	space adjoint to X, 391
X, Y	Banach spaces, 390
Y_l	spherical harmonics of degree l, 413
Z	$:= S^{n-1} \times \mathbb{R}$
Z_a	$:= S^{n-1} \times [-a, a]$
α	a unit vector or a vector in $\mathbb{R}^n$
$\alpha^\perp$	plane orthogonal to α and passing through the origin
$\Gamma(z)$	gamma function, 405
$\gamma_\pm(a)$	$:= (2\pi)^{-n}\mathrm{meas}\{\xi : \pm\tilde{a}(\xi) > 1\}$, 435
δ	delta function, 395
δ_{jl}	Kronecker symbol; $\delta_{jl} = 1$ if $j = l$ and $\delta_{jl} = 0$ if $j \neq l$
θ	angle, $0 \le \theta < 2\pi$
Θ	$:= (\cos\theta, \sin\theta)$
$\Theta^\perp$	$:= (-\sin\theta, \cos\theta)$
$\lambda_j(A)$	eigenvalues of a linear operator, 426
μ	attenuation coefficient, 57
$\Phi(x,y,\xi)$	phase function of a PDO, 400
$\Phi(x,\alpha,p)$	positive measure which defines the generalized Radon transform, 63
$\chi_D(x)$	characteristic function of the domain D
ψ_γ	function with properties (i)–(iii), 177
$\binom{n}{l}$	$:= \frac{n!}{l!(n-l)!}$ (binomial coefficient)
Δ	Laplacian, 412
Δ_α	angular part of the Laplacian, 413
∂	$:= (\partial_1, \ldots, \partial_n)$
∂_{x_k}	$:= \frac{\partial}{\partial x_k}$

∂_x^j $:= \frac{\partial^{j_1}}{\partial x_1^{j_1}} \cdots \frac{\partial^{j_n}}{\partial x_n^{j_n}}$

$*$ convolution in $\mathbb{R}^n$, 14

$\circledast$ convolution in $\mathbb{R}^1$ (with respect to the second argument), 15

$\|\cdot\|_s$ norm in $H^s(\mathbb{R}^n)$, 41

$\|\|\cdot\|\|_s$ norm in $H^s(Z)$, 42

$< \cdot, \cdot >$ inner product in $L^2(\mathbb{R}^n)$, 16

$(\cdot, \cdot)$ inner product in $L^2(Z)$, 16

$\varnothing$ empty set

Index

Milton Keynes UK
Ingram Content Group UK Ltd.
UKHW021913071024
449327UK00022B/1657